CLASTIC
DIAGENESIS

CLASTIC DIAGENESIS

Edited by

David A. McDonald
Petro-Canada
Calgary, Alberta

and

Ronald C. Surdam
University of Wyoming
Laramie, Wyoming

Published by
The American Association of Petroleum Geologists
Tulsa, Oklahoma 74101, U.S.A.

Clastic diagenesis.

    Bibliography: p.
    Includes index
    1. Rocks, Sedimentary. 2. Diagenesis. I. McDonald,
David A. II. Surdam, Ronald C. III. American Association
of Petroleum Geologists.
QE471.C573 1984 552'.5 84-70676
ISBN 0-89181-314-4

Association Editor: Richard Steinmetz
Science Director: Edward A. Beaumont
Project Editors: Douglas A. White and Ronald L. Hart
Production and Design: Custom Editorial Productions, Inc.
                   Cincinnati, Ohio

Table of Contents

1

Concepts and Principles

2
Aspects of Porosity Modification

3

Applications of Clastic Diagenesis in Exploration and Production

Preface

Interest in clastic diagenesis has grown exponentially in the last ten years. During this period, clastic diagenesis has evolved from a very descriptive science to a much more process-oriented study. This evolution has been driven by the realization that many hydrocarbon reservoirs have significant diagenetic components directly affecting porosity and permeability characteristics. The prediction in time and space of reservoir characteristics affected by diagenesis can greatly reduce the risk in the search for hydrocarbon accumulations; particularly in subtle targets lacking pronounced structural expression. Predictive potential and exploration usefulness can be evaluated only if the processes controlling clastic diagenesis are understood. We believe that this memoir will serve as a milestone marking a significant point in our progress toward this understanding.

For the reader's convenience the memoir has been divided into three sections:
1. Concepts and Principles, 2. Aspects of Porosity Modification, and 3. Applications of Clastic Diagenesis in Exploration and Production. The first two sections deal with processes controlling various aspects of clastic diagenesis, whereas the third section applies these principles and observations to specific examples.

We gratefully acknowledge the many reviewers who assisted in the scientific review process. Lastly, we thank Laura J. Crossey who expedited many editorial tasks while one of us (Surdam) was on sabbatical leave.

David A. McDonald
Ronald C. Surdam

1

Concepts and Principles

Hydrogeologic Regimes of Sandstone Diagenesis*

William E. Galloway
Bureau of Economic Geology, The University of Texas
Austin, Texas

INTRODUCTION

The objective of this short paper is to bring together some general observations about the hydrogeology of sedimentary basins to establish the dynamic framework in which sandstone diagenesis occurs. The Frio Formation (Oligocene) of the northern Gulf Coast Basin has been the subject of both regional and detailed diagenetic studies (see contributions by Loucks et al, Land, and Kaiser in this volume) and of stratigraphic, hydrogeologic, and resource analysis (Galloway, 1977; Bebout et al, 1978; Galloway and Kaiser, 1980; and Galloway et al, 1982). It provides a focus for the discussion.

The subsurface environment must be regarded as a dynamic cauldron in which fluids continuously move. Static conditions rarely, if ever, characterize subsurface fluid flow systems. To further complicate interpretation of diagenetic history, both flow dynamics and fluid chemistries evolve through time. Analysis and description of basin hydrology and extant geochemistry provide a view only of the present diagenetic environment; most diagenetic features may be relicts of earlier fluid regimes. However, such a description is a starting point. Not only does it provide a framework for examining and testing active water–rock interactions, but it also expands our understanding of natural hydrologic systems in large sedimentary basins. The thesis presented here is that a more critical analysis of diagenetic processes within a realistic hydrologic context offers a major step in understanding and predicting diagenetic histories of sandstone reservoirs. Such analysis necessitates a basic understanding of the

ABSTRACT. The subsurface hydrologic system of large, actively filling sedimentary basins includes meteoric, compactional, and thermobaric regimes. Boundaries between the regimes and their contained flow systems evolve as basin filling proceeds. As a result, sands are continuously flushed by a succession of fluids of varying origins and chemistries. Careful examination of the existing hydrologic setting and a reconstruction of generalized hydrologic history and its relationship to observed diagenetic features within a depositional sequence may serve to validate interpretive diagenetic models and may explain or predict paragenetic relationships, regional diagenetic variations within the same depositional episode, and differences in diagenetic products in different episodes in the same or similar basins.

The Frio/Catahoula Formations of the Texas Coastal Plain provide an example that both illustrates the coexistence of hydrologic regimes and relates associations of diagenetic features with existing regimes or with the important mixing zones that occur between regimes. Hydrocarbon geochemistries indicate meteoric flushing to depths approaching 2000 m. The deeper geopressured section is coincident with a thermobaric regime in which clay dewatering recharges the hydrologically restricted portions of the basin fill.

hydrogeologic regimes and evolutionary pathways of sedimentary basins.

BASIN HYDROLOGIC REGIMES

The hydrologic framework of a large, depositionally active sedimentary basin consists of ground water (the term is used in this paper in its general sense to include all waters of whatever origin that lie in the subsurface environment) moving within several different regimes (Bogomolov et al, 1978; Kissin, 1978) (Fig. 1). Although the boundaries of the regimes and their terminology are sometimes difficult to reconcile, they nonetheless form end members characterized by their relative positions, hydrochemistries, and flow dynamics.

The *meteoric regime* encompasses shallow portions of the basin fill. Water

recharges by infiltration of precipitation and surface waters, moves down the topographic gradient in the direction of decreasing gravitational potential energy, and discharges vertically at hydrologic base level. Hydrologic base level is commonly determined by sea (or lake) level. Because there is no head advantage achieved by further movement once the area of base level is attained, there is little potential for additional lateral flow beneath the subaqueous basin center. The shoreline is typically a major discharge area for meteoric aquifers (Fig. 1). On a geologic time scale, meteoric circulation is rapid.

*Publication authorized by the Director, Bureau of Economic Geology, The University of Texas at Austin.

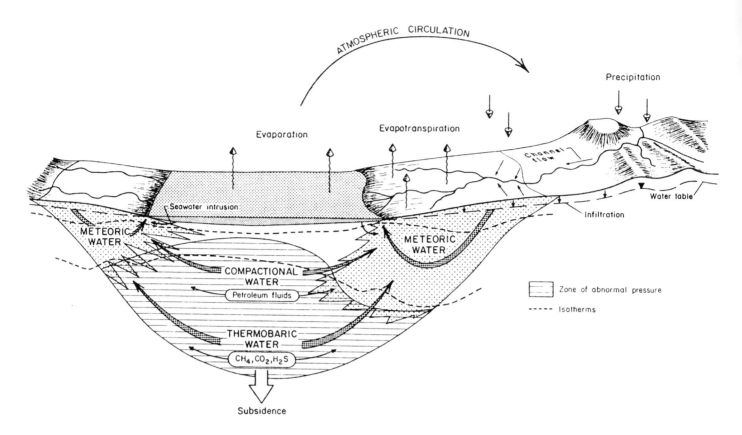

Figure 1—Hydrologic regimes of a large, subsiding depositional basin. From Galloway and Hobday. Copyright © 1983 Springer-Verlag New York. Reprinted with permission.

The *compactional* regime is characterized by upward and outward expulsion of pore waters trapped within the compacting sediment pile (Fig. 1). These waters may include evolved connate water deposited with the sediment and meteoric waters buried below the zone of active meteoric circulation. Lithostatic loading or compressive tectonic stress may generate increased pressure. Pressure head is the driving mechanism for flow. If circulation is impeded by restricted transmissivity of the sediment pile, fluid pressure/depth gradient may exceed that of a free-standing column of water of the same height.

The *thermobaric,* or abyssal, regime lies in the deepest portions of the basin fill where temperature and pressure are greatest. Significant volumes of water may be released by dehydration reactions of clay minerals or of other hydrous mineral phases. Fluids move in response to pressure gradients generated both by phase changes, such as generation of hydrocarbons or release of mineral-bound water, and by ongoing lithostatic loading. However, the extremely low permeability of the compacted sediments commonly restricts water movement over geologically significant time spans. Aquithermal pressuring generated by volume expansion of water may also play a role in raising static pressure, but is rapidly dissipated by even minor amounts of fluid leakage.

Recognition of the thermobaric and compactional regimes as important components of the hydrologic basin necessitates an expanded view of the ground-water cycle as it evolves during the filling of a sedimentary basin. Large volumes of water are diverted from the meteoric and surface flow systems and entombed within the sedimentary pile as pore or mineralogically bound water. Such waters discharge into and mix with the meteoric system slowly, if at all. Unlike the supply of water to the meteoric system, which is continually recharged, the supply of compactional and thermobaric waters, though potentially large, is finite. Processes or products related to fluid flow within these systems must recognize and accommodate this limitation.

EVOLUTION OF HYDROLOGIC BASINS

The compactional and thermobaric regimes persist as long as active basin filling continues. With cessation of subsidence or with tectonic uplift the compaction-driven flow decreases and ultimately ceases. Strata of such a hydrologically "mature" basin are increasingly flushed by meteoric waters recharged along its uplifted margins (Toth, 1980; Cousteau et al, 1975). Regional flow is directed inward toward the topographic floor of the hydrologic basin, and the sediment fill may ultimately be completely flushed by meteoric waters. Well described basins in which ambient flow is dominated by regional meteoric circulation driven by gravitational head include the Paris Basin (Korotchansky and Mitchell, 1972) and the western Canada Mesozoic foreland basin of Alberta (Hitchon, 1969a, 1969b). Long-term

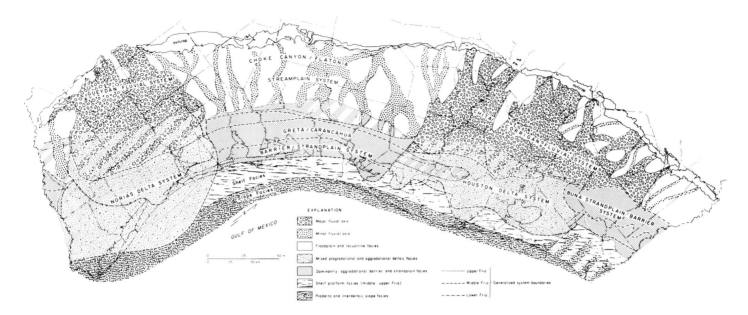

Figure 2—Depositional systems of the Frio (Oligocene) Formation, Texas Gulf Coastal Plain. From Galloway et al (1982).

flushing by regional meteoric circulation replaces residual waters with geochemically evolved meteoric ground water (Clayton et al, 1966).

With long-term tectonic stability and erosional leveling of basin topographic relief, movement of meteoric water slows, and the hydrologic basin may become effectively stagnant. Vertical pressure gradients throughout the basin would be hydrostatic (Cousteau et al, 1975), and only diffusive transfer of soluble constituents could continue. Such extreme stability is, however, rare.

HYDROGEOLOGY OF THE FRIO FORMATION

The Frio Formation, along with its updip equivalent, the Catahoula Formation, constitutes the sedimentary record of one of the major offlap depositional episodes of the Northwestern Shelf of the Gulf of Mexico Basin. Following deposition of a sand and mudstone wedge locally exceeding 3500 m in thickness during Middle to Late Oligocene and Early Miocene time, ongoing basin subsidence has buried downdip portions of the Frio under several thousand meters of younger

sediment. Updip, a thin Frio/Catahoula section crops out along a narrow belt across the inner coastal plain of Texas (Fig. 2). Basinward thickening of the section is accomplished largely by a succession of syndepositional growth faults that develop where the Frio wedge prograded beyond the underlying continental platform margin (Fig. 3).

The Frio/Catahoula depositional episode consists of several major depositional systems (Galloway et al, 1982) (Fig. 2). The Oligocene depocenter lay in the Rio Grande Embayment. Here the Gueydan bed-load fluvial system fed the enormous Norias delta system, which prograded the continental margin nearly 100 km gulfward. Along the upper Texas Coastal Plain, two or three major fluvial axes fed downdip into a broad deltaic headland, called the Houston delta system. Between the two deltaic depocenters lay an extensive wave-dominated shore-zone system that includes both well-developed barrier/lagoon and strandplain facies assemblages. Basinward, sand-rich deltaic and shore-zone facies grade into prodelta, shelf, and continental slope mudstones. The deeply buried, growth-faulted sandstones of the delta and shore-zone systems contained world class reserves of hydrocarbons and are still targets for active exploration. Variable reservoir quality of the deep exploration targets has made the

Frio the subject of extensive diagenetic study.

Hydrologic Regimes

Hydrologic regimes of the Frio as they exist today are conventionally described in terms of (1) a shallow, updip meteoric regime, (2) an intermediate depth zone characterized by normal or slightly elevated pressure gradients and saline brines, which lies below the base of relatively fresh water and above the top of geopressure, and (3) a deeply buried geopressured regime, which rises up through the Frio section basinward (Fig. 4). Examination of the characteristics of these zones shows that they broadly correspond to the more general hydrologic regimes discussed above, and thus define major environments for sandstone diagenesis. However, as the final section of this paper stresses, boundaries between regimes are both leaky and geologically ephemeral. Any selection of a fixed depth or boundary condition to define the ground-water zones is inherently arbitrary.

Meteoric Regime

Active meteoric circulation, as indicated by extent of low-chlorinity waters, extends to depths of as much as 1.5 km (Fig. 5). In portions of the south Texas Coastal Plain, however, depth of freshwater penetration is limited by shallow faulting of the aquifer or by the sand-poor character of the updip facies

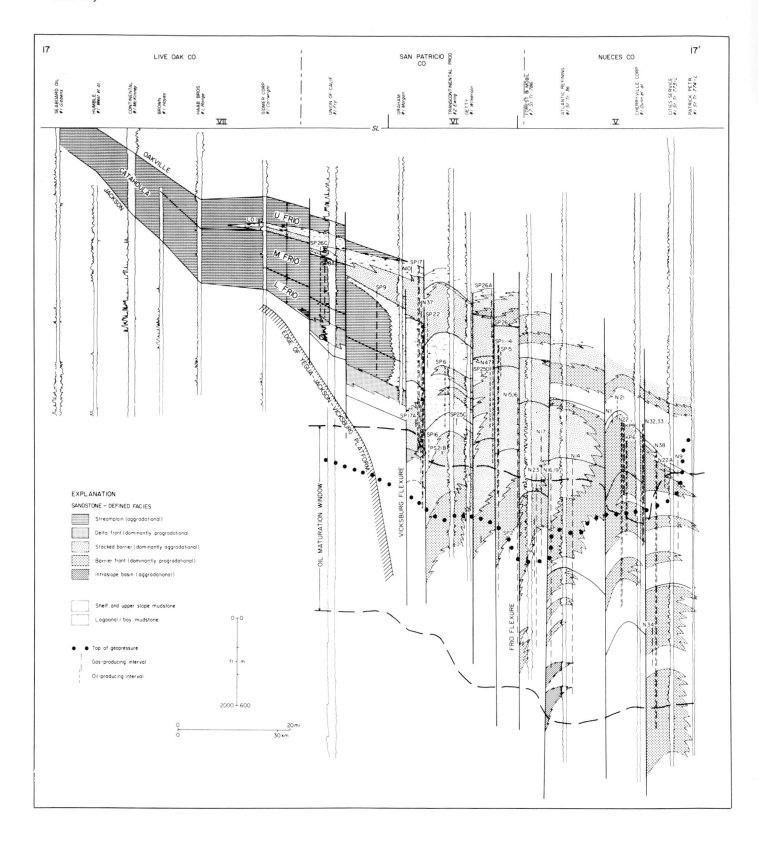

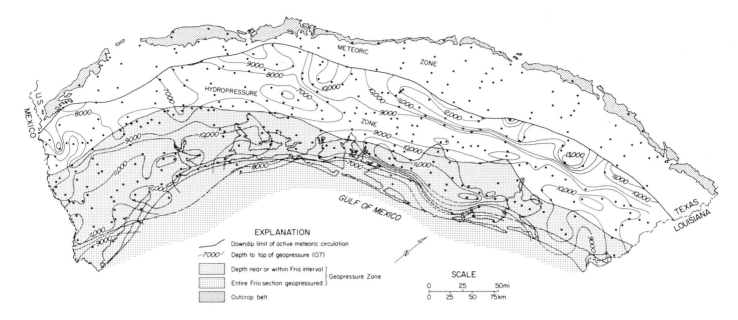

Figure 4—Lateral distribution of hydrologic zones within the Frio Formation. Depth to the "operational" top of geopressure (0.7 psi/ft or 15.8 kPa/m) is contoured in feet.

Figure 3—Stratigraphic dip cross section through the Frio Formation and its updip equivalent, the Catahoula Formation, in the middle Coastal Plain of Texas. The downdip progression of facies includes coastal stream plain, lagoon, barrier, and shelf deposits. Growth faults expand the Frio section dramatically where it offlaps the older continental platform. In the expanded section, temperatures are adequate for oil and gas generation, and much of the mud-rich barrier-front and shelf sequence is abnormally pressured. From Galloway et al (1982).

assemblage. The most active portions of the meteoric system occur primarily within axial facies of the fluvial systems; these shallow deposits have probably experienced meteoric flux throughout their depositional and burial history. Hydrocarbon geochemistry suggests that meteoric influence, as manifested by crude oil alteration by anaerobic bacteria (Fig. 6), has extended to depths of 2 km in the middle and upper coastal plain (Galloway et al, 1982). This deep meteoric penetration must have post-dated migration and accumulation of hydrocarbons. Bacterial destruction of organic acids at temperatures below 70°C (160°F) reported by Carothers and Kharaka (1980) in Miocene deposits supports the interpretation of deep meteoric penetration.

Although total dissolved solids content is characteristically low to moderate, composition of meteoric regime waters varies considerably (Galloway and Kaiser, 1980). Shallow calcium-bicarbonate waters evolve along flow paths into relatively saline sodium-chloride waters.

Diagenetic features of the Catahoula and Frio Formations produced within the meteoric regime include clay cutans, sparry calcite cement, minor amounts of kaolinite, clinoptilolite, opal, chalcedony, and feldspar overgrowths, and locally abundant leached feldspar and rock fragments.

Compactional Regime

Because muddy deltaic and marine sediments are typically deposited with high porosities (commonly exceeding 50%), physical compaction offers a source of water approximately equal in volume to the total volume of sediment. However, compaction and pore-space reduction occur rapidly with earliest burial to only a few hundred meters. Representative volumes of water lost during burial of successive thousand-meter increments of mud that might typify Frio prodelta, shelf, or slope facies are shown graphically in Figure 7. During the first kilometer of burial, the mud releases more than 21,000 cu cm of interstitial water per sq cm of surface area. At a depth of 1 km, average porosity is reduced by physical compaction to slightly more than 20%. Remaining compaction to near zero porosity at depths approaching 10 km releases a comparable volume of water in decreasing proportions with depth (Fig. 7, column B). Potential upward flux across each sq cm (column C) is no greater than the cumulative release of water by underlying sediment volumes. The illustration emphasizes the fact that the volume of available compaction water is finite.

In the Frio Formation, the depth window dominated by the compactional regime is limited by deep penetration of meteoric water (1.4–2 km) and the relatively shallow top of geopressure and associated increase in temperature and pressure gradients and resultant clay dewatering (3–4 km). Waters in the appropriate depth range are typically reducing, slightly acidic to slightly basic sodium-chloride brines containing dissolved carbon dioxide and abundant hydrocarbon gases and liquids.

Mesodiagenetic features described in recent studies that appear to be products of or associated with the compactional regime include ubiquitous quartz overgrowths, albite overgrowths, calcite-pore fill and grain replacements,

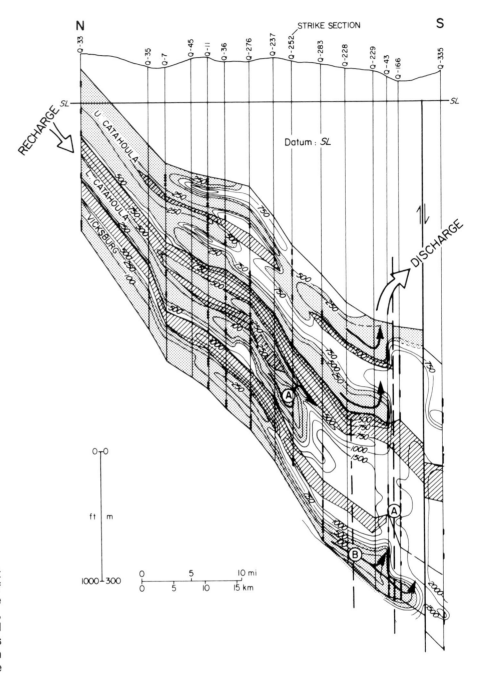

Figure 5—Calculated apparent ground-water chlorinity for a portion of the Chita–Corrigan fluvial system of the Catahoula Formation. In cross section, freshwater tongues are shown to extend into the subsurface to depths in excess of 1 km. Such tongues of relatively fresh water outline areas of most active meteoric circulation. Bunching of salinity contours at depth indicates probable areas of vertical discharge into shallower aquifers. Letters shown in circles point out complexities typical of shallow meteoric flow, such as flow focusing along more permeable channel-fill trends. From Galloway and Kaiser. Copyright © 1980 University of Texas at Austin Bureau of Economic Geology. Reprinted with permission.

EXPLANATION

	Sand body with calculated chlorinity
	Contour interval variable
	Cl⁻ ≤ 500 mg/l
(A)	Feature discussed in text

Flow Boundaries

Aquitard

Fault (dashed where inferred)

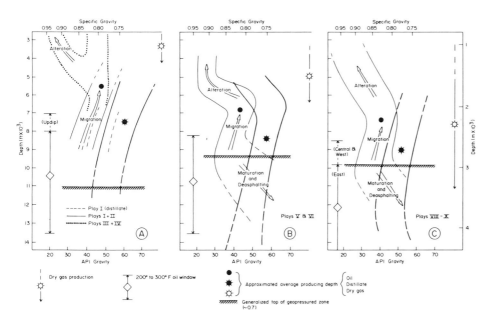

Figure 6—Generalized envelopes around Frio oil and condensate density values plotted against reservoir depth for (**A**) lower, (**B**) middle, and (**C**) upper coastal plain areas. Envelopes include more than 90% of several tens to hundreds of data points and reveal consistent vertical trends in petroleum chemistry. Well symbols show approximated average depth of all producing reservoirs. Average vertical extent of the oil-generation window and the average operational top of geopressure are also shown. Hydrocarbon liquid distributions indicate large-scale upward migration from point of origin to point of entrapment. The increasing specific gravity seen in the shallow parts of curves B and C is interpreted to reflect widespread bacterial alteration of crudes following entrapment. From Galloway et al. Copyright © 1982 University of Texas at Austin Bureau of Economic Geology. Reprinted with permission.

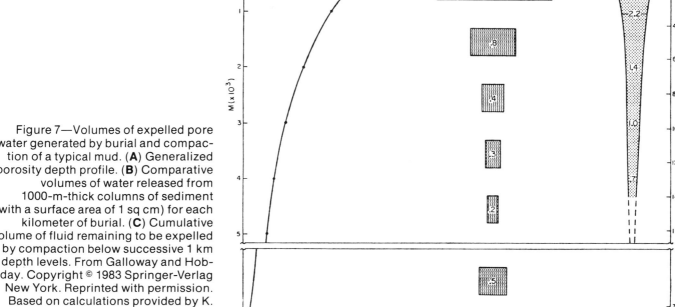

Figure 7—Volumes of expelled pore water generated by burial and compaction of a typical mud. (**A**) Generalized porosity depth profile. (**B**) Comparative volumes of water released from 1000-m-thick columns of sediment (with a surface area of 1 sq cm) for each kilometer of burial. (**C**) Cumulative volume of fluid remaining to be expelled by compaction below successive 1 km depth levels. From Galloway and Hobday. Copyright © 1983 Springer-Verlag New York. Reprinted with permission. Based on calculations provided by K. Bjorlykke.

kaolinite cement and grain replacements, authigenic chlorite, and widespread leached detrital grains and cements and partially to completely albitized plagioclase.

Thermobaric Regime

Deeper portions of the Frio contain waters exhibiting measured pressures considerably above hydrostatic. The top of this geopressured zone ideally is picked on the basis of pressure measurements at the first occurrence of such abnormally high pressures. The depth of the absolute top of geopressure, where gradients based on in-situ measurement exceed 0.465, typically lies 2.5 to 3 km below the surface. An "operational" top of geopressure that is more readily picked indirectly from electrical log measurements is commonly used for regional mapping of the geopressured zone, and is the top of geopressure used in Figures 3, 4, and 6. Fluid pressure gradients at the top of operational geopressure approach 0.7, and the underlying section is substantially overpressured. The top of operational geopressure lies within a thousand meters or so of the top of the first overpressuring.

Near the top of geopressure, temperatures in the Frio Formation are adequate for conversion of smectite layers to illite in mixed-layer clays, with consequent release of interlayer water (Freed, 1981, 1982). Clay dewatering itself may, provided bulk aquifer transmissivity is sufficiently low, be a cause of geopressuring (Burst, 1969). However, the author's observations support those of Magara (1978) and suggest that the mechanism does not fully explain associated petrophysical phenomena such as shale density decrease observed in the overpressure zone. A simpler explanation for young basins characterized by rapid deposition of deltaic and slope systems is that the rate of sediment loading exceeds the potential for fluid discharge under moderate hydraulic gradients (Bredehoeft and Hanshaw, 1968). Within the geopressure zone, decreased thermal conductivity of the sediment column results in an increased thermal gradient that accentuates further clay dewatering. Thus, regardless of the specific cause or causes of overpressuring, the deeper zone of geopressure in the Frio

Formation displays the key features that define a distinct ground-water regime.

Actual volume of released water is poorly documented (for a review, see Graf and Anderson, 1981). Nonetheless, clay dewatering offers a mechanism to "recharge" the deeper portion of the compactional regime with a pulse of relatively fresh water. At the same time, higher temperatures associated with the regime lead to thermal alteration of organic materials, producing abundant hydrocarbon liquids and gases (Figs. 3, 6). Analyses of waters of the geopressured zone indicate acidic, highly gas-charged sodium-chloride brines of variable salinity (Galloway, 1982; Kaiser, this volume).

Burial diagenetic features of Frio sandstones commonly ascribed to the geopressure zone include both ferroan and iron-poor carbonate cements, kaolinite and chlorite cements, completely albitized plagioclase feldspars, and thoroughly leached or albitized K-feldspars.

Regime Interactions

The three basin hydrologic regimes can be recognized and described within the Frio Formation today. However, their present distribution reflects only one point in the hydrologic evolution of the northern Gulf Coast Basin. Most, or perhaps all of the diagenetic events recognized in the Frio record early stages in this hydrologic evolution. Distribution and paragenesis of diagenetic features provide indirect evidence of fluxes and interactions between fluids of different regimes. In addition, data such as ambient formation water chemistry, in-situ pressure or temperature trends, and patterns of hydrocarbon distribution and physical properties all provide information on active or recent fluid fluxes.

(1) Pervasive upward flux of water through and out of the geopressured zone into the overlying normally pressured compactional zone is indicated directly by pressure distributions around fault zones, which may act as conduits for cross-stratal discharge (Berg and Habeck, 1982), and by thermal, pressure, and salinity anomalies observed within intermediate-depth portions of the Tertiary section (Harrison, 1980).

(2) Upward fluid migration is indirectly indicated by the vertical displacement of the center of mass of hydrocarbons contained in commercial reservoirs by more than 1 km above their zone of thermal generation (Fig. 6). Upward migration is further suggested by the evidence of chromatographic segregation of hydrocarbon liquids between their point of origin within the thermobaric regime and their point of residence in the compactional regime (Galloway et al, 1982) (Fig. 6).

(3) Indirect evidence for considerable mixing and interaction of waters of the compactional and thermobaric regime includes the pervasive zone of leached porosity development that characterizes intermediate depths (2.2–3.0 km) within the Frio and other Tertiary formations (Loucks et al, 1979).

(4) Intrusion of sulfide-rich thermobaric waters derived from deepest portions of the geopressured zone have played a critical role in the origin of major uranium deposits within the Gueydan fluvial system (Galloway and Kaiser, 1980). These waters, which have discharged all the way upsection into shallow portions of the meteoric regime, have produced tongues of pyritic sandstone within the syndepositionally oxidized shallow aquifers (Fig. 8). Such pyritic zones, which may record multiple pulses of intrusion and sulfidization, are centered around deep-seated growth faults (Galloway and Kaiser, 1980; Galloway, 1982). As in deeper examples of interpreted vertical fluid migration, fault zones appear to have served as principal conduits for discharge from deeper into shallower ground-water regimes. The resultant mixing zone is a focus for diagenetic alteration.

In summary, cursory examination of the Frio/Catahoula Formation as an example of a thick clastic wedge in an actively subsiding basin reveals the continuing presence of each of the three major basin hydrologic regimes. Further, reactions caused by upward flux and mixing of pressure-driven waters from the thermobaric regime continue (Fig. 9), though probably on a reduced scale because the depocenter has shifted basinward. Meteoric waters continue to invade the shallow and

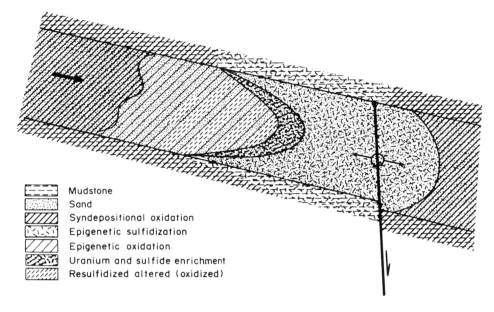

Figure 8—Schematic cross section showing the development of a pyrite-rich tongue within a shallow, oxidized aquifer by intrusion of sulfide-rich thermobaric waters along deep-seated growth faults. Such alteration of the shallow Catahoula sandstones was a prerequisite for uranium mineralization. Multiple pulses of deep-water discharge into the meteoric regime are recorded in many south Texas uranium districts. Arrows indicate the opposed, but temporally alternating counterflows of meteoric and thermobaric waters responsible for the superimposed alteration zones. From Galloway. Copyright © 1982 University of Texas at Austin Bureau of Economic Geology. Reprinted with permission.

Legend:
- Mudstone
- Sand
- Syndepositional oxidation
- Epigenetic sulfidization
- Epigenetic oxidation
- Uranium and sulfide enrichment
- Resulfidized altered (oxidized)

intermediate subsurface, expanding their influence into portions of the Frio originally dominated by upward flux of compactional waters and hydrocarbons.

Finally, it should be reemphasized that fluid flow and mixing are not limited to individual depositional wedges. The hydrologic basin encompasses the entire sediment pile. Cross-cutting growth-fault zones and diapirs provide avenues for vertical fluid movement and mixing. The extent of this mixing and the importance of the deep fluids remain topics of speculation and investigation.

In addition, this discussion has focused on description of hydrologic zones and their interactions within the basin fill. Recently, thermally driven convection of waters within aquifers has been proposed as a possible mechanism of fluid mass-transfer within sedimentary basins (Wood and Hewett, 1982). Convection has the attraction of removing some of the perplexing mass-balance limitations implicit in the finite volumes of water ultimately available from compactional and thermal dewatering of the sediment pile. Whether convection can reasonably operate at a scale adequate to explain the requisite vertical displacement of solids proposed in current diagenetic models remains to be tested in the Gulf Coast Tertiary Basin. Factors in a depositional sequence, such as the Frio Formation, that must be accommodated by a convection model include the lateral segmentation of aquifers by growth faults, the vertical pressure differentials and overall upward-directed pressure head, and the lithologic and permeability heterogeneity of the aquifers themselves. Pressures, temperature, and salinity data may provide evidence of the existence and scale of convection cells if examined with this new model in mind.

A FOCUS ON HYDROLOGIC EVOLUTION OF SANDSTONES

From the time of deposition, all but a thin veneer of terrestrial sediment is bathed in subsurface waters. The history of evolving fluxes and hydrochemistries largely determines the diagenetic history of the sediment. Carbonate sedimentologists have recognized this fact and emphasized the hydrologic regime of diagenesis. Its emphasis has been less explicit in most regional models of clastic diagenesis. As indicated in Figure 10, however, the hydrologic pathways experienced by a regional depositional sequence, such as the Frio Formation, may be complex and diverse. Depending on depositional environment, initial waters may be either meteoric or depositional connate. With ongoing deposition, portions of the sediment volume may be swept by semi-confined to confined meteoric waters, which react with the aquifer and evolve chemically (meteoric flush, Fig. 10). With burial, sands contained in mud-rich sequences, particularly in portions of the sediment pile that lie below subaqueous portions of the basin (and therefore seaward of the coastal discharge zone of the meteoric system), are flushed by chemically evolved pore waters squeezed from compactable muds (compaction flush). As the compaction flush wanes, a second pulse of water derived from thermal dehydration of hydrous mineral phases, and most likely characterized by significantly different chemistry, pervades deeply buried portions of the aquifer and discharges up into the overlying section (thermobaric flush). Finally, as subsidence, depositional loading, and thermal alteration run their course, meteoric penetration increasingly dominates the basin hydrology, and sediments may again be flushed by chemically evolved meteoric waters.

As the three hydrologic zones evolve, boundaries between them, which are better described as ill-defined mixing zones, are likely sites of active diagenetic alteration of the aquifer plumbing, the sandstones. Thermally driven convection cells may evolve within the compactional and thermobaric systems, further complicating the dynamics of this already restless system.

The three major hydrogeologic regimes persist within the Frio. Diagenetic products, within both the deep and shallow portions of the Frio/Catahoula section, record the hydrologic evolution of this portion of the Gulf Basin fill. The new data included in several companion contributions in this

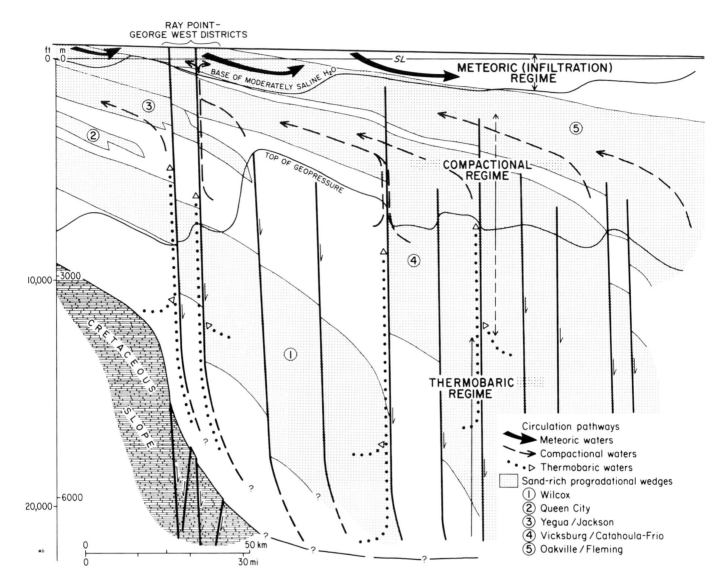

Figure 9—Ground-water regimes and circulation pathways within the offlapping Tertiary fill of the Texas Coastal Plain. Growth fault zones provide conduits for fluid discharge from deeper hydrologic regimes. Major mixing zones lie along the boundary between the geopressured and hydropressured zones and adjacent to shallow growth faults in the meteoric regime. From Galloway. Copyright © 1982 University of Texas at Austin Bureau of Economic Geology. Reprinted with permission.

volume provide a basis for a thorough evaluation of the hydrodynamics of sandstone diagenesis. Study of the modern waters and their hydrochemistries is a starting point for the reconstruction of diagenetic history.

ACKNOWLEDGMENTS

This paper was originally prepared for oral presentation at a symposium on diagenesis at the 1982 International Sedimentological Congress. Comments and discussion by several attendees added to the ideas presented here. The manuscript was reviewed by W. R. Kaiser, G. E. Fogg, and R. A. Morton. S. Doenges edited the draft manuscript. Drafting of figures was under the supervision of D. Scranton and J. Macon.

SELECTED REFERENCES

Bebout, D. G., R. G. Loucks, and A. R. Gregory, 1978, Frio sandstone reservoirs in the deep subsurface along the Texas Gulf Coast: University of Texas, Bureau of Economic Geology Report of Investigations 91, 92 p.

Berg, R. R., and M. F. Habeck, 1982, Abnormal pressures in the lower Vicksburg, McAllen Ranch field, South Texas: Gulf Coast Association of Geological Societies Transactions, v. 32, p. 247–262.

Bogomolov, Y. G., A. V. Kudelsky, and N. N. Lapshin, 1978, Hydrology of large sedimentary basins, in Hydrology of great sedimentary basins: Publication of the International Association of Scientific Hydrology, No. 120, p. 117–122.

Bredehoeft, J. D., and B. B. Hanshaw, 1968, On the maintenance of anomalous fluid pressures, 1, Thick sedimentary sequences: Geological Society of America Bulletin, v. 79, p. 1097–1106.

Burst, J. F., 1969, Diagenesis of Gulf Coast clayey sediments and its possible relation to petroleum migration: Bulletin of the

Figure 10—Generalized pathways for hydrologic evolution of a major stratigraphic unit such as the Frio/Catahoula Formation. Through its depositional and subsequent burial history, a volume of sediment may be successively swept by meteoric, compactional, and thermobaric waters. Old, depositionally moribund basins are commonly again permeated by meteoric water. The successive fluxes of these chemically differing waters and evolution of their mixing zones are recorded in the diagenetic history of the sandstone aquifers.

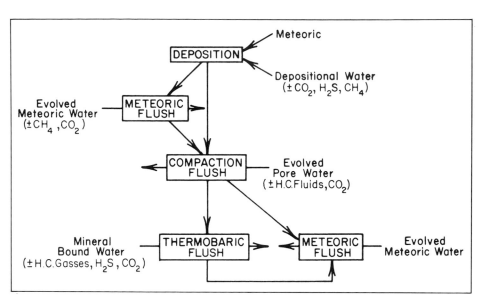

American Association of Petroleum Geologists, v. 53, p. 73–93.

Carothers, W. W., and Y. K. Kharaka, 1980, Stable carbon isotopes of HCO_3 in oil field waters—implications for the origin of CO_2: Geochimica et Cosmochimica Acta, v. 44, p. 323–332.

Clayton, R. N., I. Friedman, D. L. Graf, T. K. Mayeda, W. F. Meents, and N. F. Shimp, 1966, The origin of saline formation waters: Journal of Geophysical Research, v. 71, p. 3869–3882.

Cousteau, H., L. Rumeau, and C. Sourisse, 1975, Classification hydrodynamique des bassins sedimentaires: utilisation combinee avec d'autres methodes pour rationaliser l'exploration des bassins nonproductifs: Proceedings of the 9th World Petroleum Congress, v. 2, p. 105–119.

Freed, R. L., 1981, Shale mineralogy and burial diagenesis of Frio and Vicksburg Formations in two geopressured wells, McAllen Ranch area, Hidalgo County, Texas: Gulf Coast Association of Geological Societies Transactions, v. 31, p. 289–293.

————, 1982, Clay mineralogy and depositional history of the Frio Formation in two geopressured wells, Brazoria County, Texas: Gulf Coast Association of Geological Societies Transactions, v. 32, p. 459–463.

Galloway, W. E., 1977, Catahoula Formation of the Texas Coastal Plain: depositional systems, composition, structural development, ground-water flow history, and uranium distribution: University of Texas Bureau of Economic Geology Report of Investigations 87, 59 p.

————, 1982, Fluid flow histories and resultant epigenetic zonation of uranium-bearing fluvial aquifer systems, South Texas uranium province: University of Texas Bureau of Economic Geology Report of Investigations 119, 31 p.

Galloway, W. E., and D. K. Hobday, 1983, Terrigenous clastic depositional systems: application to petroleum, coal, and uranium exploration: New York, Springer-Verlag.

Galloway, W. E., D. K. Hobday, and K. Magara, 1982, Frio Formation of the Texas Gulf Coastal Plain—depositional systems, structural framework, and hydrocarbon distribution: Bulletin of the American Association of Petroleum Geologists, v. 66, p. 649–688.

Galloway, W. E., and W. R. Kaiser, 1980, Catahoula Formation of the Texas Coastal Plain: origin, geochemical evolution, and characteristics of uranium deposits: University of Texas Bureau of Economic Geology Report of Investigations 100, 81 p.

Graf, D. L., and D. E. Anderson, 1981, Geochemical inputs for hydrological models of deep-lying sedimentary units: loss of mineral hydration water: Journal of Hydrology, v. 54, p. 297–314.

Harrison, III, F. W., 1980, The role of pressure, temperature, salinity, lithology and structure in hydrocarbon accumulation in Constance Bayou, Deep Lake, and Southeast Little Pecan Lake fields, Cameron Parish, Louisiana: Gulf Coast Association of Geological Societies Transactions, v. 30, p. 113–127.

Hitchon, B., 1969a, Fluid flow in the western Canada sedimentary basin, 1, Effect of topography: Water Resources Research, v. 5, p. 186–195.

————, 1969b, Fluid flow in the western Canada sedimentary basin, 2, Effect of

geology: Water Resources Research, v. 5, p. 460–469.

Kissin, I. G., 1978, The principal distinctive features of the hydrodynamic regime of intensive earth crust downwarping areas, in Hydrogeology of great sedimentary basins: Publication of the International Association of Scientific Hydrology, no. 120, p. 178–185.

Korotchansky, A. N., and J. Mitchell, 1972, Rejects de dechets dans les nappes profondes: 24th International Geological Congress, Sect. 11, p. 282–295.

Loucks, R. G., M. M. Dodge, and W. E. Galloway, 1979, Importance of secondary leached porosity in lower Tertiary sandstone reservoirs along the Texas Gulf Coast: Gulf Coast Association of Geological Societies Transactions v. 29, p. 1–8.

Magara, K., 1978, Compaction and fluid migration—practical petroleum geology: Amsterdam, Elsevier, 319 p.

Toth, J., 1980, Cross-formational gravity-flow of groundwater: a mechanism of the transport and accumulation of petroleum (the generalized hydraulic theory of petroleum migration): American Association of Petroleum Geologists, Studies in Geology no. 10, p. 121–167.

Wood, J. R., and T. A. Hewett, 1982, Fluid convection and mass transfer in porous sandstones—a theoretical model: Geochimica et Cosmochimica Acta, v. 46, p. 1707–1713.

Regional Controls on Diagenesis and Reservoir Quality in Lower Tertiary Sandstones along the Texas Gulf Coast

Robert G. Loucks
Cities Service Company
Tulsa, Oklahoma

Marianne M. Dodge
Heritage Oil Corporation
San Antonio, Texas

William E. Galloway
Bureau of Economic Geology
Austin, Texas

INTRODUCTION

General Statement

Prediction of reservoir quality (as controlled by porosity and permeability) of a target sandstone is a valuable exploration tool. Along the Texas Gulf Coast, hydrocarbon production in shallowly buried sandstones is not commonly limited by reservoir quality; but in deeply buried sandstones reservoir quality can be quite variable because of extensive diagenesis. An understanding of the factors in a given basin that control diagenesis and, hence, reservoir quality, can enhance reservoir-quality prediction and improve exploration success ratios.

A regional assessment of the factors that control diagenesis in the Tertiary stratigraphic section of the onshore Texas Gulf Coast has delineated porosity and permeability trends in the major sandstone units. The study addressed the Wilcox Group and the Vicksburg and Frio Formations (Fig. 1), which are areally extensive and have thick sandstone sections.

Objectives and Scope of Study

The major objective of this regional investigation was to document the sandstone diagenetic history and delineate reservoir quality trends of the onshore Lower Tertiary stratigraphic section along the Texas Gulf Coast. Emphasis was placed on describing, quantifying, and interpreting the formation, preservation, and vertical and lateral distribution of sandstone porosity and permeability in order to develop a predictive capability for identifying favorable reservoir areas.

ABSTRACT. Reservoir quality trends in Lower Tertiary sandstones along the Texas Gulf Coast are a product of regional variations in intensity of diagenesis. The major controls on diagenesis were detrital mineralogy and regional geothermal gradient. Porosity and permeability in sandstones shallower than 3350 m (11,000 ft) are generally adequate for hydrocarbon production, whereas reservoir quality in deeper sandstones in the onshore Lower Tertiary section is highly variable. Many of these sandstone reservoirs have permeability values of less than 1 millidarcy (md), but in a few areas permeability values are higher than 1000 md.

Wilcox sandstones are poorly to moderately sorted, fine-grained, quartzose lithic arkoses, becoming more quartz-rich from the upper to the lower Texas Gulf Coast. Most rock fragments are metamorphic or volcanic in origin. Wilcox sandstones exhibit no systematic regional reservoir quality trends. Along the lower and parts of the middle and upper Texas Gulf Coast, deep Wilcox sandstones are tight, but in other parts of the middle and upper Texas Gulf Coast, porosity exists at depth. Vicksburg sandstones are poorly sorted, fine-grained lithic arkoses. Rock fragments are mainly volcanic clasts with lesser carbonate and minor metamorphic clasts. The deep Vicksburg Formation has low-quality reservoirs. Frio sandstones range from poorly sorted, fine-grained, feldspathic litharenites to lithic arkoses along the lower Texas Gulf Coast to poorly sorted, fine-grained, quartzose lithic arkoses to subarkoses along the upper Texas Gulf Coast. Volcanic and carbonate rock fragments are common in the lower Texas Gulf Coast and decrease in abundance up the coast. Frio sandstones show a systematic improvement in reservoir quality from the lower to the upper Texas Gulf Coast that is related to grain composition and geothermal gradient. Reservoir quality trends in Tertiary sandstones have been substantiated by acoustic log analysis.

In spite of variations in composition, Lower Tertiary sandstones exhibit similar diagenetic sequences generalized as follows:
Surface-to-shallow-subsurface diagenesis (0 to 1200 m ±; 0 to 4000 ft ±) began with the formation of clay coats on framework grains, dissolution of feldspar, and replacement of feldspar by calcite. Minor amounts of kaolinite, feldspar overgrowths, and Fe-poor calcite was locally precipitated. Porosity was commonly reduced by compaction and cementation from an estimated original 40% to less than 30%.   *(continued)*

Intermediate subsurface diagenesis (1200 to 3400 m ±; 4000 to 11,000 ft ±) involved dissolution of eary carbonate cements and subsequent cementation by quartz overgrowths and later by carbonate cement. Cementation commonly reduced porosity to 10% or less, but this trend could be reversed by later dissolution of feldspar grains, rock fragments, and carbonate cements. Restoration of porosity to more than 30% occurred, but some porosity was later reduced by kaolinite, Fe-rich dolomite, and ankerite cementation.

Deep subsurface diagenesis (>3400 m ±; >11,000 ft ±) was a continuation of late Fe-rich and Fe-poor carbonate cement precipitation. Plagioclase was albitized during this stage.

Differences in intensity of diagenetic events and depths at which they first occurred correspond to the chemical and mechanical stability of the original detrital mineralogy and to regional variations in geothermal gradient.

Specific objectives were:

(1) To delineate the mineralogical composition of sandstones in the Wilcox Group and the Vicksburg and Frio Formations and to establish regional compositional trends in each of these units.

(2) To synthesize a general sandstone diagenetic sequence for the entire Lower Tertiary section.

(3) To generalize regional reservoir-quality trends.

(4) To relate interval transit time from acoustic logs to sandstone diagenesis, in order to broaden the interpretation of the diagenetic sequence and its effects on reservoir quality beyond those areas where core samples were available.

Methodology

The onshore Texas Gulf Coast was divided into six geographic areas (Fig. 2) in order to delineate regional trends. Areas 1 and 2 (lower Texas) include the Rio Grande Embayment, Areas 3 and 4 (middle Texas) straddle the San Marcos Arch, and Areas 5 and 6 (upper Texas) include the Houston Embayment.

The primary data base for the sandstone study consisted of whole cores and core plugs from 179 wells (Fig. 2), core-plug porosity and permeability analyses from 253 wells (Fig. 3), and acoustic logs from 86 wells (Fig. 4). Lithology and primary structures of the cores were described and environments of deposition were interpreted. Nine-hundred-and-sixty-one thin sections were prepared from texturally mature, matrix-poor (less than 5% mud) sand-

stones at approximately 15 m (50 ft) intervals. Only matrix-poor sandstones were selected for study because the emphasis of the investigation was on delineation of high-quality reservoirs. Five-hundred-and-forty of these thin sections were impregnated with blue-dyed epoxy, and 200 points per slide were counted for framework grain mineralogy, cement composition, and porosity types. Grain size, sorting, and packing proximity (a measurement of compaction as defined by Kahn, 1956) were also determined. All thin sections were treated with amaranth solution to stain Ca-bearing plagioclase pink and with sodium cobaltinitrite to stain potassium feldspar yellow, using a technique adapted from Laniz et al (1964). Selected thin sections containing carbonate cements were treated with alizarin red-S to stain nonferroan calcite red and with potassium ferricyanide to stain ankerite, ferroan calcite, and ferroan dolomite blue, using the method of Lindholm and Finkelman (1972). Selected samples were then analyzed with the electron microprobe for carbonate composition, with the scanning electron microscope for mineral composition, textural relationships, and diagenetic products, and with the X-ray diffractometer for mineral composition.

Porosity and permeability data were obtained from both whole core and sidewall core samples (Fig. 3). One-hundred-and-fifty-six wells with whole cores (7564 data points) and 97 wells with sidewall cores (3559 data points) composed the data base.

Interval transit time for sands and

sandstones was calculated from acoustic logs at 33 to 66 m (100 to 200 ft) depth intervals (Fig. 4). Graphs of interval transit time versus depth were prepared for each well. Interval transit times were also grouped by area and by trend. The updip trend is composed of wells aligned along a trend parallel to the coast which were drilled to the Wilcox Group; the downdip trend is composed of wells drilled to the Vicksburg and/or Frio Formations (Fig. 4).

Geological Setting

The onshore Lower Tertiary section along the Gulf Coast is composed of terrigenous clastic wedges, which thicken downdip toward the Gulf of Mexico (Fig. 1). Rapid loading of sand on water-saturated prodelta and continental slope muds resulted in contemporaneous growth faulting and subsequent vertical accumulation of great quantities of deltaic and strandplain sands and muds. Equivalent sediments updip remained in the relatively shallow subsurface, whereas sediments downdip were subjected to more rapid subsidence and deeper burial. Continuous movement along growth faults isolated thick wedges of sand and mud and trapped connate fluids, which created an overpressured zone (fluid pressure greater than hydrostatic pressure of 0.465 psi/ft).

GENERAL RESERVOIR QUALITY TRENDS

Sources of Data

Core analysis data from 253 wells were examined in this study (Fig. 3). Such direct measurements are the best measure of reservoir quality, short of production tests. The principal drawback of core analysis is that porosity and permeability measurements are made at atmospheric pressures and temperatures, and thus do not represent true in-situ permeability to the original pore fluids. Values are often an order of magnitude too high. Of these 253 wells, only the 156 wells with analyses from whole cores were used to determine regional porosity and permeability trends along the Texas Gulf Coast. Core plugs, taken by drilling a cylinder into a whole core, do not disturb the fabric of consolidated sediments. In contrast, a sidewall core is taken by blasting a small hollow metal

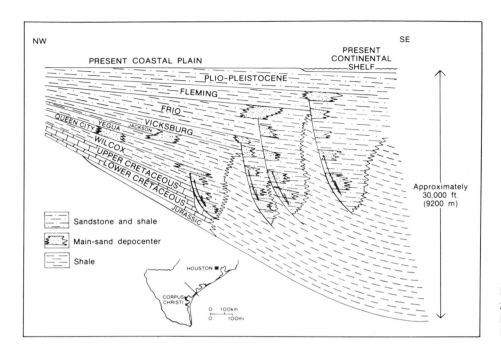

Figure 1—Depositional style of Cenozoic strata along the Texas Gulf Coast. From Bebout et al, 1979. United States Department of Energy.

Figure 2—Area of investigation showing division of lower, middle, and upper Texas Gulf Coast areas and location of wells with whole core samples. Numbers refer to well names listed in Loucks et al (in press). Copyright © University of Texas at Austin Bureau of Economic Geology. Reprinted with permission.

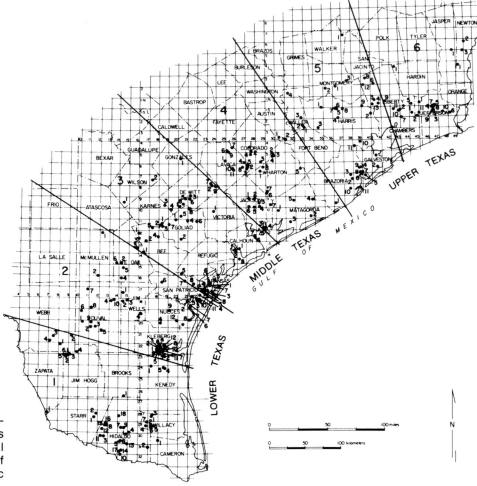

Figure 3—Location of wells with porosity and permeability data. Numbers
refer to well names listed in Loucks et al
(in press). Copyright © University of
Texas at Austin Bureau of Economic
Geology. Reprinted with permission.

cylinder horizontally into the side of
the well. The explosive impact of the
cylinder into the rock often fractures
the sample. Thin sections made from a
sidewall core commonly show numerous fine fractures. A sidewall core,
therefore, tends to give a much higher
porosity value than a core plug. Below
a depth of 1500 m (5000 ft), average
porosity values from sidewall cores
deviate significantly and systematically
from those of whole cores, because the
sediments have begun to lithify and
fracturing occurred when sidewall cores
were taken. This error increases with
depth (Fig. 5). Permeability as well as
porosity readings are affected. Therefore, only porosity and permeability
values from whole core analyses were
used in this investigation.

Porosity and Permeability Trends

A variety of graphic displays are used
to show the patterns of reservoir quality along the Texas Gulf Coast. Best-fit
lines were drawn by visual inspection,
and correlation coefficients were not
calculated because of the large number
of variables influencing these patterns.

Sandstone porosity and permeability
generally decrease with depth through
compaction and cementation, although
this trend may be reversed by dissolution of grains and cements (Chepikov et
al, 1959, 1961; Savkevich, 1969; Yermolova and Orlova, 1962; Schmidt et
al, 1977). There is a general decrease in
porosity and permeability with depth in
the Texas Gulf Coast section (Figs.
6–8), but there is a wide range of values
at any given depth (Figs. 7, 8), indicating the complexities involved in understanding controls on reservoir quality.
Variation in porosity and permeability
at depth is a function of cementation
and dissolution.

A comparison of permeability and
porosity data for the Wilcox Group

and the Vicksburg and Frio Formations shows a general relationship be
tween permeability and porosity (Fig.
9). In general, permeability increases as
porosity increases.

By superimposing plots of porosity
versus depth for each formation, differences in reservoir quality among formations can be compared (Fig. 10). The
Vicksburg Formation stands out as the
unit with the poorest reservoir quality.
Plots for Wilcox, Queen City, Yegua/
Jackson, and Frio sandstones tend to
coincide. However, if Frio values are
grouped by Areas 1 through 3 and
Areas 4 through 6 (Figs. 2, 10), Frio
sandstones of the upper Texas Gulf
Coast exhibit the best reservoir quality
of any unit.

Porosity-versus-depth plots for six
areas of the Wilcox Group and five
areas of the Frio Formation indicate no
strong systematic regional porosity
trend in the Wilcox (Fig. 11) and a

Figure 4—Location of wells with acoustic logs. Numbers refer to well names listed in Loucks et al (in press). Copyright © University of Texas at Austin Bureau of Economic Geology. Reprinted with permission.

strong regional porosity trend in the Frio (Fig. 12). The Frio Formation displays a systematic increase in reservoir quality from Area 1 northward to Area 5. This increase in reservoir quality correlates with changes in rock composition, thermal gradient, and diagenesis.

REGIONAL CONTROLS ON RESERVOIR QUALITY

General Statement

Regional influences on reservoir quality include depositional environment, initial composition of the sand and associated muds, texture, time (geologic age), subsidence rate, pressure, thermal gradient, pore-fluid composition, and diagenetic history.

Of the above factors, only three were readily correlated with reservoir quality: (1) original mineral composition of the rock, (2) geothermal gradient, and (3) sequential diagenetic changes in-

cluding cementation, replacement, and dissolution.

Mineralogy

Sandstone Classification

Folk's (1968) sandstone classification, based on three end-members— quartz, feldspar, and rock fragments (Fig. 13)—was used in this report for two reasons. The compartmentalization of the classification triangle emphasizes feldspar and rock fragments, the more chemically unstable grains, making the classification useful in grouping sandstones as to their probable degree of reaction during diagenesis. It was also useful because chert is grouped with rock fragments instead of with quartz. Chert in the Texas Lower Tertiary strata was commonly difficult to distinguish from silicified volcanic rock fragments. The term quartzose has been added to Folk's classification to delineate sandstone types with 50% or more quartz but less than 75% quartz.

Wilcox Group

The majority of samples used in this study are from the upper Wilcox Group as defined by Bebout et al (1979). These Wilcox sandstones are typically poorly to moderately well-sorted, fine-grained quartzose lithic arkoses (Fig. 14). Quartz content increases slightly from upper to lower Texas Gulf Coast; orthoclase increases in the opposite direction. There is also an increase in sorting and a slight decrease in grain size from north to south. These observations suggest transport from upper to lower Texas, but sandstone distribution and rock-fragment composition indicate a more complicated situation.

The upper Wilcox Group was deposited in a series of deltaic systems along the Texas Gulf Coast. In middle and upper Texas Gulf Coast areas, upper Wilcox deltas prograded over stable, sandy lower Wilcox deposits and only relatively small quantities of sand were transported to the lower Wilcox shelf margin. In lower Texas, upper Wilcox

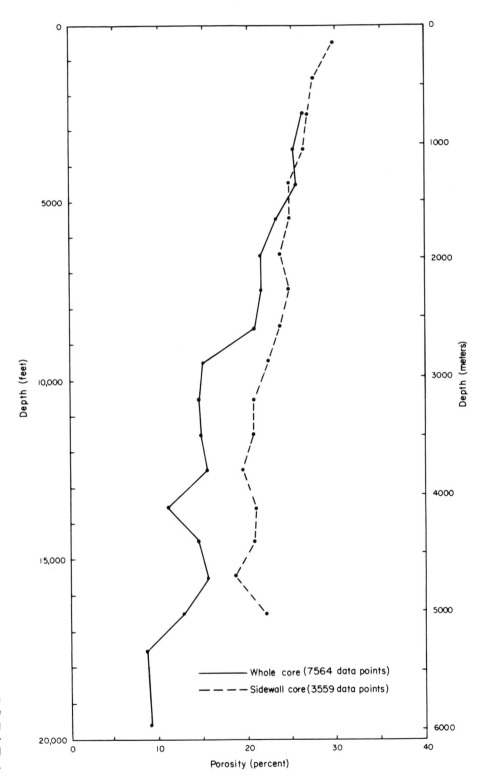

Figure 5—Mean porosity versus depth from both whole core and sidewall core from Lower Tertiary sandstones along the Texas Gulf Coast. Data were averaged over 1000-ft intervals and plotted at the midpoints of those intervals.

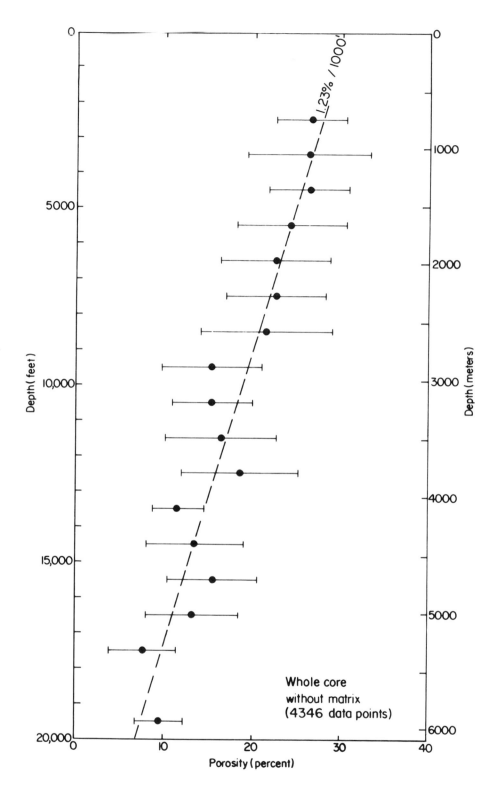

Figure 6—Mean sandstone porosity versus depth from whole-core analyses for Lower Tertiary units along the Texas Gulf Coast. Only values from sandstones without matrix were used. Raw data (Fig. 7) were averaged over 1000-ft intervals and plotted at the midpoints of those intervals. Brackets indicate standard deviation.

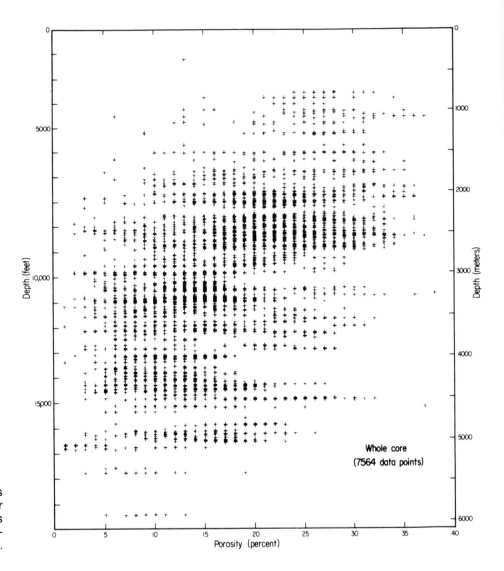

Figure 7—Sandstone porosity versus depth from whole-core analyses for Lower Tertiary units along the Texas Gulf Coast. Sandstones with and without matrix were included.

deltas prograded over unstable, muddy sediments and were subjected to extensive growth faulting. Sandstone depocenters were geographically restricted along strike indicating that the upper Wilcox was essentially dip fed, and significant transport of sediments along strike did not occur (Bebout et al, 1979).

The composition of Wilcox rock fragments suggests source areas to the north for middle and upper Texas deltas and source areas to the west for lower Texas deltas. Metamorphic and volcanic rock fragments are the major lithic debris in the Wilcox sandstones. Metamorphic rock fragments are low to medium grade, ranging from slates to quartzites and muscovite schists.

Muscovite is a common accessory mineral. Most volcanic rock fragments are highly altered and typically silicified. Many grains previously identified as detrital chert may actually be silicified volcanic rock fragments. Unaltered volcanic rock fragments are very similar to those in the Frio Formation identified as being predominantly rhyolitic in composition (Lindquist, 1977). Wilcox sandstones along the lower Texas Gulf Coast have approximately equal amounts of volcanic and metamorphic rock fragments; along the middle and the upper Texas Gulf Coast, metamorphic rock fragments predominate. The volcanic rock fragments were probably derived from west Texas, southern New Mexico, and/or

northern Mexico (Cook and Bally, 1975). A source for the metamorphic rock fragments is still in dispute. Todd and Folk (1957) and Boggs (1978) proposed the southern Appalachian Uplift and the Ouachita Fold Belt as the metamorphic source, whereas Storm (1945) and Murray (1955) postulated that the Rocky Mountains and the Central Interior supplied metamorphic rock fragments. Carbonate rock fragments are a minor constituent in Tertiary strata along the upper Texas Gulf Coast and probably were locally derived from Cretaceous rocks from the central Texas platform.

Trends in grain size, sorting, and quartz content must be considered separately for the upper and middle

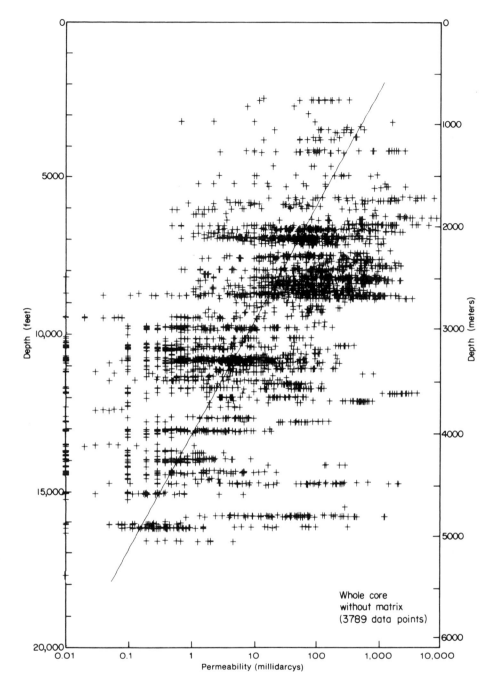

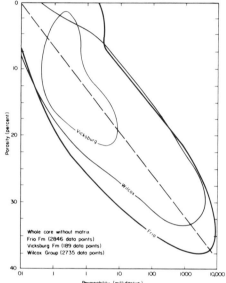

Figure 8—Permeability versus depth from whole-core analyses for Lower Tertiary units along the Texas Gulf Coast.

Figure 9—Relationship of porosity to permeability for the Wilcox Group and the Vicksburg and Frio Formations.

Gulf Coast and for the lower Gulf Coast of Texas. Wilcox sandstones are slightly more mature in middle Texas than in upper Texas because of greater distance of transport and limited strike transport. The seemingly greater maturity of Wilcox sandstones in lower Texas is probably an artifact of a different source area.

Vicksburg Formation
Vicksburg sandstones were sampled from the lower Texas Gulf Coast region (Areas 1 and 2) where the only deep, massive sandstones in this formation occur (Loucks, 1978). These sandstones are poorly sorted, fine-grained lithic arkoses (Fig. 15). Rock fragments are mainly volcanic clasts, reflecting more

extensive volcanism in west Texas and in Mexico during Vicksburg time; lesser amounts of both carbonate and metamorphic rock fragments are present. The ancient Rio Grande transported this volcanic sediment into the rapidly subsiding Rio Grande Embayment in the lower Texas region. Carbonate rock fragments are eroded caliche clasts similar to those in the Frio Formation and indicate arid conditions along the lower Texas Gulf Coast. The Rocky Mountain area may have been the source of the metamorphic rock fragments.

Frio Formation
Of the Lower Tertiary units, the Frio Formation shows the greatest regional

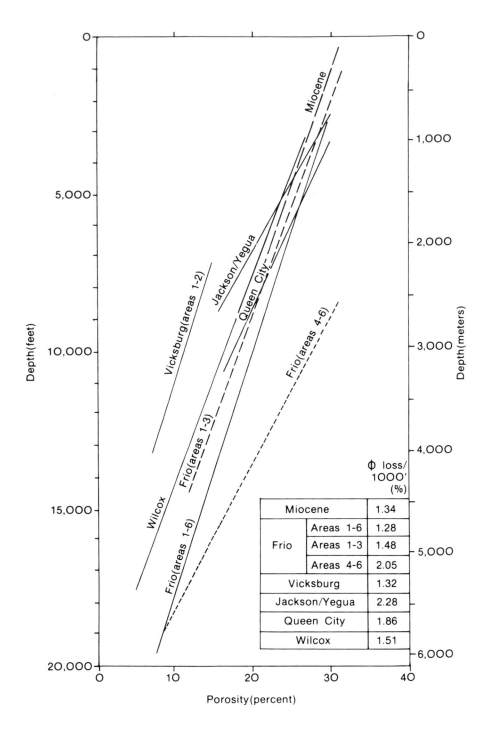

Figure 10—Mean sandstone porosity versus depth by unit for Lower Tertiary strata along the Texas Gulf Coast. Table in lower right hand corner shows porosity loss per 1000 ft for each formation.

		φ loss/1000' (%)
Miocene		1.34
Frio	Areas 1-6	1.28
	Areas 1-3	1.48
	Areas 4-6	2.05
Vicksburg		1.32
Jackson/Yegua		2.28
Queen City		1.86
Wilcox		1.51

variation in mineral composition. Along the lower Texas Gulf Coast (Areas 1 and 2), Frio sandstones are poorly sorted, fine-grained, feldspathic litharenites to lithic arkoses (Fig. 16). Middle Texas Gulf Coast (Areas 3 and

4) Frio sandstones are moderately to well-sorted, fine-grained, quartzose lithic arkoses. The Frio sandstones of the upper Texas Gulf Coast (Areas 5 and 6) are poorly sorted, fine-grained, quartzose lithic arkoses to subarkoses.

Lower, middle, and upper Texas Frio sandstones have distinct rock fragment populations. Along the lower Texas Gulf Coast the sandstones are extremely rich in volcanic rock fragments, and carbonate rock framents are

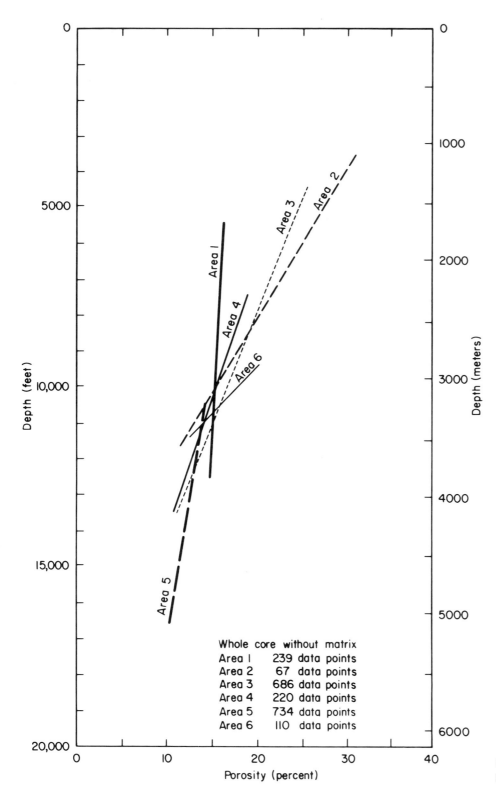

Whole core without matrix
Area 1   239 data points
Area 2   67 data points
Area 3   686 data points
Area 4   220 data points
Area 5   734 data points
Area 6   110 data points

Figure 11—Mean Wilcox sandstone porosity by area.

common. Volcanic rock fragments predominate along the middle Texas Gulf Coast, but some samples contain significant percentages of metamorphic rock fragments. In this area carbonate rock fragments are also present, but such fragments are much less common than along the lower Texas Gulf Coast. Rock fragments in upper Texas Gulf Coast sandstones are mainly volcanic rock fragments.

Sandstones of the lower Texas Gulf Coast contain the greatest abundance of rock fragments because of drainage from active volcanic areas in Mexico and west Texas into the ancient Rio Grande Basin. These volcanic rock fragments are dominantly rhyolites and

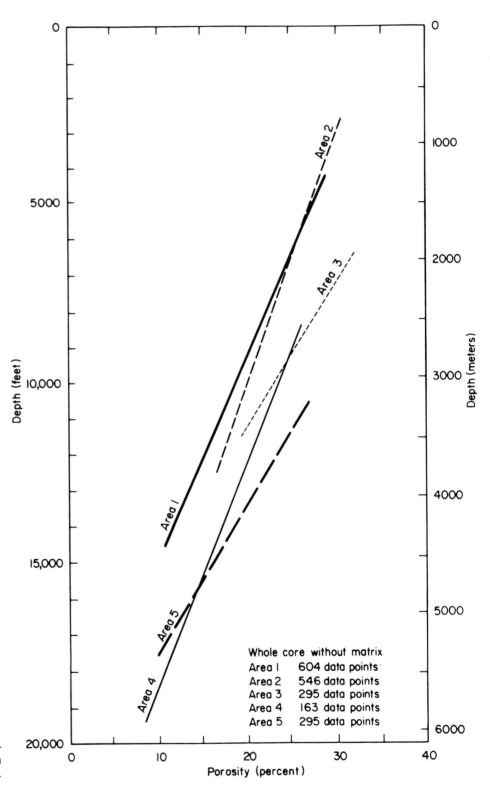

Figure 12—Mean Frio sandstone porosity by area. No data were available from
Area 6.

trachytes (Lindquist, 1977) and are
normally silicified or altered to chlorite.
Lesser amounts of these volcanic fragments survived transport to the middle
and the upper areas of the Texas Gulf
Coast. The abundant carbonate rock

fragments along the lower Texas Gulf
Coast are caliche clasts, locally derived
from caliche soils which were the result
of an arid climate (Lindquist, 1977).
Caliche soils did not form in the more
humid climate of the upper Texas Gulf

Coast areas, and the abundance of carbonate rock fragments decreases in that
direction. The source of abundant
metamorphic rock fragments in middle
Texas Frio sandstones is debatable. If
they were related to drainage from the

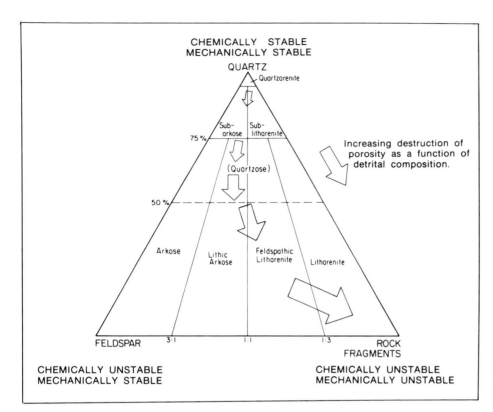

Figure 13—Sandstone classification from Folk (1968). "Quartzose" modifier added by this report. Stability poles adapted from Hayes (1979).

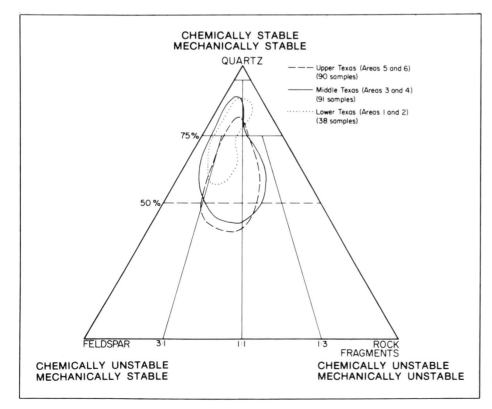

Figure 14—Wilcox sandstone composition.

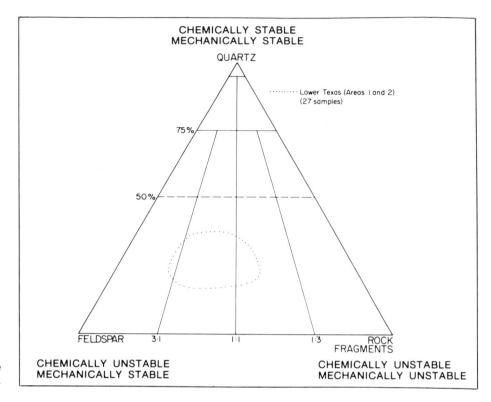

Figure 15—Vicksburg sandstone composition.

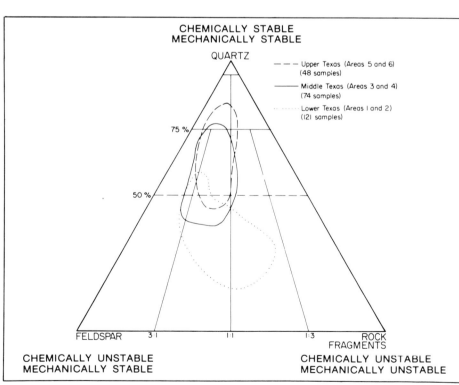

Figure 16—Frio sandstone composition.

Rocky Mountains, Frio sandstones along the upper Texas Gulf Coast area should contain more metamorphic rock fragments as well. Frio sandstones along the middle Texas Gulf Coast area also contain more orthoclase, suggesting a western source of metamorphic

rock fragments such as the Llano Uplift. Galloway (1977) reported a major Late Oligocene–Early Miocene river that crossed northern Karnes County that may have drained the Llano Uplift area.

Effect of Temperature on Porosity

Temperature is a major control on diagenesis in some sandstone suites and, hence, on porosity preservation (Galloway, 1974, 1979; Loucks et al, 1981). Porosity in the Wilcox Group and Frio Formation decreases with

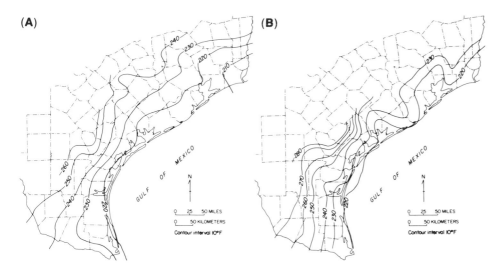

Figure 17—(**A**) Corrected temperature (° F) at 3050 m (10,000 ft) in the hydropressured zone. (**B**) Corrected temperature (° F) at 3050 m (10,000 ft) in the geopressured zone. Temperatures are corrected according to Kehle (1971).

TABLE 1

GEOTHERMAL GRADIENT INCREASES

TREND \ AREA	LOWER TEXAS 1 & 2	MIDDLE TEXAS 3 & 4	UPPER TEXAS 5 & 6
WILCOX	2.25° F/100 Ft.	2.11° F/100 Ft.	1.88° F/100 Ft.
VICKSBURG / FRIO	2.05° F/100 Ft.	1.76° F/100 Ft.	1.79° F/100 Ft.

GEOTHERMAL GRADIENT INCREASES

Table 1—Average geothermal gradient by area for the updip Wilcox and downdip Vicksburg/Frio trends. The geothermal gradient for each area is based on an average of 330 data points from well logs.

increasing temperature. Because temperature increases with depth, this correlation is simply a restatement of the previously demonstrated relationship of decreasing porosity with depth (Fig. 6). Regional relationships between porosity and temperature are more meaningful. Temperature distribution along the Texas Gulf Coast shows two regional trends (Fig. 17 and Table 1): a decrease in temperature from the lower to the upper Texas Gulf Coast and a decrease gulfward in temperature. These trends are seen both in hydropressured and geopressured formations. The trend toward cooler temperatures (Fig. 17) and a lower geothermal gradient (Table 1) in the upper Texas Gulf Coast area correlates with higher porosities in the Frio sandstones (Fig. 12). The high temperature and higher geothermal gradient inland relative to that near the coast corresponds to lower porosities in the inland Wilcox Group compared with

that of the Frio Formation along the coast (Fig. 10).

General Diagenetic Sequence

Lower Tertiary formations along the Texas Gulf Coast have all undergone a similar diagenetic history (Fig. 18). The general diagenetic sequence, with some variations, was delineated by this study for the Wilcox, Vicksburg, and Frio sandstones based on numerous samples from these units. Depths at which diagenetic products first occur differ among formations and a few formations have unique features (Loucks et al, in press), but the general diagenetic sequence is consistent. The paragenetic sequence was delineated by noting the relative positions of the various diagenetic products to each other. All features do not occur in all samples. The depth of first occurrence of each diagenetic feature helped determine maximum depth of burial that the particular feature may have started to form. Most

diagenetic features were well developed before the rock was buried to 3000 m (10,000 ft) (Fig. 18), indicating that most diagenesis occurs above the top of the "hard" geopressured zone (pressure gradient greater than 0.7 psi/ft). Porosity, however, continues to decrease downward in the geopressured zone, indicating further cementation must be taking place.

Milliken et al (1981), using isotopic analysis of cements and pore fluids in the Frio Formation along the upper Texas Gulf Coast, have reconfirmed the paragenetic sequence presented here. Other studies that have investigated the diagenesis of the Texas Gulf Coast Tertiary section are Lindquist (1977), Loucks et al (1977, 1979a, 1979b, 1980, 1981, in press), Stanton (1977), Boles (1978), Boles and Franks (1978), Richmann et al (1980), Land and Milliken (1981), Klass et al (1981), Kaiser (1982), Land (1982), and Fisher (1982).

Although all Lower Tertiary forma-

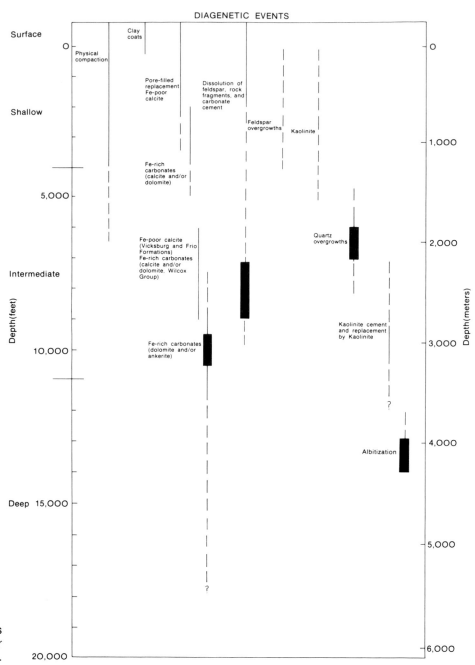

Figure 18—General diagenetic stages with increasing depth of burial for Lower Tertiary sandstones.

tions in the Texas Gulf Coast have a similar diagenetic sequence (Fig. 18), significant differences among formations in mineral composition and in importance of each diagenetic event have bearing on the existence of porosity and permeability at depth.

Lower Tertiary formations display the following general diagenetic sequence:

Surface-to-shallow-subsurface diagenesis (0 to 1200 m ±; 0 to 4000 ft ±) began with formation of clay coats (smectite?) by mechanical infiltration of colloidal clay-rich waters through the vadose zone (Burns and Ethridge, 1979; Galloway, 1974) and/or by alteration of feldspars. Although clay coats occupy only a small volume of pore space, they can be detrimental to per-

meability by reducing pore-throat diameter and causing resistance to fluid flow (Galloway, 1977).

Alteration of feldspars began in the source area and continued at the depositional surface and in the subsurface. Feldspars were either dissolved or replaced by Fe-poor calcite. Fe-poor calcite is an abundant pore-filling cement in the Catahoula (updip equi-

Figure 19—Secondary porosity as a percent of total porosity versus depth for Lower Tertiary sandstones. Note that below 3000 m (10,000 ft) most pore networks are composed predominantly of secondary porosity. Each datum point represents a thin-section analysis.

valent of the Frio Formation; Galloway, 1977) and Frio Formations, and it is a common diagenetic feature in paleosoil zones in Frio outcrop (McBride et al, 1968; Galloway, 1977). It is commonly poikilotopic.

Carbonate cement continued to precipitate with depth throughout the Lower Tertiary section (Fig. 18). At a given locality, however, precipitation with depth was discontinuous. It occurred as a series of discrete pulses separated in the stratigraphic section by as little as a few inches or as much as thousands of feet. Several different phases of carbonate cement precipitated in each unit (Fig. 18).

Dissolution occurred during the shallow and intermediate subsurface stages of diagenesis (Figs. 19–21). Dissolution commonly alternated with carbonate cementation.

Feldspar overgrowths around detrital feldspars and, less commonly, around volcanic rock fragments were also near-surface processes. Overgrowths of both plagioclase and orthoclase were noted. Overlap of early dissolution and feldspar cementation was observed in some samples. Volumetrically, this cement is insignificant.

Minor amounts of authigenic kaolinite were precipitated in the shallow diagenetic environment. Morphology of the cement can range from poorly developed, scattered plates to well-developed booklets as described by Todd and Folk (1957) from Wilcox outcrop samples. Fe-rich carbonate cement (calcite and/or dolomite) begins to precipitate by approximately 600 m (2000 ft) in pores and in molds of feldspar grains dissolved at shallower depths.

Throughout the shallow subsurface, sediments underwent relatively rapid compaction because of the lack of abundant cementation. Early-formed Fe-poor calcite was the first major compaction-arresting cement. By depths of 1200 m (4000 ft) porosity was reduced from the original 40% to approximately 30%.

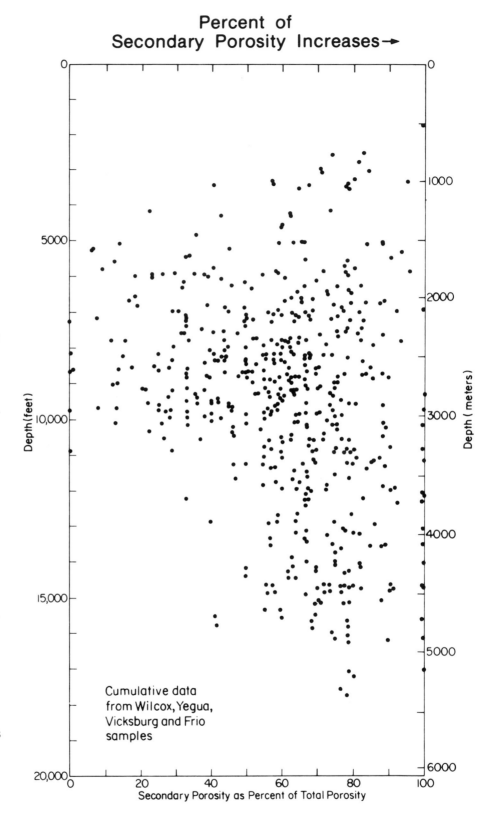

Percent of Secondary Porosity Increases →

Cumulative data from Wilcox, Yegua, Vicksburg and Frio samples

Secondary Porosity as Percent of Total Porosity

Depth (feet)

Depth (meters)

Intermediate subsurface diagenesis (1200 to 3400 m ±; 4000 to 11,000 ft ±) comprised a complex stage of cementation and dissolution. Compaction was arrested during this stage because of abundant cementation.

Carbonate cements that precipitated in the shallow subsurface, detrital feldspar grains, and early calcite that replaced feldspars underwent dissolution in the intermediate subsurface, which created secondary dissolution

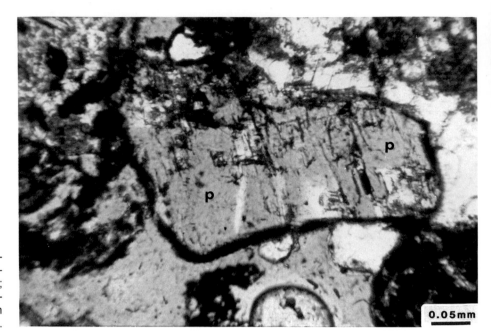

Figure 20—Formation of shallow sub-surface secondary porosity (**p**) by dissolution of feldspar. Frio Formation; Fox Minerals Corporation No. 6 Guadalupe de Garcia (920 m, 3017 ft), Jim Wells County, Texas.

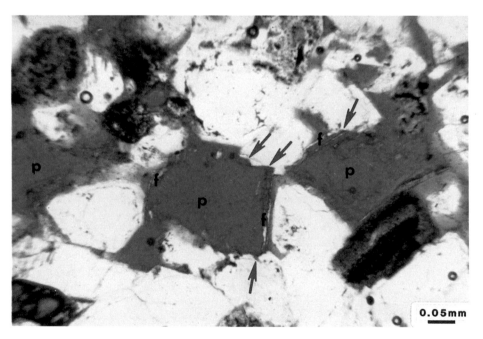

Figure 21—Formation of deep subsurface secondary porosity (**p**) by dissolution of feldspar. Note that only part of the feldspar rim is preserved (**f**) and that embayments (arrows) in the quartz overgrowth indicate post-quartz overgrowth dissolution of the feldspars.

porosity (Figs. 19, 21). Secondary porosity reached a maximum value near the end of this stage.

Quartz overgrowth cementation was an important stage of diagenesis, that appeared in the intermediate subsurface. Overgrowths arrested compaction and occluded pore space, and they were resistant to later dissolution. In the Wilcox Group in many areas, quartz overgrowths have occluded all or nearly all pore space, eliminating any potential for development of high-quality deep reservoirs. Quartz overgrowths are less abundant in other Lower Tertiary units, but locally they are still an important porosity-reducing cement.

Carbonate cementation that occurred after formation of quartz overgrowths exists in two modes. In the Vicksburg and Frio Formations, carbonate cement is Fe-poor, and in the Wilcox Group, carbonate cement is Fe-rich. This carbonate cementation was also a major porosity-reducing event.

After these last stages of quartz and carbonate cementation, porosity was commonly reduced to 10%. This reduction in porosity could be reversed, however, by intense dissolution near the base of the intermediate subsurface zone. Components that were dissolved were feldspars, volcanic rock fragments, and carbonate cements. Evidence for this stage of dissolution is the dissolution of post-quartz overgrowth carbonate cement and the formation of embayments in quartz overgrowths where grains have been dissolved (Loucks et al, 1977). Continued dissolution may resurrect porosities to more than 30%. This intermediate subsurface stage of dissolution was important in development of deep reservoirs although only 20% to 30% of the sandstone section was enhanced.

After the dissolution stage, kaolinite was precipitated as a cement and as a replacement product of feldspars. Commonly, replaced feldspars are nuclei for precipitation of the cement. Kaolinite, composed of booklets of stacked individual crystals several microns in diameter, formed a meshwork in the pore spaces that did not significantly reduce porosity but did reduce permeability. Kaolinite cement, however, is neither abundant nor widespread. Timing of this later stage of kaolinite is evidenced by growth on earlier-formed quartz overgrowths.

The last cementation phase observed in the Lower Tertiary section was the precipitation of Fe-rich dolomite and ankerite. The amount of this late cement is an important variable that controlled deep-reservoir quality in all Lower Tertiary units except the Wilcox Group where quartz overgrowths were more important in reducing porosity.

Deep subsurface diagenesis (>3400 m±; >11,000 ft ±) was a continuation of the precipitation of late Fe-rich carbonate cements. In some sections, such as in the Frio of upper Texas, late carbonate cement is minor, and high-quality reservoirs exist at depth. Also, in the deep subsurface, feldspar underwent albitization (Land and Milliken, 1981).

Reservoir development in the Lower Tertiary sandstones along the Texas Gulf Coast was controlled by a series of porosity-reducing and porosity-enhancing events (Fig. 22). In shallow reservoirs, porosity is both primary and secondary in origin (Fig. 19). Much early-formed porosity was lost through compaction and cementation. Dissolution events resurrected porosity and created reservoirs that were composed mainly of secondary porosity (Fig. 19),

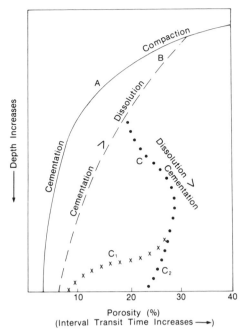

Figure 22—General pathways of porosity and interval transit time with depth. Curve A shows loss of primary porosity with depth by compaction and cementation; no secondary porosity was developed. Curve B represents loss of porosity with depth where compaction and cementation rates are greater than the production of secondary porosity. Curve C shows loss of porosity with depth by compaction and cementation followed by a major zone of secondary dissolution porosity development. Curve C_1 indicates a late stage of cementation destroying porosity, whereas, curve C_2 indicates porosity preservation with further burial.

but some of these reservoirs were destroyed by precipitation of significant quantities of late Fe-rich carbonate cement.

RESERVOIR QUALITY BY UNIT

Wilcox Group

Vertical porosity distribution in the Wilcox Group (Fig. 11) shows no regional trend. Plots of permeability values versus depth by geographic area indicate that Area 3 has the highest maximum permeability values at any depth (Fig. 23). Area 3 also has the best developed secondary dissolution porosity. It is interesting to note that Area 3 has the lowest percentage of quartz overgrowths.

Secondary dissolution porosity, resulting from dissolution of feldspars and carbonate cements, is the dominant porosity type in the Wilcox Group in the intermediate and deep subsurface (Fig. 24). Analysis of primary and secondary porosity with depth showed that at 3500 m (11,500 ft), primary porosity is less than 4%, whereas secondary porosity is as high as 10%.

Quartz cement is the major factor controlling reservoir quality in the Wilcox Group. It can be abundant enough to occlude pore spaces totally, and because it is less susceptible to dissolu-

tion than carbonate cement, destruction of reservoir quality is permanent. Carbonate cement, which does not necessarily permanently destroy a reservoir, is not responsible for as much porosity loss.

No strong regional trends in distribution of quartz and carbonate cements were observed in the Wilcox Group. This is probably because there is no significant regional variation in grain composition (Fig. 14). Locally, porosity depends upon the extent of dissolution of feldspars and carbonate cement. More unstable components and less quartz result in development of more secondary porosity and, hence, greater total porosity.

Vicksburg Formation

The deep Vicksburg Formation exhibits the poorest reservoir quality of any Lower Tertiary unit (Figs. 10, 25). This is the result of fine grain size, poor sorting, abundant unstable rock fragments, pervasive carbonate cement, high compaction from rapid subsidence, and a high geothermal gradient along the lower Texas Gulf Coast.

Most porosity beneath 2440 m (8000 ft) in depth is of secondary origin (Fig. 26). Primary porosity is reduced to less than 5%, whereas secondary porosity may be as much as 15%.

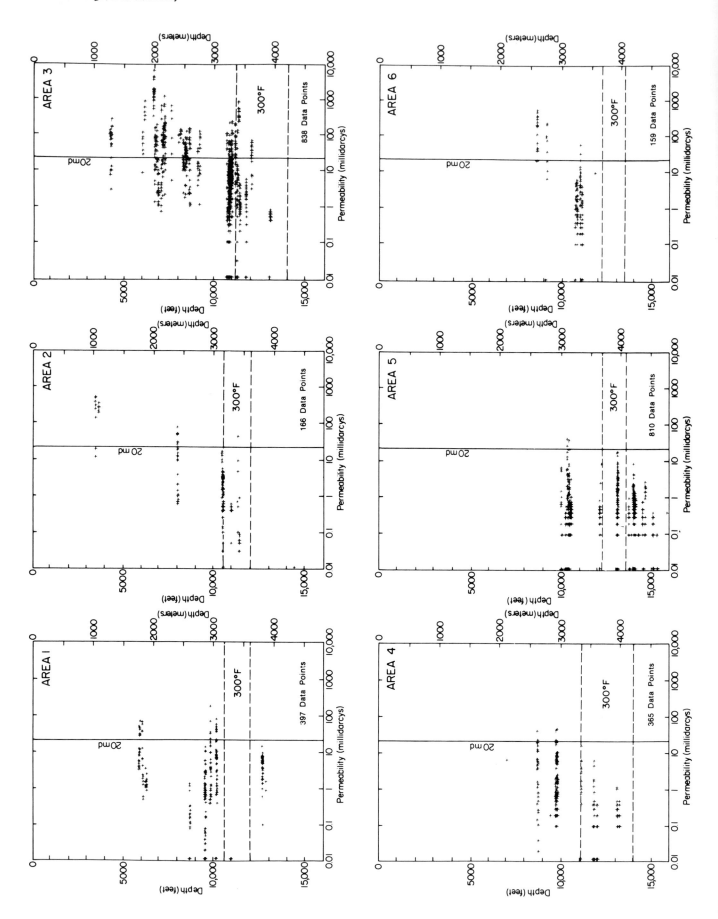

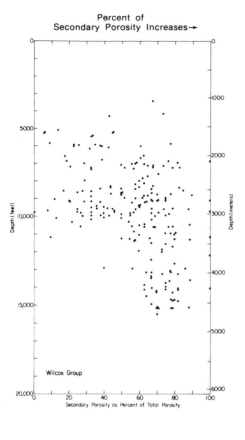

Percent of
Secondary Porosity Increases→

Wilcox Group

Secondary Porosity as Percent of Total Porosity

Figure 24—Secondary porosity as a
percent of total porosity versus depth
for Wilcox sandstones.

Figure 25—Permeability (core plugs
from whole core) versus depth by area
for Vicksburg Formation.

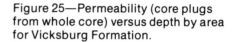

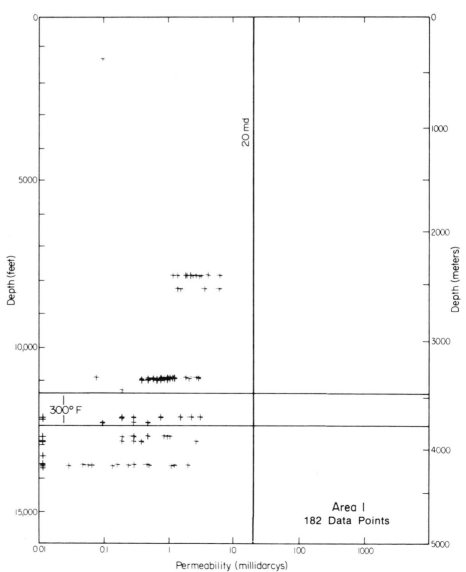

20 md

300° F

Area I
182 Data Points

Depth (feet)

Depth (meters)

Permeability (millidarcys)

Figure 23—Permeability (core plugs
from whole core) versus depth by area
for Wilcox Group.

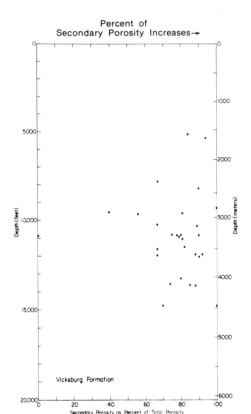

Figure 26—Secondary porosity as a percent of total porosity versus depth for Vicksburg sandstones.

Frio Formation

The Frio Formation displays the best deep-reservoir quality in the Lower Tertiary section. Porosity-versus-depth plots, however, indicate that this high reservoir quality, especially at depth, is restricted to Area 4 of the middle Texas Gulf Coast and Areas 5 and 6 of the upper Texas Gulf Coast (Figs. 10, 12, 27). Sandstones from Areas 4 and 5 from depths greater than 4500 m (15,000 ft) may have permeability values greater than 1000 millidarcys (md). In the northern part of Area 1 some permeability readings near 10 md are recorded at depths around 4500 m (15,000 ft), but the majority of permeability values are less than a few millidarcys. Porosity in the Frio Formation in the deep subsurface is dominantly secondary dissolution porosity (Fig. 28).

The increase in reservoir quality from the lower to the upper Texas Gulf Coast (Fig. 12) corresponds to other trends in rock composition, climate, and geothermal gradient. The change in Frio rock composition (Fig. 16) is probably the most important of these.

Along the lower Texas Gulf Coast, where reservoir quality is poor, Frio sandstones are low in quartz and rich in volcanic and carbonate rock fragments. Along the upper Texas Gulf Coast, where reservoir quality is good, Frio sandstones are rich in quartz, lower in volcanic rock fragments, and lacking in carbonate rock fragments. The abundance of chemically and mechanically unstable volcanic and carbonate rock fragments along the lower Texas Gulf Coast favors diagenetic processes that destroy porosity.

The geothermal gradient decreases from the lower to the upper Texas Gulf Coast (Table 1). The higher geothermal gradient (2.05° F/100 ft, Table 1) typical of the lower Texas Gulf Coast resulted in a shallower onset of thermally controlled diagenetic events. This trend was further accentuated by the unstable mineral assemblages in this area. The Rio Grande Embayment also, underwent more rapid subsidence relative to the Houston Embayment, resulting in greater compaction of sediments. A packing-proximity (a measure of number of grain contacts;

Kahn, 1956)-versus-depth plot for Area 1 shows greater compaction relative to Area 5 (Fig. 29). Frio sand grains in Areas 5 and 6 have fewer grain contacts than Area 1 even though sandstones in Areas 5 and 6 are more deeply buried.

Variations in the post-Eocene climate along the Texas Gulf Coast, from arid conditions in the south to subhumid and humid conditions in the north, indirectly influenced reservoir quality by preserving and creating unstable rock fragments. The arid climate in the south produced caliche soils that were a source of carbonate rock fragments. Caliches did not form along the more humid upper Texas Gulf Coast and carbonate rock fragments are rare. Also, the more arid climate in the lower Texas region preserved volcanic rock fragments and allowed them to be incorporated into the sandstones.

Summary of Reservoir Quality in Lower Tertiary Gulf Coast Sandstones

Diagenetic processes, which in turn determined reservoir quality in the Vicksburg and Frio Formations, were a function of rock composition, geothermal gradient, paleoclimate, and subsidence rate. Sandstone composition and geothermal gradient were probably the most important of these factors. As the geothermal gradient and amount of unstable components decreased northward along the Gulf Coast, less cementation occurred and reservoir quality remained high. The Wilcox Group, however, does not show such a trend. There was little variation in rock composition in the Wilcox Group along the Gulf Coast, and improved reservoir quality appears to be determined by local rather than regional conditions. However, the best reservoirs in the Wilcox Group were found in Area 3 where the fewest quartz overgrowths exist. Land (personal communication, 1979) has suggested that this may be due to the influence of the San Marcos Arch. Less shale was available, so consequently there were less silica-transporting fluids from shale compaction. In the Frio Formation a higher percentage of quartz typically correlates with a lower percentage of unstable rock fragments and, therefore, less carbonate and laumontite cements. Locally in the Wilcox Group, more quartz generally correlates with less

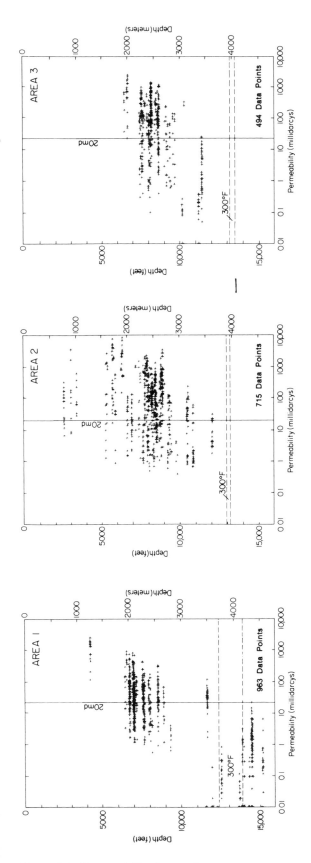
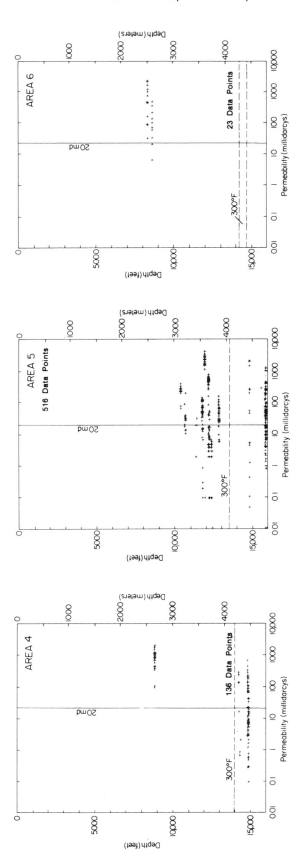

Figure 27—Permeability (core plugs
from whole core) versus depth by area
for Frio Formation.

feldspars, and, therefore, less secondary porosity. These local differences in reservoir quality may be predicted by detailed regional investigations.

RELATIONSHIP BETWEEN ACOUSTIC LOGS AND DIAGENESIS

Rock Consolidation Gradient from Interval-Transit-Time Plots

Interval transit time is the reciprocal of velocity of the compressional sound wave traveling through a rock. With increasing depth, velocity increases as a function of effective pressure ($P_{overburden}$–P_{fluid}) and cementation at grain contacts. The rapid increase in velocity continues until a depth is reached at which the rock is well consolidated. At this depth, the increase in velocity becomes asymptotic to the time-average velocity (Gregory, 1977). According to Gregory (1977), at shallow depths interval transit time in sandstones depends on porosity, rock composition, grain sorting, clay content, and overburden pressure; but at high overburden pressures, corresponding to deeper burial, only porosity and rock composition are important.

A plot of interval transit time versus depth modified from Gardner et al (1974) for a normally pressured Louisiana Miocene sandstone sequence reflects high interval-transit-time characteristics of unconsolidated sediments in the shallowest strata (Fig. 30). At about 2130 m (7000 ft) the rocks are relatively well consolidated, and interval transit time is directly related to porosity. In this plot, divergence of true interval transit time (solid line) from interval transit time related to pressure (dotted line) is attributed to consolidation.

Interval transit times for sand and sandstones from acoustic logs were plotted against depth to show the relationship between consolidation history and interval transit time along the Texas Gulf Coast area. The data align along two trends parallel to the coast (Fig. 4): an undip trend that corresponds to wells drilled into the Wilcox Group, and a downdip trend that corresponds to wells drilled into the Vicksburg and Frio Formations. Interval-transit-time-versus-depth plots exhibit the same compaction/consoli-

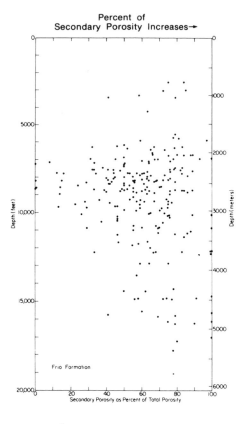

Figure 28—Secondary porosity as a percent of total porosity versus depth for Frio sandstones.

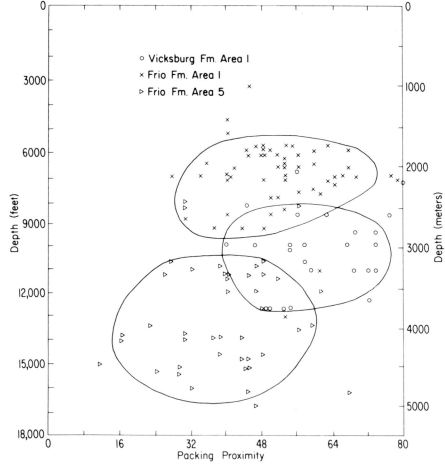

Figure 29—Packing proximity (a measure of the number of grain contacts) versus depth for Vicksburg sandstones from Area 1 and Frio sandstones from Areas 5 and 6.

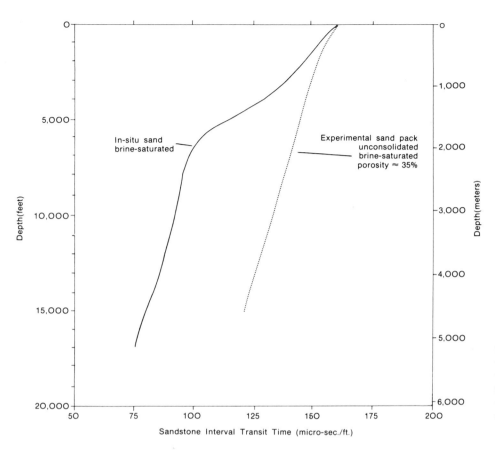

Figure 30—Interval transit time as a function of depth showing consolidation effect for Louisiana Tertiary sandstones (solid line). Modified from Gardner et al. Copyright © 1974 Society of Exploration Geophysicists. Used with permission.

dation curve as those of Gardner et al (1974) (Fig. 30), except some wells show an increase in interval transit time at the top of the geopressured zone (Fig. 31). In abnormally high-pressured (geopressured) sandstones, fluid pressure supports some of the rock column, making the effective pressure on the rocks less than for hydropressured sandstones. Thus, interval transit time increases at the top of the geopressured zone because of reduced effective pressure on the rocks. The situation in the Texas Gulf Coast is more complicated, however, because the top of the geopressured zone commonly coincides with a zone of well-developed secondary dissolution porosity (Figs. 18, 22), causing an additional increase in interval transit time. The effect of one zone on interval transit time may be inseparable from the effect of the other.

An idealized plot of porosity and interval transit time of sandstone versus depth (Fig. 22) for Tertiary sandstones of the Gulf Coast shows an initial rapid decrease in interval transit time owing to compaction of unconsolidated sedi-ments. As cementation was initiated, compaction decreased and eventually stopped, and the rate of porosity loss decreased. This corresponds to an increase in slope on the interval-transit-time plot. Also, minor dissolution porosity may have occurred. Figure 32 is an interval-transit-time-versus-depth plot that shows a curve similar to curve B in the idealized plot (Fig. 22). With major, post-quartz overgrowth dissolution, a reversal in the interval-transit-time slope may occur (Fig. 22; curve C). The interval-transit-time curve will decrease again however, if late Fe-rich carbonate cements are abundant (Fig. 22; curve C_1).

Only a few plots of interval transit time from acoustic logs along the Texas Gulf Coast display a curve similar to the idealized plot of compaction and cementation followed by dissolution (Fig. 22; curve C). One of these wells is shown in Figure 31. The possible zone of secondary dissolution porosity in this well also occurs at the top of the geopressured zone. The increase in interval transit time may be the result of both dissolution porosity and high fluid pressures. There are also slight breaks in slope in this plot (Fig. 31) at about 760 m (2500 ft) and 1520 m (5000 ft). These breaks correspond to the top of the Frio Formation and the basal Frio sandstones, respectively. Formational changes commonly show some effect on interval transit time, reflecting slight changes in rock composition or porosity, and consequently, velocity.

Regional Variation in Rock Consolidation Gradient

The acoustic logs used in this investigation were grouped by Areas 1 through 6 and further divided into an updip Wilcox trend and a downdip Vicksburg/Frio trend (Fig. 4). In each trend, all data in each area were combined to produce an average or representative sandstone interval-transit-time-versus-depth plot (Figs. 33, 34).

Sandstones from the Vicksburg/Frio trend show progressively greater consolidation gradients from the upper Texas to the lower Texas Gulf Coast (Fig. 33), further substantiating conclu-

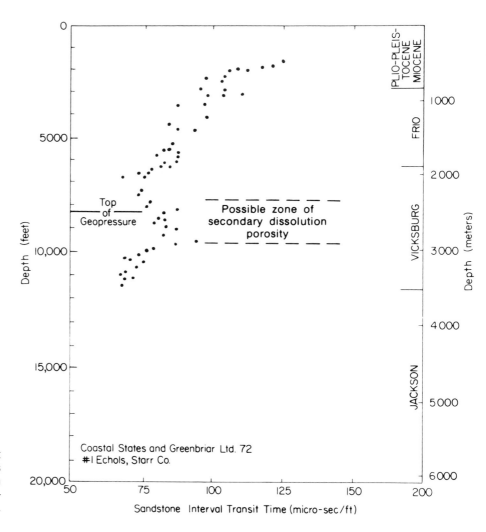

Figure 31—Sandstone interval transit time versus depth for the Coastal States and Greenbriar Ltd. 72 No. 1 Echols in Starr County. Zone of possible secondary dissolution porosity is marked.

sions based on core analyses and petrographic descriptions. Integration of reservoir data from acoustic logs greatly expanded the base on which generalizations were made. These changes correspond to high reservoir quality along the upper Texas Gulf Coast and poor reservoir quality along the lower Texas Gulf Coast. Reservoir quality improves northward at all depths.

Many of the sandstones in the updip Wilcox trend, shallower than 3050 m (10,000 ft), are younger than the Wilcox Group, and they include strata equivalent to the Vicksburg and Frio Formations (Fig. 34). They exhibit a higher consolidation gradient along the lower Texas Gulf Coast relative to the upper Texas Gulf Coast. This trend is the same for similar rocks in the downdip Vicksburg/Frio trend (Fig. 33).

At a depth greater than 3050 m

(10,000 ft), plots of interval transit time versus depth for the Wilcox trend in Areas 1, 2, 4, and 6 indicate relatively well-consolidated sandstones, but similar plots for Area 3 and possibly Area 5 show a reversal towards an apparent increase in porosity at depth (Fig. 34). Core analyses in Area 3 indicate high permeability values from samples deeper than 3350 m (11,000 ft), but core analyses in Area 5 do not support this increase in permeability (Fig. 23).

At depths shallower than 2130 m (7000 ft) both the Wilcox and the Vicksburg/Frio interval-transit-time trends are similar (Figs. 33, 34). At depths greater than 2130 m (7000 ft), Areas 1 and 2 of the Vicksburg/Frio trend exhibit interval-transit-time curves similar to those in the Wilcox trend and these curves correspond to low reservoir quality in the deep subsurface. Sandstones from Areas 3, 4,

and 5 have higher Vicksburg/Frio interval transit times below 2130 m (7000 ft) than those in the Wilcox trend except for those in Area 3 of the Wilcox trend (Figs. 33, 34). Areas 4 and 5 of the Vicksburg/Frio trend, with higher interval transit times correspond to areas with highest-quality deep reservoirs based on core analyses (Figs. 11, 12, 27). However, data for the Vicksburg/Frio trend in Area 3 are available only to a depth of 3350 m (11,000 ft); below this depth, reservoir quality may be marginal (Fig. 27).

CONCLUSIONS

Trends in reservoir quality in the Lower Tertiary section along the Texas Gulf Coast were defined by whole core analyses, grain composition of the sandstones, intensity of diagenetic features, and interval transit times for

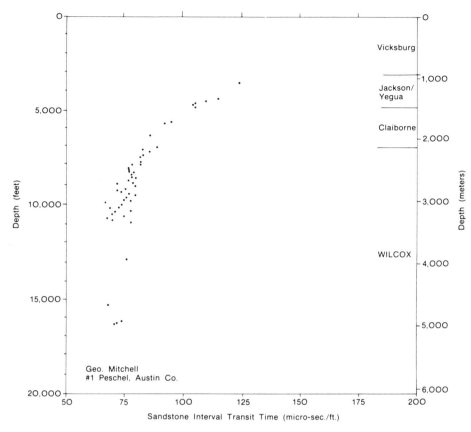

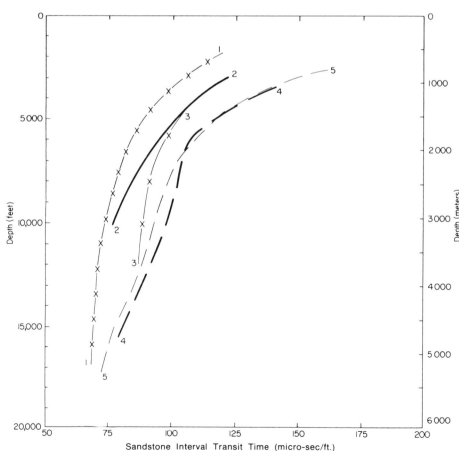

Figure 32—Sandstone interval transit time versus depth for the Geo. Mitchell No. 1 Peschel in Austin County. The curve shows no zone of secondary porosity development. Compaction and cementation occurred at a greater rate than the development of secondary porosity.

Figure 33—Sandstone interval transit time versus depth by area for the Vicksburg/Frio trend. Each curve represents a visual best-fit line drawn through data in that area.

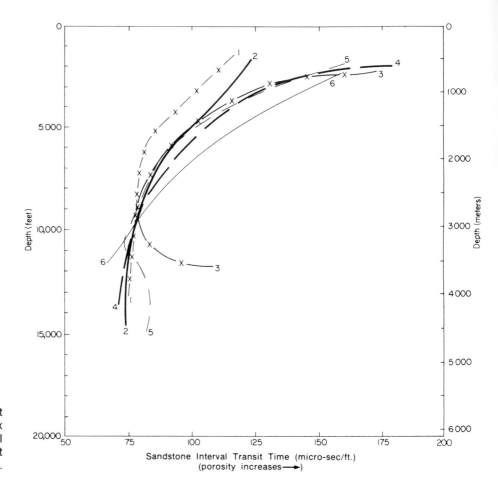

Figure 34—Sandstone interval transit time versus depth by area for the Wilcox trend. Each curve represents a visual best-fit line drawn through data in that area.

sandstones. Porosity and permeability in sandstone sections shallower than 3048 m to 3350 m (10,000 to 11,000 ft) are generally adequate for hydrocarbon production. However, reservoir quality in deep subsurface sandstones, which are the targets of deep gas condensate exploration in the Lower Tertiary section of the onshore Texas Gulf Coast, is highly variable. Most of these deep sandstone reservoirs have permeability values of less than 1 md, but in a few areas permeability values are higher than 1000 md. The *potential* for high-quality reservoirs to occur in the deep subsurface (approximately below 3350 m [11,000 ft] of burial) is summarized in Figure 35.

The Wilcox Group has good deep subsurface porosity and permeability in Area 3 and possibly in the adjacent part of Area 4. Other areas in the Wilcox Group have marginal reservoirs at depth. A few high-quality sandstone reservoirs possibly formed in marginal areas, but these sandstones would be rare and would not accumulate to any appreciable thickness.

The Vicksburg Formation in Area 1 has very poor porosity and permeability at depth. Predictions of reservoir quality in Areas 2 through 6 of the Vicksburg Formation were not made because of the lack of deeply buried sandstones.

Reservoir quality in the deep Frio Formation increases from very poor in the southern two-thirds of Area 1 to marginal through Area 3 and to good in Areas 4, 5, and 6. The Frio Formation in Area 5 has the best deep-reservoir quality of any onshore unit in any area.

Porosity and permeability were controlled by a complex series of diagenetic events consisting of compaction, cementation, and dissolution. Many physical and chemical parameters influenced these diagenetic events. Each formation along the onshore Texas Gulf Coast exhibits a similar general diagenetic history. Most dia-

genesis occurred in the hydropressured and "soft" geopressured zone (fluid pressure gradient less than 0.7 psi/ft), but some carbonate cementation continued into the "hard" geopressured zone (fluid pressure gradient greater than 0.7 psi/ft), as indicated by continued loss of porosity.

Primary porosity predominates in the shallow subsurface, but secondary dissolution porosity is dominant in the deeper subsurface (Fig. 19). Loucks et al (1977, 1979a) pointed out that the zone of well-developed secondary porosity occurs at depths and ambient temperatures that place it well within the Paleogene liquid window of hydrocarbon generation and preservation (Fig. 36) as defined by Pusey (1973). The liquid window encompasses the temperature/depth range within which major oil fields occur, unless there is significant vertical or lateral migration or post-accumulation changes in the thermal regime. The liquid window brackets much of the oil production in

Figure 35—Map displaying *potential* for high-quality, deep reservoirs (>3350 m, 11,000 ft) in Lower Tertiary strata along the Texas Gulf Coast. *Good* indicates permeability values commonly greater than 20 md. *Marginal* indicates permeability values generally less than a few md. *Poor* indicates permeability values are generally less than a md.

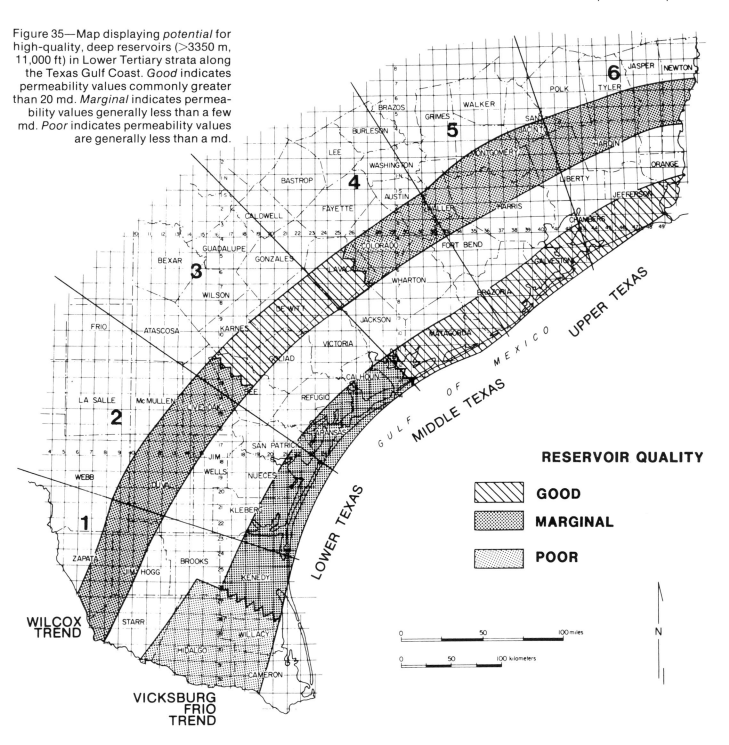

Tertiary basins such as the Gulf Coast. The window, which extends from 66°C (150°F) to 149°C (300°F) includes the minimal temperature 66°C (150°F) for generation of petroleum from source kerogen in older Eocene strata and the maximum temperature 149°C (300°F) of liquid preservation (LaPlante, 1972; Pusey, 1973). At temperatures above 149°C (300°F), only dry gas or gas with minor amounts of liquids is typically found (Klemme, 1972). The high porosity produced by dissolution within a similar depth range of the liquid window suggests that much oil and essentially all deep gas and gas-plus-condensate production from the Lower Tertiary is mainly from secondary porosity (Fig. 36).

Substantial enhancement of deeply buried sandstones by dissolution porosity generally only affects 20% to 30% of the sandstone section. The sandstones that are enhanced are commonly those that had high initial primary porosity, which was controlled by depositional environment. Depositional environment, therefore, is an indirect but important factor in searching for high-quality reservoirs composed of secondary porosity.

An understanding of the variables that control porosity and permeability

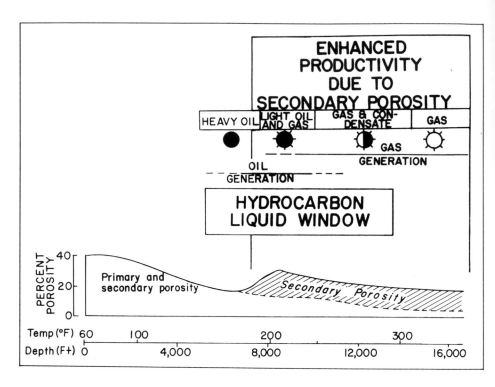

Figure 36—Schematic porosity versus depth/temperature curve for Lower Tertiary sandstones showing relative significance of primary versus secondary porosity. The lower three-quarters of the hydrocarbon liquid window, which is characterized by the production of light liquids and distillate, as well as all of the deep gas production section, lies within the zone of secondary porosity and permeability. From Loucks et al, 1979a. Copyright © 1979 Gulf Coast Association of Geological Societies. Reprinted with permission.

along the Texas Gulf Coast permits more insight into exploring for deep hydrocarbon reservoirs. Conclusions concerning controls on reservoir quality can be applied to other similar sandstone sequences.

ACKNOWLEDGMENTS

Many companies have provided various types of subsurface data for this study. These companies are gratefully acknowledged. Charles Ehlers, Turner Williamson, and Mark Yelverton were Research Assistants on this project. We thank them for their contributions.

Sincere appreciation is expressed to the following individuals who provided special assistance: Don Bebout and Ray Gregory, Bureau of Economic Geology; Tom Hill, Center for Energy Studies; Lynton Land, Department of Geological Sciences; Gus Sanders, Petroleum Engineering Department, The University of Texas at Austin; and Dusty Marshall, Department of Geology, Texas A&M University.

An extended version of this paper was reviewed by Lyle Baie, Don Bebout, Jim Bishop, Shirley Dutton, Mark Edwards, Eric Eslinger, Ken Helmold, Lynton Land, Bill Kaiser, Mark Presley, and Roger Slatt. This shorter version was reviewed by Robert Handford, Ken Helmold, Ellen Naiman, Steve Franks, and Donald McCubbin.

Initial funding for this project was provided by the Bureau of Economic Geology, The University of Texas at Austin, with subsequent funding by the Division of Geothermal Energy, U.S. Department of Energy. Results of this project were presented to D.O.E. under Contract No. EG-77-5-05-5554 ("Sandstone Consolidation Analysis to Delineate Areas of High-Quality Reservoirs Suitable for Production of Geopressured Geothermal Energy Along the Texas Gulf Coast"). Revision and updating of this investigation were completed at Cities Service Exploration and Production Research Laboratory.

SELECTED REFERENCES

Bebout, D. G., B. R. Weise, A. R. Gregory, and M. B. Edwards, 1979, Wilcox sandstone reservoirs in the deep subsurface along the Texas Gulf Coast, their potential for production of geopressure geothermal energy: Report to U.S. Department of Energy, Division of Geothermal Energy, Contract No. DEAS05-76ET 28461, 219 p.

Boggs, A. S., 1978, Petrology of the lower Eocene sandstones in south central Colorado compared to their time equivalent in Texas: Unpublished Master's Thesis, The University of Texas at Austin, 167 p.

Boles, J. R., 1978, Origin of late-stage iron-rich carbonate cements in sandstones—active ankerite cementation in Wilcox of southwest Texas (abs.): Bulletin of the American Association of Petroleum Geologists, v. 62, p. 498.

Boles, J. R., and S. G. Franks, 1978, Clay diagenesis in Wilcox sandstones of southwest Texas—implications of smectite-illite reaction for sandstone cementation (abs.): Bulletin of the American Association of Petroleum Geologists, v. 62, p. 497.

Burns, L. K., and F. G. Ethridge, 1979, Petrology and diagenetic effects of lithic sandstones: Paleocene and Eocene Umpqua Formation, southwest Oregon, in P. A. Scholle and P. R. Schluger, eds., Aspects of diagenesis: Society of Economic Paleontologists and Mineralogists Special Publication 26, p. 307–317.

Chepikov, K. P., Y. P. Yermolova, and N. A. Orlova, 1959, Epigenetic minerals as indicators of the time of entry of petroleum into commercial sandy reservoirs: Doklady of Academy of Sciences of the USSR, Earth Science Sections, v. 125, p. 288–289.

———, 1961, Corrosion of quartz grains and examples of the possible effect of oil on the reservoir properties of sandy rocks: Doklady of Academy of Sciences of the USSR, Earth Science Sections, v. 140, p. 1111–1113.

Cook, T. D., and A. W. Bally, 1975, Stratigraphic atlas of North and Central America: Princeton, Princeton University Press, 272 p.

Fisher, S. R., 1982, Diagenetic history of Eocene Wilcox sandstones and associated formation waters, south-central Texas: Unpublished Ph.D. Dissertation, The University of Texas at Austin, 120 p.

Folk, R. L., 1968, Petrology of sedimentary rocks: Austin, Texas, Hemphill Publishing Company, 182 p.

Galloway, W. E., 1974, Deposition and diagenetic alteration of sandstone in Northeast Pacific arc-related basins: implications for graywacke genesis: Geological Society of America Bulletin, v. 85, p. 379–390.

————, 1977, Catahoula Formation of the Texas Coastal Plain—depositional systems, mineralogy, structural development, ground-water flow history, and uranium distribution: University of Texas Bureau of Economic Geology Report of Investigations 87, 59 p.

————, 1979, Diagenetic control of reservoir quality in arc-derived sandstones: implications for petroleum exploration, in P. A. Scholle and P. R. Schluger, eds., Aspects of diagenesis: Society of Economic Paleontologists and Mineralogists Special Publication 26, p. 251–262.

Gardner, G. H. F., L. W. Gardner, and A. R. Gregory, 1974, Formation velocity and density—the diagnostic basics of stratigraphic traps: Geophysics v. 39, p. 770–780.

Gregory, A. R., 1977, Aspects of rock physics from laboratory and log data that are important to seismic interpretation, in C. E. Payton, ed., Seismic stratigraphy—applications to hydrocarbon exploration: American Association of Petroleum Geologists Memoir 26, p. 15–46.

Hayes, J. B., 1979, Sandstone diagenesis—the hole truth, in P. A. Scholle and P. R. Schluger, eds., Aspects of diagenesis: Society of Economic Paleontologists and Mineralogists Special Publication 26, p. 127–139.

Kahn, J. S., 1956, The analysis and distribution of the properties of packing in sand-size sediments, 1, On the measurement of packing in sandstones: Journal of Geology, v. 64, p. 384–395.

Kaiser, W. R., 1982, Predicting diagenetic history and reservoir quality in the Frio Formation (Oligocene) of Texas (abs.): International Association of Sedimentologists Abstracts, p. 119–120.

Kehle, R. O., 1971, Geothermal survey of North America: 1971 annual progress report: Unpublished report, Research Committee, American Association of Petroleum Geologists, Tulsa, 31 p.

Klass, M. J., D. G. Kersey, R. R. Berg, and T. T. Tieh, 1981, Diagenesis and secondary porosity in Vicksburg sandstones, McAllen Ranch Field, Hidalgo County, Texas: Gulf Coast Association of Geological Societies Transactions, v. 31, p. 115–123.

Klemme, H. D., 1972, Geothermal gradients: Pt. 1, Oil Journal, v. 69, p. 136, 141–144.

Land, L. S., 1982, Diagenesis of Texas Gulf Coast Frio and Wilcox sandstones (abs.): International Association of Sedimentologists Abstract, p. 119.

Land, L. S., and K. L. Milliken, 1981, Feldspar diagenesis in the Frio Formation, Brazoria County, Texas Gulf Coast: Geology, v. 9, p. 314–318.

Laniz, R. V., R. E. Stevens, and M. B. Norman, 1964, Staining of plagioclase feldspar and other minerals with F. D. and C. Red No. 2: Geological Survey Professional Paper 501, p. B152–B153.

LaPlante, R. E., 1972, Petroleum generation in Gulf Coast Tertiary sediments (abs.): Bulletin of the American Association of Petroleum Geologists, v. 56, p. 365.

Lindholm, R. C., and R. B. Finkelman, 1972, Calcite staining: semiquantitative determination of ferrous iron: Journal of Sedimentary Petrology, v. 42, p. 239–242.

Lindquist, S. J., 1977, Secondary porosity development and subsequent reduction, overpressured Frio formation sandstone (Oligocene), South Texas: Gulf Coast Association of Geological Societies Transactions, v. 27, p. 99–107.

Loucks, R. G., 1978, Sandstone distribution and potential for geopressure geothermal energy production in the Vicksburg Formation along the Texas Gulf Coast: Gulf Coast Association of Geological Societies Transactions, v. 28, p. 239–271.

Loucks, R. G., D. G. Bebout, and W. E. Galloway, 1977, Relationship of porosity formation and preservation to sandstone consolidation history—Gulf Coast Lower Tertiary Frio Formation: Gulf Coast Association of Geological Societies Transactions, v. 27, p. 109–120.

————, 1979a, Importance of secondary leached porosity in lower Tertiary sandstone reservoirs along the Texas Gulf Coast: Gulf Coast Association of Geological Societies Transactions, v. 29, p. 164–171.

————, 1979b, Sandstone consolidation analysis to delineate areas of high-quality reservoirs suitable for production of geopressured geothermal energy along the Texas Gulf Coast: Report to U.S. Department of Energy, Division of Geothermal Energy, EG-77-5-05-5554, 97 p.

————, in press, Factors controlling porosity and permeability of hydrocarbon reservoirs in lower Tertiary sandstones along the Texas Gulf Coast: University of Texas Bureau of Economic Geology Report of Investigations.

Loucks, R. G., D. L. Richmann, and K. L. Milliken, 1980, Factors controlling porosity and permeability in geopressured Frio sandstone reservoir, General Crude Oil/Department of Energy Pleasant Bayou Test Wells, Brazoria County, Texas, in M. H. Dorfman and W. L. Fisher, eds., Fourth Geopressured Geothermal Energy Conference Proceedings: University of Texas Center for Energy Studies, v. 1, p. 46–82.

————, 1981, Factors controlling reservoir quality in Tertiary sandstones and their significance to geopressured geothermal production: University of Texas Bureau of Economic Geology Report of Investigations 111, 41 p.

McBride, E. F., W. L. Lindeman, and P. S. Freeman, 1968, Lithology and petrology of The Gueydan (Catahoula) Formation in South Texas: University of Texas Bureau of Economic Geology Report of Investigations 63, 122 p.

Milliken, K. L., L. S. Land, and R. G. Loucks, 1981, History of burial diagenesis determined from isotopic geochemistry, Frio Formation, Brazoria County, Texas: Bulletin of the American Society of Petroleum Geologists, v. 65, p. 1397–1413.

Murray, G. E., 1955, Midway Stage, Sabine Stage, and Wilcox Group: Bulletin of the American Association of Petroleum Geologists, v. 39, p. 671–696.

Pusey, W. C., 1973, Paleotemperatures in Gulf Coast using the ESR-kerogen method: Gulf Coast Association of Geological Societies Transactions, v. 23, p. 195–202.

Richmann, D. L., K. L. Milliken, R. G. Loucks, and M. M. Dodge, 1980, Mineralogy, diagenesis, and porosity in Vicksburg sandstones, McAllen Ranch Field, Hidalgo County, Texas: Gulf Coast Association of Geological Societies Transactions, v. 30, p. 473–481.

Savkevich, S. S., 1969, Variation in sandstone porosity in lithogenesis (as related to the prediction of secondary porous oil and gas reservoirs): Doklady of the Academy of Sciences of the USSR, Earth Science Sections, v. 184, p. 161–163.

Schmidt, V., D. A. McDonald, and R. L. Platt, 1977, Pore geometry and reservoir aspects of secondary porosity in sandstones: Bulletin of Canadian Petroleum Geologists, v. 25, p. 271–290.

Stanton, G. D., 1977, Secondary porosity in sandstones of the lower Wilcox (Eocene), Karnes County, Texas: Gulf Coast Association of Geological Societies Transactions, v. 27, p. 147–207.

Storm, L. W., 1945, Resume of fact and opinions on sedimentation in Gulf Coast region of Texas and Louisiana: Bulletin of the American Association of Petroleum Geologists, v. 29, p. 1304–1335.

Todd, T. W., and R. L. Folk, 1957, Basal Claiborne of Texas, record of Appalachian tectonism during Eocene: Bulletin of the American Association of Petroleum Geologists, v. 41, p. 2545–2566.

Yermolova, Y. P., and N. A. Orlova, 1962, Variation in porosity of sandy rocks with depth: Doklady of the Academy of Sciences of the USSR, Earth Science Sections, v. 144, p. 55–56.

Frio Sandstone Diagenesis, Texas Gulf Coast: A Regional Isotopic Study

Lynton S. Land
The University of Texas
Austin, Texas

INTRODUCTION

Extensive study of the diagenesis of Texas Gulf Coast sandstones conducted by the Bureau of Economic Geology, University of Texas at Austin, as part of the geothermal energy program (Loucks et al, 1979a, 1980) established regional petrographic, diagenetic, and porosity trends. Detailed isotopic study in one localized area (Milliken et al, 1981) constrained the diagenetic scenario for that area. In Brazoria County, quartz and calcite cementation occurred relatively early in the burial history of the sands, followed by extensive secondary porosity development, hydrocarbon migration, continued calcite cementation, and dissolution or albitization of all detrital feldspars. Other diagenetic phases such as kaolinite, chlorite, and zeolites were of volumetrically minor and/or local significance.

As a result of the extensive regional collections amassed by the Bureau of Economic Geology, the isotopic study has been extended beyond Brazoria County, albeit on a less detailed scale. The data base for this study includes 204 Frio thin sections outside Brazoria County, 63 epoxy-impregnated thin section heels, 367 thin sections from Brazoria County, and point counts of all thin sections made by various Bureau personnel. These samples were supplemented by extensive samples from five wells examined in great detail by Lindquist (1976, 1977), mostly in south Texas.

METHODS

Because only epoxy-impregnated thin section heels were available for

ABSTRACT. Burial diagenesis of Frio sandstones deduced from detailed study of one small area of the northern Texas Gulf Coast (Brazoria County, Milliken et al, 1981) is regionally valid with only minor modifications.

Quartz is most commonly the first cement of volumetric significance to form, and constitutes 2.5% of the average sandstone volume. The average $\delta^{18}O$ of quartz cement is +31 o/oo ± 1.5 o/oo (SMOW), indicating precipitation at considerably cooler temperatures than most clay mineral transformation takes place. Calcite is the dominant cement in Frio sandstones, constituting about 5% of the total sandstone volume, and most commonly postdates quartz precipitation. Calcite more depleted than −10 o/oo (PDB) is uncommon, and most calcite has a $\delta^{18}O$ of −7.2 ± 2 o/oo (PDB). $\delta^{13}C$ values cluster closely around −4 ± 2 o/oo (PDB). Because of relatively constant isotopic composition, and relatively invariant iron and manganese content in calcite, both areally and with depth, both quartz and calcite cements appear to have been emplaced under relatively invariant chemical conditions prior to hydrocarbon migration. Detrital K-feldspar is essentially absent below 12,000 ft, and the zone of plagioclase albitization extends between about 9000 and 12,000 ft. Virtually no unaltered detrital feldspars are present below 12,000 ft in any of the samples examined, K-feldspar having been mostly dissolved and plagioclase albitized.

The volume of water required to precipitate quartz and calcite cements far from the apparent sources of material, generate secondary porosity and alter all detrital feldspars regionally in this thick sandstone sequence far exceeds the volume of pore water deposited with, near, or beneath the sands. Active thermally driven convection is a plausible (though unproven) mechanism for moving such large masses of dissolved components (and hydrocarbons) through the sandstones.

analysis, only quartz and calcite cements were examined isotopically in this study, together with carbonate and feldspar compositions. Authigenic clay minerals, especially kaolinite and chlorite, and zeolites rarely constitute more than about 2% of the rock volume and can be isolated for isotopic analyses

from unimpregnated rock samples only with great difficulty and not at all from impregnated samples. My emphasis here is on the *volumetrically* most important reactions that lead to the formation of solid phases, quartz cementation, calcite cementation, and feldspar alteration. The volumetrically

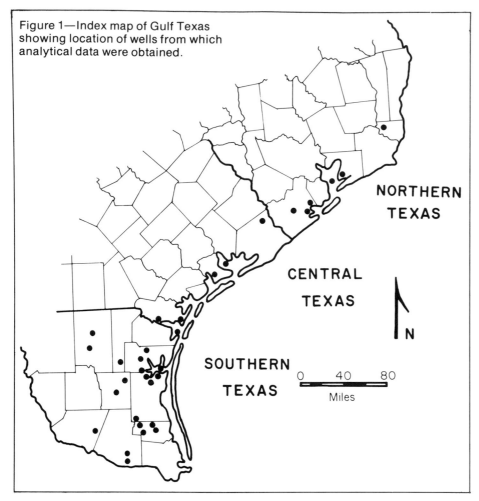

Figure 1—Index map of Gulf Texas showing location of wells from which analytical data were obtained.

NORTHERN TEXAS

CENTRAL TEXAS

N

SOUTHERN TEXAS

0   40   80
Miles

pure quartz in NaHSO₄ and H₂SiF₆, and analyzing the whole-rock quartz after oxidation with BrF₅.

Most samples isotopically analyzed for carbonate, together with additional samples spanning the entire depth and areal range of the available Frio samples, were prepared for electron-probe analysis by making polished sections from the thin section heels. Feldspars were "point counted" on the electron probe by driving the stage along a random linear direction and analyzing all feldspars intersected by the cross-hair. In this way the data obtained not only quantitatively indicate the compositions of the feldspars but their abundances as well. Carbonates were analyzed for Mg, Mn, and Fe by wavelength dispersive analysis, and simultaneously for Ca by energy dispersive analysis.

Figure 1 shows the distribution of wells from which analytical data were obtained for this study. (Previously published data [Milliken et al, 1981] are not included in Figures 3, 9, 10, and 11.)

QUARTZ CEMENTATION

On the basis of 576 thin-section point counts of Frio sandstones collected by the Bureau of Economic Geology (biased toward Brazoria County), average Frio sandstones contain 2.3% by volume quartz overgrowth cement, with sandstones toward northern Gulf Texas generally better quartz-cemented than those toward the south. Lindquist's (1976) study of 100 thin sections, mostly from south Texas, yielded an average of 3% quartz overgrowths. Thus the two studies are not entirely in agreement, Lindquist's values being somewhat higher than the Bureau's. Both biased sampling and point-counting errors by the various individuals involved are probably the reasons for the discrepancy. Frio sandstones are rarely extensively and massively quartz cemented, and only about 5% of all sands contain more than 10% quartz cement (Fig. 2). Sands containing more than 20% quartz cement are extremely rare. Recognizing the limitations of these averages, a regional value of 2.5% quartz cement will be assumed.

It is difficult to obtain the isotopic composition of quartz overgrowths

important reaction leading to secondary porosity is also discussed.

Although it would have been desirable to have chosen samples via some statistical technique that would have avoided bias, this was not possible. The samples available to Loucks and his group were already highly biased toward potential petroleum-producing areas and geothermal prospects. For example, very few samples shallower than 6000 ft were obtained. Although wells were sampled at approximately 50-ft intervals, core recovery clearly creates a bias toward "hard" samples, and sands lacking extensive mud matrix and having high secondary porosity were somewhat preferentially selected (Loucks, 1981, personal communication). Therefore the data base on which my conclusions are based may be somewhat unrepresentative, but it is very difficult to assess the degree to which I have either underestimated or overestimated the abundance of var-

ious components regionally in the Frio.

After choosing samples having more than 5% carbonate and varying amounts of quartz cement based on the thin-section point counts, approximately one gram was broken from the end of the thin section heel and ground to fine sand-size. The bulk powder was subject to X-ray powder diffraction to determine carbonate mineralogy, approximate carbonate content, and feldspar composition. Samples containing carbonate were reacted with anhydrous H₃PO₄ to release CO₂, and the gas analyzed isotopically according to standard techniques (McCrea, 1950). Several carbonate-rich but unimpregnated sandstones were analyzed with various additions of pure, powdered epoxy in order to ensure that the epoxy did not interfere with the analysis. No significant or systematic effects were found. Samples for quartz were prepared as previously described (Milliken et al, 1981) by reducing the samples to

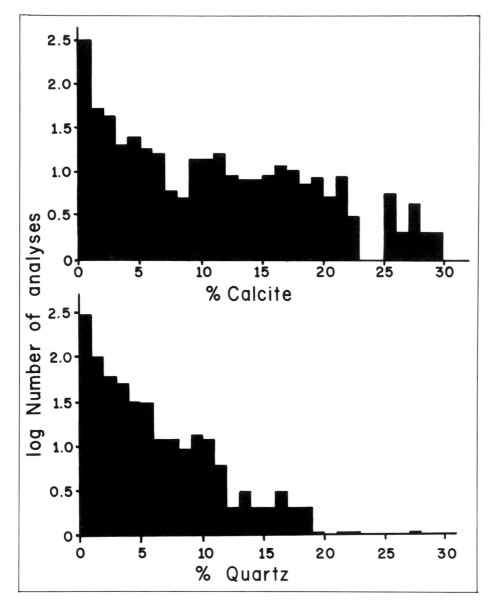

Figure 2—Percent calcite and quartz cements from all available thin-section point counts (676 thin sections) plotted against the log of the number of thin sections counted. Most sandstones contain no cement, and progressively better-cemented sands are progressively less common.

when they constitute less than about 10% of the rock (see discussion in Milliken et al, 1981). Figure 3 presents $\delta^{18}O$ values for whole-rock quartz plotted against percent quartz overgrowths of the total quartz fraction as determined from thin-section point counts. Linear regression of the data points extrapolates to a $\delta^{18}O$ for quartz overgrowths of +30.7 o/oo (correlation coefficient = 0.82). Combining these data with those reported in Milliken et al (1981) yields an "average" $\delta^{18}O$ for quartz overgrowths of about +31 ± 1.5 o/oo (SMOW).

Examination of quartz overgrowths using 15 Kv electrons, a 200 μm spot and 0.5 microamps (μA) of sample current through the reflected light microscope of the electron probe shows most overgrowths to be inhomogeneous. Fine, complex zones of orange luminescence alternate with less luminescent zones and document a complex history of nucleation and coalescence into large overgrowths (Ramseyer, 1982). Thus the overgrowths may not be isotopically homogeneous and since some abut pore space today, some could have had a long history of precipitation. Isotopic data must be interpreted with this complexity in mind.

Based on petrographic data, most quartz cementation preceded most calcite cementation, although contemporaneous precipitation and reciprocal relations are not uncommon. Since cementation by insoluble phases like quartz must result from essentially "open system" precipitation, a plot of the "average" value of quartz overgrowths as a function of temperature and $\delta^{18}O_{water}$ constrains the conditions under which the overgrowths might have formed (Fig. 4). The quartz and calcite fields do overlap somewhat, in agreement with paragenetic relationships determined petrographically. Water more positive than about +8 o/oo is unknown from Tertiary clastics in the present Gulf Coast and apparently cannot be generated by clay mineral reactions at observed temperatures (Suchecki and Land, 1983). More

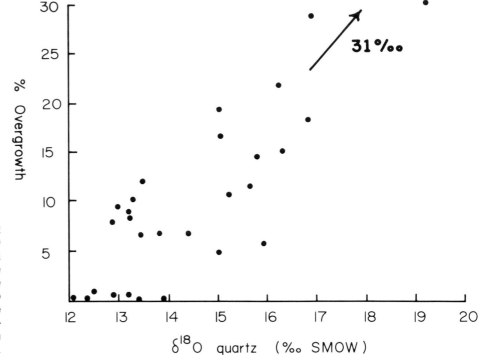

Figure 3—Scatterplot of percent quartz overgrowths of total quartz versus $\delta^{18}O$ of whole-rock quartz. Extrapolation of a regression line to +30.7 o/oo for pure quartz overgrowths is shown. The spread in values for sands with no authigenic quartz is at least partly due to inclusion of variable amounts of silicified volcanic rock fragments (Milliken et al, 1981).

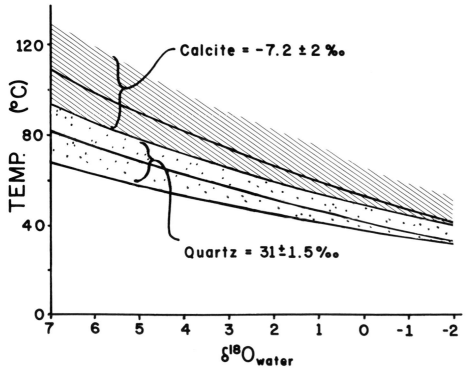

Figure 4—Locus of equilibrium values for $\delta^{18}O$ water and temperature for average quartz overgrowths ±1.5 o/oo and average calcite ±2 o/oo. Quartz usually precedes calcite petrographically and must have precipitated at temperatures considerably cooler than the smectite-to-illite transformation.

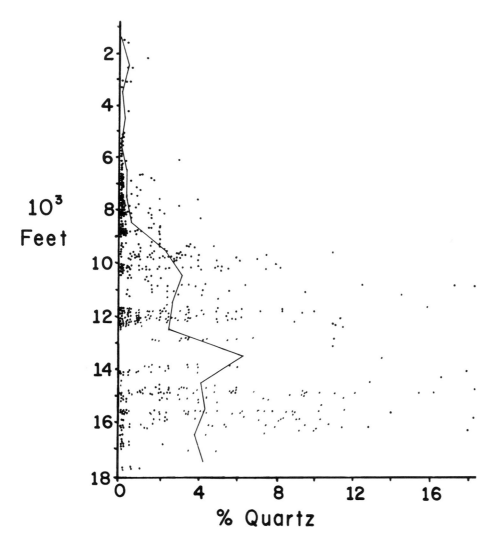

Figure 5—Percent quartz cement as determined from thin-section point counts as a function of depth. The line is a moving average. Quartz cementation is quite uncommon and not volumetrically significant above about 8000 ft. Only some sands below about 10,000 ft contain significant amounts of quartz cement. The change in degree of cementation corresponds to the type of geopressure.

enriched water is common in the underlying Cretaceous carbonates, however (Land and Prezbindowski, 1981), but no evidence for interaction with Cretaceous fluids has been documented (see below). Therefore, quartz precipitation must have occurred at relatively cool temperatures, certainly well below the temperature of approximately 100°C needed for the smectite-to-illite transition, a commonly cited source for silica. Water at equilibrium with smectite at 100°C should have a $\delta^{18}O$ of about +5 o/oo (Suchecki and Land, 1983), leading to a "best-guess" temperature of about 60°C (Fig. 4) for quartz cementation if the clay reaction was indeed the source of silica, and if the isotopic composition of the water was buffered by the clay reaction. As stated by Milliken et al, (1981), the lack of quartz overgrowths in outcrop sam-

ples, and the general rarity of quartz overgrowths shallower than about 8000 ft (Fig. 5), makes precipitation at Earth surface conditions from descending ground water (25°C from a water having a $\delta^{18}O$ of about −3 o/oo) very unlikely, although not inconsistent with the isotopic data. Precipitation at approximately 60°C from ascending water having a $\delta^{18}O$ of about +4 to +5 o/oo is consistent with all available data, but apparently requires massive, long-distance silica transport. Precipitation at intermediate temperatures from only slightly enriched water (mixed meteoric water and isotopically modified sea water), utilizing amorphous aluminosilicates as a silica source (Fisher, 1982) is another alternative.

If quartz precipitated as a result of fluids rising and cooling while basinal

shales supplied silica, the volume of fluid required raises a serious problem, as was noted by Bjorlykke (1979) and Land and Dutton (1979). If the average Frio sandstone contains 2.5% quartz overgrowth cement by volume, then about 66 mg SiO_2 must be added to each average cu cm of sandstone in the Texas Gulf Coast Frio (2.65 g/cu cm × 0.025). Figure 6 depicts the dissolved silica content of modern Frio formation water. Although the formation water today need not even closely resemble the water responsible for quartz cementation millions of years ago in most dissolved components, silica concentrations, ultimately buffered by quartz solubility and the kinetics of quartz precipitation may have been not too different. Many waters today are oversaturated with quartz by at most 50 ppm. Thus, modern water suggests the

Figure 6—Scatterplot of ppm SiO_2 versus temperature for Frio Formation waters sampled by Kharaka et al (1977) and Morton et al (1981). Line is theoretical quartz solubility, so points below the line are either valid and undersaturated with quartz or invalid because of dilution by condensed water vapor during collection of the water sample. X's are water having in situ density less than 0.990 and heavy dots are water having in situ density greater than 1.010 calculated from the algorithm of Phillips et al (1981). Low-density (buoyant) water is less saline (less than 50,000 ppm TDS) than high-density water. Small dots are intermediate in both salinity and density. It is tempting to invoke thermal convection (Wood and Hewett, 1982) to account for the density distribution, but the reason for the increased salinity in the denser (sinking) limbs of presumed convection cells is presumably salt dome dissolution.

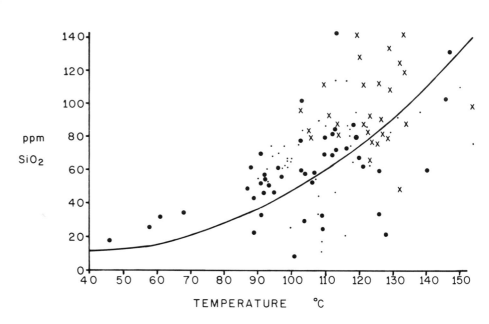

limit to the degree of oversaturation with quartz that occurs under formation conditions before precipitation actually occurs is about 50 ppm. In order to add 66 mg SiO_2 to each cu cm of Frio sandstone (on the average), about 1.3 liters of water are required, assuming: (1) no quartz is locally derived by grain interpenetration or stabilization of amorphous silica or amorphous aluminosilicates, (2) each pore volume of water loses 50 ppm SiO_2, and (3) silica transport by diffusion is negligible.

Petrographers almost uniformly agree that local grain interpenetration can account for but a small percentage of quartz overgrowth cements. Extensive grain interpenetration has never been documented from the Gulf Coast. Amorphous aluminosilicates, abundant in soils (Jackson, 1969) but not documented in shales could represent a significant source for silica. In the Frio there is little evidence for an extensive early generation of clay cement which could act as a sink for aluminum, and therefore this source is unlikely. No positive or negative evidence exists that amorphous silica (alternately marine diatoms or Radiolaria from shales, or opal phytoliths) acted as a silica source.

On the scale of a few meters, diffusion can be a significant transport mechanism (Wood and Hewett, 1982),

but diffusion cannot be the major process transporting silica in these sands. The most intensely quartz-cemented sandstones are not necessarily found immediately adjacent to shales: they are mostly within thick sandstone sequences. It is not likely that diffusive transport would almost completely bypass sands closest to the presumed source of silica. Quartz-cemented sands are distributed quite irregularly, apparently related more to permeability differences in the sands than to proximity to shales.

Many authors have noted that large amounts of silica are potentially released in the water liberated by the conversion of smectite-to-illite. If water to transport the silica is also liberated from the clay transformation, then such water presumably is at least quartz saturated (as any water released from shales by any mechanism must be). Oversaturation may occur because of buffering by more soluble silicates or by opal. What is critical is the mechanism to move the silica from the source (say the smectite-to-illite reaction) into the sands. The volume of silica released by either the reaction proposed by Boles and Franks (1979) (equation 1):

$$\text{smectite} + K^+ \longrightarrow \text{illite} \qquad (1)$$
$$+ H_4SiO_4 + \text{cations (Ca, Mg, Na, Fe)}$$

or the reaction proposed by Hower et al (1976) (equation 2):

$$\text{smectite} + \text{K-feldspar} \longrightarrow \text{illite} \qquad (2)$$
$$+ \text{quartz} + \text{chlorite}$$

may be irrelevant to sandstone cementation if significant amounts of silica are precipitated locally in the shales as Yeh and Savin (1977) have shown to occur. Dissolved silica must be continually transported out of the shales as it is being generated but before local nucleation within the shales can occur if this source of silica is important. Because isotopic data apparently preclude precipitation at high temperature, then large vertical transport of water (about 40°C, or at least a kilometer), appears to be involved.

Alternatively, the reaction of smectite-to-illite is not necessary to provide silica. Because any water expelled from shales at any temperature must be at least quartz saturated, any rising (cooling) water should precipitate quartz. If this mechanism alone operates, the 1300 cu cm of water required to cement each average cubic centimeter of sandstone in the sequence exceeds by at least a factor of 100 the available water in basinal Frio shales. Underlying Cretaceous and Paleocene sediments are also apparently of insufficient thickness to supply such large

Figure 7—Percent calcite plus dolomite cement as determined from thin-section point counts as a function of depth. The line is a moving average. In contrast to quartz (Fig. 5), no trends in degree of cementation with depth are apparent. This, together with the lack of trends in trace element content (Fig. 8) and isotopic composition (Figs. 9, 10), suggests emplacement under relatively invariant chemical conditions.

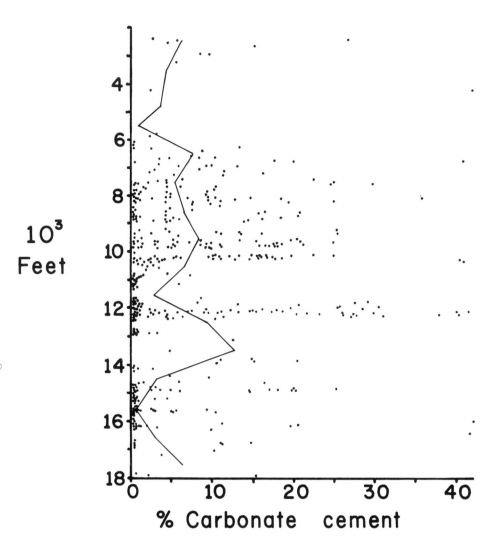

volumes of water since it is unlikely that the total volume of sediment beneath the Frio exceeds the volume of Frio sand by more than a factor of 20.

CARBONATE CEMENTATION

Point counts of calcite cement in Frio sandstone yield an average of 5.3% by volume (Figs. 2, 7), with carbonate cements and grain replacements (and carbonate rock fragments) more abundant toward south Texas. Calcite is the dominant cement; dolomite occurs only rarely. Electron-probe analysis demonstrates that the slightly luminescent calcite typically contains less than 2 mole % $MnCO_3$, $FeCO_3$, or $MgCO_3$ (Fig. 8). With the exception of a slight increase in manganese and iron below about 15,000 ft, no statistically significant correlation between calcite purity and any measured variable (percent cement, depth, $\delta^{18}O$, $\delta^{13}C$) was found.

Carbon and oxygen isotopic values for 68 samples are presented in Figures 9 and 10. Despite the large depth range and wide geographic distribution of the samples, and the fact that south Texas rocks unavoidably include samples containing carbonate rock fragments in addition to cements (accounting for many of the $\delta^{18}O$ values heavier than −6 o/oo), the data show little variability. $\delta^{13}C$ values average −4 o/oo (±2.6 o/oo) and the data-point distribution suggests a "base value" of around −2 o/oo, insensitive with respect to depth, which is modified by depleted carbon values at shallow depths. Surprisingly, these regional data are somewhat more tightly clustered than those from Brazoria County (Milliken et al, 1982; Fig. 7). Until the systematics of dissolved inorganic carbon and organic acids in

the subsurface can be documented, little can be concluded from the carbon isotope data except to be sure that if skeletal $CaCO_3$ (coccoliths, Foraminifera) mobilized from shales and having a $\delta^{13}C$ of about +2 o/oo is the ultimate source of $CaCO_3$, then ^{13}C-depleted (organic) carbon has been added to the reservoir of skeletal carbon. Because few cement values approach the $\delta^{13}C$ of organic matter (approximately −23 o/oo) or even −12 o/oo (50% organic carbon CO_2 + 50% coccolith carbon; equation 3), oxidation or decarboxylization of organic matter cannot be the dominant process of carbonate mobilization. Incorporation of slightly depleted CO_2 remaining after the generation of very ^{13}C-depleted methane from organic matter is an alternative (see below).

Like the carbon isotopic data, oxygen isotopic data (Fig. 10) are surpris-

ingly clustered considering the areal and vertical distribution of the samples, even with respect to the Brazoria County data base. The average $\delta^{18}O$ value is −7.2 o/oo ± 1.1 o/oo. Unlike Brazoria County, few samples are more depleted than −9 o/oo, probably reflecting the paucity of samples deeper than 13,000 ft. Samples shallower than about 10,500 ft are even more tightly clustered than deeper ones, consistent with petrographic data which indicate less secondary-pore-filling calcite above about 10,500 ft. Most calcite cementation preceded petroleum migration and the development of secondary porosity (Lindquist, 1977). Isotopically and in trace element content this calcite appears to reflect a reasonably homogeneous "event" resulting in calcite emplacement under considerably more restricted ranges of $\delta^{18}O$-water and temperature than are present in the

Figure 8—Scatterplot for mole percent $MgCO_3$, $FeCO_3$, and $MnCO_3$ in calcite as a function of depth. Each point represents an individual analysis. Except for slight increase in Fe and Mn below about 15,000 ft, no trends are apparent.

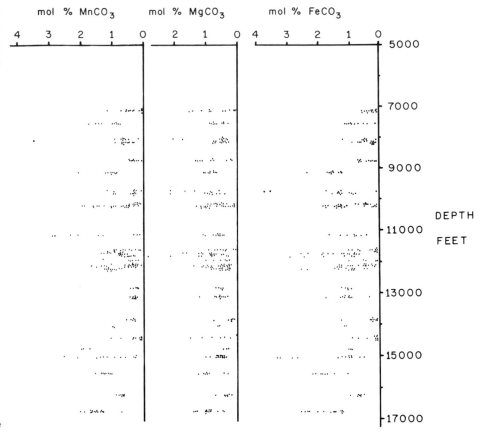

Figure 9—Scatterplot of $\delta^{13}C$ versus depth for calcite (dots) and dolomite (X's). The mean and standard deviation for calcite are also shown. Because the powdered samples were not size-separated, the "dolomite" and "calcite" values from the same samples are certainly mutually contaminated and nothing should be concluded about the isotopic differences between the two minerals.

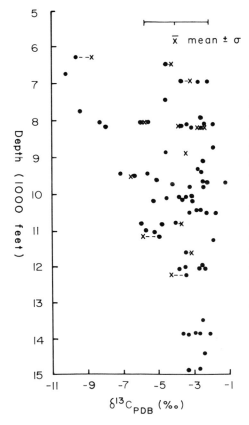

formation today. When typical $\delta^{18}O$ values for calcite are plotted as a function of temperature and $\delta^{18}O$-water (Fig. 4), precipitation at higher temperatures than quartz or from ^{18}O-depleted water is indicated. Thus precipitation may have occurred at greater burial depths than the quartz, and possibly closer to the source of material for the cements. As suggested by Milliken et al (1981), precipitation immediately above the clay transformation zone (say at 80°C from a water having a $\delta^{18}O$ of +4 o/oo) is consistent with the isotopic and petrographic data.

Calcite deeper than about 10,000 ft is typically subject to extensive dissolution, and both primary and secondary pores can be filled with later generations of calcite. Deeper calcites typically contain more depleted $\delta^{18}O$ values and some contain more iron and mang-

anese, but the relatively uniform chemistry is still quite remarkable. Based on X-ray diffraction results, dolomite constitutes between 10 and 15% of the carbonate cement in Frio sandstones, and no simple reason for its presence has been ascertained. In at least some samples, dolomite can be interpreted as a co-precipitate with calcite (Milliken et al, 1981).

Four $^{87}Sr/^{86}Sr$ ratios strongly suggest that calcite in Frio sandstones is derived from contemporaneously deposited shales (Table 1). The $^{87}Sr/^{86}Sr$ of sea water increased rapidly between the Eocene-Oligocene boundary and the present (Burke et al, 1982). Since the $^{87}Sr/^{86}Sr$ of calcite cement in the Frio corresponds closely to the ratio expected for Late Oligocene-Early Miocene marine calcite, remobilization of contemporaneously deposited skele-

TABLE 1

Sample no.	County	Depth (ft)	$\delta^{13}C$ (PDB)	$\delta^{18}O$ (PDB)	$^{87}Sr/^{86}Sr$
1671	Calhoun	7734	−9.4	− 8.1	0.70782
1414	Aransas	8151	−2.5	− 7.7	0.70849
1672	Nueces	9480	−5.7	− 7.9	0.70814
1091	Brazoria	14821	−2.9	−10.3	0.70834

Table 1—Strontium isotope composition of four calcite samples.

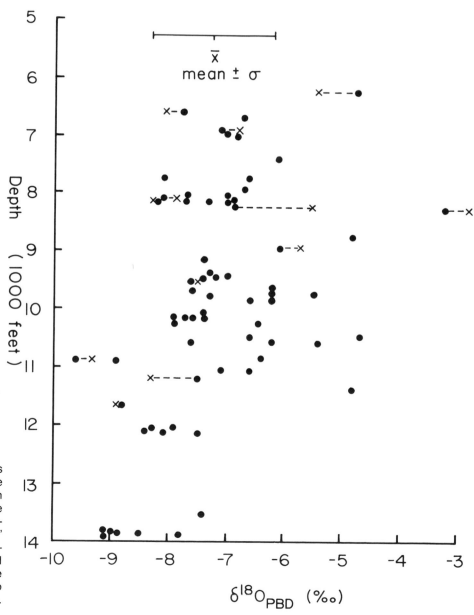

Figure 10—Scatterplot of $\delta^{18}O$ versus depth for calcite (dots) and dolomite (X's). The mean and standard deviation for calcite are also shown. Because the powdered samples were not size-separated, the "dolomite" and "calcite" values from the same samples are certainly mutually contaminated and nothing should be concluded about the isotopic differences between the two minerals.

tal carbonate prior to feldspar stabilization reactions (which would release considerable [87]Sr) is the most reasonable explanation for the data.

The solubility of calcite as a function of temperature and pressure is still imperfectly known. The decrease in solubility that occurs as a result of increased temperature is at least partially offset by an increase in solubility with increased pressure. Dissolution or precipitation of calcite related to changes in temperature and/or pressure could only achieve changes of volumetric significance if vast quantities of water were involved. It is much more likely that calcite mobilization is achieved because of changes in the amount of carbon dioxide in the system just as in the dissolution/precipitation of calcite at the Earth's surface (Holland et al, 1964; Hanor, 1978). Data on the subsurface calcite–CO_2 system are very few at the present time. If the calcium content of modern formation water is at all representative of earlier water (probably initially near sea-water composition), then calcium contents on the order of 50 to 800 ppm are not unreasonable (Kharaka et al, 1977; Morton et al, 1981). Changing the volume fraction of CO_2 in the gas phase from, say, 10^{-1} to 10^{-2} might cause the precipitation of about one millimole of calcite per liter of water.

The average calcite content of Frio sandstones today is 5.3 volume percent, or 1.5 millimoles of $CaCO_3$ per cubic centimeter of sandstone, and calcite may have been somewhat more abundant prior to the development of secondary porosity. Several liters of water were therefore probably required to cement each cubic centimeter of "average" Frio sandstone. Clearly there is no local source for $CaCO_3$ in northern and central Texas Gulf sandstones (south Texas sandstones contain carbonate rock fragments), so transport into the sands is required. Skeletal carbonate in contemporaneous shales is the most likely source for $CaCO_3$, and is supported by strontium isotopic data (Table 1). The nonradiogenic nature of the calcite cements supports their emplacement prior to significant K-feldspar destruction, and the strontium isotopic data do not support a deeper (Eocene or Cretaceous) origin for the $CaCO_3$, or the involvement of clays.

Nannofossil carbonate is absent in shales from some wells below about 10,000 ft (Hower et al, 1976; Freed, 1980), but present throughout other wells. If shales below about 10,000 ft are indeed the source of calcite deposited in the sands, then extensive vertical transport is required. It is possible that local gradients in both calcium and carbonate content could drive diffusive transport as organic maturation in the shales provides CO_2 to mobilize skeletal $CaCO_3$. But such a mechanism is very difficult to quantify because the co-diffusion of several components is required and the importance of dissolved organic acids (Carothers and Kharaka, 1978) is difficult to assess. So although the data are considerably more complex, the volume of water required for calcite cementation is at least of the same order of magnitude as for quartz cementation. It is possible that the same water transported both calcite and quartz, losing CO_2 to sandstone pore waters closer to the source of Ca^{+2} and HCO_3^- (say at about 80°C), and then precipitating quartz further updip and at cooler temperatures (say about 60°C) when sufficient oversaturation was attained.

Frio shales from several locations seem to have lost about 10% calcium carbonate below about 10,000 ft (Hower et al, 1976; Freed, 1980). Since the volume of Frio shale exceeds the volume of Frio sand by about a factor of ten, and since Frio sands contain only 5.3 volume percent calcite, considerable $CaCO_3$ may have been lost from the section. The source of sufficient acid to mobilize such a large volume of $CaCO_3$ is not at all clear.

SECONDARY POROSITY

Secondary porosity in Frio sandstones is volumetrically significant (Loucks et al, 1979b). Carbonate cements (and grains in south Texas), feldspars, and rock fragments are all affected. The volume of secondary porosity is difficult to quantify, and certainly involves considerable value judgment on the part of the (often biased) observer. Loucks et al (1979a) estimated that in Frio sandstones about 60% of the porosity was secondary (their Fig. 68, p. 77). Assuming about 20% porosity at 10,000 ft (their Fig. 16,

p. 23), 12% of the rock volume could thus be secondary porosity. I will arbitrarily assume half the secondary porosity is due to removal of calcite and half due to K-feldspar dissolution. Therefore (0.12 cu cm $CaCO_3$/36.94 cu cm $CaCO_3$ mole^{-1} $CaCO_3$ =) 1.6 millimoles of $CaCO_3$ might have been removed from each cubic centimeter of sandstone. One mole of CO_2 is required to dissolve each mole of $CaCO_3$ according to equation 3 (assuming H$^+$ is derived by $CO_2 + H_2O \rightarrow H^+ + HCO_3^-$):

$$CaCO_3 + CO_2 + H_2O \rightarrow Ca^{+2} + 2HCO_3^- \qquad (3)$$

If the reaction of Type II organic matter (Tissot and Welte, 1978) can be approximated by equation 4:

$$C_{10}H_{12}O_2 \rightarrow 6C + 3CH_4 + CO_2 \qquad (4)$$

and the density of organic matter is about 0.95 g per cu cm, then approximately 4 volumes of organic matter are required to dissolve each volume of $CaCO_3$. Clearly such large volumes of organic material could not have been deposited with the sands, and therefore CO_2 transport into the sands is required. Assuming a solubility of CO_2 of about $10^{-2.5}$ moles per liter, then the creation of 6% secondary porosity by removal of 1.6 millimoles of $CaCO_3$ requires about 0.5 liter of water for each cubic centimeter of sandstone. The "water volume" problem obviously worsens, and a new problem concerning the lack of an apparently sufficient source of CO_2 has been introduced. Removal of 5% of the volume of Frio sandstones, which constitutes about 10% of the volume of Frio sediments, apparently requires (0.05 × 0.10 × 5) about 2.5% of Frio sediments to have been composed volumetrically of organic matter as written in equation 4 if CO_2 is produced solely by decarboxylation. Clearly such large volumes of organic material cannot be documented in Frio sediments. Therefore other mechanisms of CO_2 generation must operate, or another H$^+$ source must be invoked.

Morton (1983) recently determined a Rb/Sr isochron from Brazoria County Frio shale suggesting that the smectite-to-illite transformation occurred about

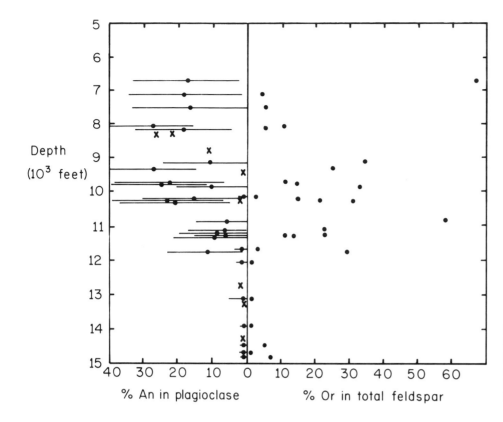

Figure 11—Feldspar compositions as a function of depth. The average An content of plagioclase for each sample is shown by a dot, and the lines through the dots include one standard deviation in An content. X's are from Boles (1982).

24 million years ago, substantially in agreement with Milliken et al's (1981, Fig. 14) estimate of the burial history of the basal Frio. If the age estimates are correct, all reactions just discussed must have taken place during the first 5 or 6 million years after deposition. Transport of the various components involved is thus constrained on this time scale. According to Dow (1978, Fig. 15), significant hydrocarbon maturation does not occur in 6 million years at 100°C. Therefore, components derived from organic maturation must have resulted from vertical transport from deeper in the section.

FELDSPAR STABILIZATION

Most Frio sandstones are lithic arkoses and feldspathic litharenites, feldspar and rock fragments being more abundant in south Texas, with quartz-rich sandstones (subarkoses) more common toward the north. Lindquist (1976) point counted an average of 27% feldspars from Frio sandstones in south Texas. Because many of her K-feldspar grains were dissolved, her values must be somewhat low with

respect to the original sandstone composition. Based on point counts of the Bureau of Economic Geology's extensive collections, feldspar constitutes about 25% of Frio sandstones from central and northern Gulf Texas, and about 33% of south Texas sandstones. Original feldspar contents were undoubtedly higher, both because of extensive dissolution of K-feldspars and because the point counts do not include the feldspar content of rock fragments. A value of 30% original feldspar will be taken as an "average" primary sand composition. Point counts of 29 rocks (1634 data points) on the electron probe (Fig. 11) suggest that about 30% of the original feldspar was K-feldspar, and 70% was plagioclase having an average An content of 20%.

Virtually no primary detrital feldspars (those retaining the original composition they had when deposited) are preserved in Frio sandstones below 12,000 ft anywhere in Texas. Albitization of plagioclase can occur as shallow as about 9000 ft, and no sandstones below 11,000 ft are unaffected by at least partial albitization. Virtually no

K-feldspars are preserved below 12,000 ft, and some samples as shallow as 7000 ft contain 95% plagioclase and only 5% K-feldspar. Although it is possible that such compositions might have been primary due to sorting effects and local source variations in this vast area (Galloway, 1977), such low K-feldspar contents seem peculiar.

Petrographic evidence suggests that K-feldspars are dissolved whereas plagioclase is albitized. Skeletal K-feldspars are more common than skeletal plagioclase, and relict plagioclase fabrics are often preserved by albitization. Lindquist was, in fact, unaware of the massive albitization of her deeper sandstones because primary plagioclase fabrics are so well preserved. I will assume complete dissolution of K-feldspars and complete albitization of plagioclase for the sake of this discussion, but I must caution that the evidence for such an assumption is not completely compelling, and alternate scenarios can be constructed (Land and Milliken, 1981).

If K-feldspars originally constituted about 10% by volume of the sand fraction of Frio sandstones (5% of the

TABLE 2

Reaction	Volume %	SiO_2	H_2O	CO_2		Mineral	Volume of Water Required (liters)
Quartz cementation $H_4SiO_4 \rightarrow SiO_2 + 2H_2O$	2.5	+1.1	+2.2		+1.1	quartz	1.3
Calcite cementation $Ca^{+2} + 2HCO_3^- \rightarrow CaCO_3 + H_2O$	10.0		+3.1	+3.1	+3.1	calcite	3.1
Secondary porosity							
Calcite dissolution	5.0		−1.6	−1.6	−1.6	calcite	0.5
K-feldspar dissolution (eqn. 5)	5.0	−0.92	−1.6	−0.46	−0.46	K-feldspar	0.15
					+0.23	kaolinite	
Albitization (eqn. 6)	2.0	+0.8	+1.6		+0.4	albite	0.02
		+1.44[1]	+3.7	+1.04			3.7[2]

[1]Includes 0.46 millimoles precipitated as kaolinite (eqn. 5)
[2]Assuming both calcite and quartz precipitate from the same water

Table 2—Material fluxes during Frio Sandstone diagenesis (millimoles per cu cm of sandstone).

volume of the sandstones assuming 50% initial porosity), then the most likely dissolution reaction can be written as equation 5:

$$2KAlSi_3O_8 + 2CO_2 + 7H_2O \rightarrow 2K^+ \quad (5)$$
$$+ Al_2Si_2O_5(OH)_4 + 4H_4SiO_4 + 2HCO_3^-$$

If K-feldspars constituted approximately 5% of the original volume of the sands, then 0.05 cu cm × 1 mole $KAlSi_3O_8$/ 109.5 cu cm = 0.46 millimoles of K-feldspar have been regionally destroyed in each cubic centimeter below about 12,500 ft in Frio sandstones. An equal number of moles of CO_2 were consumed, and K^+ and HCO_3^- released for each cubic centimeter of sandstone reacted. Comparisons of the CO_2 and H_4SiO_4 involved in this reaction with material requirements for secondary porosity generation and quartz cementation are presented in Table 2. The amount of CO_2 required is about one-third that required for calcite dissolution whereas the amount of SiO_2 released (0.92 millimoles per cu cm) is only slightly less than that required for quartz cementation. The

amount of kaolinite formed (0.23 millimoles × 99.5 cu cm per mole = 2% of the sandstone volume) is only slightly more than the value of 1.2% kaolinite obtained from all available point counts from the area. Because a great deal of fine pore-filling kaolinite is undoubtedly overlooked during point counting, it can almost certainly be concluded that a significant percentage of the kaolinite cement in these sandstones results from K-feldspar dissolution, and the few available isotopic analyses of kaolinite are in agreement with this conclusion (Milliken et al, 1981).

According to the above stoichiometry, about 0.46 millimoles of K^+ are released by destruction of K-feldspar for each average cubic centimeter of Frio sandstone below about 11,500 ft. If released into the local pore fluids, assuming about 30% porosity, then the fluid should contain 1.5 millimoles of K^+ per cu cm fluid. Sea water contains about 10^{-2} millimoles of K^+ per cu cm. Present-day pore fluids in Frio sandstones average about 5×10^{-2} millimoles K^+ per cu cm and only very rarely exceed 1.2×10^{-2} millimoles K^+ per cu cm fluid. Since no significant potassium sink in the sandstones can be identified, potassium must be transported out of the sandstones. If advective

transport is involved, and if potassium contents only twice that of sea water are allowed, then 1.5 millimoles K^+/10^{-2} millimoles K^+ per cu cm fluid = 150 cu cm fluid per cu cm sandstone are required. The 0.46 millimoles of K^+ released per cu cm of sandstone from feldspar destruction could provide enough K^+ to permit conversion of all the smectite in about 1.5 cu cm of shale to illite (Table 3), provided that transport into the shales takes place.

If plagioclase originally constituted about 20% of Frio sands, and had an average composition of An_{20} (Fig. 11), then about 2% of the volume of the sands consisted of the anorthite component in plagioclase. The albitization reaction is most likely:

$$CaAl_2Si_2O_8 + 2Na^+ + 4H_4SiO_4 \rightarrow \quad (6)$$
$$2NaAlSi_3O_8 + Ca^{+2} + 8H_2O$$

If 2% of the sandstone volume was initially anorthite, then 0.02 × 1 mole anorthite/ 100.79 cu cm = 0.2 millimoles of anorthite could be consumed and the same number of moles of Ca^{+2} released. 0.4 Millimoles of sodium are required for each cubic centimeter of sandstone and 0.4 millimoles of albite (× 100.2 cu cm per mole = 0.4 cu cm) formed. Thus, the albite formed by

TABLE 3

	Smectite	Illite	Al-conserv	Sm + K-spar	Milli-moles/cu cm
Si	1.002	0.960	0.229	0.315	2.75
Al	0.487	0.578	0	0	0
Fe	0.079	0.049	0.045	0.030	0.54
Mg	0.083	0.058	0.040	0.025	0.48
Ca	0.031	0.007	0.030	0.024	0.36
Na	0.027	0.009	0.023	0.018	0.28
K	0.054	0.115	−0.051	0.030	−0.61

Averaging Hower et al's (1976) shallowest 5 samples (= smectite) and deepest 6 samples (= illite) from their Table 7 (p. 732), and converting to moles yields columns "smectite" and "illite," respectively, in units of moles/100 g clay. Two methods of comparison have been suggested. For the first, one can assume aluminum conservation (eqn. 1), multiply "smectite" by 1.19, and subtract "illite" from the product to obtain a residual (Al-conserv column). For the second (eqn. 2), if one adds 0.91 × (K, Al and 3 × Si) to "smectite," assuming K-feldspar provides the required aluminum, again subtracting "illite" from the product, the residual in the Sm + K-spar column is obtained. In the latter case potassium is apparently released along with other constituents. Assuming the shale is about 50% illite/smectite, and the average density is 2.4 g per cu cm, the last column lists the millimoles of each component potentially released from each cu cm of shale using the aluminum-conservative reaction. The extent to which authigenic quartz and chlorite act as sinks for these components in the shales is critical, but not adequately known.

Table 3—Potential products of the smectite-to-illite reaction.

reaction 6 occupies more volume than the anorthite replaced, and in fact albite overgrowths are not uncommon accompanying the albitization process (for example, see Land and Milliken, 1981, cover photograph). The fact that albite overgrowths do not occupy 2% of the sandstone volume suggests that reaction 7:

$$CaAl_2Si_2O_8 + 3H_2O + 2CO_2 \rightarrow \qquad (7)$$
$$Al_2Si_2O_5(OH)_4 + Ca^{+2} + 2HCO_3^-$$

probably operates as well. As written in equation 6, and summarized in Table 2, the silica balance for the sands (quartz overgrowths − SiO_2 released by K-feldspar dissolution + SiO_2 consumed by albitization +SiO_2 precipitated as kaolinite) is positive, requiring net transport of SiO_2 into the sands. Since quartz overgrowth cementation apparently precedes feldspar stabilization, little quartz cementation seems to be accomplished as a result of feldspar stabilization reactions. The amount of calcium released by reaction 6 is small with respect to that involved in earlier reactions, but such calcium as calcite

could constitute up to (0.2 millimoles Ca^{+2} × 36.9 cu cm calcite per mole =) 0.7% of the volume of the sands, and may contribute significantly to the late, ^{18}O-depleted, Fe–Mn-rich secondary-pore-filling calcite. As written (equation 6), albitization of Frio sandstones requires 0.4 millimoles Na^+ per cu cm sandstone. Sea water contains 0.48 millimoles Na^+ per cu cm whereas Frio sandstone pore waters today contain an average of about 0.74 millimoles Na^+ per cu cm. Clearly the water responsible for albitization of Frio sandstones might have been quite different from either sea water or today's formation water, but sea water provides a reasonable reference to begin calculations. If we assume that $mNa^+ + 2mCa^{+2}$ in sea water remains constant at 0.496 moles per liter (that is, reaction 6 applies) and that albitization at 120°C can alter the mNa^+/mCa^{+2} of sea water from 46 to about 22 (Helgeson, 1972), then about 2×10^{-2} millimoles of Na^+ for albitization can be provided from each cubic centimeter of sea water. Thus a minimum of 20 cu cm of sea water is necessary to albitize each cubic centimeter of sandstone. Alternatively, assuming sodium released by the conversion of smectite-to-illite leaves the shale system

(Table 3), about 2 cu cm shale can provide sufficient sodium to albitize each cubic centimeter of sandstone. Sodium cannot, of course, be released by reaction 1 without equivalent consumption of H^+ or release of anions. A realistic charge balance of equation 1 as proposed by Boles and Franks (1979) is not at all clear.

Table 2 shows that significant amounts of both CO_2 and H_2O are released in the sandstones as a result of the various reactions. The volume of water released would be sufficient to reduce the salinity of the pore fluids about 20%. Nearly 4 liters of water are required to accomplish the various reactions in each cubic centimeter of sandstone.

SUMMARY DISCUSSION

The following scenario can account for most of the reactions observed in Frio sandstones. I must emphasize several points before proceeding, however. First, local variations have been ignored in this "broad brush" approach, and chlorite cement and authigenic zeolites can be locally important diagenetic phases, especially in south Texas. Second, as I previously pointed out, the averages used for these calculations may be somewhat incorrect because of sampling bias. And finally, underlying Wilcox sandstones and overlying Plio–Pleistocene sandstones are quite different in many aspects from Frio sandstones. Under no circumstances should the following "working hypothesis" be extended unmodified to those (or other) formations.

Early compactional dewatering appears to have accomplished little in the way of chemical diagenesis of Frio sandstones. Chemical transport first achieved important significance when basinal shales were heated to approximately 80–100° C. Release of CO_2 from organic maturation in more deeply buried shales mobilized the $CaCO_3$ in some Frio shales and together with SiO_2 released from underlying sediments and as a result of the smectite-to-illite reaction, quartz and carbonate-charged pore waters were expelled into the sands. Moving upward, the water, buffered in its oxygen-isotopic composition by the smectite-to-illite transformation, deposited calcite and quartz cements in the sandstones as it moved upward, cooled, lost CO_2, and mixed with more ^{18}O-depleted sea water, or possibly even meteoric water. Calcite cements were deposited closer to the material source and hence at higher temperatures. Small amounts of magnesium released by the clay transformation resulted in minor dolomite cementation. Active thermal convection must have occurred at this time to provide sufficient water (by re-using it) for transport. After skeletal calcium carbonate from shale was consumed, CO_2 continued to be generated as organic maturation proceeded, affecting both previously deposited $CaCO_3$ (moving it higher in the section) and feldspars once the section was buried to a depth where temperatures approached 120° C. Secondary porosity generation ceased once CO_2 evolution ceased, and late calcite and kaolinite cements were deposited until feldspars reached stable states. These latter reactions, controlled by the kinetics of feldspar reactions and the availability of reactants (Na$^+$ and CO_2) are probably still going on today.

Several problems exist with this scenario. The most disturbing (to me, at

least) is the tremendous transport distances and water volumes required for all these reactions (including hydrocarbon transport). Insufficient water is present in the sediments to accomplish the transport required on a "once-through" basis. Apparently either active convective overturn (Cassan et al, 1981; Wood and Hewett, 1982) is a viable transport process, local sources must provide components, or meteoric water-driven transport (Bjorlykke, 1979) must occur. If the transformation of smectite-to-illite is as important as some authors suppose (for example, Boles and Franks, 1979), and such reactions are indeed the dominant source for SiO_2 and carbonate cements, then large-scale convection appears inescapable because so much quartz and calcite are present far above the clay transformation zone. Isotopic data indicate that most quartz and calcite cement were emplaced at cooler temperatures than the smectite-to-illite transition. Alternative basinal sources for silica include opal (marine diatoms/Radiolaria and opal phytoliths) and amorphous aluminosilicates. None of these alternate sources have been adequately investigated. Alternative sources to contemporaneous nannofossil $CaCO_3$ and organic CO_2 in shales to provide for calcite cements and the acid that generates secondary porosity, respectively, are not clear, however. Strontium isotopic data and the distribution of calcium carbonate in contemporaneous shales strongly support a basinal origin for the calcite cement in the sandstones.

A second problem that has been raised is the source of acid for the various reactions involved. The one millimole of CO_2 involved in these reactions (Table 2) could be generated from about (1 millimole $\times$ [164 mg per millimole $\div$ 950 mg per cu cm]) 0.15 cu cm of organic material. Such a large volume of organic material approaches the total volume of organic matter deposited in the Gulf of Mexico (10 cu cm shale per cu cm sandstone $\times$ 0.01 cu cm organic matter per cu cm shale). If CO_2 is used to balance the cations released by equation 1 (that is, H$^+$ provided by $CO_2 + H_2O \rightarrow H^+ + HCO_3^-$ balances the released cations), the acid problem is considerably worsened. The "acid

volume problem" together with the relatively ^{13}C-rich Gulf Coast carbonate cements indicates that reactions other than oxidation and decarboxylation must be involved. The reaction: kerogen + $H_2O \rightarrow CO_2 + CH_4$ is an attractive possibility (Hoering, 1982). "Reverse weathering" reactions might provide an alternate source of H$^+$, but such reactions consume the same cations (Na$^+$, Mg^{+2}, Fe^{+2}) which are also required for diagenesis of the sandstones.

As a corollary to this problem, the "openness" of the shale system is not convincingly documented. Table 3 summarizes the material *potentially* available from the smectite-to-illite transition based on Hower et al's data. Hower et al (1976) clearly document the authigenesis of chlorite in Frio shales, in agreement with the low iron and magnesium content of most Frio carbonate cements shallower than 15,000 ft (and in contrast to Wilcox carbonate cements; Boles and Franks, 1974; Fisher, 1982). As was stated previously in the case of quartz, just because a product is released by the smectite-to-illite transition does not necessitate that the product leaves the shale system. In fact, Hower et al's 1976 data and Yeh and Savin's 1977 data suggest that most products (including silica) do not leave the shales. If expulsion from the shale system occurs, then equation 1 as written by Boles and Franks (1979, their reaction 2, p. 63) is unreasonable in that a realistic equivalency to the cations supposedly released is not expressed. O^{-2} is not a reasonable reaction product to achieve charge balance if aqueous expulsion of cations from the shale system occurs, and a reasonable anion to accompany cation expulsion was not proposed. If H$^+$ is used to balance Boles and Franks' reaction 2 (1979, p. 63) (or HCO$_3^-$ used as a product) then the "acid problem" previously discussed is quite considerably worsened.

Most Frio cementation apparently occurred early during its burial history, when porosity and permeability were still high, possibly in the first 6 million years. All proposed transport mechanisms suffer from significant flaws on this time scale. Diagenesis and dewatering of basinal shales cannot supply the volume of fluid required, but sccm to have supplied the CaCO₃. Deep recharge of meteoric water is inconsistent with the spatial distribution of cements, suffers from source problems (especially in the case of $CaCO_3$), and with problems of precipitation mechanism. Early thermal convection between the meteoric and overpressured regimes is certainly attractive to solve the water volume problem, but does not solve the problem of sourcing the $CaCO_3$ from the deep basinal shales. In addition, it is difficult to rationalize a convecting system with the presumed early, shallow development of overpressure, which must have extended to shallower depths than it does today. Why did the convecting system stop, and why is so little quartz cement present above geopressure today (Fig. 5)?

Much more extensive study of shale and sandstone burial diagenesis is clearly warranted. If we believe that the products of reactions (CO_2 and the liquid hydrocarbons themselves) cross formational boundaries, then that study must encompass the entire sedimentary basin. Frio shale diagenesis may not be as important to some aspects of Frio sandstone diagenesis as is the mineral and organic diagenesis of adjacent Tertiary and Mesozoic units.

ACKNOWLEDGMENTS

I wish to thank Bob Loucks and Bob Morton for amassing the samples on which this study was based and permitting me access to the data, the thin sections, and thin section heels. I have relied almost exclusively on their petrographic observations and those of their co-investigators, especially Kitty Milliken. Larry Mack relieved me of some of the drudgery of electron-probe analyses of feldspars and several undergraduate students assisted with sampling and sample preparation. I have profited greatly from discussions with and/or editorial efforts of several colleagues, including Steve Fisher, Bill Galloway, Tim Jackson, Earle McBride, Paul Lundegard, and Sam Savin. Analytical work for this study was supported by the National Science Foundation, EAR-7824081, and the Geology Foundation of the University of Texas. Mobil Oil Company provided the strontium isotopic data.

SELECTED REFERENCES

Bjorlykke, K., 1979, Discussion, cementation of sandstones: Journal of Sedimentary Petrology, v. 49, p. 1358–1359.

Boles, J. R., 1982, Active albitization of Gulf Coast Tertiary: American Journal of Science, v. 282, p. 165–180.

Boles, J. R., and S. G. Franks, 1979, Clay diagenesis in Wilcox sandstones of southwest Texas: implications of smectite diagenesis on sandstone cementation: Journal of Sedimentary Petrology, v. 49, p. 55–70.

Burke, W. H., R. E. Denison, E. A. Hetherington, R. B. Koepnick, H. F. Nelson, and J. B. Otto, 1982, Variation of seawater $^{87}Sr/^{86}Sr$ throughout Phanerozoic time: Geology, v. 10, p. 516–519.

Carothers, W. W., and Y. K. Kharaka, 1978, Aliphatic acid anions in oil-field waters—implications for origin of natural gas: Bulletin of the American Association of Petroleum Geologists, v. 62, p. 2441–2453.

Cassan, J. P., M. del C. Garcia Palacios, B. Fritz, and Y. Tardy, 1981, Diagenesis of sandstone reservoirs as shown by petrographical and geochemical analysis of oil bearing formations in the Gabon basin: Bulletin des Centres de Recherches Exploration-Production Elf-Aquitane, v. 5, p. 113–135.

Dow, W. G., 1978, Petroleum source beds on continental slopes and rises: Bulletin of the American Association of Petroleum Geologists, v. 62, p. 1584–1606.

Fisher, R. S., 1982, Diagenetic history of Eocene Wilcox sandstones and associated formation waters, south-central Texas: Ph.D. Dissertation, University of Texas at Austin, 195 p.

Freed, R. L., 1980, Shale mineralogy and burial diagenesis in four geopressured wells, Hidalgo and Brazoria Counties, Texas; Appendix A, in R. L. Loucks, D. L. Richmann, and K. L. Milliken, 1980, Factors controlling sandstone reservoir quality in Tertiary sandstones and their significance to geopressured geothermal production; Annual Report for the period May 1, 1979–May 31, 1980, DOE/ET/127111–1, University of Texas Bureau of Economic Geology, p. 111–172.

Galloway, W. E., 1977, Catahoula Formation of the Texas coastal plain: depositional systems, composition, structural development, ground-water flow history, and uranium distribution: University of Texas Bureau of Economic Geology Report of Investigations 87, 59 p.

Hanor, J. S., 1978, Precipitation of beach-

rock cements: mixing of marine and meteoric waters vs. CO_2-degassing: Journal of Sedimentary Petrology, v. 48, p. 489–501.

Helgeson, H. C., 1972, Chemical interaction of feldspars and aqueous solutions, in W. S. Mackenzie and J. Zussman, eds., The feldspars: Proceedings NATO Advanced Study Institute, Manchester, Manchester University Press, p. 184–217.

Hoering, T. C., 1982, Thermal reactions of kerogen with added water, heavy water, and pure organic substances, in Annual Report, Carnegie Institute Geophysical Laboratory, Washington, DC, No. 1880, p. 397–402.

Holland, H. D., T. V. Kirsipu, J. S. Huebner, and U. M. Oxburgh, 1964, On some aspects of the chemical evolution of cave waters: Journal of Geology, v. 72, p. 36–37.

Hower, J., E. V. Eslinger, M. E. Hower, and E. A. Perry, 1976, Mechanism of burial metamorphism of argillaceous sediments, 1, Mineralogical and chemical evidence: Geological Society of America Bulletin, v. 87, p. 725–737.

Jackson, M. L., 1969, Soil chemical analysis—advanced course, 2nd edition, published by the author, Madison, Wisconsin, 895 pp.

Kharaka, Y. K., E. Callender, and W. W. Carothers, 1977, Geochemistry of geopressured waters from the Texas Gulf Coast, in Proceedings of the Third Geopressured-Geothermal energy conference, University of Southwestern Louisiana, Lafayette, Louisiana, pp. 121–165.

Land, L. S., and S. P. Dutton, 1979, Cementation of sandstone, reply: Journal of Sedimentary Petrology, v. 49, p. 1359–1361.

Land, L. S., and K. L. Milliken, 1981, Feldspar diagenesis in the Frio formation, Brazoria County, Texas Gulf Coast: Geology, v. 9, p. 314–318.

Land, L. S. and D. R. Prezbindowski, 1981, Origin and evolution of saline formation water, Lower Cretaceous Carbonates, South Central Texas: Journal of Hydrology, v. 54. p. 51–74.

Lindquist, S. J., 1976, Sandstone diagenesis and reservoir quality, Frio formation (Oligocene), South Texas. Master's Thesis, University of Texas at Austin, 148 p.

———, 1977, Secondary porosity development and subsequent reduction, overpressured Frio formation sandstone (Oligocene), South Texas: Gulf Coast Association of Geological Societies Transactions, v. 27, p. 99–107.

Loucks, R. G., M. M. Dodge, and W. E. Galloway, 1979a, Sandstone consolidation analysis to delineate areas of high-quality reservoirs suitable for production of geopressured geothermal energy along the Texas Gulf Coast: Contract report EG-77-05-5554, University of Texas Bureau of Economic Geology, 97 p.

————, 1979b, Importance of secondary leached porosity in Lower Tertiary sandstone reservoirs along the Texas Gulf Coast: Gulf Coast Association of Geological Societies Transactions, v. 29, p. 164–171.

Loucks, R. G., D. L. Richmann, and K. L. Milliken, 1980, Factors controlling reservoir quality in Tertiary sandstones and their significance to geopressured geothermal production: Annual Report for the period May 1, 1979–May 31, 1980. DOE/ET/27111-1, University of Texas Bureau of Economic Geology, 188 p.

Milliken, K. L., L. S. Land, and R. G. Loucks, 1981, History of burial diagenesis determined from isotopic geochemistry, Frio formation, Brazoria County,

Texas: Bulletin of the American Association of Petroleum Geologists, v. 65, p. 1397–1413.

McCrea, J. M., 1950, On the isotopic chemistry of carbonates and a paleotemperature scale. Journal of Chemical Physics, v. 18, p. 849–857.

Morton, J. P., 1983, Age of clay diagenesis in the Oligocene Frio formation (abs.): Annual Convention of the American Association of Petroleum Geologists, Dallas, Texas, p. 132.

Morton, R. A., C. M. Garrett, Jr., J. S. Posey, J. H. Han, and L. A. Jirik, 1981, Salinity variations and chemical compositions of waters in the Frio formation, Texas Gulf Coast: Contract Report DOE/ET/27111-5, University of Texas Bureau of Economic Geology, 96 p.

Phillips, S. L., A. Igbene, J. A. Fain, H. Ozbek, and M. Tavana, 1981, A technical databook for geothermal energy utilization: Lawrence Berkeley Laboratory 12810, 18 p.

Ramseyer, K., 1982, A new cathodoluminescence microscope and its application to sandstone diagenesis: International Congress on Sedimentology, McMaster University, Hamilton, Ontario, p. 120.

Suchecki, R. L., and L. S. Land, 1983, Isotopic geochemistry of burial-metamorphosed volcanogenic sediments, Great Valley sequence, Northern California: Geochimica et Cosmochimica Acta, v. 47, p. 1487–1499.

Tissot, B. P., and D. H. Welte, 1978, Petroleum formation and occurrence: Springer-Verlag, 538 p.

Wood, J. R., and T. A. Hewett, 1982, Fluid convection and mass transfer in porous sandstones—a theoretical model. Geochimica et Cosmochimica Acta, v. 46, p. 1707–1713.

Yeh, H. W., and S. M. Savin, 1977, Mechanism of burial metamorphism of argillaceous sediments, 3, O-isotope evidence. Geological Society of America Bulletin, v. 88, p. 1321–1330.

Relationships Among Secondary Porosity, Pore-Fluid Chemistry and Carbon Dioxide, Texas Gulf Coast

Stephen G. Franks
ARCO Alaska, Inc.
Anchorage, Alaska

Richard W. Forester
ARCO Oil and Gas Company
Dallas, Texas

INTRODUCTION

Despite the importance of the concept of secondary porosity development in sedimentary basins, the actual physical and chemical processes for its development are poorly understood. In this paper, we present evidence from the Texas Gulf Coast that suggests a fundamental relationship between CO_2 content of natural gases and secondary porosity development in the subsurface. Although individual cases had been documented by earlier studies in specific areas (Hartman, 1968; Phipps, 1969; Heald and Larese, 1973; Roswell and DeSwardt, 1973; Parker, 1974; Hayes et al, 1976; Lindquist, 1976; Stanton and McBride, 1976; Alcock and Benteau, 1976), the ubiquitous occurrence of secondary porosity and its importance in petroleum exploration was not generally recognized until the work of Schmidt et al (1977) and Schmidt and McDonald (1979). Subsequent research has documented detrital and authigenic phases, which are related to the development of dissolution porosity at depth (see Hayes, 1979, Table 2). These minerals can be used to elucidate the processes responsible for porosity development.

Dissolution of carbonate cement during burial is volumetrically the most important type of secondary porosity in sandstones. It has been suggested (Schmidt and McDonald, 1979) that the primary agent responsible for carbonate dissolution in the subsurface is carbonic acid (H_2CO_3) formed by generation of carbon dioxide during the thermal maturation of organic mat-

ABSTRACT. Sequences of diagenetic minerals associated with secondary porosity show striking similarities. The formation of quartz overgrowths on detrital quartz grains is generally followed by carbonate cementation. The dissolution of this carbonate is the main secondary porosity-forming event, which commonly precedes kaolinite precipitation and iron-rich carbonate cementation. In the Texas Gulf Coast, oxygen isotopic data provide temperature estimates of authigenic phases that predate and postdate secondary porosity development: quartz, $\geqslant 80°$ C; kaolinite, $\geqslant 70°$ C; albite, $100–150°$ C; late carbonate, $> 100°$ C. These data suggest that secondary porosity in the Tertiary Gulf Coast forms at temperatures of about $100 \pm 25°$ C.

Correlations among calcite saturation indices in pore fluids, abnormally high permeabilities, and mole percent CO_2 in natural gases of the Eocene Wilcox Group imply a strong interrelationship between carbon dioxide and secondary porosity development in clastic reservoirs. The CO_2 content of gases varies systematically with both the reservoir age and temperature, which suggests a kinetic control on generation. The amount of CO_2 in natural gases increases rapidly at approximately $100°$ C; this coincides with a rapid increase in the ratio of secondary to total porosity in associated sandstones. Stable isotopic analyses of carbonate cements indicate a strong component of organically derived carbon and therefore cycling of carbon between inorganic and organic systems. The type, amount and distribution of organic matter, and early carbonate in both shales and sandstones control the quantity of CO_2 available for generating secondary porosity.

ter in sediments. This concept is based in large part on the work of Tissot et al (1974), who showed that carbon dioxide is one of the main by-products of thermal maturation of vitrinitic (Type III) kerogen. This explanation has been seized upon by subsequent workers and is frequently cited as the process responsible for carbonate and feldspar dissolution, even though little other evidence has been published in support of the hypothesis.

Deciphering the origin of secondary porosity in sandstones is analogous to understanding an unconformity in the stratigraphic record; our understanding of its origin is formulated largely on evidence from events recorded in strata preceding and following the hiatus. In the case of secondary porosity, the diagenetic minerals preceding and postdating dissolution yield clues to the physical and chemical conditions responsible for dissolution.

TABLE 1

Lower Tertiary, Texas Gulf Coast[1]	Oligocene Frio, South Texas[2]	Eocene Wilcox, Central Texas[3]	Eocene Wilcox, South & East Texas[4]	Pennsylvanian Strawn, North Texas[5]	Triassic Ivishak, North Alaska[4]
Clay Calcite (Fe-poor) Calcite and/or dolomite (Fe-rich)	Calcite		Clay	Clay	
Quartz overgrowths	Quartz overgrowths	Quartz overgrowths	Quartz overgrowths	Quartz overgrowths	Quartz overgrowths
Fe-poor calcite (Oligocene) Fe-rich calcite or dolomite (Eocene)	Calcite	Calcite	Calcite	Calcite	Siderite
Dissolution	Dissolution	Dissolution	Dissolution	Dissolution	Dissolution
Kaolinite	Kaolinite	Kaolinite	Kaolinite	Kaolinite	Kaolinite
Fe-rich dolomite or ankerite	Fe-carbonate	Fe-carbonate	Fe-rich dolomite or ankerite	Fe-carbonate	Fe-calcite

[1]Loucks et al., 1979.
[2]Lindquist, 1976.
[3]Stanton, 1977.
[4]ARCO, unpublished data.
[5]Dutton, 1977.

Table 1—Diagenetic sequences in sandstones with well-developed secondary porosity. Dissolution of feldspars, feldspar overgrowths, and authigenic kaolinite observed as minor effects in very shallow core samples have been omitted for simplicity. Similar observations have been made in outcrop samples and are likely attributable to meteoric ground water. As noted by the original authors, these effects are generally distinguishable from subsequent burial diagenetic effects. Dissolution involves both carbonate cement and feldspars in each case except the Ivishak where feldspars are generally absent.

PHYSICAL AND CHEMICAL CONDITIONS ASSOCIATED WITH SECONDARY POROSITY AND CARBON DIOXIDE GENERATION

Systematic Diagenetic Sequence

In numerous cases (see Table 1), the sequences of diagenetic minerals associated with secondary porosity formation show striking similarities. For example, calcite or siderite is usually the cement dissolved to create secondary porosity. Quartz cementation usually predates dissolution but may continue to form after dissolution and fill secondary pores. Kaolinite and iron-rich carbonates (Fe-dolomite or ankerite) are the most common late-stage cements found in secondary pores.

Because dissolution involves interaction with a fluid phase, the similarities of diagenetic evolution of rocks of different ages from different basins suggest a common link in the evolution of pore fluids in these basins. In actively subsiding basins, shale compaction yields large volumes of water that may be available for dissolution reactions in sandstones. The composition of these waters is controlled in large part by chemical reactions (organic or inorganic) occurring within the shales. Thermal maturation of organic matter and clay mineral reactions, such as the conversion of smectite to illite, are the most common reactions occurring in shales and are major influences on the geochemical evolution of pore fluids. Perhaps this is the link responsible for the diagenetic similarities.

Diagenetic Minerals Associated with Secondary Porosity

The mineralogy and composition of diagenetic minerals reflect the chemical conditions associated with secondary porosity development. Determining the temperature of formation of diagenetic cements associated with secondary porosity can place constraints on the depths and temperatures of dissolution events.

Early Carbonate Cement—Carbonate cement, primarily calcite, is common in shallow (<5000 ft) sandstones of the Texas Gulf Coast. In many samples this cement may be interpreted as pre-compaction, inasmuch as the detrital sand grains appear to be "floating" in the calcite. Similar textures and cements can be found in much more deeply buried sandstones and presumably are preserved during burial from shallower depths. This interpretation is consistent with the lack of later cements, such as quartz overgrowths, which are present in associated sandstones not cemented by early calcite. Assuming these textural arguments are

TABLE 2

Well Name and Depth	$\delta^{18}O$ (o/oo)	δD (o/oo)
Austral McLean[1] 10,627 ft	+17.0 ± 0.2	−35 ± 1
ARCO, Stroman Armstrong 12,824 ft	+14.1	−39

[1]Isotopic analyses are averages of 2 separate runs.

Table 2—Oxygen and hydrogen isotopic analyses of diagenetic kaolinites.

valid and assuming that quartz overgrowths generally form in abundance (at least in the Texas Gulf Coast) at temperatures of 80° C or more (see following discussion), early carbonate cements are interpreted to have formed at shallow depths where temperatures are less than 80° C.

Milliken et al (1981) present data on the isotopic composition of carbonate cements from the Frio Formation in Brazoria County, Texas. Calculations based on data from their Table 2 indicate that the mean oxygen isotopic composition of carbonate cements which predate quartz overgrowths has a $\delta^{18}O = +23.4 \pm 1.4$ (SMOW, per mil notation). Oxygen isotopic values of pore waters from the surface down to about 5000 ft will generally be more depleted than deeper waters. For example, waters from producing Tertiary sandstones in the northeast Thompsonville Field of south Texas have $\delta^{18}O$ values of +2.3 (2800–2900 ft), +2.9 (5675–5700 ft), and +6.1 (9784–9790 ft) (Kharaka, personal communication). Using the first two values, the calculated temperatures of formation of early (pre-quartz cementation) carbonate cements range from approximately 40 to 75° C, in agreement with their inferred shallow origin, although some are presently buried to almost 15,000 ft.

Quartz Cement—Petrologic evidence indicates that the smectite-to-illite transformation involves release of Si^{+4} (Hower et al, 1976; Boles and Franks, 1979). The transformation takes place over a broad temperature range, generally between 60 and 120° C. Some of the silica released may end up as quartz overgrowths on detrital quartz grains in nearby sandstones. Quartz overgrowth precipitation at approximately 80° C based on oxygen isotopic data (for example, Milliken et al, 1981; see also Yeh and Savin, 1977) is compatible with these observations and with quartz overgrowth formation before secondary porosity development.

Kaolinite—The diagenetic phase most intimately associated in space and time with the major dissolution event is kaolinite (Table 1). The conditions under which kaolinite precipitated from formation fluids, therefore, will closely reflect the conditions which prevailed during the creation of secondary porosity.

Table 2 gives oxygen and hydrogen isotopic analyses of diagenetic kaolinite from sandstones in the Gulf Coast. The Gulf Coast kaolinites are from the Lower Eocene Wilcox Group, northeast Thompsonville Field, south Texas. The oxygen and hydrogen isotopic compositions of formation waters at 9784–9790 ft in the same field are +6.1 and −14, respectively (Kharaka, personal communication). Oxygen and hydrogen isotopic compositions of waters from the Frio of south Texas have been reported by Kharaka et al (1978); typically they fall in the range of +5 ± 2 and −15 ± 5, respectively. Using the curve for oxygen isotope fractionation for kaolinite-water given by Kulla and Anderson (1978) and assuming a Wilcox formation water with $\delta^{18}O = +6$, we calculate the temperature of formation of kaolinites from two northeast Thompsonville Wilcox wells to be 100° C and 140° C. Utilizing these temperatures, the kaolinite δD values (Table 2), and the hydrogen isotope fractionation curve for kaolinite-water of Lambert and Epstein (1980), we calculate water δD values of −23 and −30, respectively. These values are in reasonable agreement with the values measured by Kharaka et al (1978), and provide a check of coherence.

The isotopic compositions of these fluids are typical of diagenetically evolved sea water. Note that Kharaka et al (1977) presented geochemical evidence that essentially eliminates meteoric ground water as a significant component in deep formation waters of the Gulf Coast. Therefore, the $\delta^{18}O$ values of the pore fluids must be greater than zero. Based on this value, the stable isotope data provide compelling and cogent evidence that these Wilcox kaolinites could not have formed below 70° C.

Precipitation of kaolinite from pore fluids and alteration of detrital feldspars to kaolinite during or immediately following carbonate dissolution is well documented (Loucks et al, 1979; Lindquist, 1978; Boles, 1982, and this volume). Stability of kaolinite relative to feldspars and common diagenetic clay minerals such as chlorite, illite, smectite, or mixed-layer clays at temperatures on the order of 100° C implies that relatively acidic and/or

TABLE 3

Reaction		Equilibrium pH Range
K-feldspar → Kaolinite	$2KAlSi_3O_8 + H_2O + 2H^+ \rightarrow Al_2Si_2O_5(OH)_4 + 4SiO_2 + 2K^+$	5.7–6.7
Albite → Kaolinite	$2NaAlSi_3O_8 + H_2O + 2H^+ \rightarrow Al_2Si_2O_5(OH)_4 + 4SiO_2 + 2Na^+$	5.3–5.6
Chlorite → Kaolinite	$Mg_5Al_2Si_3O_{10}(OH)_8 + 10H^+ \rightarrow Al_2Si_2O_5(OH)_4 + SiO_2 + 7H_2O + 5Mg$	5.0–5.7
Illite → Kaolinite	$K_{0.6}Mg_{0.25}Al_{2.3}Si_{3.5}O_{10}(OH)_2 + 0.75H_2O + 1.1H^+ \rightarrow$	5.8–6.6
	$1.15\,Al_2Si_2O_5(OH)_4 + 1.2SiO_2 + 0.6K^+ + 0.25\,Mg^{+2}$	
Muscovite → Kaolinite	$2KAl_3Si_3O_{10}(OH)_2 + 3H_2O + 2H^+ \rightarrow 3Al_2Si_2O_5(OH)_4 + 2K^+$	4.6–5.6

5.3–6.0 Mean Range

Range of Concentrations Assumed
K^+ = 100–1000 ppm
Na^+ = 5000–50,000 ppm
Mg^{+2} = 100–2000 ppm

Table 3—Range of pH associated with secondary porosity development and kaolinite precipitation in the Texas Gulf Coast. Calculations assume 350 bars fluid pressure (approximately 10,000 feet of burial) and temperature of 100° C. Alumina is conserved in the solid phases. Thermodynamic properties of all minerals except illite are from Helgeson et al (1978). Illite data is from Wolery (1978) and was obtained by the Tardy and Garrels estimation technique.

dilute pore fluids are involved in the generation of secondary porosity at depth. Direct calculation of the pH values of carbonate dissolution is dependent, for example, on the activities of HCO_3^- and Ca^{+2}, and the carbonate alkalinity. However, Carothers and Kharaka (1978) have discovered that titration alkalinity usually reported in the literature includes not only contributions from carbonate species but also from aliphatic acid anions. Thus, the chemical conditions favoring kaolinite precipitation are used to estimate the pH of waters involved in dissolution. Assuming quartz saturation and a range of pore-fluid compositions typical of the Gulf Coast Tertiary, the calculated range of pH values for kaolinite formation are shown in Table 3.

The pH range is fairly narrow (4.6 to 6.7), with most falling between 5 and 6 (recall that a neutral pH at 100° C is approximately 6.1). These values are somewhat lower than those reported by Kharaka et al (1977, 1978) from the northern Gulf of Mexico basin (5.2 to 7.3). However, Hankins et al (1978) obtained pH values of 4.1 to 5.4 from in-line monitors during flow tests of the Edna Delcambre et al No. 1 well in Tigre Lagoon Field, Louisiana. Laboratory measurements of these same waters yielded considerably higher values ranging from 6.19 to 6.62 (Hankins et al, 1978; Kharaka et al, 1978). Therefore, surface-measured pH values should be considered maximum values. If it can be shown that waters analyzed at the surface are representative of subsurface fluids (see, for example, Fournier and Truesdell, 1973; Kharaka et al, 1977), given compositional adjustments, and if it can be demonstrated that these present day pore waters are in equilibrium with respect to the most recent diagenetic mineral assemblages, then these fluids can be used to predict subsurface mineralogy. If equilibrium cannot be demonstrated, the waters can still be used to evaluate the potential for further reaction, and identify pathways of fluid flow. Demonstration of equilibrium requires the use of sound thermodynamic data in reliable models that incorporate the effects of compositional variation in minerals (for example, clays) and the aqueous phase.

Recent attempts using available thermodynamic data include Merino (1975), Boles (1982), and Arnorsson et al (1983).

Albite—Albitization of detrital plagioclase is one of the critical diagenetic reactions in feldspathic sandstones. In general, the zone of albitization overlaps with and postdates the zone of secondary porosity development (Boles, 1982, and this volume). Stratigraphic and petrologic data define the albitization "window" between 100 and 150°C (Iijima and Utada, 1972; Merino, 1975; Boles, 1982). Milliken et al (1981) determined the $\delta^{18}O$ of a diagenetic albite from the Gulf Coast Frio to be +17.2, consistent with formation from typical Gulf Coast waters at 135°C ± 15°C. The zone of albitization is, therefore, closely coincident with the zone of major carbon dioxide generation (to be discussed; see Fig. 1), but slightly skewed to the higher temperature side, consistent with its generally observed position in the diagenetic sequence (for example, Figure 13, Milliken et al, 1981). Moreover, a plausible albitization reaction consumes H^+; that is, pore fluids with relatively low pH values created by carbon dioxide production from organic matter could provide the potential for albitization reactions as well as carbonate dissolution:

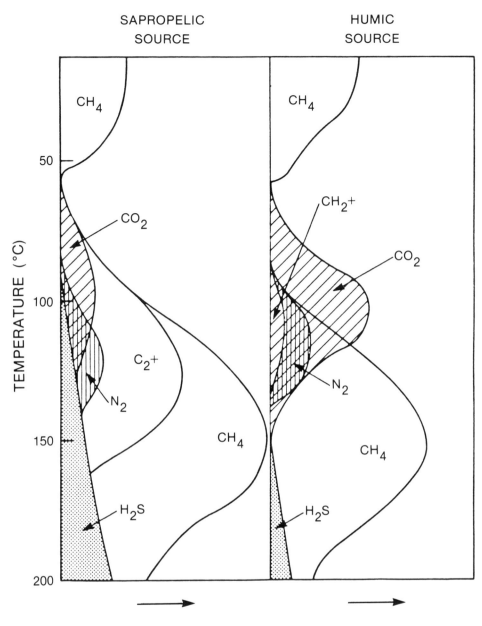

SAPROPELIC
SOURCE

HUMIC
SOURCE

Figure 1—Temperature of generation of carbon dioxide from organic matter in fine-grained sediments. Both sapropelic (Type I, II) and humic (Type III) kerogen yield maximum quantities of carbon dioxide at temperatures of about 100° C. Humic kerogen yields a much higher quantity. C_2+ represents hydrocarbons heavier than CH_4 in gas phase. N_2 is generated initially as NH_3 Modified from PETROLEUM GEOCHEMISTRY AND GEOLOGY by John M. Hunt. W. H. Freeman and Company. Copyright © 1979. Used with permission.

$$Na^+ + H^+ + \tfrac{1}{2}H_2O + 2SiO_2 \qquad (1)$$
$$+ CaAl_2Si_2O_8 = NaAlSi_3O_8$$
$$+ \tfrac{1}{2}Al_2Si_2O_5(OH)_4 + Ca^{+2}$$

Late Carbonate Cements—Carbonate cements infilling secondary porosity are commonly iron-rich (Table 1). Regional studies by Loucks et al (1979) of the Texas Gulf Coast Tertiary indicate that iron-rich calcite, dolomite, and ankerite cements typically postdate secondary porosity in the Eocene Wilcox and Yegua and the Oligocene Vicksburg and Frio Formations. Boles and Franks (1979) suggested that the occurrence of these late-stage iron-rich

carbonates is related to the release of iron and magnesium from mixed-layer illite/smectite during illitization of iron and magnesium-rich smectite layers at temperatures of 100°C or more. Boles (1978) showed that Fe-rich carbonates (for example, ankerite) should be thermodynamically more stable than calcite at these high temperatures, assuming ideal solution.

Carbon Dioxide Generation—Carbon dioxide is a significant by-product of kerogen maturation. According to Hunt (1979), its origin is primarily due to decomposition of carbonyl (C=O), methoxyl (–OCH₃),

phenolic hydroxyl (–OH), and perhaps other oxygen groups. Although all kerogen types are capable of producing carbon dioxide during thermal maturation, vitrinitic kerogen (Type III) produces by far the greatest amount per unit kerogen weight because of its high oxygen content. ARCO unpublished data (L. L. Lundell and A. Brown, personal communication) indicate that 22 weight percent of Type III kerogen is converted to carbon dioxide between vitrinite reflectance values of 0.3 and 2.4.

Peak generation of carbon dioxide from kerogen in sediments occurs at

Figure 2—Calcite saturation map for the Eocene Wilcox of Texas. Solubility indices were calculated by a chemical speciation computer program; calculations assume a uniform geothermal gradient of 31°C/km (1.7°F/100 ft) and take sample depth into account. Solubility indices have been compared with those calculated using EQ3 and WATEQ and similar results obtained. Three-hundred-and-fifty-six data points were used. Only general trends are shown on the map. Local areas or wells with undersaturated waters occur in some areas of regional oversaturation, particularly in highly faulted regions of south Texas. IAP = Ion activity product. Areas 1–6 explained in text. Stippled regions are based on 356 data points, distributed in areas 1 through 6 respectively, as follows: 11, 73, 144, 51, 13, 64. Because of the tendency for surface-measured pHs to be too high, only order of magnitude deviations from equilibrium are considered significant.

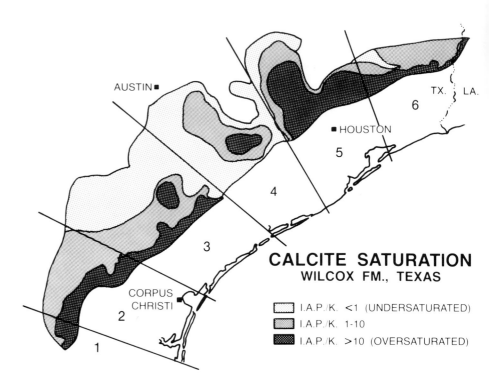

CALCITE SATURATION
WILCOX FM., TEXAS

I.A.P./K. <1 (UNDERSATURATED)
I.A.P./K. 1-10
I.A.P./K. >10 (OVERSATURATED)

approximately 100°C (Fig. 1). Although this value may vary depending on the duration of heating, it might be expected that secondary porosity formed as a result of carbon dioxide generation would also occur at temperatures close to 100°C. Thus, petroleum generated by the maturation of kerogen can accumulate in a reservoir that was created by the very same maturation processes through which the oil itself was generated.

Once formed, however, porosity may be retained during further burial if lower rates of cementation result from reduced pore-water flow and/or displacement of pore water by hydrocarbons. This can account for the presence of secondary porosity at much greater depths and temperatures (for example, the deep Tuscaloosa trend in Louisiana). Details of carbon dioxide distribution in the Texas Gulf Coast will be discussed elsewhere in this contribution.

Summary

The sequence of diagenetic events recorded by authigenic phases preceding and, in particular, following secondary porosity development in sandstones is similar in many cases. The sequence of events defined by petrographic observations is compatible with isotopically determined temperatures of mineral formation. Authigenic phases (for example, calcite and quartz) predating secondary porosity formation are estimated to have formed at temperatures generally less than 100°C. Authigenic phases postdating or forming concurrently with secondary porosity (for example, albite, kaolinite, and iron-rich, late carbonate cements) are estimated to have formed at temperatures generally higher than 100°C. These data suggest that secondary porosity in the Texas Gulf Coast Tertiary formed at temperatures of about 100°C ± 25°C. Consideration of kaolinite stability at these temperatures suggests waters with pH values of 5 to 6, possibly even lower. Lowering of pH by generation of carbon dioxide from organic matter in shales at temperatures of about 100°C is, therefore, an attractive hypothesis that seems generally compatible with a variety of data. Mass-balance calculations in regard to CO_2-generated secondary porosity are presented later in this paper.

**PORE FLUIDS,
TEXAS GULF COAST**

If carbon dioxide from organic matter is largely responsible for driving many of the diagenetic reactions in clastic reservoirs, the chemical composition of pore fluids in a basin may indicate areas undergoing active carbonate dissolution. Therefore, calcite saturation indices have been calculated for water samples from the Eocene Wilcox Group in 356 exploratory wells (Fig. 2).

The water analyses used in these calculations are from a U.S. Geological Survey compilation (Taylor, 1975) from a wide variety of sources, and the data are of highly variable quality. We have attempted to eliminate spurious analyses by deleting (1) waters from depths greater than 5000 ft which have total dissolved solids less than 5000 ppm (see Kharaka et al, 1977, 1978), and (2) waters with pH values outside of the range 3 < pH < 9.

The Texas Gulf Coast was divided into six areas, closely approximating those used by the Texas Bureau of Economic Geology in its study of regional porosity trends (see Loucks et al, 1979). Numbered from south to north, areas 1 and 2 comprise the Rio Grande Embayment, 3 and 4 approximate the San Marcos Arch, and areas 5 and 6 include the East Texas Embayment.

Figure 2 shows a trend of increasing calcite saturation index with increasing depth (temperature) in the downdip Wilcox. This trend is not surprising,

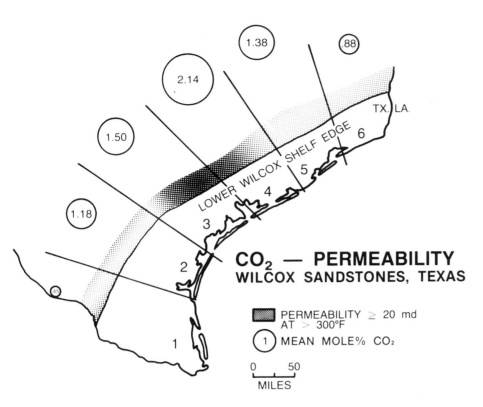

CO₂ — PERMEABILITY
WILCOX SANDSTONES, TEXAS

▓ PERMEABILITY ≥ 20 md
AT > 300°F

① MEAN MOLE% CO₂

0   50
MILES

Figure 3—Relationship between mean mole percent carbon dioxide in natural gases and regional permeability trends in deep Wilcox sandstones. Permeability trends are from Loucks et al. (1979). Mean carbon dioxide values for each area are given in Table 4. Areas 1–6 explained in text.

inasmuch as calcite solubility decreases with increasing temperature. This temperature effect is expressed in Figure 2 by a band of higher saturation indices downdip from, but more or less parallel to, an updip band of less saturated waters. A significant deviation from the expected trend is observed in central Texas (areas 3 and 4) where the updip region of low saturation indices broadens and crosscuts the band of higher saturation, suggesting that active dissolution of calcite in the deeper Wilcox may be occurring. Within this same general area, abnormally high permeabilities are found in deeply buried Wilcox sandstones (Fig. 3). Fifty percent or more of the total porosity is secondary, and an abnormally high concentration of CO_2 is present in the produced natural gases.

CARBON DIOXIDE IN NATURAL GASES, TEXAS GULF COAST

In the preceding section it was shown that the geochemistry of subsurface fluids may reveal active diagenetic reactions at depth. However, the quality of these water analyses may be questioned. Perhaps a more useful and

promising way to investigate the importance of carbon dioxide dissolved in subsurface fluids and generated during the evolution of oil and gas from kerogen is to look at carbon dioxide in the associated natural gases of hydrocarbon reservoirs.

Geographic Trends

Natural gas analyses compiled by the U.S. Bureau of Mines (Moore, 1976) provide a wealth of data on the chemistry of gases from sedimentary basins. In particular, the abundance and distribution of carbon dioxide exhibit some provocative regional trends in the Texas Gulf Coast.

Table 4 and Figure 3 show the trends in carbon dioxide content of natural gases from the Wilcox Group. The mean mole percent carbon dioxide is low in south and north Texas and increases toward central Texas (areas 3 and 4). The occurrence of high carbon dioxide content with low calcite saturation indices (compare Figs. 2 and 3) in areas 3 and 4 suggest that the low pH values responsible for low saturation indices in these areas (discussed previously) are attributable to greater abundance of dissolved carbon dioxide

in the pore fluids.

Loucks et al (1979, and this volume) documented trends in regional reservoir quality for the Lower Tertiary of the Texas Gulf Coast. They found that in the Wilcox, highest permeabilities in deep sandstones with temperatures of 150°C or more occur in central Texas (areas 3 and 4, Fig. 3). From petrographic observations they determined that more than 50% of the associated porosity is due to dissolution of calcite cement and feldspar. A similar relationship between carbon dioxide content and reservoir quality is observed in the Oligocene Frio Formation. The mean mole percent carbon dioxide in produced gases is significantly higher in north Texas (area 5, Table 4) where the highest permeabilities are found (Loucks et al, this volume). As in the Wilcox, most porosity in these sandstones is of secondary origin, resulting from the dissolution of carbonate cement and feldspar. The coincidence of pore fluids with low calcite saturation indices, high percentages of carbon dioxide in produced gases, and abundant secondary porosity in deeply buried sandstones seems more than fortuitous and supports previous suggestions

TABLE 4

Area	1	2	3	4	5	6
Frio, mole % CO_2	0.24	0.47	0.38	0.33	0.78	0.39
Wilcox, mole % CO_2	0.45	1.18	1.50	2.14	1.38	0.88

Table 4—Mean mole percent carbon dioxide for the Oligocene Frio and Eocene Wilcox by area, Texas Gulf Coast. Areas 1–6 are those shown in Figures 2 and 3.

a) **WILCOX, AREA 1**

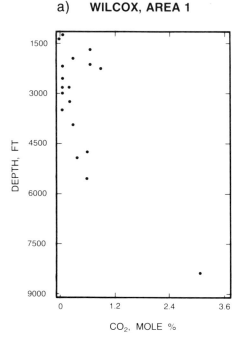

b) **WILCOX, AREA 2**

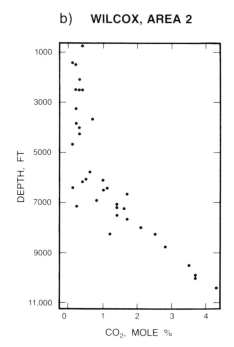

Figure 4a–b—Depth versus mole percent carbon dioxide in natural gases from the Wilcox of Texas. Areas 1–6 correspond with those shown in Figures 2 and 3. The trends are similar in each area although the depth to the inflection point of rapid carbon dioxide increase is shallower in south Texas (areas 1, 2) than in north Texas (areas 5, 6) because of changes in geothermal gradient.

that carbon dioxide generated as a by-product of petroleum generation plays an important role in sandstone diagenesis.

Depth Trends

Carbon dioxide content of Wilcox gases shows a very systematic relationship with depth of burial (Fig. 4a–f). In each area there is a fairly low and constant level at shallow depths (0 to 0.5 mole percent) down to a point at which the mole percent carbon dioxide increases rapidly to the limits of the data. Depth to the point at which the increase begins is shallower in south Texas (6000–7000 ft) than in north Texas (8000–9000 ft), probably reflecting differences in geothermal gradients, which decrease from south to north. In

areas 1 and 2 (south Texas) geothermal gradients in the Wilcox trend average about 42°C/km (assuming a mean surface temperature of 23.3°C, after Jones, 1969). In north Texas the mean gradient is approximately 29.6°C/km.

Using these gradients, the temperature of rapid increase in carbon dioxide is 100–113°C in south Texas and 95–105°C in north Texas, which is in reasonable agreement with a 100°C peak of carbon dioxide generation. The shapes of the carbon dioxide depth curves and sharp inflection near 100°C suggest these are generation curves and have not been substantially influenced by vertical migration of carbon dioxide. This is a somewhat surprising conclusion because the associated hydrocarbon gases are generally dry and

almost certainly have undergone significant vertical migration (3000–4000 ft, at least) assuming generation temperatures of 150°C or more (see Fig. 1). However, this may account for the lack of any significant correlation between carbon dioxide content and composition of the reservoired hydrocarbon gases. This lack of correlation implies that although the CO_2 and the hydrocarbon gases may both originate as by-products of the maturation of kerogen, different sources and/or migration pathways may be involved.

In Figure 5 we have compared the Wilcox depth–carbon dioxide data with the ratio of secondary to total porosity in the Wilcox from areas 1 through 6. It can be seen that the depth at which a rapid increase in carbon

c)   **WILCOX, AREA 3**

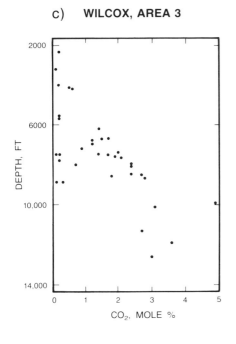

d)   **WILCOX, AREA 4**

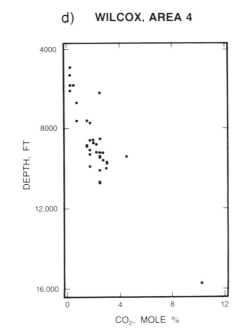

e)   **WILCOX, AREA 5**

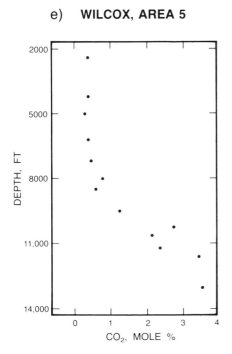

f)   **WILCOX, AREA 6**

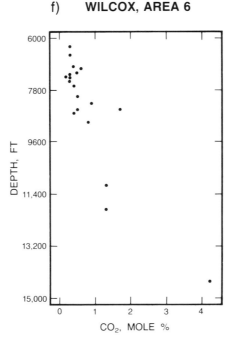

Figure 4c–f—Depth versus mole percent carbon dioxide in natural gases from the Wilcox of Texas. Areas 1–6 correspond with those shown in Figures 2 and 3. The trends are similar in each area although the depth to the inflection point of rapid carbon dioxide increase is shallower in south Texas (areas 1, 2) than in north Texas (areas 5, 6) because of changes in geothermal gradient.

dioxide content occurs corresponds closely with the depth at which a rapid increase in secondary porosity is observed.

Age Trends

Calculations of depth versus mole percent carbon dioxide for Miocene through Jurassic reservoirs show trends similar to those for the Wilcox (Fig. 6). In each case a trend of increasing carbon dioxide with increasing depth of burial is observed. As in the Wilcox, each unit exhibits a low and consistent carbon dioxide content down to a point at which the percent carbon dioxide rapidly increases. This inflection point is at a rather constant temperature across the state for a given age, but occurs at shallower depth (temperature) in the older rocks (Table 5). In general, at any given depth (temperature) the mole percent carbon dioxide is greater in older stratigraphic units (Table 6). This could be a result of the kinetics of carbon dioxide generation from kerogen.

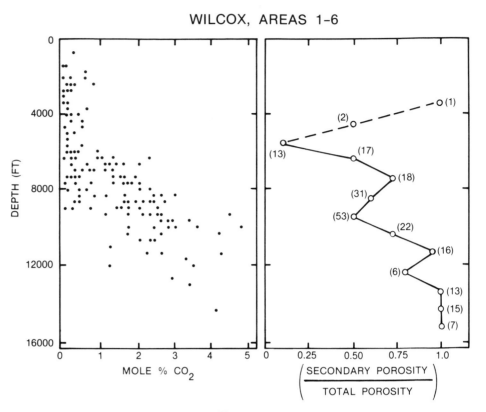

WILCOX, AREAS 1-6

Figure 5—Comparison of depth-mole percent carbon dioxide and ratio of secondary to total porosity for the Wilcox in areas 1–6, Texas. The three very shallow samples have abundant secondary porosity probably related to influx of meteoric water. However, the most rapid increase in deep subsurface secondary porosity occurs from about 6000 to 8000 ft, coincident with the depth of rapid increase in the carbon dioxide content of gases. Porosity data are from Loucks et al (1979). Numbers in parentheses are number of data points used in calculating the ratio.

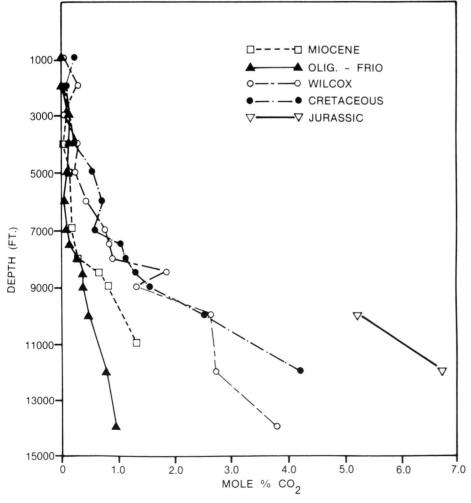

Figure 6—Depth versus mole percent carbon dioxide in Miocene through Jurassic reservoirs, Texas Gulf Coast. Carbon dioxide values were averaged at 1000 ft increments except where data were sparse and larger increments were required. Where data were especially abundant smaller increments were used to better define the inflection point. Most data points are the mean of 3 to 8 values.

TABLE 5

	Modal CO$_2$	CO$_2$ at 100° C	Inflection Temperature	n
Miocene	0.15%	0.30%	79° C	69
Oligocene–Frio	0.15	0.30	75	181
Eocene	1.0	0.91	75	160
Cretaceous	0.8	1.16	66	177
Jurassic	2.2	—	—	19

Table 5—Summary of carbon dioxide data by age. The inflection temperature is the temperature at which a rapid increase in mole percent carbon dioxide is observed. Note that it occurs at lower temperatures in older stratigraphic units. The total number of samples for the unit is "n." CO$_2$ (100° C) is the mole percent CO$_2$ in the gas at the 100° C isotherm. Temperatures are based on an average regional geothermal gradient of 31° C/km (1.70° F/100 ft). Areal variations within a unit can be significant but are averaged out in this data.

TABLE 6

Unit	n	Avg. % CO$_2$	Avg. Depth (ft)
Miocene	69	0.30 ± 0.39	4330 ± 2230
Olig–Frio	180	0.41 ± 0.70	6750 ± 2640
Wilcox	160	1.37 ± 1.34	7060 ± 2830
Cretaceous			
Olmos	5	0.20 ± 0.20	5420 ± 2290
Woodbine	16	2.1 ± 1.9	7140 ± 4220
Edwards	9	6.6 ± 2.8	11,140 ± 2560
Paluxy	8	0.4 ± 0.2	4660 ± 1800
Glen Rose	12	1.23 ± 0.8	7480 ± 2720
Travis Peak	24	0.8 ± 0.3	7220 ± 1030
Rodessa	35	1.25 ± 0.6	7440 ± 1240
Pettit	45	1.1 ± 0.4	7610 ± 1420
Cret-L	5	2.4 ± 2.4	8870 ± 64
Jurassic			
Cotton Valley	16	1.8 ± 0.6	10,070 ± 1350
Smackover	21	5.0 ± 1.9	10,890 ± 1770

Table 6—Mean mole percent carbon dioxide and depth by stratigraphic unit. Variability is expressed as ± one standard deviation. The number of samples is "n." Units except Lower Cretaceous, undifferentiated, (Cret-L) are arranged in stratigraphic order.

The data represented in Figure 6 are amenable to an Arrhenius relationship. This is given in Figure 7, which represents the carbon dioxide generated in unit time over the appropriate temperature interval. We could have presented the ordinate as lnΔG, where ΔG would be the difference in mole percent CO$_2$ between some initial value (for example, a baseline value at, say, 1000 ft) and its present value, but this would make no difference in the analysis. Essentially, Figure 7 represents an in-situ CO$_2$ generation rate; the slope corresponds to a pseudoactivation energy of 8 kcal per mole, and a frequency factor of $10^{2.9}$.

This pseudoenergy of activation is similar to those calculated for the generation of petroleum compounds from kerogen by Connan (1974; 11 to 14 kcal per mole) and Tissot (1969; 14 to 20 kcal per mole), and identical to the activation energy of decarboxylation derived from lab experiments by Kharaka et al (1983; 8.1 kcal per mole). It must be recognized that many assumptions and generalizations are implicit in this analysis. Several chemical reactions must be involved in the generation of CO$_2$, but we treat the data as representing a single "overall" reaction. This overall reaction has but one activation energy, but there is little doubt that as CO$_2$ generation from kerogen proceeds the activation energy increases (Wayne, 1969). In addition, even the types of kerogen in the Gulf Coast sediments may vary, so that the yields of CO$_2$, and the activation energies for different kerogen types, would vary. We used the age of the forma-

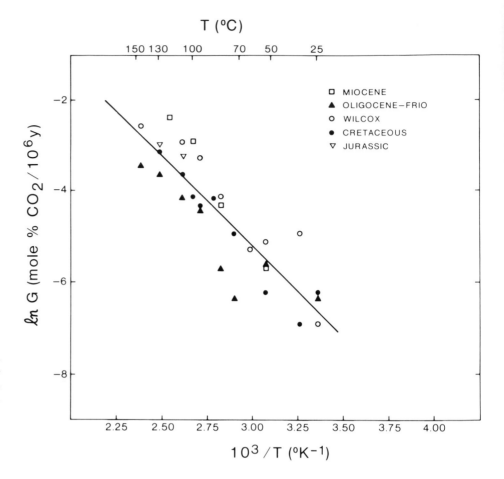

Figure 7—Arrhenius plot of mole percent carbon dioxide generated per million years versus reciprocal temperature. Calculations based on following average ages (m.y.): Miocene, 15; Oligocene-Frio, 30; Wilcox, 50; Cretaceous, 100; Jurassic, 135. Best-fit line is shown.

tions, rather than an integrated geothermal gradient–subsidence curve for the Gulf Coast sediments. However, if each formation spent a constant fraction of the sediment's age near its maximum temperature, the slope of the line (which is proportional to the activation energy) would not be affected (Waples, 1976).

Despite these oversimplifications, this model is compatible with the generation of CO_2 along with water and heavy heteroatomic compounds, followed by progressively smaller molecules and hydrocarbons (Tissot and Welte, 1978).

Lithologic Effects

Present data are not adequate to determine whether gross lithology (clastic versus carbonate) of the reservoir has any influence on carbon dioxide content. For example, Figure 8 shows that mean mole percent carbon dioxide is very similar in Cretaceous clastics (Travis Peak and Woodbine)

and carbonates (Glen Rose, Rodessa, and Pettit-Sligo) at comparable burial depths. In Jurassic rocks there is considerable difference in carbon dioxide content of the clastic Cotton Valley and the carbonate Smackover sequences (for a review of Gulf Coast stratigraphy, see Murray, 1961). The data are too limited to argue strongly for a lithologic effect, but it is possible that at higher temperatures carbon dioxide may be liberated by carbonate–clay reactions. An isotopic study of carbon dioxide from clastic and carbonate reservoirs of the Gulf Coast is presently underway to address this question.

CARBONATE CYCLING IN THE SUBSURFACE

It is clear that significant mass transfer takes place during diagenesis in a sedimentary basin. The extent to which carbonate dissolution and precipitation reactions take place depends, among other things, on solution com-

position, mineral assemblage, and fluid flux through the sedimentary rocks.

Schmidt and McDonald (1979) suggested that the zone of active carbonate dissolution generally underlies the zone of maximum active carbonate cementation. They postulate that the former is the source of carbonate for the latter and that carbonate is moving upward in aqueous solution to form a "carbonate curtain." They surmised that the chemistry of interstitial waters (influenced by clay diagenesis) or migration of waters upwards from a geopressured zone to a hydropressured zone could cause carbonate precipitation.

Certainly recycling of carbonate must occur in sedimentary basins, and in juvenile basins in which the general direction of fluid flow is simply upward and outward from the basin center, the gross transport of dissolved carbonate is likely to be in the same direction. However, at constant composition, the solubility of calcite is a function of the specific pressure and temperature gra-

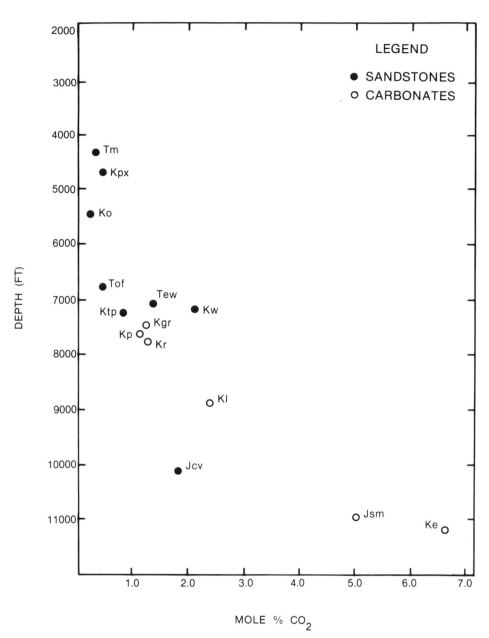

Figure 8—Reservoir lithology versus mole percent carbon dioxide. The mean production depth and mean mole percent carbon dioxide are shown for all formations studied. Means and standard deviations are shown in Table 6. Symbols are as follows: (**Jsm**) Jurassic Smackover; (**Jcv**) Cotton Valley; (**Kl**) Lower Cretaceous, undifferentiated; (**Kp**) Pettit (Sligo); (**Kr**) Rodessa; (**Ktp**) Travis Peak; (**Kgr**) Glen Rose; (**Kpx**) Paluxy; (**Ke**) Edwards; (**Kw**) Woodbine; (**Ko**) Olmos; (**Tew**) Eocene Wilcox; (**Tof**) Oligocene Frio; (**Tm**) Miocene.

dient along a given flow path. Thus, generalizations about the stability of carbonate in the subsurface based solely on the basis of upward moving fluids can be misleading. Although the thermodynamics of the calcium carbonate system is reasonably well known, the geochemical environment in the subsurface with respect to carbonate stability is poorly understood. It should be noted, contrary to the observations of Schmidt and McDonald (1979), that some sandstone reservoirs show evidence of downward mass transfer of calcite (Cassan et al, 1981). The

temperature increase due to burial may result in precipitation of calcite cements from trapped pore fluids, because sea water is near saturation with respect to calcite, and calcite exhibits retrograde solubility with respect to temperature. Note also that temperature generally dominates the effect of fluid pressure on calcite stability in sedimentary basins.

It might be expected that there are two end-member sources of carbonate available for cementation: early (shallow) marine carbonate (including recycled shell material) and late (deep)

carbonate from organic reactions in shales. For simplicity we are not considering caliche and other shallow carbonate cements of nonmarine origin. Marine carbonate derived directly from sea water or through dissolution of carbonate shelly material should have a carbon isotopic composition of approximately 0 ± 4 (PDB). Carbonate species from organically derived carbon should be considerably more depleted. During burial, these end-member components may be mixed through recycling processes. Petrographic observations and geochemical evidence

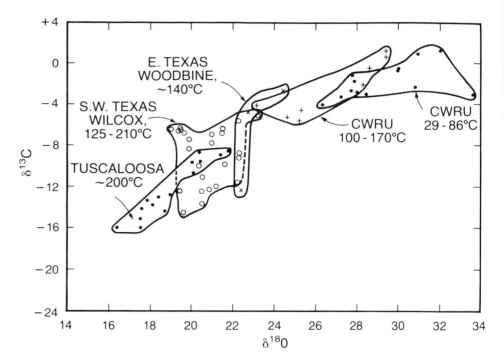

Figure 9—Carbon and oxygen isotopic composition of carbonate cements in sandstones and shales from Gulf Coast. Data points labeled (**CWRU**) are from shales as reported by Yeh (1974). Present-day temperatures are also given.

indicate that in many cases dissolution and reprecipitation of shell material are involved (Garrison et al, 1969).

Stable isotope data from Gulf Coast sandstone and shale carbonate cements (Fig. 9) are compatible with a two end-member mixing model. In spite of the geographic (Louisiana and Texas) and stratigraphic (Miocene through Cretaceous) range, the data exhibit a clear trend of increasingly negative $\delta^{13}C$ values with decreasing $\delta^{18}O$. The observed trend between the carbon and oxygen isotopic data is a result of two independent processes, one determining the $\delta^{13}C$ values, the other determining the $\delta^{18}O$ values, both of which depend on depth (temperature).

The $\delta^{18}O$ variations are a consequence of temperature-dependent fractionations in the carbonate–water system. The entire range, approximately 15 per mil, can be accounted for solely by a temperature variation of approximately 120°C. However, as pointed out previously, typical Gulf Coast formation waters have $\delta^{18}O$ values between +2 and +8 (Kharaka et al, 1977), becoming more positive at depth because of enhanced fluid–rock interac-

tion at elevated temperatures. With a typical Gulf Coast formation water $\delta^{18}O = +6$, we calculated (O'Neil et al, 1969) calcite formation temperatures of approximately 130–190°C for the deep Tuscaloosa, 100–150°C for the Wilcox of southwest Texas, and 100–110°C for the Woodbine of east Texas. Present-day temperatures are only slightly higher, and imply that either these late-stage carbonate cements formed at slightly shallower depths (Franks, unpublished ARCO Research Report), or that recrystallized carbonate has incorporated some marine carbonate oxygen, with atomic water/rock ratios of approximately two.

The carbon isotopic composition of sandstone carbonate cements shows a clear signal of an organic carbon component (Fig. 9). Although the relative contributions from thermal degradation of organic matter and acetic acid, thermally induced decarboxylation, and the oxidation of organic matter are not clear (Carothers and Kharaka, 1980; Kharaka et al, 1983; Irwin et al, 1977), maturation of organic matter is assuredly a significant source of isotopically depleted carbon. The other

source is from marine HCO_3^-, either as a component of the sediment (as marine $CaCO_3$, for example, shell material) and/or carried by formation waters. The $\delta^{13}C$ trend implies increasing contribution of organically derived carbon to marine carbonate carbon with increasing depth.

The stable isotopic evidence suggests that the subsiding sedimentary basin carries marine-derived carbonate to greater depths, and that during recrystallization and/or dissolution and reprecipitation, organically derived carbon from the maturation of kerogen is added to the system.

MASS BALANCE CONSIDERATIONS

Impressive volumes of carbonate have been dissolved out of sandstones to produce high-quality reservoirs for hydrocarbons. We now address the question of whether there is sufficient CO_2 generated from kerogen to account for secondary porosity development by carbonate dissolution.

The average shale contains about 1% organic carbon (Hunt, 1972). The original sediments very likely contained more than this because some organic carbon is lost during conversion of mud to shale. We shall assume 1 weight percent. If all this kerogen is Type III, 1 g of shale can yield up to 2.2×10^{-3} g of carbon dioxide (5×10^{-5} moles) during thermal maturation to a vitrinite reflectance value of 2.4% (ARCO unpublished data, L. Lundell and A. Brown). According to the following reaction, 1 g of shale could provide enough carbon dioxide to dissolve 5×10^{-5} moles (5×10^{-3} g) of calcite, assuming total utilization of the CO_2:

$$CaCO_3 + H_2O + CO_2 \qquad (2)$$
$$= 2HCO_3^- + Ca^{+2}$$

With sand/shale ratios of 1/10 to 1/20, typical of the Gulf Coast, approximately 5 to 10% secondary porosity could be generated by this reaction if there is no calcite in the shales to react with the carbon dioxide.

In the Texas Gulf Coast calcareous shales are common. Data from Hower et al (1976) show up to 7 weight percent calcite in their shallowest shale samples (6000 ft) and less than 0.5% in their deepest samples (18,000 ft). They interpreted the difference as loss of calcite by diagenetic reactions during burial. Although this trend is not observed in all areas (in fact, some wells we have analyzed show an increase in calcite with depth), it suggests that reactions between carbon dioxide and calcite within shales could neutralize the expelled fluids, thereby making them ineffective in generating secondary porosity in associated sandstones. Surdam et al (this volume) suggested that fluids expelled from shales flow through microfractures rather than around individual grains, reducing the volume of shale contacted by the solutions and, therefore, reducing the extent of reaction within the shales themselves. Although microfractures may serve as avenues of expulsion, shale pore water must first flow through the shales to microfractures. Therefore, shales with less than 1% organic carbon and more than about 1% calcite are unlikely to yield low pH pore fluids capable of dissolving calcite in interbedded sandstones. Obviously, the most favorable shales for secondary porosity development would be those

with very low original calcite content and high percentages of Type III kerogen. Note that Lundegard (1983) has suggested, based on material-balance considerations, that decarboxylation of organic matter cannot generate sufficient CO_2 to account for the secondary porosity in the Texas Gulf Coast Frio Formation. In order to develop predictive models of porosity distribution in sedimentary basins it will be necessary to consider not only the chemical reactions occurring during burial but also the distribution of types and amounts of organic matter, facies relationships, and distribution of early carbonate cements in both sands and shales. In addition, the variability in the amount and distribution of secondary porosity and its dependence on fluid pathways and scale of mass transport must be evaluated.

CONCLUSIONS

Several independent lines of evidence have been drawn together to evaluate the hypothesis that carbon dioxide generated as a by-product of thermal alteration of organic matter in sediments is a major factor in secondary porosity development. The data from the Texas Gulf Coast lend strong support to this concept. Specifically:

(1) The sequence of diagenetic events associated with secondary porosity development in the Texas Gulf Coast is not unique to that area. Similarities in rocks of different ages from other basins suggest common origins. The most likely common denominator is organic and inorganic reactions occurring in shales from which most of the pore fluids are derived.

(2) Carbon dioxide content of natural gases commonly shows a rapid increase at temperatures of approximately 100°C. The depth at which the increase occurs coincides with a rapid increase in the ratio of secondary to total porosity in associated sandstones.

(3) Areas of high carbon dioxide content in natural gases coincide with areas having pore fluids that are relatively undersaturated with respect to calcite. These areas also coincide with better than usual reservoir quality in deep, high temperature sandstones. More than half of the porosity in these sandstones is a consequence of the dis-

solution of carbonate cement and feldspar.

(4) Carbon dioxide content of gases is related to age and temperature of the reservoir formation, which suggests a kinetic control on generation.

(5) Stable isotope data on subsurface cements suggest recycling of carbonate through dissolution/reprecipitation reactions associated with secondary porosity formation.

ACKNOWLEDGMENTS

We thank Jim Hickey for generating many of the computer plots of CO_2 distribution, Hung Chang who wrote the program for calculating calcite saturation indices and generating an earlier version of the Wilcox pore fluid chemistry map, and Crozier Brown for helping us develop the computer programs used to manipulate the gas and water data. Discussions with Sal Bloch, Carol Bruton, Lee Lundell, Alton Brown, Keith Thompson, Hung Chang, Jim Hickey, and Chuck Vavra have been most informative. We thank Carol Bruton and Jim Boles for their careful and scholarly reviews. Two anonymous reviewers also have helped improve the manuscript. The enthusiastic support and encouragement of Dave McDonald was pivotal to the completion of this report. We gratefully acknowledge Lisa Patterson and Virginia Todd for word processing wizardry, and ARCO Oil and Gas Company for permission to publish this paper.

SELECTED REFERENCES

Alcock, F. G. and Benteau, 1976, Nipisi Field—a middle Devonian clastic reservoir, in M. M. Lerand, ed., The sedimentology of selected clastic oil and gas reservoirs in Alberta: Canadian Society of Petroleum Geology, Special Publication, 125 p.

Arnorsson, S., E. Gunnlaugsson, and H. Svavarsson, 1983, The chemistry of geothermal waters in Iceland, II, Mineral equilibria and independent variables controlling water compositions: Geochimica et Cosmochimica Acta, v. 47, p. 547–566.

Boles, J. R., 1978, Active ankerite cementation in the subsurface Eocene of Southwest Texas: Contributions to Mineralogy and Petrology, v. 68, p. 13–22.

———, 1982, Active albitization of plagioclase Gulf Coast Tertiary: American Journal of Science, v. 282, p. 165–180.

Boles, J. R., and S. G. Franks, 1979, Clay diagenesis in Wilcox sandstones of Southwest Texas: implications of smectite diagenesis on sandstone cementation:

Journal of Sedimentary Petrology, v. 49, p. 55–70.

Carothers, W. W., and Y. K. Kharaka, 1978, Aliphatic acid anions in oil-field waters—implications for origin of natural gas: Bulletin of the American Association of Petroleum Geologists, v. 62, p. 2441–2453.

———, 1980, Stable carbon isotopes of HCO_3 in oil-field waters—implications for the origin of CO_2: Geochimica et Cosmochimica Acta, v. 44, p. 323–332.

Cassan, J-P., M. Gracia Palacios, R. Fritz, and Y. Tardy, 1981, Diagenesis of sandstone reservoirs as shown by petrographical and geochemical analysis of oil bearing formations in the Gabon Basin: Bulletin des Centres de Recherches Exploration-Production Elf-Aquitaine, v. 5, p. 113–135.

Connan, J., 1974, Time-temperature relation in oil genesis: Bulletin of the American Association of Petroleum Geologists, v. 58, p. 2516–2521.

Dutton, S. P., 1977, Diagenesis and porosity distribution in deltaic sandstone, Strawn series, Pennsylvania, North Central Texas: Gulf Coast Association of Geological Societies Transactions, v. 27, p. 272–277.

Fournier, R. O., and A. H. Truesdell, 1973, An empirical Na-K-Ca geothermometer for natural waters: Geochimica et Cosmochimica Acta., v. 37, p. 1255–1275.

Garrison, R. E., J. L. Luternauer, E. V. Grill, R. D. MacDonald, and J. W. Murray, 1969, Early diagenetic cementation of recent sands, Fraser River Delta, British Columbia: Sedimentology, v. 12, p. 27–46.

Hankins, B. E., R. E. Chavanne, R. A. Ham, O. C. Karkalits, and J. I. Palermo, 1978, Chemical analysis of water from the world's first geopressured-geothermal well: Transactions of the Geothermal Resources Council, v. 2, p. 253–255.

Hartman, J. A., 1968, The Norphlet Sandstone, Pelahatchie Field, Rankin County, Mississippi: Gulf Coast Association of Geological Societies Transactions, v. 18, p. 2–11.

Hayes, J. B., 1979, Sandstone diagenesis—The hole truth, in P. A. Scholle and P. R. Schluger, eds., Aspects of diagenesis: Society of Economic Paleontologists and Mineralogists Special Publication 26, p. 127–139.

Hayes, J. B., J. C. Harms, and T. Wilson, Jr., 1976, Contrasts between braided and meandering stream deposits, Beluga and Sterling Formations (Tertiary), Cook Inlet, Alaska, in T. P. Miller, ed., Recent and ancient sedimentary environments in Alaska: Anchorage, Alaska, Alaska Geological Society, p. J1–J27.

Heald, M. T., and R. E. Larese, 1973, The significance of the solution of feldspar in

porosity development, Journal of Sedimentary Petrology, v. 43, p. 458–460.

Helgeson, H. C., J. M. Delany, H. W. Nesbitt, and D. K. Bird, 1978, Summary and critique of the thermodynamic properties of rock-forming minerals: American Journal of Science, v. 278-A, 277 p.

Hower, J., E. V. Eslinger, M. E. Hower, and E. A. Perry, 1976, Mechanism of burial metamorphism of argillaceous sediments, 1, Mineralogical and chemical evidence: Geological Society of America Bulletin, v. 87, p. 725–737.

Hunt, J. M., 1972, Distribution of carbon in crust of earth: Bulletin of the American Association of Petroleum Geologists, v. 56, p. 2273–2277.

———, 1979, Petroleum geochemistry and geology: San Francisco, W. H. Freeman Publishing Co., 617 p.

Iijima, A., and M. Utada, 1972, A critical review on the occurrence of zeolites in sedimentary rocks in Japan: Japanese Journal of Geology and Geography, v. 42, p. 61–83.

Irwin, H., C. Curtis, and M. Coleman, 1977, Isotopic evidence for source of diagenetic carbonates formed during burial of organic-rich sediments: Nature, v. 269, p. 209–213.

Jones, P. H., 1969, Hydrology of Neogene deposits in the Northern Gulf of Mexico Basin: Louisiana Water Resources Research Institute Bulletin GT-2, 105 p.

Kharaka, Y. K., E. Callendar, and R. H. Wallace, Jr., 1977, Geochemistry of geopressured geothermal waters from the Frio Clay in the Gulf Coast Region of Texas: Geology, v. 5, p. 241–244.

Kharaka, Y. K., W. W. Carothers, and P. M. Brown, 1978, Origins of water and solutes in the geopressured zones of the northern Gulf of Mexico Basin: Proceedings of the Society of Petroleum Engineers of AIME, Houston, Texas, SPE 7505, 5 p.

Kharaka, Y. K., W. W. Carothers, and R. J. Rosenbauer, 1983, Thermal decarboxylation of acetic acid: implications for origin of natural gas: Geochimica et Cosmochimica Acta, v. 47, p. 397–402.

Kulla, J. B., and T. F. Anderson, 1978, Experimental oxygen isotope fractionation between kaolinite and water: United States Geological Survey Open File Report 78-701, p. 234–235.

Lambert, S. J., and S. Epstein, 1980, Stable isotope investigations of an active geothermal system in Valles Caldera, Jemez Mountains, New Mexico: Journal of Volcanic and Geothermal Research, v. 8, p. 111–129.

Lindquist, S. J., 1976, Sandstone diagenesis and reservoir quality, Frio Formation (Oligocene), South Texas: Master's Thesis, University of Texas at Austin, 148 p.

———, 1978, How mineral content

affects reservoir quality of sands: World Oil, April 1978, p. 99–102.

Loucks, R. G., D. G. Bebout, and W. E. Galloway, 1977, Relationship of porosity formation and preservation to sandstone consolidation history—Gulf Coast Lower Tertiary Frio Formation: Gulf Coast Association of Geological Societies Transactions, v. 27, p. 109–120.

Loucks, R. G., M. M. Dodge, and W. E. Galloway, 1979, Sandstone consolidation analysis to delineate areas of high-quality reservoirs suitable for production of geopressured geothermal energy along the Texas Gulf Coast: Contract Report EG-77-5-05-5554, University of Texas Bureau of Economic Geology, 97 p.

Lundegard, P. D., 1983, Origin of secondary porosity: Frio Formation (Oligocene), Texas Gulf Coast: Geological Society of America, Abst. with Programs, p. 632.

Merino, E., 1975, Diagenesis in Tertiary Sandstones from Kettleman North Dome, California, I, Diagenetic mineralogy: Journal of Sedimentary Petrology, v. 45, p. 320–336.

Milliken, K. L., L. S. Land, and R. G. Loucks, 1981, History of burial diagenesis determined from isotopic geochemistry, Frio Formation, Brazoria County, Texas: Bulletin of the American Association of Petroleum Geologists, v. 65, p. 1397–1413.

Moore, R. J., 1976, Analyses of natural gases, 1917–1974, United States Bureau of Mines, Publ., NTIS PB-251 202.

Murray, G. E., 1961, Geology of the Atlantic and Gulf Coastal Province of North America: New York, Harper and Bros., 692 p.

O'Neil, J. R., R. N. Clayton, and T. K. Mayeda, 1969, Oxygen isotope fractionation in divalent metal carbonates: Journal of Chemical Physics, v. 51, p. 5547–5558.

Parker, C. A., 1974, Geopressures and secondary porosity in the deep Jurassic of Mississippi: Gulf Coast Association of Geological Societies Transactions, v. 24, p. 69–80.

Phipps, C. B., 1969, Post-burial sideritisation of calcite in Eocene beds from the Maracaibo Basin, Venezuela: Geological Magazine, v. 106, p. 485–495.

Roswell, D. M., and A. M. J. DeSwardt, 1974, Secondary leaching porosity in Middle Ecca Sandstones: Geological Society of South Africa Transactions and Proceedings, v. 77, p. 131–140.

Schmidt, V., D. A. McDonald, and R. L. Platt, 1977, Pore geometry and reservoir aspects of secondary porosity in sandstones: Bulletin of Canadian Petroleum Geology, v. 25, p. 271–290.

———, 1979, The role of secondary porosity in the course of sandstone diagenesis, in P. A. Scholle and P. R. Schluger, ed., Aspects of diagenesis, Society of Economic Paleontologists and Mineralogists Special Publication 26, p. 185–207.

Stanton, G. D., 1977, Secondary porosity in sandstone of the lower Wilcox (Eocene), Karnes County, Texas: Gulf Coast Association of Geological Societies Transactions, v. 27, p. 197–207.

Stanton, G. D., and E. F. McBride, 1976, Factors influencing porosity and permeability of Lower Wilcox (Eocene) sandstone, Karnes County, Texas, [abs.]: American Association of Petroleum Geologists and Society of Economic Paleontologists and Mineralogists Annual Meeting Abstracts, v. 1, p. 119.

Taylor, R. E., 1975, Chemical analyses of ground water for saline-water resources studies in Texas coastal plain stored in national water data storage and retrieval system: United States Geological Survey Open File Report 75–79, 669 p.

Tissot, B., 1969, Premieres donnees sur les mecanismes et la cinetique de la formation du petrole dans les sediments: simulation d'un schema reactionnel sur ordinateur: Revue de l'Institut Francais du Petrole et Annales des Combustibles Liquides, v. 24, p. 470–501.

Tissot, B., and D. H. Welte, 1978, Petroleum formation and occurrence: New York, Springer-Verlag, 538 p.

Tissot, B., B. Durand, J. Espitalie, and A. Combay, 1974, Influence of nature and diagenesis of organic matter in formation of petroleum: Bulletin of the American Association of Geologists, v. 58, p. 499–506.

Waples, D., 1976, Time-temperature relation in oil genesis: discussion: Bulletin of the American Association of Petroleum Geologists, v. 60, p. 884–885.

Wayne, R. P., 1969, The theory of kinetics of elementary gas phase reactions, *in* Comprehensive chemical kinetics: Amsterdam, Elsevier, v. 2, 486 p.

Wolery, T. J., 1978, Some chemical aspects of hydrothermal processes at mid-oceanic ridges—a theoretical study, I, Basalt–seawater reaction and chemical equilibrium between aqueous solutions and mineral: Ph.D. Dissertation, Northwestern University, Evanston, Illinois, 262 p.

Yeh, H. W., 1974, Oxygen isotope studies of ocean sediments during sedimentation and burial diagenesis: Ph.D. Dissertation, Case Western Reserve University, Cleveland, Ohio, 135 p.

Yeh, H. W., and S. M. Saving, 1977, Mechanism of burial metamorphism of argillaceous sediments, 3, Oxygen isotope evidence: Geological Society of America Bulletin, v. 88, p. 1321–1330.

The Role of Meteoric Water in Diagenesis of Shallow Sandstones: Stable Isotope Studies of the Milk River Aquifer and Gas Pool, Southeastern Alberta

Fred J. Longstaffe
University of Alberta
Edmonton, Alberta

INTRODUCTION

The purpose of this paper is to describe and explain the oxygen- and carbon-isotope compositions of authigenic and detrital minerals from the Milk River aquifer and the southeastern Alberta Milk River Gas Pool (Fig. 1). The investigation has three main objectives: (1) to determine the role of meteoric water in the diagenesis of sandstones within the aquifer and the Milk River Gas Pool; (2) to estimate the temperature(s) at which diagenesis has occurred; and (3) to use the stable isotope compositions of authigenic minerals to help decipher the paleohydrology of the study area.

The oxygen-isotope ratio ($^{18}O/^{16}O$) of an authigenic mineral in a clastic sedimentary rock is a function of several variables. Of primary importance is the oxygen-isotope composition of the water present during mineral formation. The Milk River area has been selected for study because the meteoric water that is recharging the aquifer is significantly depleted in ^{18}O relative to other possible formation fluids (Schwartz and Muehlenbachs, 1979). Authigenic minerals formed in equilibrium with the low-^{18}O meteoric water should have $\delta^{18}O$ values predictably lower than such minerals of detrital origin, or minerals formed earlier in the diagenetic history of the unit from more ^{18}O-rich formation fluids.

Geological Background

The Milk River Formation is composed mostly of sandstone. Three members have been recognized: (1) interbedded shale and sandstone of the lowermost Telegraph Creek Member, (2) poorly consolidated sandstone of

ABSTRACT. Oxygen- and carbon-isotope compositions have been determined for clay and carbonate minerals from the Upper Cretaceous clastic rocks of the Milk River and Lea Park Formations. These units contain the Milk River aquifer and the southeastern Alberta Milk River Gas Pool, respectively.

The stable isotope data provide important information concerning the diagenesis and paleohydrology of the study area. Authigenic minerals from sandstones in the Milk River aquifer are characterized by low $\delta^{18}O$ and $\delta^{13}C$ values: clay minerals ($<2\,\mu m$), dominated by authigenic kaolinite, $\delta^{18}O = +11.3$ to $+14.2$ (SMOW); authigenic calcite, $\delta^{18}O = +15.3$ to $+18.5$ (SMOW), $\delta^{13}C = -9.9$ to -2.6 (PDB). The authigenic minerals with the lowest $\delta^{18}O$ values occur within a zone of local recharge in the aquifer. Here the authigenic clay minerals and calcite closely approach isotopic equilibrium with existing meteoric water at low temperatures ($<+15°$ C). Low $\delta^{13}C$ values for the calcite indicate incorporation of organically derived CO_2, probably from decaying plant material in the overlying soil and till.

Close agreement between actual formation temperatures and those calculated from isotopic data disappears in downdip portions of the aquifer, mostly because of the drastic enrichment in ^{18}O of the formation water in this direction (-20 to -6, SMOW; Schwartz et al, 1981). Authigenic minerals from these locations have retained isotopic signatures characteristic of ^{18}O-poor meteoric water no longer present in the system. This water was displaced by ^{18}O-rich formation fluids that are themselves now being flushed from the aquifer by modern-day ground water.

Authigenic minerals from sandstones in the southeastern Alberta Milk River Gas Pool are more ^{18}O-rich than those from the aquifer: clay minerals, dominated by illite, $+14.7$ to $+15.8$, SMOW; calcite, $+19.3$, SMOW. Such compositions are compatible with mineral crystallization at low temperatures ($+15$ to $+20°$ C) from formation fluids similar in $\delta^{18}O$ to other Cretaceous oil and gas pools that occur in Alberta. The low-^{13}C nature of authigenic calcite and some dolomite (-7.6 to -3.0, PDB) from the Milk River Gas Pool may be related to the production of biogenic methane in this reservoir.

Of all clay minerals analyzed, the illite-dominated mixtures from argillaceous rocks of the Milk River aquifer and the Milk River Gas Pool have the highest $\delta^{18}O$ values ($+16.0$ to $+19.0$, SMOW). Such compositions reflect a detrital origin rather than diagenetic processes. Isotopic exchange between these clay minerals and formation water is insignificant.

Most dolomite from the sandstones and the argillaceous rocks is not in equilibrium with the authigenic calcite. The dolomite is much richer in ^{18}O ($+24.4$ to $+28.3$, SMOW) and ^{13}C (-2.7 to $+1.0$, PDB); such values are typical of platform carbonate rocks. No evidence for extensive isotopic exchange between formation water and the dolomite can be demonstrated.

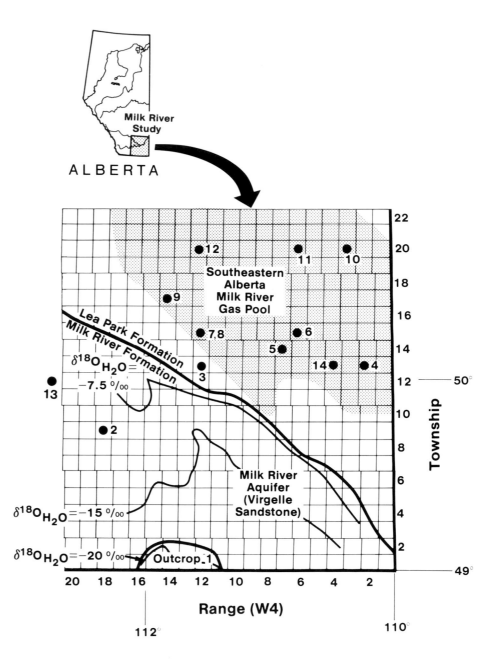

Figure 1—Location map, Milk River study area. The depositional edge of the Virgelle Sandstone demarcates the boundary between the Milk River Formation to the southwest and its argillaceous equivalent, the Lea Park Formation, to the northeast (after Meijer Drees and Myhr, 1981). The southeastern Alberta Milk River Gas Pool (stippled pattern) is contained within the Lea Park Formation. The Virgelle Sandstone is the dominant rock unit of the Milk River aquifer. Contours for the $\delta^{18}O$ of ground water in the Milk River aquifer are also shown, simplified after Schwartz and Muehlenbachs (1979).

The Virgelle Sandstone has been sampled in outcrop (1) (Milk River valley) and in the subsurface. The locations of subsurface samples are indicated by ●; well names are given by number: (2) Imperial Chin 7-10-9-18 W4; (3) Dome Cecil 6-28-13-12 W4; (4) Merland et al Medhat 6-30-13-2 W4; (5) McCulloch Medhat 6-20-14-7 W4; (6) Alberta Suffield 7-14-15-6 W4; (7) PCP Alderson 7-21-15-12 W4 (8) PCP Alderson 6-33-15-12 W4; (9) PCP Johnson 7-8-17-14 W4; (10) CRL Yan HB Atlee 10-36-20-3 W4; (11) Alberta Suffield 6-8-20-6 W4; (12) AEG Princess 6-13-20-12 W4; (13) Banner et al Iron Sp. 7-9-12-21 W4; (14) Bralorne et al Medhat 6-12-13-4 W4.

the overlying Virgelle Member, and (3) interbedded sandstone, sandy claystone, and shale of the uppermost Deadhorse Coulee Member (Meijer Drees and Myhr, 1981).

The Milk River aquifer is an artesian system that is contained mostly within the Virgelle Member (Meyboom, 1960; Schwartz and Muehlenbachs, 1979). Most recharge to the aquifer occurs outside of the study area, in northern Montana. However, limited recharge also occurs in the Milk River valley, where the Virgelle Member is exposed (Fig. 1; Schwartz and Muehlenbachs, 1979). The Virgelle Member plunges

away from its area of outcrop in a fan-shaped pattern (Meyboom, 1960).

The Milk River Formation, including the Virgelle Member, disappears in the subsurface towards the northeast owing to a facies change from sandstone to sandy shale (Fig. 1). The sandy shale unit has been named the Alderson Member of the Lea Park Formation (Meijer Drees and Myhr, 1981). The Alderson Member is stratigraphically equivalent to the Milk River Formation and formerly was termed the "Milk River Equivalent." The facies change from dominantly sandstone to dominantly shale forms a transitional

boundary between the Milk River aquifer and the Milk River Gas Pool (Fig. 1). One well sampled in this study (Dome Cecil 6-28-13-12 W4) lies within the transition zone. While formally part of the Alderson Member, it comprises the effective northeastern margin of the Milk River aquifer.

Within the southern part of the Milk River Gas Pool, the lower portion of the Alderson Member is composed of silty shale and mudstone with lenses and thin beds of very fine-grained sandstone. The upper part is comprised of thinly interbedded silty shale and sandstone. Towards the north, the upper portion of the Alderson Member becomes less sandy, grading into silty shale and mudstone similar to the lower part of the unit (Meijer Drees and Myhr, 1981).

Schwartz and Muehlenbachs (1979) and Schwartz et al (1981) have studied the stable isotope geochemistry of the water in the Milk River aquifer. They found that the ground water becomes progressively richer in ^{18}O from southeast to northwest (Fig. 1). To explain this variability in isotopic composition, Schwartz and Muehlenbachs (1979) suggested that the aquifer is being recharged in the south by meteoric water. This water is displacing preexisting, more ^{18}O-rich formation water. The flushing is incomplete, a zone of mixing remaining between the meteoric water and older formation water. The extent of the flushing depends upon the transmissivity of particular zones within the ground-water system. Fluxes of meteoric water through the highly permeable sandstone of the Virgelle Member have promoted a marked northward penetration of water depleted in ^{18}O (Fig. 1). This effect is not observed for the much less transmissive rocks of the Alderson Member. The extent to which these hydrological conditions have affected mineral diagenesis in the Milk River Formation and its equivalent is examined in this paper.

ANALYTICAL METHODS

Clay minerals, calcite, and dolomite were selected for isotopic analysis because they occur both as pore linings or pore fillings (that is, authigenic minerals) and as detrital grains in these rocks (Longstaffe, 1983; Black et al, 1980). In addition, these minerals can be separated from the rocks in quantities sufficient for analysis of their oxygen- and carbon-isotope ratios (see below).

The isotope data are presented in the usual δ notation with respect to Standard Mean Ocean Water (SMOW) for oxygen (Craig, 1961) and the *Belemnitella americana* from the Peedee Formation (PDB) for carbon (Craig, 1957). The partitioning of ^{18}O between two phases, A and B, is given by

$$\Delta_{A-B} = 10^3 \ln\alpha_{A-B} \simeq (\alpha-1)10^3 \simeq \delta_A - \delta_B$$

where α is the oxygen-isotope fractionation factor between A and B.

Clay Minerals

The rock samples were dispersed in distilled water and allowed to settle in a column for the appropriate period of time (normally 24 hr) required to obtain the <2 μm size-fraction. This procedure was repeated three times to ensure effective separation of the clay minerals. Once separated, the clay-sized material was treated with a 3% sodium hypochlorite solution for 48 hr at $+65°C$ and then thoroughly washed with distilled water, using a high-speed centrifuge.

The <2 μm size-fraction was split into three portions; the first portion was saturated with a 2 molar solution of Ca^{+2} and the second with a 2 molar solution of K^+. The third portion was set aside for isotopic analysis. The clay mixtures were thoroughly washed in distilled water and freeze dried. About 50 mg of the Ca- and K-saturated samples were then dispersed in distilled water and deposited by suction upon a ceramic disc. This procedure produces a preferred basal orientation of platy minerals. The Ca- and K-saturated samples were analyzed using a Philips X-ray diffractometer under the following conditions:

(1) Co K-alpha radiation filtered by a graphite monochromator;

(2) 1 degree divergent slit; time constant = 2;

(3) 1 degree two-theta/min at 600 mm/hr;

(4) 50 kV, 20 mA.

The following X-ray patterns were obtained (Ignasiak et al, 1983):

(1) Ca-disc at 54% relative humidity, 2-35 degrees two-theta;

(2) Ca-disc glycolated, 2-80 degrees two-theta;

(3) K-disc at 0% relative humidity, 2-25 degrees two-theta;

(4) K-disc at 54% relative humidity, 2-25 degrees two-theta;

(5) K-disc at $+300°C$, 2-25 degrees two-theta;

(6) K-disc at $+550°C$, 2-25 degrees two-theta.

For X-ray patterns 1 and 4, samples were equilibrated to 54% relative humidity over a magnesium nitrate solution for a minimum of 12 hr prior to analysis. For X-ray pattern 2, sam-

ples were vapor-solvated with ethylene glycol for 18 hr at $+65°C$ followed by >24 hr at $+22°C$. Samples analyzed at 0% relative humidity were heated to $+105°C$ for a minimum of 2 hr and analyzed while still hot. Samples treated at $+300$ and $+550°C$ were heated for 3 and 2 hr, respectively, prior to analysis. In all cases, humidity conditions were strictly controlled during sample equilibration and during the analysis of the sample on the diffractometer.

All samples consisted of a mixture of clay minerals and trace amounts of quartz. Samples containing greater than 3% quartz were rejected prior to oxygen-isotope analysis of the clay minerals.

Less than 0.2 μm size-fractions were also prepared for a representative suite of samples in order to examine the effect of grain size upon the isotopic composition of the clay minerals. The <2 μm samples selected for further separation were saturated with Li, dispersed and centrifuged for the appropriate time necessary to collect the <0.2 μm material. The <0.2 μm samples were then analyzed by X-ray diffraction prior to isotopic analysis.

Oxygen-isotope analyses of the clay minerals were performed using the BrF_5 method of Clayton and Mayeda (1963). The clay mixtures were dried at $+105°C$ for 2 hr and placed in a zero humidity environment for 24 hr prior to treatment with BrF_5. This procedure is required to remove interlayer water from the clay minerals prior to isotopic analysis.

Carbonate Minerals

Crushed bulk rock samples were analyzed by X-ray diffraction to determine the nature of carbonate minerals. For isotopic analysis, carbonate minerals were separated by reacting powdered bulk rock samples (<44 μm) in phosphoric acid using a procedure modified after McCrea (1950), Epstein et al (1964), and Walthers et al (1972). Because of possible cross contamination of gases produced from calcite and dolomite, isotope results for calcite are presented only when the calcite/dolomite ratio in the sample exceeds 0.5. Furthermore, if the calcite/dolomite ratio in the sample exceeds 0.7, only data for calcite are presented.

RESULTS

The clay mineralogy and oxygen-isotope compositions of about 100 <2 and <0.2 μm size-fractions from sandstone, siltstone, and shale samples of the Milk River Formation, the Lea Park Formation, and the underlying Colorado Group have been determined. In addition, the oxygen-and carbon-isotope ratios of calcite and/or dolomite that coexist with the clay minerals in these rocks have been measured.

The locations of outcrop and subsurface cores that have been sampled are shown on Figure 1. The oxygen-isotope results of Schwartz and Muehlenbachs (1979) and Schwartz et al (1981) for ground water are also summarized in Figure 1. Isotopic data for clay and carbonate minerals from the Milk River aquifer and the Milk River Gas Pool are presented in Tables 1 and 2. Table 3 lists the $\delta^{18}O$ values for ground water at specific locations of interest within the Milk River aquifer. The clay mineralogy of the Milk River aquifer and the Milk River Gas Pool is given in Tables 4 and 5, respectively. Isotopic data for clay and carbonate minerals from shale and sandstone samples of the underlying Colorado Group are given in Table 6.

Clay Minerals

The $\delta^{18}O$ values of <2 μm clay minerals from all samples range from +11.3 to +19.0 (Tables 1, 2). These data divide into two main groups: (1) clay-poor sandstones (<5% argillaceous material) – $\delta^{18}O$ <+16 and (2) clay-rich rocks (shale, argillaceous siltstone, and argillaceous sandstone) – $\delta^{18}O$ >+16. The $\delta^{18}O$ values of the clay minerals from the sandstones increase from about +11 for samples from outcrop in the Milk River valley in the southernmost part of the study area to about +15 for subsurface samples of the Alderson Member to the north (Tables 1, 2; Fig. 1).

Illite is the dominant clay mineral in the <2 μm material from clay-rich rocks of both the Alderson and Virgelle Members (60 to 80%; Tables 4, 5). Minor quantities of kaolinite, smectite, and chlorite are also present. The clay mineralogy (<2 μm) of sandstones from the southeastern Alberta Milk River Gas Pool (Alderson Member) is very similar to that of the enclosing argillaceous rocks (Table 5). The clay minerals from the sandstones, however, have lower $\delta^{18}O$ values (+14.7 to +15.8) than the mineralogically similar mixtures from the shales (+16.0 to +19.0; Table 2; Fig. 2). As Black et al (1980) have also noted, illite, smectite, chlorite, kaolinite, and illite/smectite line and fill the pores in the sandstone lenses; these clay minerals are dominantly authigenic in origin.

The <2 μm material from sandstones in the Milk River aquifer is composed mostly of kaolinite and/or smectite (70 to 90%), kaolinite usually being most abundant (Table 4). Chlorite and illite are also present in minor amounts. The kaolinite and smectite occur as pore linings or pore fillings typical of authigenic clay minerals (Longstaffe, 1983; Wilson and Pittman, 1977). Samples richest in authigenic kaolinite and smectite have the lowest $\delta^{18}O$ values (+11.3 to +14.2; Tables 1, 4). Illite is concentrated in clay laminae and is presumed to be detrital. This assumption is supported by the sympathetic variation between the $\delta^{18}O$ values and clay mineral compositions of <2 μm size-fractions from the Virgelle Sandstone (Fig. 3). As the sandstones become more argillaceous (that is, contain a higher percentage of clay laminae), both the relative abundance of illite in the <2 μm size-fraction and the $\delta^{18}O$ value of this material increase towards compositions typical of detrital clay minerals from shales in the study area.

Oxygen-isotope data were also obtained for <0.2 μm clay minerals from the Milk River aquifer (Table 1). The $\delta^{18}O$ results are similar to those obtained for the <2 μm materials: (1) clay-poor sandstones, +10.4 to +14.2; (2) clay-rich rocks, +15.4 to +16.5.

Some differences in clay mineralogy exist between <0.2 μm and <2 μm samples of the clay-poor sandstone (Table 4). The finer size-fraction contains more smectite and less kaolinite, illite, and chlorite than the <2 μm material. The <0.2 μm size-fraction also contains degraded illite, a phase not detectable in the <2 μm size-fraction. This illite is most probably of detrital origin, but because of its very small grain size, has become more extensively altered during weathering and leaching.

In contrast to the sandstones, little difference in clay mineralogy exists between the <0.2 μm and <2 μm fractions of argillaceous rocks from the Milk River aquifer (Table 4). Minor amounts of illite/smectite can be detected in some <0.2 μm samples.

Carbonate Minerals

Calcite and dolomite are the main carbonate minerals present in the samples; traces of siderite were also detected in rocks from the Alderson Member. Calcite occurs in abundance only in the sandstones. It occurs as a pore-lining and pore-filling cement and is authigenic (Longstaffe, 1983). A composition of Ca:Mg = 97:03 for the calcite is indicated from X-ray diffraction data ($d_{104} = 3.027 \pm 0.008$ Å). This composition does not vary significantly throughout the study area. In contrast, the $\delta^{18}O$ values of the authigenic calcite vary from +15.3 to +19.3 (SMOW) (Tables 1, 2), increasing by about 4 o/oo from south to north, a pattern similar to the authigenic clay minerals. Within the Milk River Gas Pool, the $\delta^{18}O$ value of calcite from the sandstones remains constant at +19.3 (SMOW). The $\delta^{13}C$ values of the authigenic calcite range from −9.9 to −2.6 (PDB) and average −5.5 (PDB) (Tables 1, 2; Fig. 4). No systematic variations in carbon-isotope values were detected.

Traces of calcite are present in the argillaceous rocks from the Alderson Member. Its composition (Ca:Mg = 85:15, $d_{104} = 2.992 \pm 0.003$ Å) is quite different from that of the authigenic calcite present in the sandstone. This high-Mg calcite is comprised of fossil debris. The very low calcite/dolomite ratios in the argillaceous rocks precluded accurate isotopic analysis of this calcite.

Dolomite ($d_{104} = 2.888 \pm 0.004$ Å) occurs as grains in the Virgelle Member sandstone. It is also the dominant carbonate phase in the argillaceous rocks of the Alderson Member ($d_{104} = 2.885 \pm 0.003$ Å). The dolomite has $\delta^{18}O$ values

---

Table 1—Stable isotope results for clay and carbonate minerals, Milk River Aquifer.

TABLE 1

Sample Location	Depth[1] (m)	Clays δ^{18}O <2 μm SMOW	Clays δ^{18}O <0.2 μm SMOW	Carbonates δ^{18}O SMOW	Carbonates δ^{13}O PDB
Virgelle Member, Milk River Formation					
Milk River Valley					
Twp 2, R 15 W4					
1. sandstone	0.0	+11.3	+13.6	+15.3	−3.7 c[2]
2. sandstone	7.6	+12.8	+13.0		
3. sandstone	13.7	+13.0	+13.2	+15.6	−6.7 c
4. sandstone	19.8	+12.2	+12.9	+25.2	−1.0 d[3]
5. sandstone	25.9	+11.9	+13.4	+25.6	−0.8 d
Imperial Chin					
7-10-9-18 W4					
6. sandstone	241.7	+12.4	+10.4	+25.1	−1.3 d
7. silty sandstone	264.4	+14.2	+14.4	+16.4	−3.3 c
8. sandstone	265.5			+24.7	−1.3 d
9. sandstone	266.4			+24.5	−1.4 d
10. arg. sandstone[4]	268.5	+16.0	+15.4	+24.5	−1.3 d
11. sandstone	271.0	+12.7	+11.4	+25.0	−1.1 d
Dome Cecil					
6-28-13-12 W4					
12. siltstone	231.8	+16.7	+16.1	+26.0	−1.4 d
13. silty shale	235.3			+26.1	−1.5 d
14. silty shale	237.4			+26.0	−1.5 d
15. arg. siltstone	238.4	+17.4	+16.1	+25.9	−1.6 d
16. silty shale	241.1			+24.9	−1.4 d
17. arg. siltstone	243.5			+27.4	−1.5 d
18. arg. siltstone	244.9			+26.2	−1.2 d
19. arg. siltstone	246.3	+16.8	+15.8	+25.9	−1.0 d
20. arg. siltstone	249.3			+25.7	−1.0 d
21. siltstone	250.9	+15.4		+18.5	−5.4 c
22. arg. sandstone	252.2			+15.5	−9.9 c
				+25.0	−2.0 d
23. arg. sandstone	252.4			+17.8	−7.7 c
				+25.2	−1.7 d
24. sandstone	253.6			+17.7	−2.6 c
25. sandstone	254.7	+13.0		+17.2	−3.0 c
26. silty shale	256.0			+25.7	−0.9 d
27. sandy siltstone	258.8	+15.7	+15.6	+26.5	−0.9 d
28. silty shale	259.4			+25.6	−0.9 d
29. arg. siltstone	260.8			+25.4	−1.1 d
30. arg. siltstone	262.3	+18.9	+16.2	+26.4	−0.9 d
31. silty shale	263.7			+25.8	−0.6 d
32. silty shale	265.8			+25.9	−0.7 d
33. silty shale	268.2	+16.6	+16.5	+25.0	−0.9 d

[1]Depth = depth below surface.
[2]c = calcite.
[3]d = dolomite
[4]arg. = argillaceous.

TABLE 2

Sample Location	Depth[1] (m)	Clays δ¹⁸O <2 μm SMOW	Carbonates δ¹⁸O SMOW	δ¹³C PDB
Alderson Member, Lea Park Formation				
Merland et al Medhat				
6-30-13-2 W4				
34. arg. siltstone[4]	366.0	+18.1	+26.2	−1.2 d[2]
35. arg. siltstone	371.7	+16.9	+26.4	−1.5 d
36. arg. siltstone	376.4	+17.7	+26.5	−1.3 d
37. bentonite	380.9	+18.5	+25.6	−2.4 d
McCulloch Medhat				
6-20-14-7-W4				
38. arg. siltstone	344.7	+17.4	+25.2	−1.0 d
39. bentonite	353.1	+18.1		
Alberta Suffield				
7-14-15-6 W4				
40. mudstone	266.7	+17.7	+26.3	−1.2 d
41. sandstone	269.8	+14.7	+26.8	−1.1 d
42. mudstone	270.4	+17.9	+28.3	−1.1 d
43. mudstone	272.3	+17.7	+25.5	−0.3 d
44. silty mudstone	288.3	+17.4	+26.8	−1.2 d
45. mudstone	292.2	+17.1	+26.4	+0.9 d
46. silty sandstone	292.6	+15.8	+24.2	−5.3 d
47. arg. siltstone	309.4	+17.1	+26.6	−1.1 d
48. arg. siltstone	311.7	+17.0	+27.1	−1.4 d
49. silty shale	313.9	+17.6	+26.6	−1.4 d
PCP Alderson				
7-21-15-12 W4				
50. silty shale	305.0	+16.9	+28.2	+1.0 d
51. arg. siltstone	310.0	+16.7	+27.1	−0.6 d
52. arg. siltstone	326.4	+17.1	+26.2	−1.0 d
53. arg. siltstone	329.9	+16.2	+26.0	−1.0 d
54. arg. sandstone	334.8	+15.8	+19.3	−3.0 c[3]
55. silty shale	338.3	+16.2	+26.7	−0.8 d
PCP Alderson				
6-33-15-12 W4				
56. arg. siltstone	293.2	+16.9	+27.1	−1.9 d
57. arg. siltstone	295.0	+17.5	+25.8	−0.9 d
58. arg. siltstone	299.0	+17.7	+26.1	−0.6 d
59. arg. siltstone	301.3	+16.7	+25.4	−0.9 d
60. sandstone	306.5	+14.7	+19.3	−7.6 c
			+24.4	−1.3 d
61. siltstone	310.6	+16.3	+26.8	−0.9 d

(continued)

Table 2—Stable isotope results for clay and carbonate minerals, Southeastern Alberta Milk River Gas Pool.

TABLE 2 *(continued)*

Sample Location	Depth[1] (m)	Clays δ^{18}O <2 μm SMOW	Carbonates δ^{18}O SMOW	δ^{13}C PDB
PCP Johnson				
7-8-17-14 W4				
62. arg. siltstone	370.0	+17.6	+26.0	−0.7 d
63. sandstone	372.2	+14.7	+19.3	−6.8 c
64. silty shale	373.7	+17.6	+25.6	−1.2 d
65. silty shale	377.0	+17.3	+27.1	−0.1 d
66. silty shale	382.5	+17.3	+25.8	−1.0 d
67. arg. siltstone	386.0	+16.1	+27.8	−0.7 d
68. arg. sandstone	387.1	+15.4	+27.8	+0.0 d
CRL Yan HB Atlee				
10-36-20-3 W4				
69. arg. siltstone	391.1	+17.1	+26.4	−1.2 d
70. arg. siltstone	395.6	+17.1	+25.7	−1.2 d
71. arg. siltstone	415.7	+17.0	+27.8	−5.5 d
72. arg. siltstone	419.9	+17.1	+27.4	−2.7 d
73. arg. siltstone	422.9	+16.7	+28.0	−7.0 d
74. arg. siltstone	426.7	+17.3	+27.7	−5.5 d
Alberta Suffield				
6-8-20-6 W4				
75. arg. siltstone	416.1	+17.3	+27.0	−2.4 d
AEG Princess				
6-13-20-12 W4				
76. arg. siltstone	311.5	+17.3	+26.5	−1.4 d
77. arg. sandstone	316.5	+17.0	+26.0	−1.5 d
78. arg. siltstone	320.0	+19.0	+24.7	−1.8 d
79. arg. siltstone	324.0	+16.5	+25.9	−1.9 d
80. arg. siltstone	329.8	+16.6	+27.5	−4.2 d

[1]Depth = depth below surface.
[2]d = dolomite.
[3]c = calcite.
[4]arg. = argillaceous.

of +24.2 to +28.3 (SMOW), regardless of unit or rock type (Tables 1, 2; Fig. 4). Unlike calcite, no systematic variations in the oxygen-isotope composition of dolomite were observed. The δ^{13}C values of most dolomite samples range from −2.7 to +1.0 (PDB) and average −1.1 ± 0.1 (Tables 1, 2; Fig. 4). A few dolomite samples from the Milk River Gas Pool have low δ^{13}C values (−7.0 to −4.2, PDB; Table 2, Fig. 4). However, in all cases where samples of coexisting calcite and dolomite were successfully separated during analysis, the dolomite is always significantly richer in ^{18}O and ^{13}C than the authigenic calcite (Table 1, #22, 23; Table 2, #60).

DISCUSSION

Clastic sedimentary rocks are composed of: (1) unaltered grains that nor-mally retain oxygen-isotope compositions characteristic of their source, (2) minerals formed during weathering that are usually rich in ^{18}O (for example, detrital clay minerals), and (3) minerals formed during diagenesis (for example, clay and carbonate cements and pore-filling materials) (Savin and Epstein, 1970a, 1970b, 1970c; Long-staffe, 1983). Minerals formed during weathering or diagenesis are usually rich in ^{18}O because they have crystallized at low temperatures in isotopic equilibrium with water. The oxygen-isotope fractionations for clay-water and carbonate-water are large at low temperatures (Friedman and O'Neil, 1977; Savin and Epstein, 1970a, 1970b, 1970c; Lawrence and Taylor, 1971, 1972).

How rich in ^{18}O the clay or carbonate minerals will be is controlled by fluid composition, temperature, and the effective mineral/water ratio for the element of interest. In a sandstone aquifer, the mineral/water ratio for oxygen will be low. Authigenic minerals will inherit ^{18}O/^{16}O ratios controlled by the ground water, without any notable change in the oxygen-isotope composition of the much larger water reservoir. Under such conditions, kaolinite formed in isotopic equilibrium with water of δ^{18}O = 0 at +25°C will have an ^{18}O/^{16}O ratio of about +25. Should the fluid be low-^{18}O meteoric water, the oxygen-isotope composition of the precipitating mineral will be lower, given no change in temperature. As the crystallization temperature increases, the oxygen-isotope fractionation between the mineral and water becomes smaller. For example, kaolinite precipitating at +270°C in equilibrium with 0 o/oo water, will have a δ^{18}O value of approximately +6.

An important concern is that the clay or carbonate minerals may have exchanged structural oxygen with fluids subsequent to crystallization. Exchange of oxygen isotopes between water and structural sites in clay minerals has been shown both in natural systems and by experiments to be unimportant at temperatures typical of sedimentary environments (Savin and Epstein, 1970a, 1970b; O'Neil and Kharaka, 1976; James and Baker, 1976; Yeh and Savin, 1976; Eslinger and Yeh, 1981). Yeh and Savin (1976) and Eslinger and Yeh (1981) demonstrated that

TABLE 3

Location	δ^{18}O SMOW
Twp2, R13 W4[1]	−18.9
Twp2, R16 W4[1]	−18.2
Twp3, R13 W4[1]	−19.1
Twp2, R14 W4[2]	−18.5
Twp2, R13 W4[2]	−19.2
Twp2, R10 W4[2]	−18.6
Twp1, R11 W4[2]	−19.2
Twp8, R17 W4[1]	−9.3
Twp9, R17 W4[1]	−8.7
Twp9, R17 W4[2]	−8.5
Twp9, R16 W4[2]	−9.0
Twp13, R15 W4[1]	−5.8

[1]From Schwartz and Muehlenbachs (1979) and Schwartz et al (1981).
[2]From Swanick (1982).

Table 3—δ^{18}O of ground water, Milk River Aquifer.

oxygen-isotope exchange between recent clay minerals and ocean water begins to be significant for only the very finest size-fractions (<0.1 μm). Isotopic exchange rates for clay minerals increase during burial diagenesis, finer size-fractions being significantly affected at lower temperatures than coarser-grained clay minerals. Yeh and Savin (1977) showed that significant isotopic reequilibration between pore fluids and very fine-grained material (<0.5 μm) did not occur below about +80 to +100°C during burial diagenesis.

Like clay minerals, carbonate minerals do not exchange oxygen isotopes to any significant extent except through dissolution–reprecipitation reactions (Anderson, 1969; Anderson and Chai, 1974; Land, 1980). However, such reactions can be easily achieved during recrystallization, a process which is much more common, and occurs at lower temperatures, in carbonate minerals than in silicate minerals (Clayton, 1959). Of dolomite and calcite, dolomite is much less prone to such recrystallization (Epstein et al, 1964; Land, 1980).

Detrital Minerals, Milk River Aquifer and the Milk River Gas Pool

The δ^{18}O values of clay minerals from the argillaceous rocks (+16.0 to +19.0) reflect weathering conditions at their source rather than subsequent diagenetic activity. The oxygen-isotope compositions are virtually identical to those of clay minerals from other argillaceous rocks in Alberta (Longstaffe, 1983), including shales from the underlying Colorado Group (Table 6). Such values are typical of detrital clay minerals from fine-grained sediments throughout the world (Savin and Epstein, 1970a, 1970b, 1970c; Longstaffe et al, 1982). Savin and Epstein (1970a, 1970b) predicted that detrital clay mixtures from shales should range in δ^{18}O from +16 to +22 o/oo.

It is extremely unlikely that the <2 μm clay minerals have experienced significant oxygen-isotope exchange subsequent to their formation. However, the <0.2 μm material from selected argillaceous rocks (Table 1, #10, 12, 15, 19, 30, 33) is poorer in ^{18}O than the <2 μm fraction by 0.1 to 2.7 o/oo. Part of this difference may result from minor differences in clay mineralogy between the two size-fractions (Table 4). However, the finer size-fraction contains less chlorite than the coarser-grained clay minerals. This difference should increase, rather than decrease, the δ^{18}O values of the <0.2 μm material, since chlorite is normally poorer in ^{18}O than other clay minerals. All of the <0.2 μm samples are located within the Milk River aquifer. They have been in contact with low-^{18}O ground water since the artesian system was established, probably during the Early Tertiary. While these very fine-grained clay minerals have not been heated to the temperatures required for extensive reequilibration with ground water to occur (maximum burial depths <300 m), some isotopic exchange may have taken place.

Virtually all samples of dolomite have δ^{18}O (+24.4 to +28.3, SMOW) and δ^{13}C (−2.7 to +1.0, PDB) values typical of platform dolomite (Land, 1980). Dolomite from the underlying Colorado shales has virtually identical oxygen- and carbon-isotope compositions (Tables 1, 2, 6). Similar results were reported by Hitchon and Friedman (1969) for dolomite from other

TABLE 4

Sample Location	Depth[1] (m)	Relative % Clays								
		K		Sm		I/Sm (μm)	I		Chl	
		<2	<0.2	<2	<0.2	<0.2	<2	<0.2	<2	<0.2
Virgelle Member, Milk River Formation										
Milk River Valley Twp 2, R 15 W4										
1. sandstone	0.0	80	55	10	25	15	10	5	—	t
2. sandstone	7.6	60	55	25	25	10	15	10	t	t
3. sandstone	13.7	15		60			20		5	
4. sandstone	19.8	55	10	20	40	35	20	15	5	t
5. sandstone	25.9	50	10	25	40	35	20	15	5	t
Imperial Chin 7-10-9-18 W4										
6. sandstone	241.7	60		10			20		10	
7. silty sandstone	264.4	5		50			30		15	
10. arg.[2] sandstone	268.5	5	5	40	40	10	50	45	5	t
11. sandstone	271.0	45	35	20	55	5	30	5	5	t
Dome Cecil 6-28-13-12 W4										
12. siltstone	231.8	15	10	5	5	—	65	80	15	5[3]
15. arg. siltstone	238.4	5	5	5	5	—	75	80	15	10[3]
19. arg. siltstone	246.3	35	10	5	5	—	55	75	5	10[3]
21. siltstone	250.9	15		20			55		10	
25. sandstone	254.7	70		5			20		5	
27. sandy siltstone	258.8	20	10	15	20	15	50	55	15	t
30. arg. siltstone	262.3	15	5	10	15	10	70	70	5	t
33. silty shale	268.2	15	10	5	10	5	65	75	15	t

[1]Depth = depth below surface.
[2]arg. = argillaceous.
[3]Exhibits some swelling.
K = kaolinite, Sm = smectite, I/Sm = degraded illite or illite/smectite, I = illite, Chl = chlorite, t = trace.

Table 4—Clay mineralogy, Milk River Aquifer.

reservoir rocks in Alberta.

Coexisting dolomite and calcite are not in isotopic equilibrium. The dolomite is richer in ^{18}O than the calcite by about +8 to +10 o/oo in the Milk River aquifer and by +5 to +7 o/oo in the Milk River Gas Pool (Tables 1, 2; Fig. 4). While the oxygen-isotope fractionation between dolomite and calcite at sedimentary temperatures is not well known, values of +2 to +4 o/oo at +25°C are indicated by most estimates (Land, 1980). At the even lower temperatures of the Milk River aquifer (+6 to +18°C; Meyboom, 1960; Swanick, 1982), the fractionation could be 1 or 2 o/oo larger. Even so, for most samples, the oxygen-isotope fractionation between dolomite and calcite lies outside of reasonable limits for isotopic equilibrium.

Another argument for isotopic disequilibrium between the two carbonate minerals is that the dolomite samples are significantly enriched in ^{13}C relative to coexisting calcite (by +6 to +8 o/oo; Tables 1, 2). Dolomite precipitated in equilibrium with calcite at sedimentary temperatures should be only slightly richer in ^{13}C (<2.5 o/oo) (Sheppard and Schwarcz, 1970; Friedman and O'Neil, 1977). A logical conclusion is that most of the dolomite is detrital and has retained carbon- and oxygen-isotope values characteristic of its provenance.

TABLE 5

Sample Location	Depth[1] (m)	Relative % Clays (<2 μm)			
		K	Sm	I	Chl
Alderson Member, Lea Park Formation					
Merland et al Medhat 6-30-13-2 W4					
34. arg. siltstone[2]	366.0	15	5	75	5
35. arg. siltstone	371.7	15	5	75	5
36. arg. siltstone	376.4	15	5	70	10
37. bentonite	380.9	10	85	5	t
McCulloch Medhat 6-20-14-7 W4					
38. arg. siltstone	344.7	15	5	70	10
Alberta Suffield 7-14-15-6 W4					
40. mudstone	266.7	15	5	70	10
41. sandstone	269.8	15	5	70	10
42. mudstone	270.4	15	5	70	10
43. mudstone	272.3	15	5	75	5
44. silty mudstone	288.3	20	5	70	5
45. mudstone	292.2	10	10	70	10
46. silty sandstone	292.6	20	5	70	5
47. arg. siltstone	309.4	20	5	70	5
48. arg. siltstone	311.7	20	5	70	5
49. silty shale	313.9	15	5	75	5
PCP Alderson 7-21-15-12 W4					
50. silty shale	305.0	20	5	70	5
51. arg. siltstone	310.0	15	5	70	10
52. arg. siltstone	326.4	30	15	50	5
53. arg. siltstone	329.9	20	5	65	10
54. arg. sandstone	334.8	25	5	60	10
55. silty shale	338.3	15	t	75	10
PCP Alderson 6-33-15-12 W4					
56. arg. siltstone	293.2	15	5	70	10
57. arg. siltstone	295.0	15	5	65	15
58. arg. siltstone	299.0	25	5	60	10
59. arg. siltstone	301.3	25	5	60	10
60. sandstone	306.5	20	10	60	10
61. siltstone	310.6	15	5	70	10

(continued)

Table 5—Clay mineralogy, Southeastern Alberta Milk River Gas Pool.

TABLE 5 *(continued)*

Sample Location	Depth[1] (m)	Relative % Clays (<2 μm)			
		K	Sm	I	Chl
PCP Johnson					
7-8-17-14 W4					
62. arg. siltstone	370.0	20	5	70	5
63. sandstone	372.2	10	5	80	5
64. silty shale	373.7	10	5	75	10
65. silty shale	377.0	15	5	75	5
66. silty shale	382.5	15	5	70	10
67. arg. siltstone	386.0	20		70	10
CRL Yan HB Atlee					
10-36-20-3 W4					
69. arg. siltstone	391.1	20	5	70	5
70. arg. siltstone	395.6	15	5	75	5
71. arg. siltstone	415.7	15	5	75	5
72. arg. siltstone	419.9	20	5	70	5
73. arg. siltstone	422.9	15	t	75	10
74. arg. siltstone	426.7	15	5	75	5
Alberta Suffield					
6-8-20-6 W4					
75. arg. siltstone	416.1	20	t	75	5
AEG Princess					
6-13-20-12 W4					
77. arg. siltstone	316.5	15	5	70	10
78. arg. siltstone	320.0	15	t	75	10
79. arg. siltstone	324.0	20	t	75	5
80. arg. siltstone	329.8	20	t	75	5

[1]Depth = depth below surface.
[2]arg. = argillaceous.
K = kaolinite, Sm = smectite, I = illite, Chl = chlorite, t = trace

Authigenic Minerals, Milk River Aquifer

Authigenic minerals from the sandstones of the Milk River aquifer have attained their isotopic and mineralogical compositions under quite different conditions than detrital minerals from the argillaceous rocks. Not surprisingly, pore-lining and pore-filling kaolinite and smectite have much lower $\delta^{18}O$ values than the detrital clay minerals (Tables 1, 2; Fig. 2). A logical explanation of this difference is that the authigenic clay minerals have formed in equilibrium with low-^{18}O ground water.

Kaolinite, in particular, is a common diagenetic phase in sandstones that are recharged by meteoric water. The ability of kaolinite to form at low temperatures has been well documented (Curtis and Spears, 1971; Kittrick, 1970; Linares and Huertas, 1971; La Iglesia and Van Oosterwyck-Gastuche, 1978; Van Oosterwyck-Gastuche and La Iglesia, 1978). However, the exact nature of the oxygen-isotope fractionation between kaolinite and water at low temperatures is not well known; the relation of Eslinger (1971) and Land and Dutton (1978) provides one rea-

sonable estimate (Fig. 5). The oxygen-isotope fractionation between smectite and water is somewhat better understood; the equation of Yeh and Savin (1977) has been used here (Fig. 5).

These relationships between mineral $\delta^{18}O$, fluid $\delta^{18}O$ and temperature make possible more detailed speculation concerning the formation of authigenic clay minerals in the Milk River aquifer. An average $\delta^{18}O$ of +10.9 can be calculated for authigenic kaolinite + smectite mixtures from the Milk River valley, assuming all illite and chlorite to be detrital (Fig. 5a). Ground water at this location has $\delta^{18}O$ values of -19.2 to -18.2 (Table 3). From these data, a temperature for clay crystallization of +3 to +5°C can be calculated, assuming isotopic equilibrium (Fig. 5a).

Similar calculations are possible for authigenic calcite that coexists with the kaolinite and smectite. In this instance, the appropriate equations of O'Neil et al (1969) and Tarutani et al (1969) have been used. Assuming an average calcite $\delta^{18}O$ of +15.45 (SMOW), a temperature of +2°C can be determined for equilibrium precipitation (Fig. 5a).

Such values are in reasonable agreement with the temperature of ground water at this location (+6°C, Meyboom, 1960; +9 to +11°C, Schwartz and Muehlenbachs, 1979). This concordance suggests a close approach to isotopic equilibrium between the authigenic minerals and the ground water; crystallization of the clay minerals and calcite has occurred at low temperatures in the presence of meteoric water.

Precipitation of the calcite from meteoric water is consistent with other observations. Firstly, the low-Mg character of the calcite is typical of an inorganic derivation. Secondly, the low-^{13}C nature of the calcite suggests involvement of organically derived CO_2. The probable source of such CO_2 is decaying plant material from the overlying soil and till; local recharge to the aquifer in this area occurs by infiltration of precipitation through the soil zone (Meyboom, 1960; Schwartz and Muehlenbachs, 1979). The carbon dioxide gas can be expected to have $\delta^{13}C$ values of -25 to -15 (PDB) (Pearson and Hanshaw, 1970; Deines et al, 1974). At low temperatures (+5 to +20°C), the $CO_2(g)$–calcite fractionation for carbon is about -13 to -11

TABLE 6

Sample Location	Depth[1] (m)	Clays $\delta^{18}O$ <2 μm SMOW	Carbonates $\delta^{18}O$ SMOW	$\delta^{13}C$ PDB
Banner et al Iron Sp. *7-9-12-21 W4*				
81. shale	567.8	+16.2	+24.1	−1.2 d[2]
82. sandstone	573.8	+14.3	+23.8	−1.4 d
Bralorne et al Medhat *6-12-13-4 W4*				
83. shale	404.0	+16.5	+25.4	−1.0 d
84. shale	410.2	+16.5	+25.3	−0.7 d
85. shale	416.0	+16.5	+24.9	−1.3 d
86. shale	421.1	+16.5	+27.1	−0.9 d
87. sandstone	422.4		+18.4	−4.5 c[3]
			+25.2	−1.3 d
88. shale	423.4		+26.0	−1.0 d
89. shale	425.0		+25.8	−1.1 d
90. sandstone	426.0	+15.3	+17.4	−4.9 c

[1]Depth = depth below surface.
[2]d = dolomite.
[3]c = calcite.

Table 6—Stable isotope results for clay and carbonate minerals, Colorado Group.

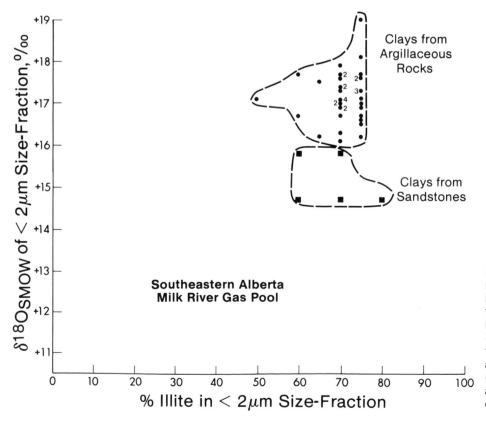

Southeastern Alberta Milk River Gas Pool

Figure 2—$\delta^{18}O$ values of <2 μm clay minerals versus the relative percent illite in the clay mineral mixture for samples from the southeastern Alberta Milk River Gas Pool. (•) argillaceous rocks (number of samples indicated adjacent to data points); (■) sandstones. All data are from Tables 2 and 5; samples #37, 39 (bentonite) are not shown. Samples #68 and 76 are not plotted owing to insufficient data for clay minerals.

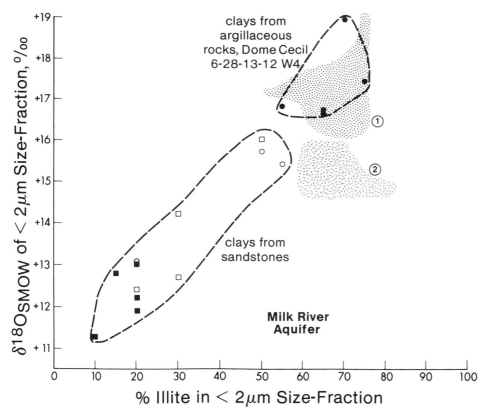

Figure 3—$\delta^{18}O$ values of $<2\ \mu m$ clay minerals versus the relative percent illite in the clay mineral mixture for samples from the Milk River aquifer. Similar data for shales (**1**) and sandstones (**2**) from the southeastern Alberta Milk River Gas Pool are also shown in the shaded areas (see Fig. 2). All data are from Tables 1 and 4. (■) Milk River valley; (□) Imperial Chin 7-10-9-18 W4; (o) Dome Cecil 6-28-13-12 W4 (sandstone); (●) Dome Cecil 6-28-13-12 W4 (shale).

(Bottinga, 1968; Friedman and O'Neil, 1977). This fractionation would produce calcite with $\delta^{13}C$ values in the observed range (−10 to −3, PDB; Table 1). Reactions involving biogenic methane found in parts of the Milk River aquifer (Swanick, 1982) are another potential source of ^{13}C-depleted carbon dioxide.

The close agreement between measured and calculated temperatures disappears for subsurface samples from the Milk River aquifer located further to the north (Figs. 1, 5b,c). The average $\delta^{18}O$ values for authigenic calcite (+16.4, SMOW) and authigenic clay minerals (+11.3; Fig. 5b) from Imperial Chin 7-10-9-18 W4 are somewhat higher than in the Milk River valley. Formation fluid at this location is considerably richer in ^{18}O (−9.3 to −8.5; Table 3). If equilibrium is assumed, isotopic temperatures of +45 to +60°C for crystallization of the clay minerals and calcite can be calculated (Fig. 5b). Such values are clearly unrealistic. Groundwater temperatures near this locality are only +15 to +18°C (Swanick, 1982); a temperature of +16°C can be predicted using the temperature–depth

relationship of Meyboom (1960). The results of such calculations for authigenic minerals from Dome Cecil 6-28-13-12 W4 are equally bizarre (+55 to +80°C; $\delta^{18}O$ of water = −5.8; Table 3, Fig. 5c), given the present formation temperature (+15 to +18°C; Meyboom, 1960).

The discrepancy between isotopic temperature estimates and measured temperatures from the Imperial Chin and Dome Cecil wells indicates that the diagenetic clay and carbonate minerals are grossly out of isotopic equilibrium with existing formation fluids. The extent of the isotopic disequilibrium is mostly a function of the radically changing $\delta^{18}O$ (−20 to −6) of the fluid (Fig. 1). Proceeding downdip in the aquifer, the formation fluids systematically deviate from the meteoric water line for oxygen and hydrogen isotopes (Schwartz and Muehlenbachs, 1979). This isotopic variation is accompanied by large-scale changes in the major ion chemistry of the formation fluids. The concentrations of Na^+, HCO_3^- and Cl^- increase, and those of Ca^{+2}, Mg^{+2}, and SO_4^{-2} decrease, in the direction of ground-water flow (Schwartz and

Muehlenbachs, 1979; Swanick, 1982).

Schwartz and Muehlenbachs (1979) proposed that most of the isotopic and chemical variations in the formation fluids of the Milk River aquifer result from dispersion involving recharging meteoric water and older, more ^{18}O-rich formation fluids. Swanick (1982) has demonstrated that the formation water in the northern portion of the aquifer is much older than the meteoric water to the south. Flushing of the aquifer by meteoric water is not yet complete.

A possible origin for the older formation fluids still present in downdip portions of the Milk River aquifer has been suggested by Schwartz et al (1981). They noted that glacial ice covered the northern half of the study area for much of the last 100,000 years, blocking the discharge areas of aquifer systems in the western Canada sedimentary basin. Formation fluids would then begin to discharge in advance of the ice front. Cross-formational flow would be greatly increased and formation water from greater depths would move into shallower units. This change in flow patterns provides a mechanism

for ^{18}O-rich formation water to be present in downdip portions of the Milk River aquifer. Support for this hypothesis is provided by the similarity in isotope chemistry between water from the northern portion of the Milk River aquifer and water from the underlying Bow Island Formation (Schwartz et al, 1981).

A logical extension of these ideas is that the ^{18}O-rich formation fluids now present in the northern portion of the aquifer represent a recent modification to an earlier ground-water system. This conclusion may help to explain the oxygen-isotope compositions of the authigenic minerals from the Imperial Chin and Dome Cecil wells. Assuming temperatures of +15 to +18°C, δ^{18}O values of −16 to −13 can be predicted for the water involved in the formation of the authigenic phases (Fig. 5b,c). Still lower δ^{18}O values (−18 to −16) for the water can be predicted if cooler subsurface temperatures, based upon reduced thickness of Pleistocene glacial debris, are used in the calculation. Such δ^{18}O values are very similar to present-day meteoric water from this location (Hitchon and Friedman, 1969; Schwartz and Muehlenbachs, 1979; Schwartz et al, 1981; Swanick, 1982). No evidence exists to indicate that the oxygen-isotope composition of meteoric water in this area has changed more than ±2 o/oo over the last 500,000 years.

Authigenic Minerals, Southeastern Alberta Milk River Gas Pool

The isotopic data for authigenic minerals in sandstones of the Milk River Gas Pool suggest that the fluids involved in their formation also contained a significant component of meteoric water. However, the occurrence of illite as an authigenic mineral suggests that it has precipitated from water containing a much higher K^+/H^+ ratio than is typical of meteoric water (Montoya and Hemley, 1975).

Detailed evaluation of the isotopic data for the illite-dominated clay mixtures is difficult, as the amount of detrital material is not easily estimated. However, clay minerals from the sandstones with the lowest clay content consistently have δ^{18}O values of +14.7 (Table 2, #41, 60, 63). Assuming all of this material to be authigenic, and

choosing temperatures appropriate for the present depth of the reservoir (+15 to +20°C), fluid compositions of about −10 can be calculated (the oxygen-isotope fractionation for illite-water of Eslinger and Savin [1973] has been used). Calculations using the coexisting authigenic calcite (Table 2, #54, 60, 63; δ^{18}O = +19.3, SMOW) produce δ^{18}O values of −12 to −11 for the formation water. Even if a somewhat higher temperature of about +30°C is assumed (that is, greater depth of burial in the past), comparable fluid compositions (about −8) result. Such oxygen-isotope compositions are typical of water from other Upper Cretaceous oil and gas fields in Alberta (−13 to −9; Hitchon and Friedman, 1969). These formation fluids are derived by mixing of meteoric water and evolved sea water rich in dissolved solids (Clayton et al, 1966; Hitchon and Friedman, 1969). Such water can have the appropriate chemical composition for the formation of illite.

Authigenic calcite from the Milk River Gas Pool has low δ^{13}C values, as do a few samples of dolomite that may also be diagenetic in origin (Fig. 4). Low-^{13}C biogenic methane is found in the Milk River Gas Pool (Rice and Claypool, 1981). Its presence suggests the availability of ^{13}C-depleted CO_2 for incorporation into diagenetic carbonate minerals.

CONCLUSIONS

The low δ^{18}O values of authigenic minerals from the Milk River aquifer reflect their crystallization at low temperatures (pehaps as low as +5°C) from meteoric water. Samples from areas of local recharge closely approach oxygen isotopic equilibrium with the existing ground water. At depth, where the recharging meteoric water has not yet completely displaced older, more ^{18}O-rich formation fluids, the authigenic minerals are grossly out of equilibrium with the existing fluid. These minerals retain oxygen-isotope compositions diagnostic of still earlier formation fluids dominated by meteoric water. It is believed that a flow system of this kind is now being reestablished within the Milk River aquifer, following deglaciation.

The higher δ^{18}O values of authigenic

clay minerals and calcite from sandstones within the southeastern Alberta Milk River Gas Pool are consistent with crystallization at temperatures of +15 to +20°C from formation fluids derived by mixing of meteoric water and evolved sea water. Such formation waters are typical of many oil and gas pools in the Alberta Basin.

Detrital clay minerals and dolomite from both the Milk River aquifer and the Milk River Gas Pool have δ^{18}O values typical of such materials throughout the world. Except for the finest-grained materials, significant isotopic exchange between detrital phases and formation water has not occurred.

Low-^{18}O clay minerals and calcite occur in other shallow sandstone units in Alberta (Longstaffe, 1983), including those that directly underlie the Milk River aquifer (for example, Table 6). Such minerals should be characteristic of clastic sedimentary basins where meteoric water has displaced earlier formation fluids. In addition to recording an important stage of diagenesis in such rocks, these minerals, with their diagnostic isotopic signatures, have value as pathfinders when deciphering the paleohydrology of sedimentary basins.

ACKNOWLEDGMENTS

I wish to thank Dr. J. F. Lerbekmo for providing some of the samples used in this study and Dr. F. W. Schwartz for first suggesting the problem. D. Caird and E. Toth provided valuable assistance with the analyses. K. Muehlenbachs provided access to his stable isotope laboratory. This research was supported by grants from Imperial Oil Ltd. (Canada) to Longstaffe and the Natural Sciences and Engineering Research Council of Canada (NSERC) to Longstaffe and Muehlenbachs. F. W. Schwartz, D.A. McDonald, E. Olson, and B. J. Tilley provided helpful comments on earlier drafts of this manuscript.

SELECTED REFERENCES

Anderson, T. F., 1969, Self-diffusion of carbon and oxygen in calcite by isotope exchange with carbon dioxide: Journal of Geophysical Research, v. 74, p. 3918–3932.
Anderson, T. F., and B. H. T. Chai, 1974, Oxygen isotope exchange between calcite and water under hydrothermal conditions, *in* A. W. Hoffman, B. J. Gilletti, H.

S. Yoder, Jr., and R. A. Yund, eds., Geochemical transport and kinetics: Carnegie Institution of Washington Publication 634, p. 3918–3932.

Black, H. N., H. E. Ripley, and W. H. Beecroft, 1980, Fracturing fluid improvements for gas wells in Canada: Journal Canadian Petroleum Technology, April–June, p. 58–64.

Bottinga, Y., 1968, Calculation of fractionation factors for carbon and oxygen exchange in the system calcite–carbon dioxide–water: Journal of Physical Chemistry, v. 72, p. 800–808.

Clayton, R. N., 1959, Oxygen isotope fractionation in the system calcium carbonate–water: Journal of Chemical Physics, v. 30, p. 1246–1250.

Clayton, R. N., and T. K. Mayeda, 1963, The use of bromine pentafluoride in the extraction of oxygen from oxides and silicates for isotopic analysis: Geochimica et Cosmochimica Acta, v. 27, p. 43–52.

Clayton, R. N., I. Friedman, D. L. Graf, T. K. Mayeda, W. F. Meents, and N. F. Shimp, 1966, The origin of saline formation waters, 1, Isotopic composition: Journal of Geophysical Research, v. 71, p. 3869–3882.

Craig, H., 1957, Isotopic standards for carbon and oxygen and correction factors for mass-spectrometric analysis of carbon dioxide: Geochimica et Cosmochimica Acta, v. 12, p. 133–149.

———, 1961, Standards for reporting concentrations of deuterium and oxygen-18 in natural waters: Science, v. 133, p. 1833–1834.

Curtis, C. D., and D. A. Spears, 1971, Diagenetic development of kaolinite: Clays and Clay Minerals, v. 19, p. 219–227.

Deines, P., D. Langmuir, and R. S. Harmon, 1974, Stable carbon isotope ratios and the existence of a gas phase in the evolution of carbonate groundwater: Geochimica et Cosmochimica Acta, v. 38, p. 1147–1164.

Epstein, S., D. L. Graf, and E. T. Degens, 1964, Oxygen isotope studies on the origin of dolomite, *in* H. Craig et al, eds., Isotopic and cosmic chemistry. Amsterdam, North Holland Publishing Company, p. 169–180.

Eslinger, E. V., 1971, Mineralogy and oxygen isotope ratios of hydrothermal and low-grade metamorphic argillaceous rocks: PhD dissertation, Case Western Reserve University, 205 p.

Eslinger, E. V., and S. M. Savin, 1973, Mineralogy and oxygen isotope geochemistry of the hydrothermally altered rocks of the Ohaki-Broadlands, New Zealand geothermal area: American Journal of Science, v. 273, p. 240–267.

Eslinger, E. V., and H. Yeh, 1981, Mineralogy, O^{18}/O^{16}, and D/H ratios of clay-rich sediments from Deep Sea Drilling Project site 180, Aleutian Trench: Clays and Clay Minerals, v. 29, p. 309–315.

Friedman, I., and J. R. O'Neil, 1977, Compilation of stable isotope fractionation factors of geochemical interest, *in* M. Fleischer, ed., Data of geochemistry, 6th ed.: United States Geological Survey Professional Paper 440-KK, 12 p.

Hitchon, B., and I. Friedman, 1969, Geochemistry and origin of formation waters in the western Canada sedimentary basin, 1, Stable isotopes of hydrogen and oxygen: Geochimica et Cosmochimica Acta, V. 33, p. 1321–1349.

Ignasiak, T. M., L. Kotlyar, F. J. Longstaffe, O. P. Strausz, and D. S. Montgomery, 1983, Separation and characterization of clay from Athabasca asphaltene: Fuel, v. 62, p. 353–362.

James, A. T., and D. R. Baker, 1976, Oxygen isotope exchange between illite and water at 22°C: Geochimica et Cosmochimica Acta, v. 40, p. 235–239.

Kittrick, J. A., 1970, Precipitation of kaolinite at 25°C and 1 atm: Clays and Clay Minerals, v. 18, p. 261–267.

La Iglesia, A., and M. C. Van Oosterwyck-Gastuche, 1978, Kaolinite synthesis, 1, Crystallization conditions at low temperatures and calculation of thermodynamic equilibria. Application to laboratory and field observations: Clays and Clay Minerals, v. 26, p. 397–408.

Land, L. S., 1980, The isotopic and trace element geochemistry of dolomite: the state of the art, *in* D. H. Zenger, J. B. Dunham, and R. A. Ethington, eds., Concepts and models of dolomitization: Society of Economic Paleontologists and Mineralogists Special Publication 28, p. 87–110.

Land, L. S., and S. P. Dutton, 1978, Cementation of a Pennsylvanian deltaic sandstone: isotopic data: Journal of Sedimentary Petrology, v. 48, p. 1167–1176.

Lawrence, J. R., and H. P. Taylor, Jr., 1971, Deuterium and oxygen-18 correlation: clay minerals and hydroxides in Quaternary soils compared to meteoric waters: Geochimica et Cosmochimica Acta, v. 35, p. 993–1003.

———, 1972, Hydrogen and oxygen isotope systematics in weathering profiles: Geochimica et Cosmochimica Acta, v. 36, p. 1377–1393.

Linares, J., and F. Huertas, 1971, Kaolinite: synthesis at room temperature: Science, v. 171, p. 896–897.

Longstaffe, F. J., 1983, Stable isotope studies of diagenesis in clastic rocks: Geoscience Canada, v. 10, p. 43–58.

Longstaffe, F. J., B. E. Nesbitt, and K. Muehlenbachs, 1982, Oxygen-isotope geochemistry of shales hosting the Pb-Zn-Ba mineralization at the Jason prospect, Selwyn Basin, Yukon, *in* Current Research, Part C. Geological Survey of Canada, Paper 82-1C, p. 45–49.

McCrea, J. M., 1950, On the isotopic chemistry of carbonates and a paleotemperature scale: Journal of Chemical Physics, v. 18, p. 849–857.

Meijer Drees, N. C., and D. W. Myhr, 1981, The Upper Cretaceous Milk River and Lea Park Formations in southeastern Alberta: Bulletin of Canadian Petroleum Geology, v. 29, p. 42–74.

Meyboom, P., 1960, Geology and groundwater resources of the Milk River Sandstone in southern Alberta: Research Council of Alberta Memoir 2, 89 p.

Montoya, J. W., and J. J. Hemley, 1975, Activity relations and stabilities in alkali feldspar and mica alteration reactions: Economic Geology, v. 70, p. 577–583.

O'Neil, J. R., and Y. F. Kharaka, 1976, Hydrogen and oxygen isotope exchange reactions between clay minerals and water: Geochimica et Cosmochimica Acta, v. 40, p. 241–246.

O'Neil, J. R., R. N. Clayton, and T. K. Mayeda, 1969, Oxygen isotope fractionation in divalent metal carbonates: Journal of Chemical Physics, v. 51, p. 5547–5558.

Pearson, F. J., and B. B. Hanshaw, 1970, Sources of dissolved carbonate species in groundwater and their effects on carbon-14 dating, *in* Isotope hydrology 1970 (Proceedings of the Symposium in Vienna, 1970), IAEA, Vienna, p. 271–286.

Rice, D. D., and G. E. Claypool, 1981, Generation, accumulation, and resource potential of biogenic gas: Bulletin of the American Association of Petroleum Geologists, v. 65. p. 5–25.

Savin, S. M., and S. Epstein, 1970a, The oxygen and hydrogen isotope geochemistry of clay minerals: Geochimica et Cosmochimica Acta, v. 34, p. 25–42.

———, 1970b, The oxygen and hydrogen isotope geochemistry of ocean sediments and shales: Geochimica et Cosmochimica Acta, v. 34, p. 43–63.

———, 1970c, The oxygen isotopic chemistry of coarse grained sedimentary rocks and minerals: Geochimica et Cosmochimica Acta, v. 34, p. 323–329.

Schwartz, F. W., and K. Muehlenbachs, 1979, Isotope and ion geochemistry of groundwaters in the Milk River aquifer, Alberta: Water Resources Research, v. 15, p. 259–268.

Schwartz, F. W., K. Muehlenbachs, and D. W. Chorley, 1981, Flow-system controls of the chemical evolution of groundwater: Journal of Hydrology, v. 54, p. 225–243.

Sheppard, S. M. F., and H. P. Schwarcz, 1970, Fractionation of carbon and oxygen isotopes and magnesium between coexisting metamorphic calcite and dolomite: Contributions to Mineralogy and

Petrology, v. 26, p. 161–198.

Swanick, G. B., 1982, The hydrochemistry and age of the water in the Milk River aquifer, Alberta, Canada: Unpublished Master's Thesis, University of Arizona, 103 p.

Tarutani, T., R. N. Clayton, and T. K. Mayeda, 1969, The effect of polymorphism and magnesium substitution on oxygen isotope fractionation between calcium carbonate and water: Geochimica et Cosmochimica Acta, v. 33, p.

987–996.

Van Oosterwyck-Gastuche, M. C., and A. La Iglesia, 1978, Kaolinite synthesis, II, A review and discussion of the factors influencing the rate process: Clays and Clay Minerals, v. 26, p. 409–417.

Walthers, Jr., L. J., G. E. Claypool, and P. W. Choquette, 1972, Reaction rates and δO^{18} variation for the carbonate–phosphoric acid preparation method: Geochimica et Cosmochimica Acta, v. 36, p. 129–140.

Wilson, M. D., and E. D. Pittman, 1977, Authigenic clays in sandstones: recognition and influence on reservoir properties and paleoenvironmental analysis: Journal of Sedimentary Petrology, v. 47, p. 3–31.

Yeh, H., and S. M. Savin, 1976, The extent of oxygen isotope exchange between clay minerals and sea water: Geochimica et Cosmochimica Acta, v. 40, p. 743–748.

Reservoir Diagenesis and Convective Fluid Flow

J. R. Wood
Chevron Oil Field
   Research Company
La Habra, California

T. A. Hewett
Chevron Oil Field
   Research Company
La Habra, California

INTRODUCTION

Diagenesis constitutes a rather broad field of study that includes not only inorganic reactions and processes involving minerals, but also organic transformations. Although these two aspects tend to be considered exclusively of one another, the processes that lead to cementation of mineral grains, for example, must influence gas and oil generation and transport to some extent and vice versa. Since organic and inorganic reactions and transport occur in the same rocks at the same time, it is reasonable to assume that some of the same processes are involved. In particular, organic and inorganic diagenesis share the same temperature–pressure environment and surely respond to changes in these parameters in ways that are more similar than dissimilar.

Since inorganic diagenetic products compete with gas and oil for the available pore space in a reservoir, the timing of peak hydrocarbon generation can influence subsequent inorganic diagenesis and vice versa. Inorganic diagenesis thus should not be considered independently of hydrocarbon generation and migration. Elevated temperatures are known to drive kerogen maturation, and recent work (Wood and Hewett, 1982) suggests that temperature gradients are responsible for much fluid and mass transfer in porous rocks.

The arguments presented here are extensions of earlier work (Wood and Hewett, 1982), which assumes the presence of convective fluid currents in porous media. The chemical consequences of cycling saturated fluid parcels around the convection cell are examined for several relevant cases and the possibility of generating hydrocarbon accumulations by exsolution from a convecting medium is discussed.

ABSTRACT. A diagenetic model based on convective fluid flow has been analyzed for typical reservoir conditions. Calculations based on the model suggest that a significant fraction of the inorganic diagenesis observed in sandstone reservoirs can be attributed to the presence of slowly circulating aqueous fluids. Stability considerations indicate that static pore fluids do not exist in porous bodies of geologic dimensions and that pore fluids will convect at a rate of about 10^{-8} m per sec ($\sim$1 m per yr) in the presence of a normal geothermal gradient (25° C per km).

If it is assumed that the pore fluid maintains chemical equilibrium with the rock matrix, it follows that mass must be transferred as the fluid crosses isotherms. Minerals such as quartz, which have prograde solubilities under normal reservoir conditions, will move from hot source zones to cooler sinks. Minerals such as calcite, which have retrograde solubilities, will move from cool sources to hot sinks. The net effect is a continuous transfer of rock matrix in the reservoir for as long as the fluid circulates.

Because the temperature field can change sharply along a streamline, convection can localize precipitation and dissolution zones. In anticlinal structures, the fluid flow is most likely a modified torus in which warm fluid flows up the base of the ascending limb while cooler fluid flows down along the upper surface. The regions of most rapid heating and cooling of the fluid occur at the synclinal troughs and at the anticlinal crests. This flow pattern will produce zones of intense diagenesis at the crests and troughs of the structure. Zones of secondary porosity produced by the dissolution of framework grains or previously deposited cements are also predictable and the model provides explicit conditions for isomorphic replacement.

Since hydrocarbon solubilities are similar to quartz solubility in the temperature range 60–150° C, hydrocarbon transfer and accumulation should closely approximate that of quartz. Calculations suggest that convection can transfer significant quantities of hydrocarbons in molecular solution and exsolve them in traps in relatively short geologic times. The convection model thus links inorganic and organic diagenesis and provides reasonable explanations for such observed phenomena as secondary porosity and thermal anomalies.

STATEMENT OF THE PROBLEM

The purpose of this paper is to examine the effect of a weakly convecting fluid in a porous reservoir rock with respect to mass transfer involving the framework minerals. As a conceptual model, we can take a slab of porous, granular material with lateral dimensions very large relative to the vertical (thickness) dimension. In a geologic context, the scale of the lateral dimensions is on the order of 10–100 km, while the vertical dimension is on the order of 10–100 m, an aspect ratio of 100–10,000. The porous medium itself can be considered to consist of grains with average diameters on the order of 0.1 to 1 mm (medium to coarse sand) with no interstitial cement and/or clay. To simplify analysis, we can further assume the grains to be monomineralic, say quartz, with a simple composition (SiO_2). The body is thus homogeneous both in composition and in grain-size distribution and will have high values for porosity, ϕ, on the order of 20–30%, and permeability, k, on the order of 1–10 darcies. The pore fluid can be taken to be pure water saturated with SiO_2 that is in equilibrium with the quartz matrix at the local temperature and pressure.

Hydrodynamic, or forced, fluid flow is not considered here. We wish to consider the question of fluid stability in the absence of external boundary forces since it is not clear that fluid movement will occur when the only driving force is thermally induced bouyancy. However, the diagenetic model considered here is applicable to both free and forced convective flow. Although hydrodynamic flow is certainly involved in many cases, this contribution is concerned principally with the free convective component of the flow.

Two basic questions that arise are: (1) Under what conditions is the fluid static? and (2) What is the fluid path for nonzero fluid velocities? For those conditions under which fluid flow is possible, the problem of matrix mass transfer can be related to the fluid movement by the simple expedient of assuming local chemical equilibrium between the fluid and the rock matrix.

THE CONVECTIVE MODEL— FREE CONVECTION

Fluid Stability and Flow Velocity

The problem of fluid stability in a porous medium heated from below has been treated extensively (see review by Combarnous and Bories, 1975) in terms of the fluid mechanics involved and also with respect to natural systems (McKibbin and O'Sullivan 1981, Straus, 1974; Straus and Schubert, 1977, 1978, 1979; Schubert and Straus, 1979).

The fluid motion considered here is a type of thermal convection commonly

referred to as Benard or Rayleigh-Benard convection. It arises where a fluid contained between two horizontal plates is heated below and cooled above. The basic driving force is the slight density difference in the warmer fluid at the bottom relative to the cooler fluid at the top. In terms of temperature, the density contrast is given by

$$\Delta\rho/\rho = \alpha\Delta T$$

where α is the coefficient of expansion for the fluid and ΔT is the temperature difference between the plates. $\Delta\rho$ is the density difference between a bouyant fluid element and its surrounding fluid at the lower boundary, and ρ is the density of the fluid (see Turner, 1973, p. 208). Since α is approximately $2 \times 10^{-6} K^{-1}$ for an aqueous fluid under normal (60–150° C) reservoir temperatures, the buoyant force $\Delta\rho$ is obviously very small for a ΔT on the order of 1–2° K. Viscous forces will act to retard any buoyant motions, and there is a critical value for ΔT below which no convective motion is possible. Under these conditions, marginal stability corresponds to a particular value of a parameter, the Rayleigh number, defined as

$$Ra = g\alpha H^{3} (\rho C)_f \Delta T / v\lambda$$

where g is the gravitational constant, λ is the thermal conductivity, $(\rho C)_f$ is the heat capacity of the fluid, and v is the kinematic viscosity. H is a characteristic length usually taken as the distance between the two plates. If the fluid is contained in a porous medium, then the Rayleigh number must be modified to the filtration Rayleigh number, which includes the permeability

$$Ra = k g\alpha(\rho C)_f H\Delta T / v\lambda$$

(Combarnous and Bories, 1975).

Both theory and experiment show that the Rayleigh number must be greater than 40 for fluid flow to occur in a horizontal layer. For H = 100 and $\Delta T = 2.5°$ K, Ra is approximately 35, slightly less than required for convective flow. However, this assumes that the isotherms are horizontal. If there is any slight departure of the isotherms from the horizontal, then the fluid is

unconditionally unstable and flow will occur regardless of the value of the Rayleigh number.

Wood and Hewett (1982) examined this problem from the point of view of reservoir diagenesis and concluded that mean flow velocities on the order of 10^{-8} m per sec can be expected. This flow results from the inability of a natural system to satisfy the (necessary) condition for fluid equilibrium

$$\overline{\nabla}\rho\times\overline{g} = 0$$

where ρ is the fluid density, $\overline{g}$ is the gravity vector, and $\overline{\nabla}\rho$ represents the density gradient in the pore fluid. If the fluid density is a function of temperature only, then the necessary condition for a static pore fluid is that all isotherms be horizontal. Similar conclusions regarding fluid stability have been reached for cases including permeable boundaries, for boundary conditions that specify a heat flow (Straus and Schubert, 1979), and on purely geologic grounds (Roberts, 1981). The theoretical basis for steady and pervasive fluid movement in reservoir rocks under normal geologic conditions thus seems well founded, even in the absence of hydrodynamic flow.

However, fluid movement does not necessarily result in major diagenetic alteration. To explain many diagenetic observations it is necessary to show that sufficient solid mass can be transferred from point to point in reasonable geologic times.

The estimate of 10^{-8} m per sec (approximately 1 m per yr) for the fluid velocity was obtained from the following expression for flow in a large convection cell

$$U = Kg \alpha \sin \theta \, \Delta T / v \qquad (1)$$

where U is the pore fluid velocity, K is the permeability, g is the acceleration of gravity, α is the (volumetric) thermal expansion coefficient, ΔT is the temperature difference across a layer, v is the kinematic viscosity, and θ is the angle between the isotherms and a horizontal datum. Similar estimates have been observed for fluid flow in oceanic sediments, derived from field temperature measurements (Anderson et al, 1979). Calculations of the rates of vertical ground-water movement

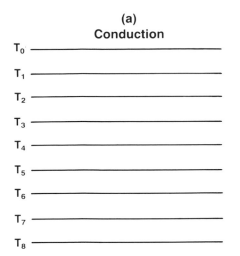

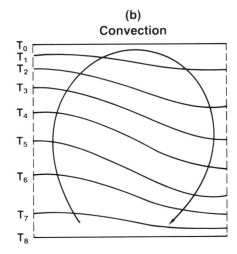

Figure 1—Schematic diagram showing isotherms for (**a**) conductive heat transport and (**b**) convective heat transport. Arrow indicates sense of fluid rotation in convection cell for T_0 less than T_8. Dashed vertical lines indicate boundaries of convection cell.

(Bredehoeft and Papadopulous, 1965) also suggest that a fluid velocity of 10^{-8} m per sec is the correct order of magnitude. Additionally, model studies of hydrothermal systems (Cathles, 1977; Norton, 1978; Norton and Cathles, 1979) suggest that fluid velocities of 10^{-6} m per sec are reasonable estimates for fluid velocities in these systems, which would be expected to have a more vigorous fluid circulation.

Role of the Temperature Field

If thermal convection occurs in rocks under conditions of chemical equilibrium, it follows that mass transfer of the matrix will occur if the matrix phases have temperature-dependent solubilities and fluid-rock equilibrium is maintained. For any volume element along the streamline, the change in porosity is related to the fluid flow and temperature by the relation (Wood and Hewett, 1982)

$$\frac{d\phi}{dt} = U(\rho_w/\rho_s)\frac{dC}{dS} \qquad (2)$$

where S represents distance along a streamline, C is the equilibrium concentration of SiO_2 in the fluid, ρ_w is the density of the aqueous fluid and ρ_s is the density of the solid (quartz in the present discussion). Since C is a function of T and P, the total derivative is

$$dC = \left(\frac{\partial C}{\partial T}\right)_P dT + \left(\frac{\partial C}{\partial P}\right)_T dP \qquad (3)$$

and

$$\frac{dC}{dS} = \alpha_T\frac{dT}{dS} + \alpha_P\frac{dP}{dS} \qquad (4)$$

where

$$\alpha_T = \left(\frac{\partial C}{\partial T}\right)_P$$

and

$$\alpha_P = \left(\frac{\partial C}{\partial P}\right)_T$$

so that

$$\frac{d\phi}{dt} = U(\rho_w/\rho_s)\left[\alpha_T\frac{dT}{dS} + \alpha_P\frac{dP}{dS}\right] \qquad (5)$$

where t is time. α_T and α_P represent the temperature and pressure coefficients for quartz solubility and should not be confused with α, the volumetric thermal expansion coefficient in equation 1.

The second term on the right-hand side of equation 5, which involves the pressure coefficient, α_P, and the (hydrostatic) pressure gradient, dP/dS, is much smaller than the corresponding temperature term. Consequently, we will ignore the pressure term in the subsequent discussion, but there may be cases (Bruton and Helgeson, in press) where this approximation is not justified.

The spatial locations of the dissolution and precipitation zones are functions of the variations in the temperature field along a streamline. Consider a simple case of closed-loop fluid flow in which streamlines cross isotherms (Fig. 1). It is helpful to imagine a streamline removed from the convection cell and straightened out. This streamline can then be used as an ordinate and chemical and thermal variations along this streamline can be plotted on the abscissa. Figure 2 shows qualitatively the temperature distribution for the convection cell of Figure 1. This temperature distribution will be used to illustrate precipitation/dissolution along nonisothermal flow paths.

The character of the temperature gradient in the cell can be approximated by the derivative behavior of this temperature field. In Figure 2, we sketch dT/dS for this function and note that the maximum and minimum occur at the inflection points of the function. These points correspond to the loci of maximum precipitation and dissolution (depending on the sign of α_T) and show that these processes are largely confined to the regions in the reservoir where the temperature field changes rapidly.

The relationship between the temperature gradient and mass transfer is one of the principal features of the diagenetic convective model and it is worthwhile exploring briefly two limiting cases: (1) uniform deposition/dissolution, and (2) point deposition/dissolution. Uniform deposition/dissolution occurs when the mass of material delivered to a unit volume of rock or removed from a unit volume of rock is the same at all points along the streamline. In this case, there is no heterogeneity in either accumulation of cement

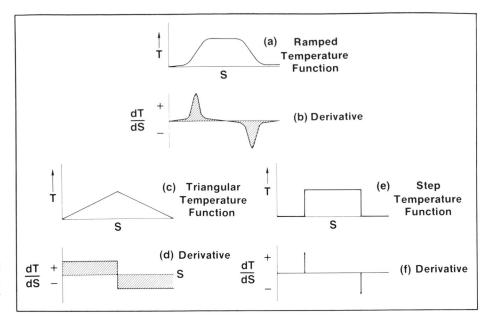

Figure 2—Schematic diagram of temperature functions plotted along streamline. Hatched areas are proportional to mass precipitated or dissolved.

or rock dissolution. This is clearly the least favorable case for generation of mineral deposits, but (probably) leads to minimal pore closure and/or obstruction and would be a favorable condition for hydrocarbon transport and accumulation.

In terms of the convective model, this case requires linear temperature gradients (Fig. 2b), that is, $dT/dS = \pm$ constant. Under these conditions, the change in porosity is uniform. ΔM mass units are removed from all points in the source zone and ΔM units are deposited in the sink zone. Since the total mass removed from the source equals the total mass deposited in the sink, the area under the derivative curves must be equal.

For point deposition/dissolution the temperature field along an isotherm must be a step function (Fig. 2d) such that $dT/dS = \pm \infty$ at the inflection points. This leads to an infinite source or sink at a point. Again, mass balance requires that the total and incremental mass changes at the source and sink be identical.

In nature, neither the linear or step temperature function is realized, but these temperature distributions do represent limiting behavior, and correspond to extremes in the distribution of diagenetic assemblages. The ramped step function represents an intermediate case between these two and shows how the system changes as the tempera-

ture gradients are either increased or decreased.

These arguments are also intended to show that the change in porosity with time does not depend on the concentration of SiO_2 in the pore fluid but rather on the temperature coefficient for the solubility of a phase at a particular temperature and the temperature derivatives along the streamline. Many arguments regarding cementation and gas and oil transport are based on the solubilities of the phases involved (Jones, 1981; Price, 1981a). However, equation 5 shows that only derivative quantities are involved and, for small changes in T, α_T can be assumed to be essentially constant. The principal mechanism for concentrating hydrocarbon and mineral accumulations in convecting systems is a change in the temperature field along fluid streamlines.

Time Scales

In addition to showing how the temperature gradients and solubility coefficients control the diagenesis, equation 5 is also useful for providing estimates of the time required to achieve certain levels of pore filling (or mass removal). Assuming constant values for all parameters, equation 5 can be integrated to provide a time-dependent expression for porosity,

$$\phi = \phi_o(1-t/\tau) \tag{6}$$

where

ϕ_o = original porosity

and τ is a time constant defined by

$$\tau = -[U/\phi_o(\rho_w/\rho_s)\alpha_T \ dT/dS]^{-1}.$$

For

U	$= 10^{-8}$ m-sec^{-1}
ρ_w	$= 1000$ Kg-m^{-3}
ρ_s	$= 2600$ Kg-m^{-3}
ϕ_o	$= 0.25$
α_T	$= 5 \times 10^{-6}$ K^{-1}
dT/dS	$= 0.025$ K-m^{-1}

we find that $\tau = 5$ million years.

Equation 6 is a good approximation during the early stages of pore filling, but the permeability and hence fluid velocity can be expected to change as porosity decreases. The dependence of permeability on porosity is difficult to predict. However, if we assume proportional changes in porosity, permeability, and fluid velocity such that

$$U/U_o = K/K_o = \phi/\phi_o \tag{7}$$

where the subscripted variables refer to the original values, and substitute (7)

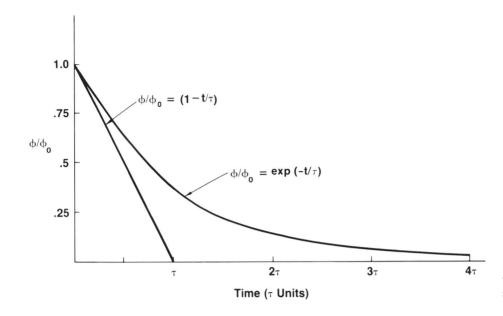

Figure 3—Porosity reduction as a function of time in thermally convecting system.

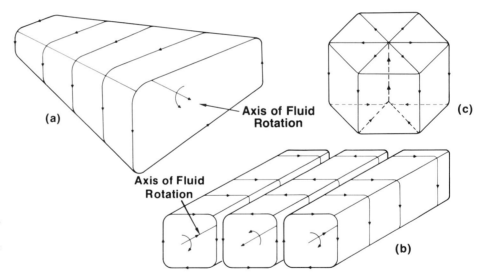

Figure 4—Types of convection cells: (**a**) sheet roll, (**b**) longitudinal rolls, (**c**) polyhedral cell. Solid arrows show direction of fluid flow on cell surface. Dashed arrows in (**c**) indicate paths of fluid ascent.

into (2) we have

$$\frac{d\phi}{dt} = U_o(\phi/\phi_o)(\rho_w/\rho_s)\ \alpha_T\frac{dT}{dS} \qquad (8)$$

$$= -\phi/\tau \ .$$

Integration then yields

$$\phi = \phi_o\exp(-t/\tau) \ . \qquad (9)$$

These two functions are plotted in Figure 3 in terms of τ (time) units. The time required for a porosity reduction of 50% of the original porosity is 0.5τ for the linear model and about 0.69τ

for the exponential model. However, the differences rapidly become larger. When the porosity reaches zero on the linear model, it is still 37% of ϕ_o on the exponential model and increasingly longer times are required for further porosity reduction.

Spatial Distribution and Shape of Convection Cells

One aspect of the convection model relevant to exploration strategy is the size, shape, and distribution of the convection cells in the porous body. The general shape of the convection cells is

fairly well known for some system geometries and conditions (Davis, 1967; Combarnous and Bories, 1975; Turner, 1973). Basically, the cells can be divided into three classes: sheet rolls (Fig. 4a), longitudinal or countra-rotating rolls (Fig. 4b), and polyhedral cells (Fig. 4c). For the sheet rolls, a hot current of fluid flows up the bottom and cooler fluid flows along the top. Longitudinal rolls can be visualized as a sheet roll which has broken into two-dimensional cells parallel to the axis of fluid rotation. The polyhedral cells are generally 5 6 sided and circulate fluid

$$f_q = \frac{(1-\phi_o)f_q^o + \phi_o(t/\tau_q)}{(1-\phi_o) + \phi_o(t/\tau)} \qquad (18)$$

where (6) has been used to substitute for ϕ. A similar equation is obtained for calcite

$$f_c = \frac{(1-\phi_o)f_c^o + \phi_o(t/\tau_c)}{(1-\phi_o) + \phi_o(t/\tau)} \qquad . \qquad (19)$$

The volume ratio of the two phases is

$$\gamma_{c,q} = \frac{f_c}{f_q} = \frac{[(f_c^o(1-\phi_o) + t/\tau_c]}{[f_q^o(1-\phi_o) + \phi_o(t/\tau_q)]} \qquad (20)$$

The ratio of the two phases depends on the initial amount of the two phases in the rock and thus requires knowledge of the original distribution of quartz and calcite in the reservoir. This information is rarely, if ever, available and, thus, requires some assumptions. However, if we are concerned only with the ratio of the authigenic phases and not the ratio of the authigenic + initial phases, then setting $f_c^o = f_q^o = 0$ yields

$$\gamma_{c,q}' = f_c'/f_q' = \tau_q/\tau_c \qquad . \qquad (21)$$

The primes indicate that the quantities are relative to a "zero-level" defined as the total amount of a phase minus the initial mass. This ratio is a function of the individual time constants for each phase, which reduces to

$$\gamma_{c,q} = (\alpha_{T,c}/\rho_c)/(\alpha_{T,q}/\rho_q) \qquad . \qquad (22)$$

Thus, the ratio of the authigenic phases is a function of the α_T's and the densities and is essentially a constant. γ will generally be negative for quartz-calcite since $\alpha_{T,c}$ is negative and $\alpha_{T,q}$ is positive under common reservoir conditions.

It is interesting, and possibly quite significant geologically, that the magnitude of $\alpha_{T,q}$ for quartz is very close to that of $\alpha_{T,c}$ for calcite in the temperature range 50–150°C. Since quartz and calcite densities are also approximately equal (2.65 for quartz, 2.71 for calcite), it would not be surprising for $\gamma_{c,q}$ to be close to −1 in a geologic environment, implying that calcite will replace quartz in the hotter regions of the convection cell without any change in porosity. Quartz will replace calcite, with no net porosity change, in the cooler parts of the cell.

The development of secondary por-

osity in a case such as this depends on the relative values of the temperature coefficients for the solubilities of the two phases. For $\alpha_{T,q} = -\alpha_{T,c}$ there would be no net porosity reduction for all points along the streamline where this condition is met. For $\alpha_{T,q} > -\alpha_{T,c}$ more quartz will be removed in the hotter regions than calcite will be precipitated. However, more quartz will be deposited in the cooler regions than calcite is dissolved. This will result in a net porosity increase at depth, but a porosity decrease at shallower levels. The reverse will occur for $\alpha_{T,q} < -\alpha_{T,c}$.

However, the solubility of calcite depends on more parameters than temperature alone. The solution pH exerts a strong effect that can act to reverse the retrograde solubility behavior of calcite. The presence of dissolved salts, particularly Ca-salts, will also influence the precipitation/dissolution behavior. Thus, no general rule can be given regarding the porosity relations. However, for any particular case, the porosity change resulting from chemical reaction can be deduced from equations like 10–22.

Feldspar Dissolution

As another example, we can look qualitatively at the case of an inhomogeneous initial distribution of phases, such as the feldspars in a quartz-rich reservoir (see Boles, 1982). For brevity, we can consider a mixture of potassium feldspar and plagioclase grains dispersed along a certain horizon in a quartz-rich rock. Assume that this feldspar layer occurs in the lower, hotter section of the convection cell (Fig. 6) and is itself a homogeneous mixture of feldspar and quartz. The fluid can saturate with the K-feldspar, but since the Ca component of the plagioclase has no stability field under normal reservoir conditions, this phase must continuously dissolve, adding material to the fluid at some steady rate. The fluid concentrations will increase until a state of saturation with incongruent product phases is reached (Helgeson, 1979), after which material will be removed at some steady rate. For the sake of argument, we can assume these product phases to be, say, albite and some Ca-bearing clay. The identity of these phases is important, of course, but the arguments regarding their

transfer and distribution in the convecting system will be largely the same regardless of their actual composition.

As a fluid parcel starting at point A (Fig. 6) moves down the descending limb of the convection cell, it enters the feldspar horizon (point B) and increases in temperature. Since solid feldspar is present and feldspar solubilities increase with increasing temperature, the fluid concentrations increase slightly until the fluid leaves the feldspar horizon (point C). Between points C and D, the fluid is no longer in contact with solid feldspar, but no precipitation occurs until the fluid parcel cools to the temperature at point D. Once that temperature is reached, precipitation will occur and will continue until the fluid parcel begins to heat again. Maximum precipitation will occur where the temperature gradient is steepest (F–G), producing, in effect, a transfer of quartz and K-feldspar from the source region accompanied by albite and a Ca-bearing clay which results from the presence of plagioclase in the source region.

As before, we can define volume fractions for the authigenic phases similar to (10a, b). Mass conservation equations similar to (12) and (13) can also be written and summed as in (14). The result will be a porosity–time relation of the form

$$d(1-\phi)/dt = \Sigma 1/\tau_i \quad i = 1 \text{ to } p \qquad (23)$$

where p is the number of authigenic phases. A composite time constant, τ, can also be defined as

$$1/\tau = \Sigma 1/\tau_i \qquad . \qquad (24)$$

An integrated expression for the porosity variation can be obtained which is identical to equation (6) and the individual volume fractions can be expressed as

$$f_i = \frac{(1-\phi_o)f_i^o + \phi_o(t/\tau_i)}{(1-\phi_o) + \phi_o(t/\tau)} \qquad . \qquad (25)$$

In this case, where the authigenic phases (quartz, albite, orthoclase, Ca-clay) all have positive temperature coefficients, the composite time constant is less than any individual time constant. However, this assumes that none of the phases possess a retrograde solubility.

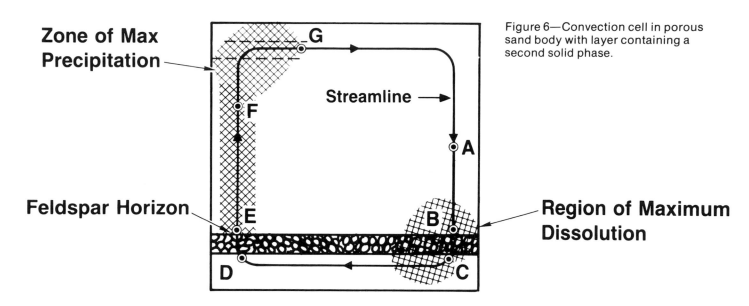

Zone of Max Precipitation

Feldspar Horizon

Streamline →

G

F

A

E

B

D

C

Region of Maximum Dissolution

Figure 6—Convection cell in porous sand body with layer containing a second solid phase.

If calcite is involved in addition to, or instead of, the Ca-clay, the composite time constant can increase. Other phases,* in addition to calcite, may show a retrograde solubility in a certain P–T–X space, so that no general rule can be given in terms of time to complete porosity reduction and number of authigenic phases.

This relatively simple picture has a number of consequences, among which is the suggestion that authigenic phases appearing in one section of a reservoir may be in response to minerals that are many hundreds of meters to kilometers removed from them. The local mineralogy may have no influence on the authigenic assemblage, and in particular, may not be supplying any mass at all to the precipitating phases. In a convecting system, the diagenetic patterns scale with the convection cells and attempts to achieve a local mass balance will be fruitless.

The actual phases precipitated may be complicated by the presence of other minerals in any real system, as well as by the original composition of the pore fluid. But, in general, the processes must transfer material from the hot zones to the cold zones, or vice versa,

and the regions of maximum dissolution and precipitation must correspond to extremes in the temperature gradients. As in the case of quartz and calcite considered earlier, no general rule can be given regarding porosity shifts resulting from chemical reaction.

Hydrocarbon Transport and Accumulation

It obviously does not matter in terms of the convective model if the pore-filling phase(s) are inorganic or organic. Provided that they are in true solution, hydrocarbons dissolved in a circulating pore fluid should accumulate by exsolution in zones where the fluid cools during its cycle. Equation 5 can be easily modified for the case of hydrocarbon transport by denoting the amount of oil accumulation in a unit volume by $\rho_o \phi_o S_o$ where ρ_o is the oil density and S_o is the fraction of pore space occupied by oil. For constant density,

$$\frac{d(\phi S_o)}{dt} = -U\left(\frac{\rho_w}{\rho_o}\right)\alpha_T \frac{dT}{dS} = \phi_o/\tau_{HC} \quad (26)$$

where α_T refers to the temperature coefficient for the dissolved hydrocarbon components. Published data (Price, 1981b) suggest that the temperature coefficients for the solubility of quartz and many hydrocarbon components are similar, so the primary differences in the rate of pore filling by hydrocarbons and quartz will be due to differences in the densities of the exsolved

phases. Since

$$\rho_q/\rho_{HC} \cong 2.5,$$

$$\tau_{HC} \cong \tau_q/2.5 \cong 2\times10^6 \text{ yr} \quad .$$

When hydrocarbons and quartz cement are being exsolved simultaneously, the rate at which the original pore space is filled can best be followed by defining a new variable, $\beta = (\phi_o-\phi) + \phi S_o$, which includes the contributions of both mechanisms. Combining (5) and (26), we see that

$$\frac{d\beta}{dt} = \frac{-d\phi}{dt} + \frac{d(\phi S_o)}{dt} =$$
$$\phi_o\left(\frac{1}{\tau_q} + \frac{1}{\tau_{HC}}\right) = \phi_o/\tau_\beta \quad .$$

The time scale for pore filling by both mechanisms is reduced from that for filling by quartz alone by a factor of

$$\frac{\tau_q}{\tau_\beta} = 1 + \rho_q/\rho_{HC} \cong 3.5 \quad .$$

Because deposition/dissolution in this model depends on the temperature coefficient for the solubility of a phase, all phases that have the same temperature dependence will be deposited, or will dissolve, in the same proportions regardless of the absolute solubilities. This means that the molecular composition of exsolved hydrocarbons in a sink will not reflect the solubility of the source phase as much as it reflects the temperature dependence of the compo-

---

*All carbonates and most (all) sulfates exhibit retrograde solubility behavior in at least some region of T–P space (Helgeson, 1969). Most of these correspond to reservoir conditions. But pressure and the presence of dissolved salts have a marked effect (Holland and Malinin, 1979).

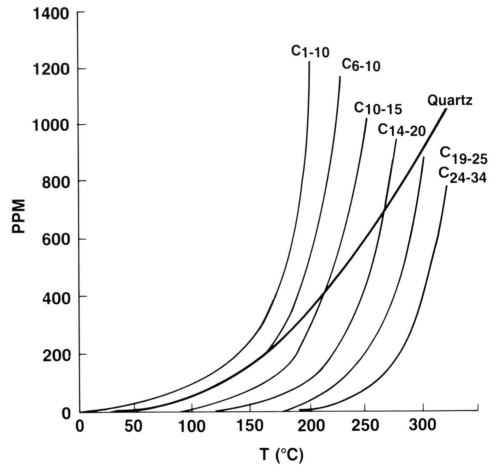

Figure 7—Solubility of quartz and 6 distillation fractions of petroleum as a function of temperature. Data from Price (1981b) and Morey et al (1962).

nents of that phase.

Price (1981b) has recently presented some data on hydrocarbon solubilities as a function of temperature and pressure. He shows that the solubilities of the C_1–C_{34} components from 50 to 400°C plot as a family of roughly parallel straight lines (on a semi-log plot), implying similar temperature dependence for all of the components considered. These data have been replotted (Fig. 7), together with some data for quartz solubility (Morey et al, 1962), on a linear scale. This plot shows that the α_T values are a function of the hydrocarbon fractions. But, in the temperature range 60–150°C, which is generally thought to be the temperature range for peak hydrocarbon generation (Tissot and Welte, 1978), the slopes of all hydrocarbon fractions are similar, particularly for the lighter components. This implies that the α_T values are also similar and suggests that an exsolved hydrocarbon acccumulation would have a composition reasonably similar

to the parent phase.

It is also apparent that the temperature dependence of hydrocarbon solubilities is remarkably similar to that for quartz in this temperature range. This observation justifies the approximation made above, at least for the temperature range of 60–150°C. In this temperature window, which coincides with peak kerogen maturation, a convecting fluid would be expected to exsolve hydrocarbons at the same time and in the same place that it is precipitating quartz.

Unfortunately, this model will predict that the most intense clay mineral diagenesis will also occur in the hydrocarbon exsolution zones. This is an apparently inescapable consequence of the fact that nearly all common authigenic clays have prograde solubilities.* Thus, they will precipitate in the cooler

---

*However, clay solubilities are also strongly pH dependent. This could lead to effectively retrograde behavior in acid pore fluids.

regions of the convection cell along with the exsolving hydrocarbons.

A brief sequential summary of these processes would be: (1) a toroidal convection cell is established around an anticlinal dome, (2) authigenic quartz and clays precipitate in the vicinity of the crest, (3) nearby source rocks mature and provide hydrocarbons to the convecting fluid, (4) the hydrocarbons exsolve at the crest together with the quartz and clays, and (5) the hydrocarbon source exhausts itself and the system returns to inorganic diagenesis.

SUMMARY

The convective model says that a mineral moves from hot to cold regions if the temperature coefficient for its aqueous solubility is positive, and from cold to hot if it is negative. The picture can be complicated if pressure gradients also contribute to the mass flow (Bruton and Helgeson, 1982), but pressure effects are undoubtedly small unless the pressure gradient is large. Fluid leaking from an overpressured zone may be such a case, but for convecting systems, the pressure terms can be safely ignored.

The convective model is unique in its ability to describe diagenesis in reservoir rocks. The fact that the fluid flow is driven by normal geothermal gradients and is capable of moving large quantities of mass in relatively short times geologically makes convective processes very attractive candidates for large-scale diagenetic alteration. The model accommodates such phenomena as secondary porosity since convective mass transfer can lead naturally to the development of nonuniform porosity distributions. The model is also capable of describing hydrocarbon transfer and accumulation, provided molecular solution in an aqueous phase is involved.

Calculations based on a fluid convection model suggest that dissolved hydrocarbons can be transported long distances and exsolved to produce substantial accumulations in short geological times. While these calculations do not prove that any commercial deposits originated by a solution–exsolution process, the fact remains that convecting fluids have the capacity to transfer comparable quantities of organic and inorganic material in reasonable times. This leads to conclusions that are in substantial agreement with the hypothesis of "deep water discharge" that Roberts (1981) has suggested and supports ideas regarding molecular hydrocarbon transport (Bonham, 1980; Price, 1981a). In addition, some of the objections to a molecular solution mechanism for hydrocarbon transport disappear if convection (or forced fluid flow) is considered. In particular, the restriction to high temperatures (275°C or greater) as suggested by Price (1981a) is eliminated, as is the requirement for large volumes of water.

It would seem, in fact, that the problem(s) of accounting for large volumes of pore-filling cements is roughly parallel to accounting for the transport and accumulation of hydrocarbons. In both cases, a viable solution would seem to require the circulation of large quantities of aqueous pore fluids, which are ostensibly connate waters. Conceptually, it would be very satisfying to have a single mechanism that is capable of accommodating hydrocarbon, as well as inorganic phases.

SELECTED REFERENCES

Anderson, R. N., M. A. Hobart, and M. G. Langseth, 1979, Geothermal convection through oceanic crust and sediments in the Indian Ocean: Science, v. 204, p. 828–832.

Boles, J. R., 1982, Active albitization of plagioclase, Gulf Coast Tertiary: American Journal of Science, v. 282, p. 165–180.

Bonham, L. C., 1980, Migration of hydrocarbons in compacting basins, in Problems of petroleum migration: American Association of Petroleum Geologists Studies in Geology, v. 10, 69–88.

Bredehoeft, J. D., and I. S. Papadopulous, 1965, Rates of vertical groundwater movement estimated from the earth's thermal profile: Water Resources Research, v. 1, no. 2, p. 325–328.

Bruton, C. J., and H. C. Helgeson, in press, Calculation of chemical and thermodynamic consequences of differences between fluid and geostatic pressure in hydrothermal systems: American Journal of Science.

Cathles, L. M., 1977, An analysis of the cooling of intrusives by ground-water convection which includes boiling: Economic Geology, v. 72, p. 804–826.

Combarnous, M. H., and S. A. Bories, 1975, Hydrothermal convection in saturated porous media: Advances in Hydroscience, v. 10, p. 231–307.

Davis, S. H., 1967, Convection in a box: linear theory: Journal of Fluid Mechanics, v. 30, no. 3, p. 465–478.

Hayes, J. B., 1979, Sandstone diagenesis—the hole truth: Society of Economic Paleontologists and Mineralogists Special Publication 26, p. 127–139.

Helgeson, H. C., 1969, Thermodynamics of hydrothermal systems at elevated temperatures and pressures: American Journal of Science, p. 729–804.

———, 1979, Mass transfer among minerals and hydrothermal solutions, in H. L. Barnes, ed., Geochemistry of hydrothermal ore deposits, 2nd ed.: New York, Wiley-Interscience.

Holland, H. D., and S. D. Malinin, 1979, The solubility and occurrence of non-ore minerals, in H. L. Barnes, ed., Geochemistry of hydrothermal ore deposits, 2nd ed.: New York, Wiley-Interscience.

Jones, R. W., 1981, Some mass balance and geological constraints on migration mechanisms: Bulletin of the American Association of Petroleum Geologists, v. 65, no. 1, p. 103–122.

McKibbin, R., and M. J. O'Sullivan, 1981, Heat transfer in a layered porous medium heated from below: Journal of Fluid Mechanics, v. 111, p. 141–173.

Morey, G. W., R. O. Fournier, and J. J. Rowe, 1962, The solubility of quartz in water in the temperature interval from 25°C to 300°C: Geochimica et Cosmochimica Acta, v. 26, p. 1029–1043.

Norton, D., 1978, Source lines, source regions and pathlines in hydrothermal systems related to cooling plutons: Economic Geology, v. 73, p. 21–28.

Norton, D., and L. M. Cathles, 1979, Thermal aspects of ore deposition, in H. L. Barnes, ed., Geochemistry of hydrothermal ore deposits, 2nd ed.: New York, Wiley-Interscience.

Pittman, E. D., 1979, Porosity, diagenesis and productive quality of sandstone reservoirs: Society of Economic Paleontologists and Mineralogists Special Publication 26, p. 159–173.

Price, L. C., 1981a, Primary petroleum migration by molecular solution: consideration of new data: Journal of Petroleum Geology, v. 4, no. 1, p. 89–101.

———, 1981b, Aqueous solubility of crude oil to 400°C and 2000 bars pressure in the presence of gas: Journal of Petroleum Geology, v. 4, no. 2, p. 195–223.

Roberts, W. H., 1981, Some uses of temperature data in petroleum exploration, in B. Gottlieb, ed., Unconventional methods in exploration for petroleum and natural gas, symposium II: Dallas, Southern Methodist University Press.

Schmidt, V., and D. A. McDonald, 1979, The role of secondary porosity in the course of sandstone diagenesis: Society of Economic Paleontologists and Mineralogists Special Publication 26, p. 175–207.

Schubert, G., and J. M. Straus, 1979, Three-dimensional and multi-cellular steady and unsteady convection in fluid-saturated porous media at high Rayleigh numbers: Journal of Fluid Mechanics, v. 94, p. 25–38.

Straus, J. M., 1974, Large amplitude convection in porous media: Journal of Fluid Mechanics, v. 64, p. 51.

Straus, J. M., and G. Schubert, 1977, Thermal convection of water in a porous medium: effects of temperature- and pressure-dependent thermodynamic and transport properties. Journal of Geophysical Research, v. 82, no. 2, p. 325–333.

———, 1978, On the existence of three-dimensional convection in a rectangular box containing fluid-saturated porous material: Journal of Fluid Mechanics, v. 87, p. 385–394.

———, 1979, Three-dimensional convection in a cubic box of fluid-saturated porous material: Journal of Fluid

Mechanics, v. 91, p. 155–165.

Tissot, B., and D. H. Welte, 1978, Petroleum formation and occurrence: New York, Springer-Verlag, 530 p.

Turner, J. S., 1973, Bouyancy effects in fluids: Cambridge Monographs on Mechanics and Applied Mathematics. New York, Cambridge University Press.

Wood, J. R., and T. A. Hewett, 1982, Fluid convection and mass transfer in porous sandstones—a theoretical approach: Geochimica et Cosmochimica Acta, v. 46, no. 10, p. 1707, 1713.

Interpretation of Methanic Diagenesis in Ancient Sediments by Analogy with Processes in Modern Diagenetic Environments

Donald L. Gautier
U.S. Geological Survey
Denver, Colorado

George E. Claypool
U.S. Geological Survey
Denver, Colorado

INTRODUCTION

For a number of years, investigators have known that the mineralogy and isotopic composition of authigenic carbonates and sulfides can, in principle, be used to reconstruct the early diagenetic environment of marine mudstones (for example, Murata et al, 1969; Curtis et al, 1972; Raiswell 1976; Irwin et al, 1977; Hudson, 1978; Curtis, 1978; Coleman and Raiswell, 1981). In practice, however, ancient mudstones are far removed in time from their early diagenetic environment, and reliable quantification and predictive capability for interpretation of even relatively simple diagenetic systems in ancient sediments remain elusive. Nevertheless, a great deal is now known about processes in modern methanic marine sediments and much of this information can be used in interpreting diagenesis from authigenic species in ancient methanic sediments.

Early diagenesis in modern organic carbon-rich marine muds is approximated by a depth zonation of processes, driven by microbial oxidation of organic matter. In descending order, the three principal zones of early diagenesis (aerobic respiration, sulfate reduction, and carbonate reduction/ methane generation) reflect the ecological succession of progressively less efficient modes of metabolism within the water and sediment column (Claypool and Kaplan, 1974). The transition from one metabolic zone to the next is the ecological response to biogeochemical changes in interstitial waters that are caused by the microbial population. Thus, sulfate-reducing bacteria replace aerobes once free oxygen has been depleted from pore waters. Likewise,

ABSTRACT. Methanic diagenesis commonly dominates pore-water chemistry in organic carbon-rich sediments from depths of tens of centimeters to 1000 m or more, and is coincident with the depths of principal sediment dewatering. As a result, methanic diagenesis in organic carbon-rich mudstones may control early diagenesis in adjacent sand and sandstone and can result in the accumulation of economic quantities of biogenic methane. Chemical aspects of methanic diagenesis in ancient marine sediments can be reconstructed by analogy with processes in modern diagenetic environments, on the basis of the mineralogy, texture, and isotopic composition of concretionary carbonate cements and other related authigenic minerals.

This sort of diagenetic reconstruction is well illustrated by means of examples from the Upper Cretaceous Gammon Shale from southeastern Montana. Bioturbated mudstones of the Gammon accumulated in oxic, open marine waters, but dissolved oxygen was probably depleted from pore waters a few tens of centimeters beneath the sediment/water interface. Sulfate reduction took place beneath this depth and was most important in a zone of mixing at the base of bioturbation, where isotopically light ($\delta^{34}S \simeq -25$ o/oo) iron sulfides accumulated. Organic matter oxidized during sulfate reduction gave rise to isotopically light calcite ($\delta^{13}C \simeq -21$ o/oo) that formed discrete concretions and that formed the interior portions of zoned calcite-siderite concretions. Sulfate was exhausted at depths of about 5–10 m, and CO_2 reduction (methanogenesis) became the dominant form of anaerobic respiration. Carbonate precipitation accelerated as pH increased because of CO_2 removal, while continued anaerobic oxidation of organic matter maintained bicarbonate activity at high levels. In the absence of dissolved sulfide, increased Fe^{+2} activity favored siderite over calcite as the principal authigenic carbonate. During the early stages of methanogenesis, kinetic fractionation caused $\delta^{13}C$ of CH_4 to change from −90 to −70 per mil and $\delta^{13}C$ of bicarbonate to change from −22 to approximately zero per mil over a depth interval of a few meters in the sediment column.

Interpretation of methanic diagenesis in the Gammon Shale illustrates only part of a single diagenetic pathway for one type of organic carbon-rich mudrock. And yet, the implications are clear: Early diagenesis of muds is dominated by processes involving organic matter and by the products of organic matter decomposition. Because of the economic significance of organic carbon-rich mudrocks as source beds for hydrocarbons and because their diagenesis probably controls mineral precipitation and dissolution in many reservoir rocks, it is of the utmost importance that diagenesis in ancient mudstones be understood.

when sulfur species have been removed from solution by sulfate reducers, methanogens become the dominant life-form in the sediment. In contrast to both aerobic oxidation and sulfate reduction, the methane-generating ecosystem (in the presence of excess organic matter) is not limited by availability of electron acceptors, either initially present or available as products of downward diffusion from overlying ecologic regimes. Rather, methanogens apparently utilize two principal sources of dissolved CO_2. The first is CO_2 that is already present when methane generation begins. This CO_2 is derived mainly from organic carbon that was oxidized during sulfate reduction. The second source of CO_2 is that added by fermentation reactions within the methanic zone itself. In addition, some evidence exists that methanogens can also use CO_2 released during early thermal alteration of organic matter (Claypool and Kaplan, 1974; Rice and Claypool, 1981). Methane generation persists within an accumulating sediment column until precluded by temperature (75°C) and/or by depletion of metabolizable organic matter (Rice and Claypool, 1981). Methanic sediments, then, are defined (Berner, 1981) as those in which the principal products of early diagenesis (commonly, iron-bearing carbonate minerals) formed (or are forming) during microbial methane generation, subsequent to the depletion of O_2 and SO_4^{-2} from solution.

The depletion of pore-water sulfur and the onset of methane generation in an organic carbon-rich marine sediment generally occurs a few tens of centimeters (Goldhaber and Kaplan, 1980) to several tens of meters (Claypool and Threlkeld, 1983) beneath the sediment/interface and may persist to depths of a kilometer or more. Thus methanic diagenesis grades downward into the realm of early thermal alteration of organic matter and is generally coincident with the main episode of sediment dewatering. In the upper one kilometer of water-laden mud, the volume of water expelled may equal or exceed one-half of the entire volume of the sediment (for example, Rieke and Chilingarian, 1974). In sedimentary sequences that contain both organic carbon-rich muds and lesser amounts of sand, expelled water is initially transmitted upward. After burial to a few hundred meters, vertical permeability is reduced (Burst, 1976) and most expelled water moves laterally out of the subsiding sediments through the more permeable lithologies. It is axiomatic that these waters exert primary controls on sandstone diagenesis (Hayes, 1979), and that depth-related diagenetic zones in shales are directly linked to processes of lithification and cementation in sands and sandstones (Curtis, 1978). This interpreted linkage of sandstone and shale diagenesis is consistent with and supported by the observation of cements in sandstones that seem to have formed before the main stage of thermal decarboxylation of organic matter in adjacent mudstones (for example, Gautier, 1981a). These are referred to as "eogenetic carbonate" cements by Schmidt and McDonald (1979), and their dissolution is thought to play a significant role in the development of secondary porosity. Thus, early diagenesis in methanic marine mudstones probably controls, to a large degree, the water chemistry in many marine sands and sandstones during their early history and sets the stage for diagenesis during later thermal transformations of organic matter and clay minerals.

In addition to their effects on nearby sand bodies, the processes and products of methanic diagenesis are themselves economically significant. The principal product of methanic diagenesis, isotopically light methane ($\delta^{13}C = -50$ to -90 o/oo, PDB relative to the Chicago standard), is a valuable hydrocarbon resource, accounting for more than 20% of the world's discovered reserves of natural gas (Rice and Claypool, 1981) and perhaps a higher percentage of undiscovered gas resources. Further, the potential of organic carbon-rich mudstones to be source beds for biogenic methane and hydrocarbons generated during increasing thermal maturation depends, in large measure, on the degree of preservation of organic matter during early oxic diagenesis (for example, Tourtelot, 1979; Coleman et al, 1979).

Processes of early diagenesis in organic-rich terrigenous marine sediments have been studied in some detail and are comparatively well understood in modern diagenetic environments. Sulfate reduction, in particular, has been investigated from a variety of observational, experimental, and theoretical perspectives (for example, Goldhaber and Kaplan, 1974, 1980; Berner, 1981; as well as may others and references therein). Similarly, carbonate reduction/methane generation has also been investigated from a number of standpoints, but with particular success by means of analysis of gases and pore waters squeezed from sediments collected during coring of the Deep Sea Drilling Project (DSDP) (Claypool and Kaplan, 1974; Claypool and Threlkeld, 1983; and references therein). Authigenic carbonates from deep-sea sediments have been examined by a number of workers (for example, Hein et al, 1979; Kelts and McKenzie, 1980, 1982; Matsumoto and Iijima, 1980; Pisciotto and Mahoney, 1981; Wada et al, 1982). However, interpretation and quantitative evaluation of authigenic carbonates from the DSDP cores with respect to analyses of pore waters and sediment gases remain fertile fields for future research.

CONCRETIONS: RECORDS OF EARLY DIAGENESIS

Since the 1920s, it has been recognized that some concretions form during the early burial history of fine-grained sediment (Tarr, 1921; Tomkeieff, 1927). These concretions provide at least three types of information useful in interpreting early diagenesis: (1) sediment textures, (2) sediment porosity, and (3) geochemistry and mineralogy.

Textural Information

The first students of concretions recognized that concretions in mudstones display textural relationships useful in timing the growth of concretions (Tarr, 1921; Tomkeieff, 1927; Weeks, 1957; Lippmann, 1955; Pantin, 1958; Woodland, 1964; Hallam, 1969; Raiswell, 1971; Oertel and Curtis, 1972). Most of these studies were concerned with determining the degree of sediment compaction during concretion growth by tracing continuous but deformed bedding laminae through concretions into compacted adjacent sediment (Tomkeieff, 1927; Raiswell, 1971); by

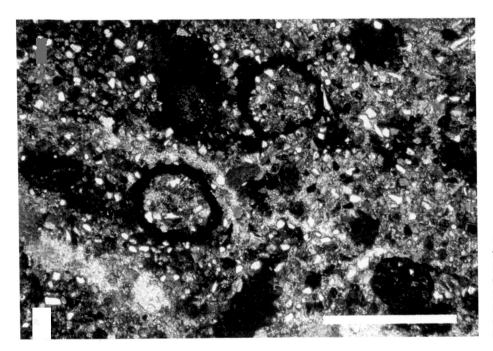

Figure 1—Clay-lined burrows (ring-like forms) and claystone clasts in intensely burrowed clay-rich siltstone of the Gammon Shale from Porcupine Dome, Montana. These delicate features are preserved in a septarian, siderite concretion formed during early diagenesis. Scale bar is 1 mm long.

examining the degree of compaction of delicate fossils preserved in the concretions (for example, Weeks, 1957); or by quantifying radial variation in the degree of preferred orientation of clay minerals and, therefore, sediment compaction from concretion center to outside edge (Oertel and Curtis, 1972).

Concretions generally consist of a microcrystalline carbonate mosaic, the crystal size of which commonly reflects the prevailing grain size of the original sediment. Concretions, therefore, commonly preserve the most delicate sedimentologic features, such as burrow structures (Fig. 1) or fecal pellets (Fig. 2). When such excellent preservation of original sediment textures is observed in conjunction with bedding planes that continue through the concretion and into the adjacent sediment and that curve around the concretions, growth is indicated to have occurred prior to at least some sediment compaction and before maximum burial. The fine-scale preservation of sediment textures is evidence that the concretions formed mainly as pore-filling cement, subsequent to sediment deposition but prior to extensive dewatering. Preservation of uncompacted sedimentary structures such as burrows or fecal pellets that display evidence of compaction in adjacent sediment is a further indicator of early formation. Similarly,

the presence of septaria is an excellent criterion of early concretion growth, because septaria apparently only develop in concretions that began forming in virtually uncompacted, water-laden sediment (see arguments by Lippmann, 1955, and Raiswell, 1971).

Conversely, petrographic evidence of extensive disruption of original sediment textures by concretionary carbonate should be taken as a warning that analytical data may not reflect early diagenetic conditions. Destruction of original sediment textures may be observed within concretions that otherwise seem to have formed very early in the burial history. Figure 3 illustrates the partial obliteration of sediment textures in a limestone concretion from the Upper Cretaceous Carlile Shale in a core from Bowdoin Dome, north-central Montana. The process of textural destruction shown in Figure 3 appears to be recrystallization of the original microcrystalline, carbonate pore-filling cement to form coarser crystals in a manner analogous to aggrading neomorphism of micritic carbonate as observed and described by carbonate petrologists (Folk, 1965; Bathurst, 1976). The timing of neomorphism is not known, although recrystallization has been observed in concretions of recent sediments (K.

Pye, written communication, 1982). Wherever recrystallization has occurred, measurements that require preservation of original textures or composition for reliability, such as volume percent carbonate data or stable isotope ratios, especially oxygen isotope ratios, must be considered suspect. For example, in the concretion illustrated in Figure 3, the relationship of the concretion to the adjacent shales and its septarian structure suggests early formation. However, the oxygen isotope values for the carbonate of the concretion suggest equilibration during later diagenesis (that is, $\delta^{18}O = -5$ o/oo, PDB rather than 0 to -2 o/oo).

Porosity Information

Concretions formed during early diagenesis are interpreted by most investigators to develop mainly through the precipitation of carbonate cements that occupy available porosity without significant displacement or replacement of the host sediment (Lippmann, 1955; Seibold, 1962; Raiswell, 1971; Oertel and Curtis, 1972; Raiswell, 1976; Gautier, 1982). Thus, the volume percent carbonate of a concretion closely approximates the porosity of the sediment during the time the concretion grew.

While these concretions are forming, the enclosing water-laden muds

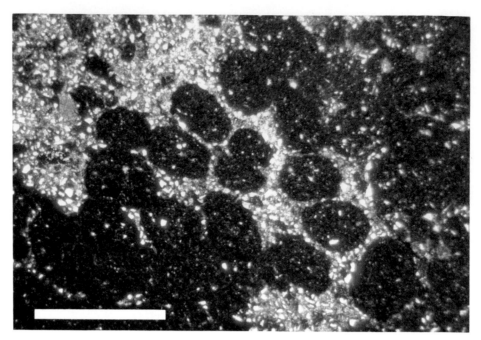

Figure 2—Fecal pellets in silty mudstone from the Upper Cretaceous Belle Fourche Shale in the subsurface of Bowdoin Dome, north-central Montana. The pellets are preserved and enclosed in an early diagenetic concretion of septarian limestone. Scale bar is 1 mm long.

undergo dramatic changes in porosity and water content. In concretions that formed during early diagenesis, the effects of compaction and dewatering are commonly manifested as septaria (Raiswell, 1971; Lippmann, 1955), as progressive center-to-outside radial enhancement of clay-mineral orientation (Oertel and Curtis, 1972) and by a decrease in volume percent carbonate (increasing acid-insoluble residue) from center to outside edge (for example, Raiswell, 1971; Oertel and Curtis, 1972; Gautier, 1982). These radial trends in volume percent carbonate and clay-mineral orientation are probably indicative of the degree to which the sediment was compacted and dewatered during the period of concretion growth. Indeed, if the volume percent carbonate values of the concretions correspond closely to true sediment porosities, then radial trends such as these provide a measure of the amount of water lost from the sediment during concretion growth.

Although most early diagenetic concretions display such center-to-edge trends in volume percent carbonate, variations within the concretions are small when compared with differences between concretions. Analysis of differences among concretions from various localities and within populations at the same localities may provide a

standard measure of the degree of sediment compaction and dewatering at the time of concretion growth (Gautier, 1982).

Geochemistry and Mineralogy

The mineralogical, elemental, and stable isotopic compositions of the authigenic minerals provide independent evidence of the composition of interstitial waters at the time of mineral precipitation. The carbon isotope composition of the carbonate phases in particular provides a sensitive indicator of biogeochemical processes (Galimov and Girin, 1967), owing to the large-scale swings in carbon isotopic composition of interstitial waters during early diagenesis.

The equilibrium isotopic composition of oxygen in carbonate minerals is determined mainly by the temperature and by the isotopic composition of the water from which the minerals precipitate (Urey, 1947; McCrea, 1950; Epstein et al, 1953). In principle, calcite can be used as a paleothermometer if the composition of the waters from which it precipitated is known. However, the isotopic composition of ancient marine waters is not generally known (Tourtelot and Rye, 1969), nor are all the mechanisms of oxygen isotope fractionation during diagenesis (Hudson and Friedman, 1976).

Further, some of the minerals of most interest in diagenetic reconstructions, siderite, for example, have uncertain fractionation factors (α) with respect to water and phosphoric acid (Gautier, 1982). For these reasons, oxygen isotopes are probably best used as a relative scale to measure the degree to which interstitial pore waters have departed from ideal sea-water composition.

Generally, authigenic carbonates in marine sediments become progressively depleted in $\delta^{18}O$ in direct proportion to the depth and time after burial at which precipitation occurs (Kelts and McKenzie, 1982). This trend results mainly from increasing temperatures in the sediment column. Accordingly a relationship is to be expected between $\delta^{18}O$ and volume percent carbonate (sediment paleoporosity) in early diagenetic concretions. This relationship was recently investigated in a number of siderite samples from methanic sediments of the Cretaceous in the Western Interior (Gautier, 1982). The expected relationship was strongly supported by a large and significant positive correlation coefficient (0.89, n = 51) between volume percent carbonate and $\delta^{18}O$ of the carbonates.

The various types of textural, mineralogical, and isotopic evidence provide information by which the relative depth

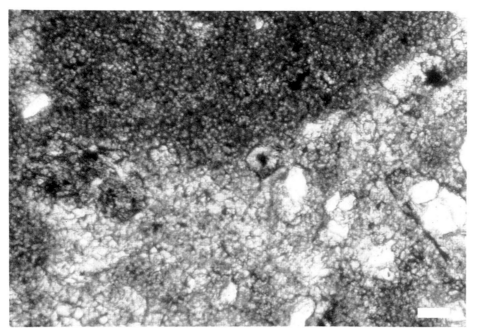

Figure 3—Example of partial destruction (or incomplete preservation) of sediment textures resulting from carbonate crystal growth. Process of textural destruction resulting from crystal growth seems analogous to processes of aggrading neomorphism of micritic carbonate in limestones. Darker area is a clay-filled burrow within lighter-colored, silty mudrock. Limestone concretion, Upper Cretaceous Carlile Shale, in core from Bowdoin Dome, north-central Montana. Scale bar is 0.1 mm long.

of concretion growth can be estimated for concretionary carbonates formed during early diagenesis in marine sediments (for discussion see Gautier, 1982). These relationships are illustrated in Figure 4. Concretions that form at shallow depths in the sediment column, shortly after sediment deposition (upper left in Fig. 4), preserve uncompacted sediment textures (Figs. 1, 2, 4), contain a large volume percent carbonate (75–80%), and have oxygen isotope ratios similar to those of carbonates precipitated in equilibrium with sea water ($\delta^{18}O = 0$ to -2 o/oo, PDB). Concretions that formed later in the burial history and at greater depths in the sediment column (toward lower right, in Fig. 4) have lower volume percent carbonate values and more negative $\delta^{18}O$ values, and display textural evidence of clay-mineral alignment and squashing of sedimentary structures.

EXAMPLE FROM THE GAMMON SHALE

The Upper Cretaceous (Lower Campanian) Gammon Shale of the northern Great Plains of the United States provides a fine example to illustrate the interpretation of early diagenesis in an ancient methanic marine mudrock by analogy with processes in modern diagenetic environments. The Gammon is a sequence of mudstone that accumulated offshore during a major regression of the epeiric sea in the Western Interior (Gill and Cobban, 1973; Gautier, 1981b). Within most of eastern Montana, North Dakota and South Dakota, and northeastern Wyoming, time–temperature conditions have never been intense enough for oil or thermal gas generation in the Gammon Shale, or even for extensive decarboxylation of organic matter (Gautier, 1981b). Carbon dioxide reduction and associated methane generation is the last diagenetic event for which significant evidence exists.

The products of the zone of carbon dioxide reduction and methane generation (methanic zone) characterize the Gammon Shale. Gammon outcrops were originally identified and characterized by the conspicuous oxidation products of siderite weathering (Rubey, 1931), and cores of the Gammon contain abundant siderite in nodules and disseminated crystals (Gautier, 1981b, 1982). Although dolomite is more familiar than siderite as a product of methanic diagenesis in organic carbon-rich marine sediments (for example, Murata et al, 1969; Pisciotto and Mahoney, 1981; Kelts and McKenzie, 1982), siderite is common and abundant in many marine mudstones, including the Cretaceous System in the Western Interior and many rapidly accumulated methanic sediments cored by Deep Sea Drilling Project. Occurrence of siderite probably reflects availability of reactive iron during early diagenesis. The Gammon and its stratigraphic equivalents, the Eagle Sandstone of central Montana (Gautier, 1981a) and the Alderson Member of the Lea Park Formation of southeastern Alberta in Canada, an equivalent of the Milk River Formation of south-central Alberta (Meijer-Drees and Mhyr, 1981), contain economic accumulations of natural gas that are interpreted to be biogenic on the basis of molecular composition and carbon isotope ratio (Rice and Shurr, 1980; Rice and Claypool, 1981). These accumulations of biogenic gas were apparently generated within the Gammon Shale itself during early diagenesis and have been entrapped in the fine-grained sediments ever since (Gautier, 1981b; Rice and Claypool, 1981).

Generation of biogenic methane requires interstitial waters that contain abundant dissolved CO_2 and that are free of dissolved sulfate and molecular oxygen (Claypool and Kaplan, 1974). Siderite precipitation requires similar chemical conditions, being restricted to waters of high bicarbonate activity and extremely low concentrations of molecular oxygen and dissolved sulfide

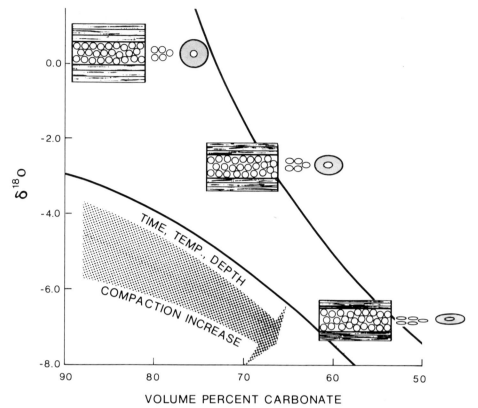

Figure 4—Relationship of $\delta^{18}O$ and volume percent carbonate data for concretions formed during the initial subsidence of marine mudstones. Solid lines bound range of concretion composition known to us. Time, temperature, depth, and degree of sediment compaction at the onset of concretion growth increase to the lower right of the diagram.

(Garrels and Christ, 1965; Curtis, 1967). The siderite concretions in the Gammon formed, for the most part, early in the burial history (Gautier, 1982), as did the biogenic methane (Gautier, 1981b; Rice and Claypool, 1981). The Gammon Shale thus contains two of the principal products of early methanic diagenesis and has been only slightly affected by later diagenetic events. The following sections will provide an interpretation of the diagenetic history of the Gammon Shale from the sediment/water interface to the zone of methanic diagenesis.

CONCRETIONS FROM THE GAMMON SHALE AT OWL CREEK

Diagenesis of the Gammon Shale is particularly well illustrated by means of concentrically zoned calcite/siderite concretions such as those collected from outcrops in the vicinity of Owl Creek along the northern flank of the Black Hills in southeastern Montana and northwestern South Dakota. Sample localities and mineralogic and isotopic data were previously tabulated for some Owl Creek and lower Owl Creek samples including those discussed herein (Gautier, 1982). At Owl Creek the Gammon is thick (>300 m) and sedimentation was relatively rapid, probably exceeding 100 m per m.y.; siderite concretions in the vicinity formed at shallow depths in the sediment column (Gautier, 1982). Typical zoned calcite/siderite concretions from this locality are illustrated in Figure 5, and an idealized zoned concretion drawn by generalization from analyses of several such concretions is illustrated in Figure 6. Stable carbon isotope and volume percent carbonate data as measured and averaged on the actual concretions are generalized on the idealized concretion.

The paragenetic sequence of mineral formation is obvious in zoned concretions: The interior calcitic nodules formed prior to the surrounding siderite-cemented sediment. Both the calcitic and the sideritic portions of the concretions contain disseminated framboids and crystals of pyrite (Fig. 7). Original sediment textures preserved in the concretion and in adjacent rocks indicate that the Gammon Shale at this locality was mottled owing to intense bioturbation prior to concretion formation.

In both core materials and insoluble residues of siderite and calcite, sulfide sulfur is present in amounts of 0.5 to 1.5%, thus exceeding that which could have been available by simply burying the sea-water sulfate originally contained in pore waters at the time of deposition ($\simeq$ 28 mmoles/liter) (Gautier, 1982). Most of this sulfide sulfur occurs as framboids that are depleted in the heavy isotope ^{34}S ($\delta^{34}S = -8$ to -32 o/oo). Petrographic evidence indicates that these pyrite framboids are the earliest formed authigenic species so far identified in the Gammon Shale (Fig. 7). Pyrite with these characteristics and paragenetic placement is typical of iron sulfides observed forming in the upper part of the sulfate-reduction zone in recent diagenetic environments (for example, Goldhaber and Kaplan, 1974, 1980).

The calcitic nodules in the interiors of the zoned concretions are petrographically, mineralogically, and iso-

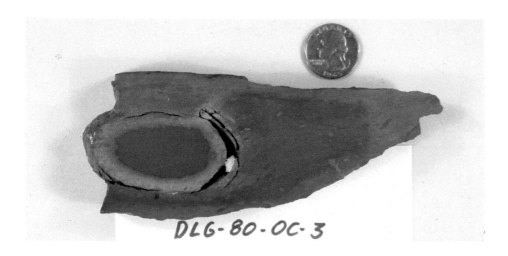

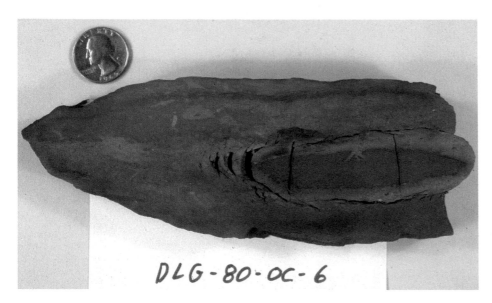

Figure 5a and b—Examples of zoned calcite (interior nodule) and siderite (main body) concretions from the Owl Creek localities, southwestern Montana and northwestern South Dakota. Note burrow mottling.

topically similar to discrete calcite concretions and baculite shell fillings that are found throughout the upper part of the Gammon Shale in the vicinity of Owl Creek and at other localities with relatively high sedimentation rates. This calcite preserves textures of included sediment in great detail and closely mimics the original grain texture. The calcite is nonferroan, makes up 80% or more of the concretionary nodules by volume and displays $\delta^{18}O$ values of approximately zero per mil, PDB, as would calcite precipitated in equilibrium with sea-water bicarbonate at reasonable surface temperatures.

These calcites are composed of carbon that is significantly depleted in the heavy isotope ^{13}C, and can be subdi-

vided into two groups on the basis of their carbon isotope composition (see Fig. 6). The first and largest group, including most of the nodules from the interiors of the zoned concretions, yields $\delta^{13}C$ values of around −20 per mil, suggesting that the carbon was derived directly from the oxidation of organic matter. The second group of calcites displays $\delta^{13}C$ values of −30 per mil or less.

Surrounding the interior calcitic nodules in the zoned calcite/siderite concretions is siderite. The siderite from the zoned concretions is otherwise similar mineralogically, compositionally, and texturally to the siderite concretions that are present throughout the Gammon Shale at Owl Creek and that

characterize the Gammon in the study area (Gautier, 1982). The paragenetic sequence is clear: The siderite formed subsequent to the calcite. The siderite concretions, like the calcitic nodules, consist of micritic carbonate that generally preserves original sediment fabric, and has a large ratio of volume percent carbonate to volume percent acid-insoluble residue (volume percent carbonate >80). The $\delta^{18}O$ of the siderite (0 to −2 o/oo) is similar to that of the calcitic nodules and thus also similar to carbonates precipitated in equilibrium with sea water.

Although $\delta^{18}O$, volume percent carbonate, sulfide sulfur content, and sediment textures preserved by the sideritic concretions are similar to the pre-

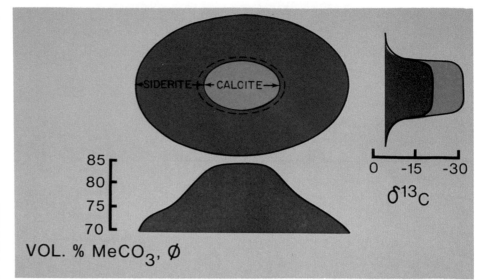

Figure 6—Cross-sectional view, looking parallel to bedding, of an idealized zoned calcite/siderite concretion showing trends of volume percent mineral carbonate and $\delta^{13}C$, relative to PDB, for calcite and siderite.

viously formed calcitic nodules, they differ significantly in their carbon isotope composition. The $^{13}C-^{12}C$ equilibrium fractionation between dissolved bicarbonate and siderite, and siderite and calcite is so small that it can be ignored in geochemical interpretations such as this. Consequently, variation in carbon isotope composition between the interior calcitic nodules and the exterior main body of siderite reflects true changes in the isotopic composition of dissolved bicarbonate in the interstitial pore waters during concretion growth. In contrast to the ^{13}C-depleted calcites, the siderites display increasingly less negative $\delta^{13}C$ values from interior to exterior (that is, through time), with most values being about −4 per mil. This implies that an isotopic shift of at least 16 per mil had occurred in aqueous bicarbonate during the time elapsed from the cessation of calcite precipitation until the last siderite was formed.

The carbonates of the Gammon Shale record dramatic shifts in the carbon isotope composition of interstitial waters within the upper 10 m or so of the sediment column. These shifts begin with open-ocean $\delta^{13}C$ values near zero per mil, indicated by analyses of biogenic aragonite in baculite shells, to −20, −30 per mil or less in early formed calcitic nodules, and approaching zero again in the later formed siderite concretions. These trends in conjunction with mineralogic and textural relations

form the basis of the interpretation that follows.

INTERPRETATION AND DISCUSSION

Figure 8 provides a general interpretation of the chemical and isotopic evolution of pore waters in organic carbon-rich terrigenous marine muds during early diagenesis under conditions of oxygenated bottom waters and relatively rapid sedimentation. Central to the reconstruction is the $\delta^{13}C$ profile on the right side of the diagram. Labeled arrows indicate chemical species undergoing diffusion and principal directions of diffusion imposed by the concentration gradients. Authigenic minerals and the corresponding carbon isotope compositions that were used in reconstructing early diagenesis in the Gammon Shale are shown near the middle of the diagram (diagenetic products). For the discussion on the following pages, refer to Figure 8.

Setting the Stage: The Geochemical/Diagenetic Baseline

The Gammon Shale has been intensively bioturbated by deposit-feeding macrofauna. The vigorous activity of these organisms is a certain indicator that bottom waters were oxygenated and it is likely that interstitial waters in the upper few centimeters of freshly deposited parts of the Gammon sediments were oxic as well. Oxygen iso-

topic analyses of shell material from marine mollusks, such as aragonite from shells of the ammonite *Baculites* sp. (smooth), suggest that sea water above the Gammon sediments was not only of normal marine salinity, but had a temperature similar to that of modern subtropical waters ($\delta^{18}O_{ARAG} = 0$ to -2 o/oo).

This isotopic, paleontological, and sedimentologic evidence is the basis for the first assumption required for the reconstruction of early diagenesis in an ancient sediment: that the waters in which the Gammon sediments accumulated were of normal marine composition. It is further assumed that the sediments and interstitial waters were well mixed to about the base of bioturbation, below which the sediments became anoxic. For the purposes of this discussion, the base of bioturbation is assigned a depth of 0.5 m below the sediment/water interface. Although this depth is selected somewhat arbitrarily, it is a reasonable one based on observations of bioturbation in recent sediments deposited below storm wave base from oxygenated bottom waters (for example, Howard, 1975). The starting value for $\delta^{13}C$ of pore-water bicarbonate is fixed at approximately zero per mil by isotopic analyses of aragonite from shells of marine mollusks collected from outcrops of the Gammon Shale at the Owl Creek sections and collected from cores of the Joseph J. C. Paine and Associates

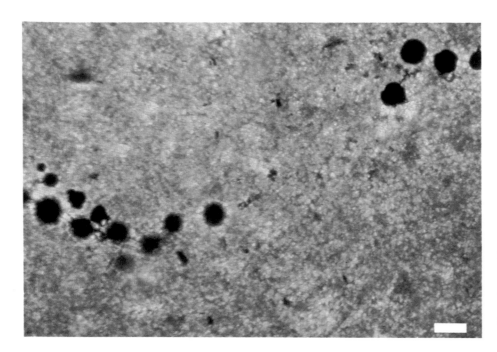

Figure 7—Pyrite framboids in micro-crystalline siderite. Crystals of siderite radiating from framboids indicate that pyrite formation preceded siderite precipitation. Siderite concretion from the Gammon Shale, Owl Creek locality, southwestern Montana. Scale bar is 0.1 mm long.

Aasen 1–9 well from Bowman County, North Dakota (sec. 9, T. 129 N., R. 106 W.). Volume percent carbonate values, $\delta^{18}O$ data, and textural evidence suggest that the entire diagenetic history represented by the sideritic concretions at the Owl Creek localities occurred in the upper part of the sediment column, almost certainly within a few tens of meters of the sediment/water interface, and probably within the upper ten meters of the sediment column (Gautier, 1982). Accordingly, a depth of -10 m is chosen as the lower boundary for interpretation. This choice is also somewhat arbitrary, but small adjustments in this depth would not seriously affect interpretations of early diagenesis.

Immediately beneath the zone of aerobic oxidation, sediment pore waters were anoxic and sulfate reduction was the dominant metabolic activity. The principal products of sulfate reduction are reduced sulfur, mostly as H_2S and HS^-, and bicarbonate derived from the oxidation of carbon in organic compounds (for example, Goldhaber and Kaplan, 1974). The Gammon Shale contains evidence of both of these products.

The abundance of sulfide sulfur and the mixed but generally ^{34}S-depleted isotopic composition ($\delta^{34}S = -8$ to -32 o/oo) of these early formed sulfides

does not permit an unambiguous interpretation. However, if the pyrite were simply derived from the complete (closed-system) reduction of buried sea-water sulfate, it would display a $\delta^{34}S$ value identical to the sulfate from which it was derived. During the early part of the Campanian Stage such sulfate sulfur had a $\delta^{34}S$ of about 18 per mil relative to the Canyon Diablo Troilite (CDT) (Claypool et al, 1979). Conversely, if the pyrite in the Gammon formed entirely in the zone of mixing, in communication with the overlying "infinite" reservoir of sea-water sulfate, the resulting iron sulfides would be depleted in ^{34}S by about 54 per mil relative to the sea-water sulfate (that is, 18 o/oo $-$ 54 o/oo $= -36$ o/oo) (Goldhaber and Kaplan, 1980). Thus the isotopic data suggest that much of the pyrite formed from sea-water sulfate that was added by mixing and diffusion to sea-water sulfate originally buried with the sediment. However, some of the pyrite probably formed beneath the zone of rapid interchange with sea water by processes that approximate closed-system sulfate reduction.

In recent diagenetic environments, H_2S concentration attains a maximum near the base of the zone of bioturbational sediment mixing; it is here where most iron sulfides precipitate (for example, Berner, 1980). In sediment

pore waters below the H_2S maximum, both dissolved sulfide and sulfate decline dramatically. In contrast with the sulfur species, bicarbonate concentration increases steadily with depth and attains a maximum near the base of the sulfate-reduction zone while the carbon isotope composition becomes continually more depleted in ^{13}C with depth, reaching its minimum ($\delta^{13}C = -22$ o/oo) at the base of the sulfate-reduction zone. This increasing concentration and isotopic shift results from the continual addition of carbon derived from the anaerobic oxidation of organic matter by sulfate-reducing bacteria. Because of the continuing (albeit slow) precipitation of iron sulfides within the entire sulfate-reduction zone, the concentration of iron in solution is buffered to very low values, similar to those of waters in equilibrium with pyrite (Garrels and Christ, 1965). Carbonates precipitating in this zone are free of iron.

The calcite cement that composes discrete concretions, baculite fillings, and interior nodules of zoned concretions in the Gammon Shale has all the characteristics of carbonates formed in the lower part of the sulfate-reduction zone. The principal evidence for this interpretation is:

(1) The calcite formed early in the burial history, probably within a few

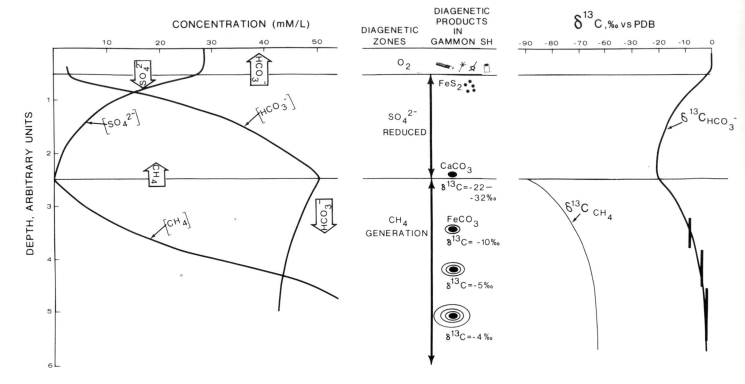

Figure 8—Principal trends (SO_4^{-2}, HCO_3^-, CH_4) of chemical and isotopic evolution of interstitial waters in terrigenous marine muds under conditions of oxygenated bottom waters, excess organic matter, and rapid sedimentation. Arrows denote major diffusion gradients. Center of diagram shows major diagenetic zones and examples of diagenetic products from zoned calcite/siderite concretions in the Gammon Shale at the Owl Creek localities of southwestern Montana and northwestern South Dakota. Although $\delta^{13}C$ trend on right of diagram shows a minimum value of −22 o/oo, where anaerobic oxidation of CH_4 is important, $\delta^{13}C_{HCO_3^-}$ may be as light as −35 o/oo at the base of the sulfate-reduction zone.

meters of the sediment/water interface, as indicated by its volume percent carbonate values, its preservation of sediment fabric, and by its $\delta^{18}O$ values, which are similar to those of carbonates precipitated in equilibrium with sea water of normal surface temperature.

(2) Calcite precipitation occurred subsequent to precipitation of most pyrite framboids, as indicated by petrographic observations of crystals of calcite within the nodules, radiating away from pyrite framboids.

(3) The calcite from the nodules, fillings, and concretions, is virtually free of iron, even though the Gammon sed-

iments subsequently produced iron carbonates (that is, siderite concretions).

(4) The isotopic composition of the early formed carbonates is extremely depleted in ^{13}C.

Whereas most of the calcite cements display $\delta^{13}C$ values similar to the organic matter from which they were derived (that is, −20 to −23 o/oo), some of these early calcites display values that are much more depleted in ^{13}C (that is, $\delta^{13}C < -30$ o/oo). These carbonates did not result from the direct oxidation of sedimentary organic matter, but must have been derived from a more ^{13}C-depleted carbon. The likely source of such carbon is biogenic methane ($\delta^{13}C = -50$ to −90 o/oo) that formed beneath the sulfate-reduction zone and subsequently moved upward. In modern diagenetic environments, especially under shallow (<100 m) water, a considerable fraction of the methane formed in the carbonate-reduction (methanic) zone diffuses and bubbles upward to be consumed by anaerobic oxidation in the sulfate-reduction zone (Barnes and Goldberg, 1976; Reeburgh and Heggie, 1977). Such consumption results in addition of extremely ^{13}C-depleted carbon to the pore waters and in an accompanying negative shift in the $\delta^{13}C$ of total dissolved carbon. Such negative carbon

isotope values are occasionally observed in the sulfate-reduction zone in modern diagenetic environments. For example, carbon isotope values of less (more negative) than −30 per mil were observed for total dissolved carbon in pore waters from DSDP Site 533, Leg 76, at depths below about 15 m near the base of the sulfate-reduction zone (Claypool and Threlkeld, 1983). In the Owl Creek concretions, this isotopically light, iron-free calcite in which pyrite framboids are contained, is strong evidence for the upward diffusion of biogenic methane and of methane consumption in the sulfate-reduction zone. This calcite is the basis for the upward-pointing CH_4 arrow on Figure 8.

At the termination of sulfate reduction in the Gammon sediments at the Owl Creek locality and other areas of relatively rapid sedimentation, pore waters were probably saturated with respect to calcite and had carbon isotopic composition of $\delta^{13}C = -20$ o/oo or less. For a wide range of typical marine environments, the depth of the base of the sulfate-reduction zone can be estimated from the sedimentation rate (Berner, 1980; Toth and Lerman, 1977). Such estimates suggest a depth of 30 m or less, although siderite concretions that formed in the underlying methanic zone seem to have formed in the upper ten meters or so of the sediment column (Gautier, 1982).

As in modern anoxic diagenetic environments, the activity of the methanogenic archaebacteria controlled pore-water chemistry below the sulfate-reduction zone. Methanogenesis preferentially removed isotopically light CO_2 to form CH_4 with a $\delta^{13}C$ of -80 to -90 per mil. Removal of ^{12}C-enriched CO_2 by methanogens caused the pH to rise slightly and caused the $\delta^{13}C$ of residual aqueous bicarbonate to shift dramatically to less negative values. By analogy with modern diagenetic environments, the isotopic composition of bicarbonate behaved as a reservoir undergoing Rayleigh distillation. This relationship has been described and illustrated in some detail by Claypool and Kaplan (1974) and by Claypool and Threlkeld (1983). With continuing methane production, both dissolved bicarbonate and cumulative methane became isotopically heavier. The concentration of dissolved iron increased because H_2S was no longer being generated by sulfate reduction, and siderite precipitated copiously in the zone of methanic diagenesis. The earliest siderite to precipitate, as indicated by analyses of the centers of pure siderite concretions and of the innermost layer of siderite in zoned concretions, had a $\delta^{13}C$ of about -10 per mil or heavier. However, the vast majority of siderite in the concretions display $\delta^{13}C$ values of -5 to -4 per mil. Inasmuch as the siderites apparently formed at shallow depths soon after deposition, these values suggest that pore-water $\delta^{13}C$ shifted toward less ^{13}C-depleted compositions rapidly after the onset of methane generation. Carbonate $\delta^{13}C$ apparently soon stabilized at values slightly depleted relative to seawater bicarbonate (-5 to -4 o/oo).

The asymptotic approach of carbonate $\delta^{13}C$ values to within ±5 per mil of zero is a commonly observed phenomenon (for example, sediments in the South Guaymas Basin [Goldhaber, 1974], and in DSDP sites 102, 147, 174A, and 180 [Claypool and Kaplan, 1974]). Although evidence exists that pore-water bicarbonate and associated carbonate minerals can attain much more ^{13}C-enriched values in the methanic zone (for example, Murata et al, 1969; Nissenbaum et al, 1972; Curtis et al, 1972; Pisciotto and Mahoney,

1981), such isotopically heavy carbonates or ^{13}C-enriched dissolved bicarbonate are rarely encountered. It seems reasonable that the near-zero values reflect some commonly achieved dynamic balance between isotopic fractionation resulting from methane generation and the continued addition of isotopically light carbon through fermentation reactions.

Judging from the large quantities of biogenic methane currently in the Gammon and equivalent rocks, and by analogy with observations in modern environments, methane generation probably continued within the Gammon sediments to depths of hundreds of meters. However, in regions of relatively rapid sedimentation, such as Owl Creek, siderite precipitation apparently ceased at shallow depths. Inasmuch as the Gammon Shale is not enriched in iron except for the siderite (Schultz et al, 1980), siderite precipitation was probably limited by the availability of reactive iron in the solid sediments.

CONCLUSIONS AND FUTURE RESEARCH

The interpretation of early diagenesis with a prominent phase of methane generation presented above (Fig. 8) illustrates a common sequence of diagenetic processes for one type of organic carbon-rich sediment accumulating at a relatively rapid rate. The implications seem clear enough: Early diagenesis of the muds is dominated by redox processes involving organic matter and by the products of organic matter decomposition. Furthermore, processes involving organic matter probably continue to dominate mineral diagenesis throughout the low-temperature ($<70°C$) burial history of mudstone. Because of the economic significance of the organic carbon-rich mudrocks as source beds of biogenic and thermogenic hydrocarbons and because diagenesis of mudrocks influences mineral precipitation and dissolution in adjacent sandstone reservoirs, it is of the utmost importance that diagenesis in ancient organic-rich rocks be understood.

When compared with the state of knowledge of processes in modern diagenetic environments, our efforts to reconstruct those ancient diagenetic environments are primitive indeed. However, much of the diagenetic his-

tory of marine mudstones is recorded in authigenic minerals that can be used in diagenetic reconstructions by analogy with processes that have been documented in recent diagenetic environments. The qualitative approach illustrated here is only a first step. Quantitative description and predictive interpretation are the ultimate goal. Such a quantitative approach may be possible for certain aspects of early diagenesis. In principle, the abundance of each critical authigenic mineral used in interpreting the burial history of an ancient marine mudstone can be determined. The depths of mineral precipitation may be inferred, as illustrated in the preceding discussion, and the isotopic and chemical evolution of pore waters can be interpreted. This information can then be treated mathematically.

One approach that seems particularly fruitful is the application of mass-balance considerations in the form of the diagenetic equation (Berner, 1980). In this approach, changes in concentration of components of sediments undergoing diagenesis are related to pertinent fluxes and reaction rates. In principle, stable isotopic species are amenable to the same type of treatment (Craig, 1969; Cline and Kaplan, 1975), although published attempts at application to diagenetic processes are still quite limited in number (Goldhaber and Kaplan, 1980). Generalization of the rigorous mathematical approach may be justified because of evidence that diagenetic processes affecting dissolved components (fluxes and reaction rates) are interrelated in such a way that knowledge of a single sedimentation variable, such as sedimentation rate, allows prediction of apparent biological reaction rates (Toth and Lerman, 1977) and concentration gradients (Berner, 1978), at least within a certain range of sedimentary environments. Similar generalizations may be possible for gradients of stable isotope ratios.

Studies of early diagenetic processes in modern sediments commonly focus on characterization of reactants and ephemeral products (for example, SO_4^{-2}, HCO_3^-, CH_4). In contrast, the only materials available for reconstructing the early diagenetic environment in ancient rocks are the stable authigenic minerals. More complete knowledge of the partitioning of stable

2

Aspects of Porosity Modification

The Chemistry of Secondary Porosity

Ronald C. Surdam
University of Wyoming
Laramie, Wyoming

Steven W. Boese
University of Wyoming
Laramie, Wyoming

Laura J. Crossey
University of Wyoming
Laramie, Wyoming

INTRODUCTION

The discovery of secondary porosity in sandstone has been the most significant advance in the study of clastic diagenesis in the past decade. Only recently has the widespread development of secondary porosity in sandstones been fully appreciated (Hayes, 1979; Schmidt and McDonald, 1979; McBride, 1980). Initially it was thought that most of the secondary porosity in sandstones resulted from carbonate dissolution, but it is now recognized that silicate dissolution also is an important mechanism in the formation of secondary porosity (Fig. 1). Obviously, the discovery of secondary porosity was an essential step to a more comprehensive understanding of progressive diagenesis of clastic rocks.

ESSENTIAL PROBLEM

How can the explorationist predict secondary porosity (porosity enhancement) in potential hydrocarbon reservoirs? At present, the chemical mechanisms responsible for the development of secondary porosity are poorly understood, making it difficult to evaluate the process and impossible to predict its occurrence.

While the existence of secondary porosity in clastic rocks is no longer a matter of debate, its contribution to reservoir porosity is still a subject of concern. The current understanding of secondary porosity results mainly from the effort to develop criteria to distinguish secondary from primary porosity. As a consequence, porosity is commonly thought of as either primary or secondary, when in most rocks the porosity and permeability are a result of

ABSTRACT. The development of secondary porosity (porosity enhancement) in many sandstones is the result of aluminosilicate and/or carbonate dissolution. The dissolution of aluminosilicate minerals and subsequent porosity enhancement is a problem of aluminum mobility. Our experimental data demonstrate that it is possible to increase significantly the mobility of aluminum and to transport it as an organic complex in carboxylic acid solutions. These same carboxylic acid solutions have the capability of destroying carbonate grains and cements.

Carothers and Kharaka have shown that concentrations of carboxylic acid anions range up to 5000 ppm over a temperature range of 80–200° C in some oil field formation waters. Our experiments show that acetic acid solutions at the same concentrations and over the same temperature range can increase the solubility of aluminum by one order of magnitude, whereas oxalic acid solutions increase the solubility of aluminum by three orders of magnitude. The textural relations observed in the experiments are identical to those observed in sandstones containing porosity enhancement as a result of aluminosilicate dissolution.

A natural consequence of the burial of sedimentary prisms is the maturation of organic material. These maturation reactions result in the evolution of significant amounts of organic acids and carbon dioxide. The experiments suggest that the enhancement of porosity in a sandstone as a result of aluminosilicate or carbonate dissolution is the natural consequence of the interaction of organic and inorganic reactions during progressive diagenesis. The degree to which porosity enhancement develops depends on the ratio of organic to inorganic matter, the initial composition of the organics, the sequences, rates and magnitude of diagenetic reactions, fluid flux, and sand/shale geometry.

both primary and secondary processes. By investigating processes capable of significantly enhancing porosity and permeability, and which are compatible with the progressive diagenesis of both organic and inorganic phases in a prism of sedimentary rocks, it is possible to avoid attributing all porosity to a single process. This integrated approach enables the explorationist to predict the distribution of enhanced porosity no matter what its origin.

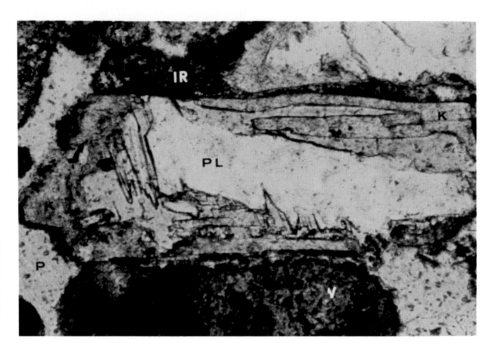

Figure 1 — Photomicrograph of a plagioclase-rich sandstone. **(PL)** plagioclase; **(K)** K-feldspar; **(IR)** clay and iron oxide; **(V)** volcanic lithic grains; and blue is porosity. Center of photomicrograph is an etched and dissolved plagioclase grain, which is rimmed by authigenic K-feldspar. Magnification = 100X.

APPROACH

The approach consists of three phases: (1) to explore the chemistry of fluids present during progressive diagenesis of a typical sand/shale sequence; (2) to examine some of the porosity-enhancing mineral reactions in the context of this fluid chemistry; and (3) to document some of the mechanisms of mass transfer responsible for porosity and permeability enhancement.

The dissolution of plagioclase framework grains is a common element in many sandstones characterized by porosity enhancement (Fig. 2). This study reports on the chemistry of plagioclase dissolution under diagenetic conditions (75 to 200°C) and in a variety of fluid compositions that simulate those present during diagenesis. In addition, the results of a series of experiments designed to investigate the progressive diagenesis of type I (sapropelic) and III (humic) kerogen in a water-wet system are presented. These two sets of experiments, (1) plagioclase dissolution in a simulated diagenetic environment, and (2) kerogen maturation in an aqueous environment, allow an evaluation of the roles of organic and inorganic reactions in the

enhancement of porosity during progressive burial of sediments. More specifically, when the results of these experiments are combined with oil field water observations, a general integrated model for porosity and permeability enhancement in sandstones is developed. This model is a first step and will need much refinement. The complex nature of the interactions precludes the development of a unique model that explains every case of so-called secondary porosity, but the model presented is viable in that it is consistent with the processes characterizing progressive diagenesis of sediments.

ALUMINUM MOBILITY

Plagioclase dissolution during diagenesis is a problem of aluminum mobility (Fig. 2). Petrologists have long used aluminum conservation as a basic tenet in evaluating mineral reactions; however, porosity enhancement as a result of aluminosilicate framework dissolution requires some mobility of aluminum. Skeletal textures (see Fig. 2), kaolinite veins, and pore fillings common in many sandstones, zeolite veins in volcanogenic sandstones, and plagioclase dissolution in some soils are

just some of the observations that document aluminum mobility in a variety of diagenetic conditions.

Plagioclase Dissolution

Considerable laboratory work has been done on the dissolution of plagioclase (see among others, Seifert, 1967; LaGache, 1976; Petrovic, 1976a, 1976b; Holdren and Berner, 1979). Dissolution occurs at sites of excess surface energy (Holdren and Berner, 1979), and judging from the similarity between etched textures characterizing both artifically and naturally weathered plagioclase grains, it is apparent that similar surface reactions are operative in nature. Although similar surface reactions may characterize both the experimentally and naturally weathered plagioclase, the analogy ends at this level. The previous plagioclase dissolution experiments have been run in a variety of solutions (that is, NaOH), none of which simulate the natural waters characterizing clastic diagenetic environments. In these experiments the aluminum solubility is low and is thought to be controlled by $Al(OH)_3$ saturation (Holdren and Berner, 1979). If aluminum solubilities were this low during sandstone diagenesis, unreasonably large volumes of fluid would be

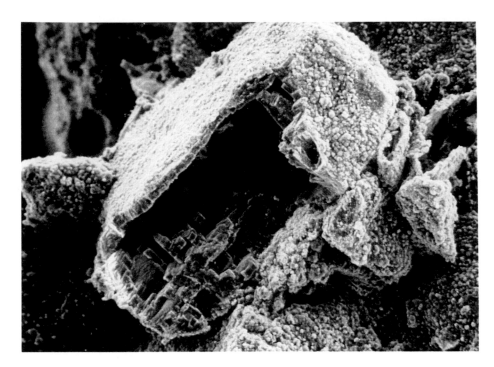

Figure 2 — SEM photomicrograph of grain from same sandstone as Figure 1. Plagioclase grain almost completely dissolved and rimmed by authigenic K-feldspar. Figure width = 350 microns.

required to create significant secondary porosity as a result of plagioclase dissolution. These experiments yield general descriptive information about the process, but little information regarding the fluids responsible for secondary porosity.

Organic Complexes in Soils

Recent observational work in the field of soil chemistry and mineralogy has yielded important information pertaining to plagioclase dissolution and aluminum mobility. Dissolution textures characterizing plagioclase grains in soils are very similar to the textures seen in plagioclase grains dissolving in diagenetic terranes characterized by secondary porosity (see Berner and Holdren, 1979; Boles, 1982). In addition, in soils it has been shown that dissolution of plagioclase occurs extensively in horizon A and that aluminum is transported in solution from horizon A to C, where it is precipitated with iron (Holdren et al, 1977). The observed presence of calcium oxalate in soils indicates the mechanism of aluminum transport (Graustein et al, 1977). More recently, Antweiler and Drever (1982) isolated oxalate fom the interstitial waters of a soil profile. Even small amounts of oxalate in solution increase the effective solubility of aluminum and

iron by several orders of magnitude (Lind and Hem, 1975). The general nature of the interaction of calcium oxalate and aluminum in soils can be illustrated as follows:

$$3H^+(aq) + Al(OH)_3(s) + 2CaC_2O_4(s) \quad (1)$$
$$= Al(C_2O_4)_2^-(aq) + 2Ca^{+2}(aq) + 3H_2O$$

$$\text{where } [Al(C_2O_4)_2^-] = Keq\left[\frac{(H^+)^3}{(Ca^{+2})^2}\right].$$

According to equation 1 the amount of complexed aluminum in solution is sensitive to both pH and activity of Ca^{+2}. The reaction of oxalate with an aluminum hydroxide or silicate to form a soluble trivalent metal complex consumes H^+. If the ligand is destroyed, the liberated metal will remove OH^- from solution as it precipitates. If the process is accurately described by equation 1, the effect of transport is to reduce the acidity in the zone of ligand production and to increase it in the zone of ligand destruction (Graustein et al, 1977). Thus in weathering profiles plagioclase dissolution and aluminum mobility has been documented; aluminum migrates as an organic complex [that is, $Al(C_2O_4)_2^-$]. In this setting, as a result of the organic complexation of aluminum, there is sufficient mass transfer to create secondary porosity. This same type of process may be operative during

the progressive diagenesis of a prism of sediments.

ORGANIC ACIDS IN DIAGENETIC ENVIRONMENTS

Carothers and Kharaka (1978) report that concentrations of aliphatic acid anions (acetate) range up to 5000 ppm in 95 formation-water samples from 15 oil and gas fields in the temperature range 80–200° C. Their work also shows that the aliphatic acid anions decrease with increasing subsurface temperatures and age of the reservoir rocks. It should be noted that Carothers and Kharaka (1978) did not analyze for the presence of the oxalate anion.

The hypothesis that organic complexation is critical in the mobility of aluminum during diagenesis may be tested by investigating the relation between organic and inorganic reactions during diagenesis. A hypothetical diagenetic reaction can be formulated as follows:

$$CaAl_2Si_2O_8 + 2H_2C_2O_4 + 8H_2O \quad (2)$$
$$+ 4H^+ = 2H_4SiO_4$$
$$+ 2(AlC_2O_4 \cdot 4H_2O)^+ + Ca^{+2}.$$

If this type of reaction is operative during diagenesis, then it will be possible to begin to quantify the processes control-

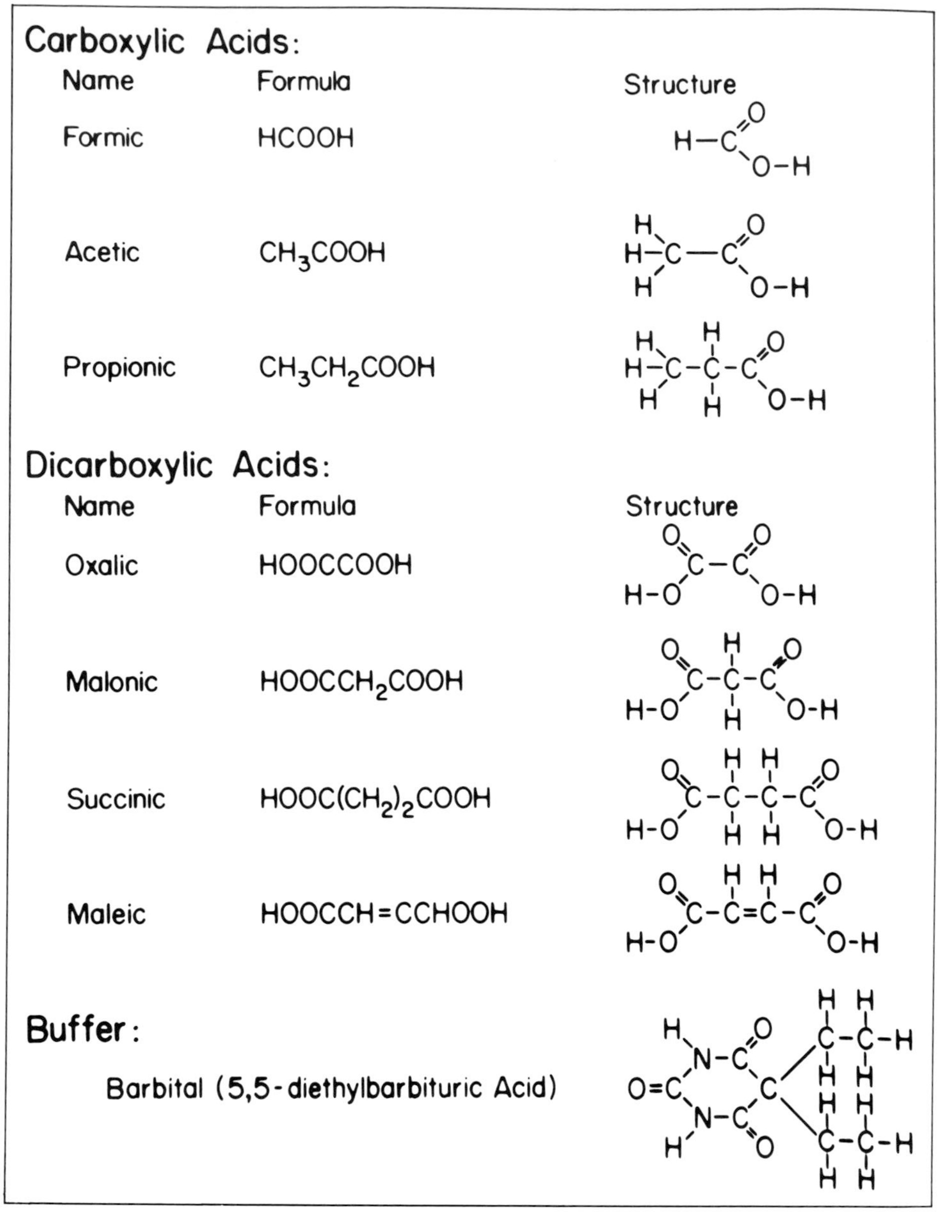

Carboxylic Acids:

Name	Formula	Structure
Formic	HCOOH	
Acetic	CH₃COOH	
Propionic	CH₃CH₂COOH	

Dicarboxylic Acids:

Name	Formula	Structure
Oxalic	HOOCCOOH	
Malonic	HOOCCH₂COOH	
Succinic	HOOC(CH₂)₂COOH	
Maleic	HOOCCH=CCHOOH	

Buffer:

Barbital (5,5-diethylbarbituric Acid)

Figure 3—Formula and structures of carboxylic acids discussed in this paper.

ling the distribution of enhanced porosity in sandstones.

Silica also may be released as an organic complex. If so, it would resolve the current debate between Bjorlykke (1979) and Land and Dutton (1979), and Boles and Franks (1979) concerning silica cementation in sandstones. At present this debate is irreconcilable on the basis of conventional inorganic geochemistry.

EXPERIMENTAL WORK

Apparatus

The experiments were conducted using stainless steel containers with Teflon liners in conjunction with a fluidized bath (SBL-2). The containers in the fluidized bath were constantly agitated.

Temperature and Time

The experiments were run at 100°C for a duration of 2 weeks. Preliminary experimental runs also have been made at 50 and 75°C and for periods of 1 to 4 weeks. None of these reconnaissance runs have changed any of the interpretations made from the 100°C and 2-week runs.

Fluid Composition

Two types of fluids were used in these experiments: (1) acetate-rich ($HC_2H_3O_2$) and (2) oxalate-rich ($H_2C_2O_4$). Guided by the work of Carothers and Kharaka (1978), these solutions were chosen because they represent simple straight-chain monofunctional and difunctional organic acids. Figure 3 illustrates the structure of several of the carboxylic acids discussed in this paper. The total carbon content of the solutions is the same as that reported by Carothers and Kharaka (1978) in oil field formation waters. In addition, Na-acetate and Na-oxalate

TABLE 1

	Initial Solution Concentration (ppm)	ppm Al	ppm SiO$_2$	[1]ppm Ac$^-$ (Final)	Final pH
Andesine	10,000 Ac$^-$	24	221	9640	3.40
Albite	10,000 Ac$^-$	14	198	—	3.20
Labradorite	10,000 Ac$^-$	72	195	9490	3.35
Microcline	10,000 Ac$^-$	21	152	—	3.15
Blank	10,000 Ac$^-$	<2	<1	9590	2.90

[1]Ac$^-$ = acetate anion.

Table 1—Acetic acid dissolution experiment (2 weeks at 100°C).

solutions were used in some initial experiments.

Solid Material

The starting material in most, but not all, of the runs was andesine plagioclase. Albite, labradorite, anorthite, microcline, laumontite, clinoptilolite, and mixtures of laumontite and calcite were also used in some of the runs. The solid material was sized, cleaned, and washed thoroughly before use. All of the starting materials were characterized both chemically and mineralogically. The experiments were run at fluid/solid ratios of 1000/1.

Analytical Procedures

The fluids were analyzed as follows: (1) pH measurements were made in a nitrogen environment with a glass electrode; (2) cations and anions were analyzed with a Dionex unit (ion chromatograph) and atomic absorption spectroscopy (Si was analyzed both by atomic absorption and colorimetrically); and (3) alkalinity was determined by titration.

The solids were analyzed by (1) scanning electron microscopy and (2) where feasible by atomic absorption and electron microprobe techniques.

Experimental Results

Tables 1, 2, and 3 show the results of the dissolution experiments. Figure 4

compares selected total aluminum values and the theoretical solubility curve for gibbsite assuming only aluminum hydroxyl species. It is apparent that acetic acid solutions (Table 1, Fig. 4) at the same concentrations and over the same temperature range as natural oil field waters can increase the total aluminum in solution by an order of magnitude (depending on pH) above its value based on inorganic equilibria. Because of its difunctional behavior, oxalic acid (Table 2, Fig. 4) is even more effective (for example, three orders of magnitude increase in total Al at pH 5). The textural relations observed in the experiments are identical to those observed in sandstones containing secondary porosity as a result of aluminosilicate dissolution (see Figs. 5, 6, 7). A diagrammatic view of the aluminum oxalate complex is shown in Figure 8. In addition to the effect on aluminum mobility, the carboxylic acids also may affect the behavior of dissolved silica (Fig. 9 shows some preliminary results). There is little question that the presence of organic acids, particularly the difunctional forms, can significantly increase the mobility of aluminum. Also, by tying up Al^{+3}, the organic acids effectively destabilize alumino-silicate framework grains.

Aluminosilicate framework grain dissolution has been emphasized, but carboxylic acids also have pronounced effects on carbonate dissolution. Mix-

tures of laumontite and calcite were run in buffered solutions over a pH range of 5 to 9 with and without oxalic acid. Results are compiled on Table 4. The interaction of oxalic acid, calcite, and laumontite at fixed pH values is shown in Figure 9b. Laumontite without calcite exhibits a trend of decreasing SiO$_2$ as pH increases. The trend is exaggerated in the presence of oxalate

$$\left(\frac{SiO_2 \text{ at pH 5}}{SiO_2 \text{ at pH 9}} = 9.3 \text{ with oxalate, and} \right.$$

2.8 without oxalate). When calcite is added to the system, the effect of oxalic acid on the laumontite is nullified; silica is maintained at a constant low value (11 ppm) independent of pH. The activity of Ca^{+2} is increased owing to the presence of calcite (particularly at low pH) and laumontite (CaAl$_2$Si$_4$O$_{12}$ · 4H$_2$O) is stabilized. The oxalate curves (Fig. 9b, dashed lines) demonstrate that calcite reduces the concentration of oxalic acid most effectively at low pH

$$\left(\frac{Ox^{-2} \text{ at pH5}}{Ox^{-2} \text{ at pH9}} = 0.0065 \text{ with calcite} \right.$$

present, and 0.73 without calcite). Thus, regardless of the Ca^{+2} activity, at low to neutral pH values oxalate concentrations would be too low in the presence of calcite to significantly increase the solubility of any aluminosilicate unless the calcite could be stabilized by another mechanism (that is, increased P$_{CO_2}$ at fixed pH, discussed below).

TABLE 2

	Initial Solution Concentration	ppm Al	ppm SiO$_2$	[1]ppm Ox^{-2}	Final pH
Andesine	Distilled H$_2$O	<1	67	nd[2]	7.70
Andesine	10 ppm Ox^{-2}	3.2	55	8	6.65
Andesine	100 ppm Ox^{-2}	3.9	143	26	5.95
Andesine	1000 ppm Ox^{-2}	130	421	600	4.30
Andesine	10,000 ppm Ox^{-2}	1300	413	7600	2.50
Albite	1000 ppm Ox^{-2}	75	349	570	3.60
Albite	10,000 ppm Ox^{-2}	790	506	7140	1.70
Labradorite	1000 ppm Ox^{-2}	125	376	880	4.65
Labradorite	10,000 ppm Ox^{-2}	1400	374	7900	2.70
Anorthite	1000 ppm Ox^{-2}	80	376	490	4.25
Microcline	10,000 ppm Ox^{-2}	150	540	4890	1.75
Laumontite	1000 ppm Ox^{-2}	137	596	690	3.50
Clinoptilolite	1000 ppm Ox^{-2}	137	441	970	4.50
Blank	10,000 ppm Ox^{-2}	<1	<1	8200	1.50

[1]Ox^{-2} = oxalate anion.
[2]nd = not detected.

Table 2—Oxalic acid dissolution experiments (2 weeks at 100°C).

TABLE 3

	Initial Solution Concentration	ppm Al	ppm SiO$_2$	[1]ppm Ox^{-2} (Final)	Initial pH	Final pH
Andesine	1000 ppm Ox^{-2}	112	479	690	3.15	3.65
Andesine	1000 ppm Ox^{-2}	80	382	580	4.00	4.20
Andesine	1000 ppm Ox^{-2}	72	275	640	5.00	5.15
Andesine	1000 ppm Ox^{-2}	25	104	860	6.05	6.15
Andesine	HCl	67	348	0	2.00	2.80

[1]Solutions buffered with 10,000 ppm Ac$^-$ (NaAC + Hac).

Table 3—Buffered oxalic dissolution experiment (andesine, 2 weeks at 100°C).

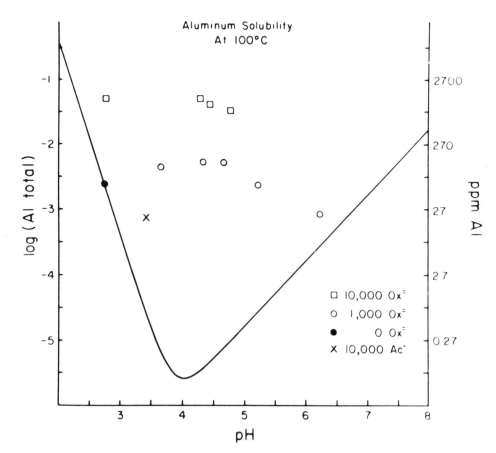

Figure 4—Gibbsite solubility curve at 100°C. Circles and squares represent Al data from the experiments. Ox^{-2} is equal to the amount of oxalate in solution in ppm.

Figure 5—SEM photomicrograph of starting material (andesine) before experiment.

Figure 6—SEM photomicrograph of surface of andesine fragment after experiment. Experiment conditions: 100° C at 10,000 ppm oxalate for 2 weeks. Note gel precipitated on grain surface.

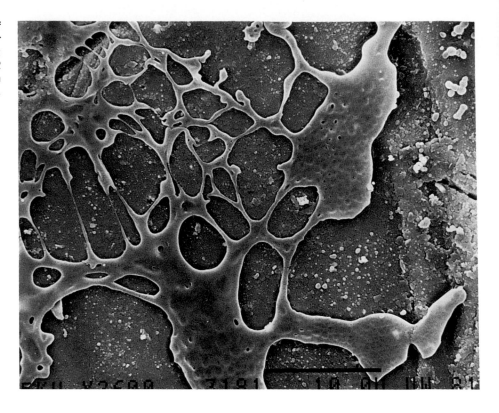

Figure 7—SEM photomicrograph of andesine fragment after experiment. Experimental conditions same as Fig. 6.

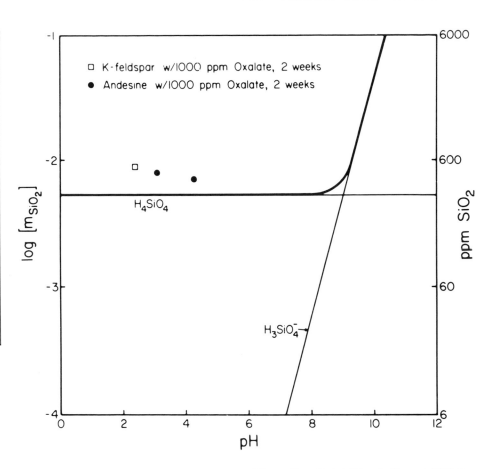

Figure 8—Schematic diagram showing complexation of Al^{+3} by oxalic acid.

Figure 9a—Solubility of silica at 100° C. The square box and black dots are results from this study.

KEROGEN MATURATION

Generative Potential

A key element in building diagenetic models that include organic diagenesis is the determination of the organic acid generative potential of kerogen. To our knowledge there is no experimental evidence available that allows this determination to be made directly. Much of what is known about kerogen structure has come from a variety of chemical methods (Victorovic, 1980). One of the most productive methods has been the *carbon balance* permanganate oxidative degradation technique. In this technique the kerogen is subject to stepwise oxidation at a constant temperature in an aqueous solution; at each oxidation step the products are analyzed. Table 5 presents the amounts and types of products obtained from several organic-rich shales. Large variations in the amount of oxidation products are observed, but the products remained relatively constant. Up to 31 wt % oxalic acid is produced. One model of the kerogen molecule that emerges from these studies is shown in Figure 10. Mono- and difunctional carboxylic acids are bonded peripherally to the core of the molecule.

Because these short chains are attached by only one bond, they are likely to be broken off in early stages of thermal cracking.

The oxidation work of Victorovic involves pervasive oxidation in which no hydrocarbons are produced; in the natural system oxidation potential will be limited by the availability of mineral oxidants. Examples of such mineral oxidants may be ferric iron released from clay mineral reactions (Boles and Franks, 1979) and amorphous material (Curtis, 1978), or sulfur in the disulfide pyrite (that is, early formed pyrite from bacterial sulfate reduction could be reduced further to form H_2S and siderite or chlorite). The reduction of the ferric iron or disulfide could be balanced by the oxidation of organic matter. Thus, carboxylic acids may be produced from thermal cracking of peripheral aliphatic chains and/or oxidation of the peripheral groups as a result of the reduction of mineral oxidants.

The maturation of two types of kerogen was examined in order to evaluate their potential to produce carboxylic acid in an aqueous system. Kerogen from oil shale out of the Green River Formation (specifically from a layer in the upper Laney Member containing fossil catfish [Buchheim and Surdam, 1977]), was used as a source for the type I kerogen. Pyrolysis showed this kerogen to be characterized by no S_1 peak (thermal extract) and a very large S_2 peak (pyrolysate); the sample is interpreted to be very immature. Material from a peat bog near Myrtle Beach, South Carolina was used as a source for the type III kerogen experiments.

Type I Kerogen

Two types of type I kerogen samples were used: (1) finely ground oil shale samples with no pretreatment, and (2) kerogen extracted from finely ground oil shale samples with 50% v/v HCl at

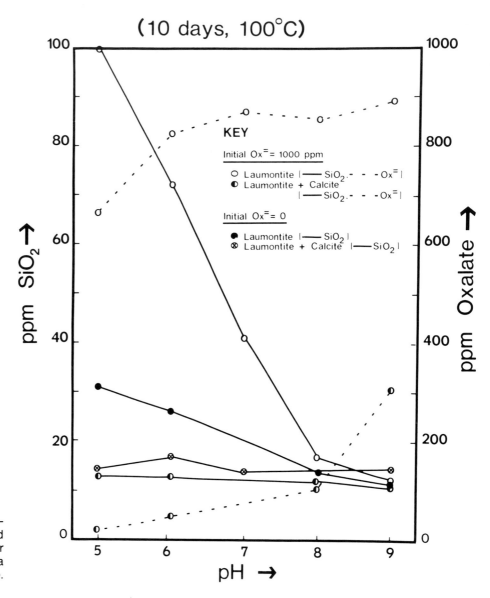

Figure 9b—Experimental results of dissolution of mixtures of laumontite and calcite (20% calcite). Dashed lines refer to oxalate scale, solid lines to silica scale.

80°C for 4 hr followed by extraction with 30% v/v HF at 80°C for 4 hr. The residue was washed with distilled water to remove soluble material and excess acid. The resulting material was essentially free of carbonate and silicate materials. Approximately 50 mg of each type of sample were placed in 10 ml Teflon containers and fitted with a tightly sealing threaded lid. To each type of kerogen sample 5.0 ml of one of three different solutions were added: (1) distilled water; (2) acetic acid–sodium acetate solution of pH = 3.0 and total acetate concentration of 1000 ppm; (3) same as (2) but with pH = 6.0. Each kerogen–solution mixture was prepared in duplicate. Duplicate samples of solution alone were also carried through the experiment as controls. The Teflon containers were capped and sealed in a stainless steel vessel containing a small amount of distilled water. The vessel was immersed in a fluidized sand bath at a temperature of 100°C and rotated slowly with the axis of rotation inclined approximately 60° from horizontal. This arrangement allowed accurate temperature control and continuous mixing of the container contents. The samples were heated and mixed in this manner for 2 weeks, after which the vessel was removed from the fluidized bath and allowed to cool. The resulting solutions were pipetted off the settled residue and placed in air-tight polyethylene containers until analyzed. The solutions were analyzed for carboxylic acids using a Dionex Ion Chromatograph and the pH of each was measured with a glass electrode. Oxalic acid and acetic acid were the only carboxylic acids detected. The acetic acid, having been added initially as a buffer, was of limited interest. The results are listed in Table 6.

From the results, it can be seen that significant amounts of oxalic acid were

TABLE 4

Solids (% by wt)	Buffered pH	Initial Oxalate Concentration (ppm)	Final Oxalate Concentration (ppm)	ppm Al	ppm SiO_2
laumontite (80) calcite (20)	5	1000	18	<1	13
laumontite	5	1000	664	26	102
laumontite (80) calcite (20)	6	1000	49	<1	13
laumontite	6	1000	827	18	72
laumontite	7	1000	872	8.6	41
laumontite (80) calcite (20)	8	1000	106	<1	12
laumontite	8	1000	858	1.0	17
laumontite (80) calcite (20)	9	1000	310	<1	11
laumontite	9	1000	894	1.0	12
laumontite	5	0	—	<1	31
laumontite + calcite	5	0	—	<1	14
laumontite	6	0	—	<1	26
laumontite + calcite	6	0	—	<1	17
laumontite + calcite	7	0	—	<1	14
laumontite	8	0	—	<1	14
laumontite	9	0	—	<1	12
laumontite + calcite	9	0	—	<1	14

Table 4—Laumontite and calcite dissolution experiment (10 days at 100°C).

TABLE 5

Shale	CO_2	Volatile Acids	Oxalic Acid	Nonvolatile "Nonoxalic" Acids	Unoxidized Carbon	Oxidation Period (hrs)
Kimmeridge	49.8	8.3	31.0	10.5	0.6	57
Pumpherston	21.0	4.5	10.7	13.1	51.4	125
Gdov	98.0	2.2	3.8	0.5	0.0	500
Volga	88.1	3.9	5.2	4.1	0.0	500
St. Hilaire	35.3	0.2	6.6	10.1	48.5	100

Table 5—Carbon distribution in oxidation products of kerogen (modified from Victorovic, 1980).

TABLE 7

Sample Number	T°C	Time (Weeks)	TOC	Formic	Acetic	ppm Propionic	Succinic	Oxalic
				Waters (1 ft)				
1 (f)[1]	100	2	50 ppm	3.9	3.1	nd	tr	6.7
2(u)	100	2	50 ppm	6.4	3.8	nd	0.8	8.4
3(f)	57	2	50 ppm	0.1	nd	nd	nd	3.6
4(u)	57	2	50 ppm	nd	nd	nd	nd	nd
5(f)	100	1	50 ppm	2.2	2.3	nd	nd	nd
6(u)	100	1	50 ppm	3.3	1.9	nd	nd	nd
7(u)	57	1	50 ppm	nd	nd	nd	nd	nd
				Acid Ext. (layer 3)				
(pH = 1) 1	100	2	50 ppm	5.8	2.3	4.3	nd	nd
(pH = 1) 2	57	2	50 ppm	1.9	2.0	nd	nd	nd
(pH = 1) 3	100	1	50 ppm	3.3	2.2	2.1	nd	nd
(pH = 1) 4	57	1	50 ppm	1.5	1.9	nd	nd	nd
(pH = 7) 5	100	2	50 ppm	4.3	3.8	14.0	2.2	nd
(pH = 9) 6	100	2	50 ppm	6.6	12.2	34.4	4.9	nd
				Base Ext. (layer 3)				
(pH = 6.5) 1	100	2	2500+ ppm	37	18.0	nd	5.8	nd
(pH = 6.5) 2	57	2	2500+ ppm	15.0	7.6	nd	nd	nd
(pH = 6.5) 3	100	1	2500+ ppm	27	13.0	nd	nd	9.0
(pH = 6.5) 4	57	1	2500+ ppm	6.0	nd	nd	nd	nd
(pH = 4)	100	2	2500+ ppm	1.6	0.7	nd	nd	nd
(pH = 9)	100	2	2500+ ppm	58	33	39	15.0	57

[1](f) = filtered   (u) = unfiltered   nd = not detected   tr = trace

Table 7—Analytical results of maturation study of humic material.

4.9 ppm succinic. No oxalic acid was present in the acid-extracts.

Base-extractable Humic Material

About 10 ppm oxalate was present in the unheated base extract. No other carboxylic acids were observed. The TOC for the base extract was about 1500 ppm, and the initial pH was 6.5.

After 1 week at 57°C, 6 ppm formic acid was present and no oxalic acid was observed. After 2 weeks at 57°C, formic acid increased to 15 ppm, acetic acid to 7.5, and still no oxalic acid was observed.

For the 100°C experiments, both Layer 2 and Layer 3 extracts were prepared. After 1 week, Layer 2 and Layer 3 contained (respectively) 19 and 27 ppm formic, 6 and 13 acetic, and Layer 3 contained 9 ppm oxalic. After 2 weeks at 100°C, the Layer 2 extract contained 22 ppm formic, 18 ppm acetic and 11 ppm oxalic acid. Layer 3 contained 37 ppm formic, 18 ppm acetic, and 6 ppm succinic acid (no oxalate). The Layer 3 extract was buffered to a lower pH (4) and higher pH (9) with Na-barbital and run for 2 weeks at 100°C. At the lower pH, fewer carboxylic acids were produced: 2 ppm formic and 1 ppm acetic acid are observed. At the higher pH, however, significant increases in carboxylic acid production is noted: 60 ppm formic, 33 ppm acetic, 40 ppm propionic, 15 ppm succinic, and 60 ppm oxalic acid are present.

Solid Peat Material and Distilled Water

A quantity of 50 mg of solid Layer 3 material and 5 ml of distilled water were run at 57° and 100°C for 1- and 2-week time periods. The final pH of the samples after thermal decomposition ranged from 3.4 to 3.7.

Heating at 57°C produced no detectable carboxylic acids after 1 week and only a trace of formic acid after 2 weeks. At 100°C and 1 week only minor amounts of carboxylic acids were present. After 2 weeks, 13 ppm formic, 7 ppm acetic, 4 ppm succinic, and 13 ppm oxalic were observed.

Interpretation

Thermal degradation of dissolved organic material in water samples from the peat bog results in the generation of significant quantities of carboxylic acids (Table 8). The acids produced after 2 weeks at 100°C account for an average of 9.3% of the TOC for the samples. The unfiltered samples show a slight increase in the ratio of carboxylic

TABLE 8

Waters

| | 1 Week | | | | 2 Weeks | | | |
| | 57°C | | 100°C | | 57°C | | 100°C | |
	Filtered	Unfiltered	Filtered	Unfiltered	Filtered	Unfiltered	Filtered	Unfiltered
1 ft	nd[1]	tr	3.0	3.3	2	tr	8.3	11.7
2 ft	tr	nd	5.7	5.7	2.8	nd	8.8	8.5
3 ft	tr	nd	1.9	1.8	3.4	nd	8.9	nd
4 ft	tr	nd	2.1	2.3	2.1	nd	8.7	7.9
5 ft	1.0	nd	1.8	2.8	3.0	nd	10.8	9.9

Extractions

| | Acid | | | | Base | | | |
| | 1 Week | | 2 Weeks | | 1 Week | | 2 Weeks | |
Sample	57°C	100°C	57°C	100°C	57°C	100°C	57°C	100°C
Layer 2		5.1		9.3		0.3		0.6
Layer 3	2.3	5.5	2.6	9.0	0.1	0.6	0.3	0.8
pH = 7 Layer 3				10.0				
pH = 9 Layer 3				25.0				2.7
pH = 4 Layer 3								>0.1

[1]nd = not detected   tr = trace

Table 8—Fraction of total organic carbon (TOC) converted to carboxylic acids (C) during humic material maturation.

acid to total organic carbon (C/TOC) over the filtered samples. It appears that dissolved carbon undergoes significant thermal degradation. The pH values of the water samples are all <4; as discussed below this is not the most productive pH range for carboxylic acid generation. If these waters were to mix with other waters of higher pH as the material heated with increasing burial, greater concentrations of carboxylic acids could be generated.

The extraction removed some carboxylic acids already present in the soil (geochemical fossils). In terms of TOC content, these extractions are similar to the water samples. The lower pH of the extracts (< 1 final pH) explains their relatively low content of carboxylic acids. When buffered to higher pH values, the extractions generate up to 25% of the TOC content as carboxylic acids. The effects of the barbital buffers interacting with organic matter are not known. However, when the barbital is heated alone at 100°C for 2 weeks it shows no decomposition to simple carboxylic acids. The presence of propionic acid in some unbuffered acid

extracts (albeit in small quantities) indicates that generative potential for propionic acid is present. Potentially a substantial portion of the TOC may be converted to carboxylic acids on heating at a high pH.

Again, a geochemical fossil (oxalic acid) is present in the unheated base extractable humic material. Previous experiments have shown that oxalic acid converts to formic acid at low pH. The initial and final pH of Layer 2 material is lower than that of Layer 3 and this may explain the absence of oxalic in the heated extracts after 1 week at 100°C.

In summary, humic material is capable of generating significant quantities of carboxylic acid during early maturation. Distilled water heated in contact with peaty material can produce carboxylic acids at relatively low (<4) pH values. At higher pH there is a considerable increase in carboxylic acid production. Lastly, it is most important to note that the carboxylic acid-producing reaction, with respect to the humic material, appears to be very sensitive to the solution pH. Thus, it is speculated that this sensitivity to solution pH could result in a wide range of organic acid conditions in a natural diagenetic environment containing humic material.

Conclusion from Kerogen Maturation Experiments

Both algal and humic materials are capable of generating carboxylic acids during maturation. The reactions involving both humic and algal material appear to be particularly sensitive to pH and to the presence of carbonates. In conclusion, it appears from these experiments that sediments containing organic debris have the potential to generate significant amounts of carboxylic acid during diagenesis. The degree to which this carboxylic acid would be available for porosity enhancement would depend on the following: pH of the interstitial solution; mineralogy of framework grains and preexisting cements; ratio of inorganic to organic matter (sand/shale); the composition of the organic material; the sequence, rates and magnitude of diagenetic reactions; fluid flux; and sand/shale geometry. Depending on variations of these parameters and probably others, a wide range of porosity effects could be explained by carboxylic acids.

OIL FIELD WATERS

The effectiveness of carboxylic acid, particularly the difunctional forms, in increasing the solubility of aluminosilicate framework grains has been demonstrated. Furthermore, the experimental work suggests that significant amounts of carboxylic acid can be generated by the diagenesis of kerogen.

As already mentioned, Carothers and Kharaka (1978) showed that over the temperature range of 80 to 200°C there can be up to 5000 ppm acetate

present in oil field waters. Moreover, in 95 formation waters they studied from 15 relatively young (Eocene–Miocene) oil and gas fields it was shown that the aliphatic or carboxylic acid anions generally contribute 50 to 100% of the measured alkalinity over the 80–200°C temperature range. They suggested that the carboxylic acids that are highly soluble in water are generated as a result of thermocatalytic degradation of kerogen in the source rocks. The carboxylic acids are destroyed at higher temperatures as a result of thermal decarboxylation according to the following reaction (Carothers and Kharaka, 1978):

$$CH_3COO^- + H_2O \rightarrow CH_4 + HCO_3^- \quad (3)$$

(see Fig. 11).

Thirteen oil field waters have been obtained and analyzed for both mono- and difunctional carboxylic acids by ion chromatography (see Table 9 and Fig. 3). Acetic acid (monofunctional) is present in concentrations up to 10,000 ppm. Difunctional carboxylic acids (such as maleic and malonic acids) are present in 11 out of the 13 oil field water samples. As these waters have undergone extensive diagenetic reaction in the course of migration, the presence of the highly reactive difunctional carboxylic acids is more significant than their concentration.

ORGANIC–INORGANIC INTERACTIONS

A standard explanation of secondary porosity is based on the well-known fact that with increasing P_{CO_2} the solubility of calcite increases (Fig. 12). Thus, it is typically stated that secondary porosity resulting from dissolution of carbonate grains and cements can be attributed to elevated P_{CO_2} values resulting from decarboxylation reactions (Al-Shaieb and Shelton, 1981). This is an important process and may well explain some cases of secondary porosity, particularly in environments rich in humic material (Fig. 13). However, the presence of carboxylic acids and their effect on carbonate solubility cannot be overlooked. As already mentioned, carbonates are readily dissolved in carboxylic acid solutions over a range of pH (Fig. 9b). Thus, the presence of these organic acids alone can

explain some cases of secondary porosity resulting from carbonate dissolution. Elevated P_{CO_2} cannot be the cause of dissolution of aluminosilicates. The only aqueous species in the carbonate system that has even a weak ability to complex metals is CO_3^{-2}. As can be seen from Figure 14, CO_3^{-2} is significant only at elevated pH (>10.3). Although the presence of carbonic acid may cause dissolution of carbonates, it will have little effect on the aluminosilicate framework grains. Another important consideration is the interaction of carboxylic acids, CO_2, and carbonate minerals. As mentioned above, Carothers and Kharaka (1978) found that in the oil field waters they examined, 50 to 100% of the alkalinity was provided by carboxylic acid. Thus, in these solutions the principal control on pH would be the carboxylic acid anions. In oil field waters containing 5000 to 10,000 ppm acetate, the carboxylic acid could easily buffer the pH of the pore fluid (acetate's maximum buffering capacity is at a pH of approximately 5). If the system is buffered externally, that is, by acetate, then with increasing P_{CO_2} the solubility of calcite decreases. For example, in Figure 15 at a pH of 5 the solution composition would be confined to a diagonal line (pH = 5) and would move to the lower right as P_{CO_2} increases (decreased carbonate solubility). P_{CO_2} increases from left to right along the abcissa, as does ΣCO_3 (see Fig. 16).

Before outlining possible diagenetic pathways, a brief discussion of the mechanism of expulsion of the organic acids from the source rocks must be made. It is evident that hydrocarbons and aqueous phases do not migrate through 'impermeable' shales in the same manner as through more porous reservoir rocks. Carboxylic acids are water soluble, and their transport will largely be influenced by water-releasing reactions in the source rocks (smectite-to-illite reaction) and dehydration reactions (opal CT to quartz transformation). The migration of the aqueous solution along microfractures rather than tortuously around individual grains will tend to reduce the volume of rock contacted by the solution, hence reducing the extent of reaction in the fluids. Also, the fluids will be transported more rapidly as they may be confined to a migration system of relatively small volume relative to that of

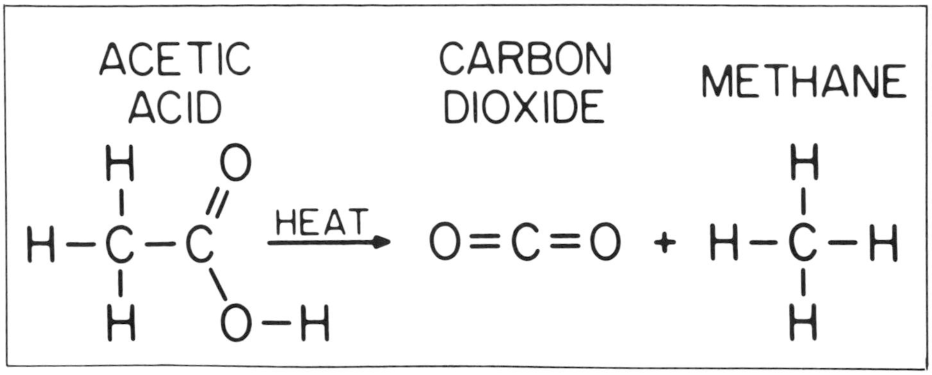

ACETIC ACID   CARBON DIOXIDE   METHANE

$$\text{H-C-C} \xrightarrow{\text{HEAT}} \text{O=C=O} + \text{H-C-H}$$

Figure 11—Schematic diagram showing decarboxylation of acetic acid.

TABLE 9

Sample	ppm Acetic Acid	ppm Propionic Acid	ppm Maleic Acid	ppm Malonic Acid
OB1	tr[1]	30	8	9
OB2	580	nd	20	11
OB3	510	nd	12	11
OB4	120	60	12	4
OB5	380	90	5	9
OB6	560	nd	26	tr
OB7	50	60	8	tr
OB8	480	tr	7	7
OB9	450	tr	8	5
OB10	140	120	nd	1
OB11	nd	nd	nd	4
OB12	nd	nd	nd	nd
OB13	10,000	nd	nd	nd

[1]tr = trace   nd = not detected

Table 9—Carboxylic acid analysis of oil field brines.

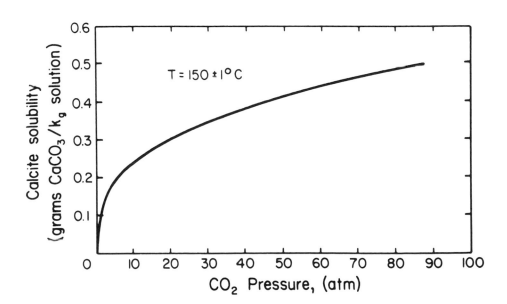

Figure 12—Calcite solubility as a function of P_{CO_2} at 150° C from Holland and Borcsik, 1976. Used with permission of author.

Figure 13—Relative yield of gas from organic matter. Note that humic material generates considerably more CO_2 than does sapropelic material. C_{2+} represents hydrocarbons heavier than CH_4 in gas phase. N_2 is generated initially as NH_4. Modified from PETROLEUM GEOCHEMISTRY AND GEOLOGY by John M. Hunt. W. H. Freeman and Company. Copyright © 1979. Used with permission.

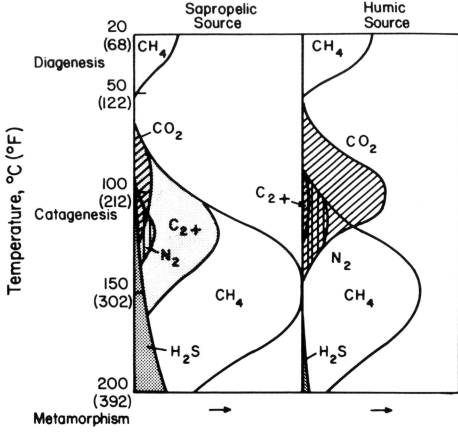

Figure 14—Activities of different species in the carbonate system as a function of pH. Only CO_3^{-2} has even weak ability to complex aluminum; it is dominant species at pH > 10.3. The diagram assumes $\Sigma CO_3 = 10^{-2}$ and temperature = 25°C. Activities of H^+ and OH^- are defined by pH. After James I. Drever, THE GEOCHEMISTRY OF NATURAL WATERS, © 1982, p. 38. Used with permission of Prentice-Hall, Inc., Englewood Cliffs, NJ.

Relative Yield of Gas from Organic Matter in Fine-Grained Sediments

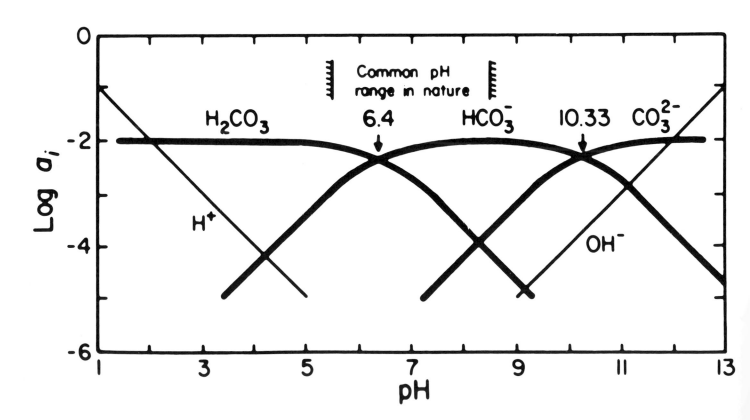

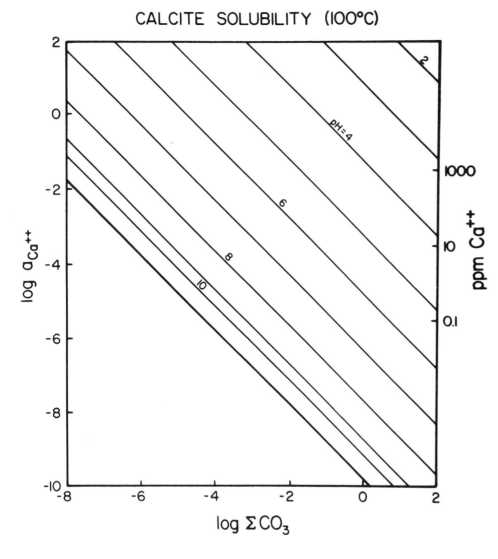

CALCITE SOLUBILITY (100°C)

Figure 15—Calculated equilibrium surface for calcite at 100° C projected onto the $\log a_{Ca^{-2}} - \log \Sigma CO_3$ ($H_2CO_3 + HCO_3^- + CO_3^{-2}$) plane. Solid lines represent contours of constant pH. Note that if the system is externally buffered (restricted to a constant pH value) and ΣCO_3 is increased by increasing P_{CO_2} (see Fig. 18), calcite solubility is decreased. Values were calculated from data in Helgeson, 1969.

the reservoirs, reducing reaction times. Thus the reactivity of the fluids may be reduced to a certain extent depending on: (1) geometry of migration system, (2) transport velocity, (3) mineralogy of the source rocks, and (4) overall distance of transport.

Considering the interaction of only four diagenetic components (aluminosilicate minerals, carbonate minerals, carboxylic acids, and CO_2) several diagenetic pathways are possible. A schematic flow diagram of possible reaction pathways is presented in Figure 17.

MODEL FOR POROSITY ENHANCEMENT

The model that we propose to explain enhanced porosity is presented in Figures 18 and 19. This model requires modification of diagrams published by Tissot and Welte (1978) and Carothers and Kharaka (1978, 1980). It is essential to include the short chain carboxylic acids shown in Figures 18 and 19 in any diagenetic scenarios involving porosity and permeability evaluation. As can be seen in Figure 18 the short chain carboxylic acids occur in the maturation scheme between the long chain humic acids and the generation of hydrocarbons. In detail, the presence and abundance of the short chain carboxylic acids are determined on the low temperature side by bacterial degradation and on the high temperature side by thermal degradation (Fig. 19). The position of the carboxylic acids in Figures 18 and 19 is most important. The reduction of min-

eral oxidants and consequent oxidation of organic matter may be more effective in releasing peripheral difunctional carboxylic acid groups than thermal degradation. The coincidence in time, temperature, and space of smectite/illite ordering (release of Fe^{+3} from the octahedral layers) and the carboxylic acid maxima shown in Figure 19 suggest the potential for the highly soluble organic acids, including difunctional forms, to be swept through the adjacent sandstones just prior to the generation of hydrocarbons. Thus, an ideal mechanism is available for dissolving carbonates and/or aluminosilicates out of pores and pore throats and thereby enhancing porosity and permeability prior to hydrocarbon generation. Porosity and permeability enhancement according to this model becomes a nat-

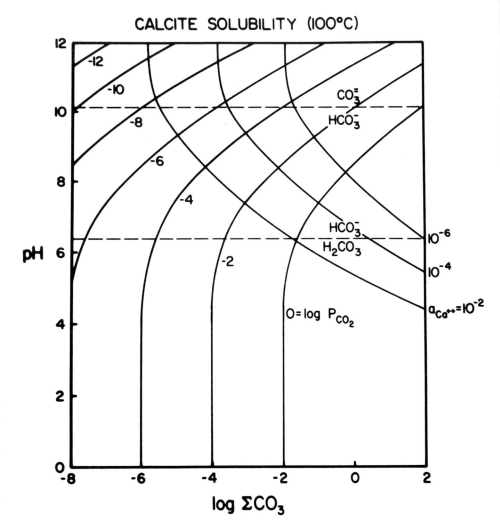

Figure 16—Calculated equilibrium surface for calcite at 100° C projected into the pH – log ΣCO_3 plane. Solid lines represent f_{CO_2} isopleths and contours of constant Ca^{+2} activity (see labels). Light dashed lines separate the pH regions where different carbonate species (H_2CO_3, HCO_3^-, and CO_3^{-2}) are dominant. Note the relationship between ΣCO_3 and P_{CO_2}.

ural consequence of the interaction between progressive maturation of kerogen and diagenesis of inorganic cements and framework grains. If organic-rich rocks are in the diagenetic system of interest and if the rocks are hydraulically connected to the sandstones, then porosity enhancement should occur to some degree.

The timing of the organic and inorganic reactions, the availability of a fluid flux, the development of migration paths, and the initial porosity and permeability characteristics of the potential reservoir sandstone will be determinative to porosity enhancement. The effectiveness of the organic acids in both silicate and carbonate dissolution will be determined by the reaction history of the solutions as they migrate from source rock to reservoir. The organic acids are so reactive (par-

ticularly the difunctional carboxylic acids) that if the migration path is long and the fluids react with the conduit, the organic acids could be partly or wholly consumed before reaching the reservoir rocks. The ideal conditions for the development of porosity enhancement are to have the organic-rich source rocks adjacent to the reservoir rocks, and to have some primary porosity remaining (for the reactive fluids, like any fluid, will follow the path of least resistance). This may explain why in many cases secondary porosity is best developed where the reservoir originally had the best primary porosity.

The development of porosity enhancement or secondary porosity requires mass transfer. If the dissolution of framework grains and/or cements is to significantly improve the

reservoir, the removed material must be transported beyond the immediately adjacent pore or pore throat. What happens to the material in those sandstones where the porosity has been significantly enhanced? In cases where the enhancement is the result of alumino-silicate dissolution, some of the transferred material eventually may form kaolinite. The late kaolinite that is common in many sandstone reservoirs may be the result of mass transfer as a result of the dissolution of alumino-silicate minerals. It is speculated that the kaolinite forms as the aluminum complexes that are responsible for the aluminum mobility are destabilized. The complexes are probably destabilized by variations in pH. It is interesting to note that the only evidence of the aluminum complex left in the rock would be the kaolinite.

CLASTIC REACTION PATHWAYS

$\left[\text{CO}_2,\text{ carboxylic acids, carbonates, aluminosilicates}\right]$ Initial Components

~80° C

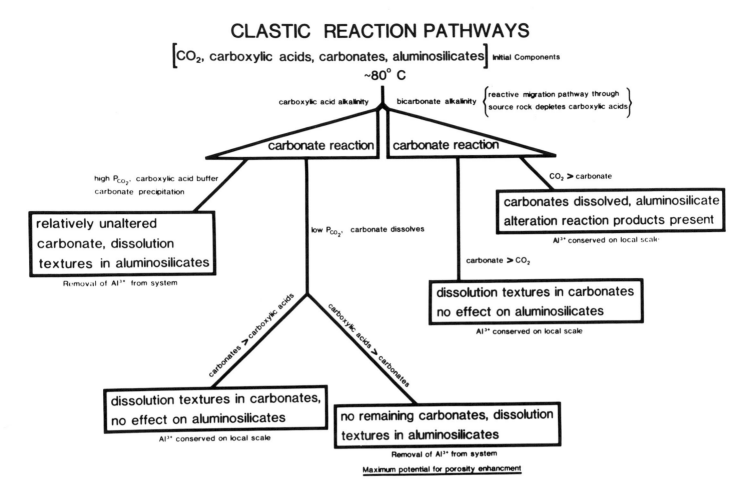

Figure 17—Flow diagram combining aspects of organic and inorganic diagenesis. The initial components are carboxylic acids, CO_2, carbonate minerals, and aluminosilicate minerals. The system is designed to represent simplified reservoir conditions at about 80° C.

The model as proposed explains many of the observations made relative to "secondary porosity," and it involves rock/fluid parameters characterizing most diagenetic systems. The proposed model is sensitive to the details of the interactions of organic and inorganic diagenesis and in that way it differs from earlier models. The model offers the explorationist a process-oriented framework in which to evaluate porosity enhancement. The details of the model need much refinement, but in its present form it should be of use conceptually. The key to understanding and applying the model, and thus predicting porosity enhancement, will be found in the integration of organic and inorganic diagenesis.

ACKNOWLEDGMENTS

Bob Siebert of CONOCO was a constant source of ideas and inspiration to the work; we gratefully acknowledge his significant contribution. CONOCO provided financial support for this work; we are thankful for this support. This work also benefited from discussion with R. Antweiler, T. Dunn, and J. Fahy. T. Dunn also contributed photomicrographs used in the paper. We would like to thank J. Brown of CONOCO and T. Dunn of Amoco for providing us with the oil field water samples.

SELECTED REFERENCES

Al-Shaieb, Z., and J. W. Shelton, 1981, Migration of hydrocarbons and secondary porosity in sandstones: Bulletin of the American Association of Petroleum Geologists, v. 65, no. 11, p. 2433–2436.

Antweiler, R. C., and J. I. Drever, 1983, The weathering of a Late Tertiary volcanic ash: importance of organic solutes: Geochimica et Cosmochimica Acta, v. 47, p. 623–629.

Berner, R. A., and G. R. Holdren, Jr., 1979, Mechanism of feldspar weathering, II,

Observations of feldspars from soils: Geochimica et Cosmochimica Acta, v. 43, p. 1173–1186.

Bjorlykke, K., 1979, Cementation of sandstones: Journal of Sedimentary Petrology, v. 49, p. 1358–1359.

Boles, J. R., 1982, Active albitization of plagioclase, Gulf Coast Tertiary: American Journal of Science, v. 282, p. 165–180.

Boles, J. R., and S. G. Franks, 1979, Reply cementation of sandstones: Journal of Sedimentary Petrology, v. 49, p. 1362.

Buchheim, H. P., and R. C. Surdam, 1977, Fossil catfish and the depositional environment of the Green River Formation, Wyoming: Geology, no. 5, p. 196–198.

Carothers, W. W., and Y. K. Kharaka, 1978, Aliphatic acid anions in oil-field waters—implications for origin of natural gas: Bulletin of the American Association of Petroleum Geologists, v. 62, p. 2441–2453.

———, 1980, Stable carbon isotopes of HCO_3 in oil field waters—implication for the origin of CO_2: Geochimica et Cosmochimica Acta, p. 323–332.

Curtis, C. D., 1978, Possible links between sandstone diagenesis and depth related

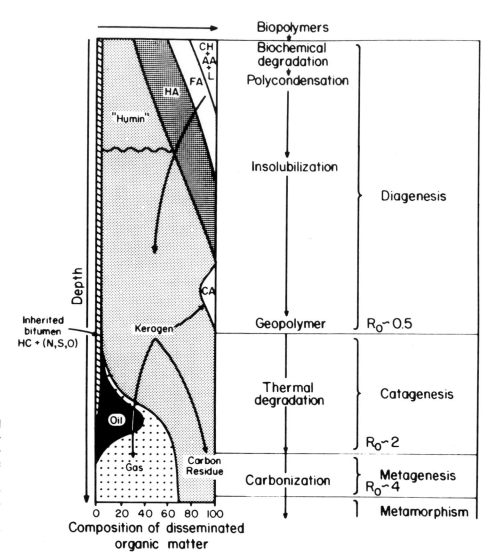

Figure 18—Model of carboxylic acid (**CA**) generation during thermal maturation of organic material. (**CH**) carbohydrates, (**AA**) amino acids, (**FA**) fulvic acid, (**HA**) humic acids, (**L**) lipids, (**HC**) hydrocarbons, (**N, S, O**) N, S, O compounds (nonhydrocarbons), (**CA**) carboxylic acids. Modified from Tissot and Welte. Copyright © 1978 Springer-Verlag New York. Used with permission.

geochemical reactions occurring in enclosing mudstones: Quarterly Journal of the Geological Society of London, v. 135, p. 107–117.

Drever, J. I., 1982, The geochemistry of natural waters: Englewood Cliffs, NJ, Prentice-Hall, 388 p.

Graustein, W. C., J. Cromack, Jr., and P. Sollins, 1977, Calcium oxalate: occurrence in soils and effect on nutrient and geochemical cycles: Science, v. 198, p. 1252–1254.

Hayes, J. B., 1979, Sandstone diagenesis—the hole truth: Society of Economic Paleontologists and Mineralogists Special Publication 26, p. 127–140.

Helgeson, H. C., 1969, Thermodynamics of hydrothermal systems at elevated temperatures: American Journal of Science, v. 267, p. 729–804.

Holdren, Jr., G. R., W. C. Graustein, and R. A. Berner, 1977, Chemical weathering in soils: evidence from surface compositions: Geological Society of America Abstracts with Programs, v. 9, p. 1020–1021.

Holdren, Jr., G. R., and R. A. Berner, 1979, Mechanism of feldspar weathering, I, Experimental studies: Geochimica et Cosmochimica Acta, v. 43, p. 1161–1171.

Holland, H. D., and M. Borcsik, 1965, On the solution and deposition of calcite in hydrothermal systems: Symposium on the problems of postmagmatic ore deposition, Prague, no. 2, p. 364–374.

Hunt, J. M., 1979, Petroleum geochemistry and geology: San Francisco, W. H. Freeman, 617 p.

LaGache, M., 1976, New data on the kinetics of the dissolution of alkali feldspar at 200°C in CO_2 charged water: Geochimica et Cosmochimica Acta, v. 40, p. 175–181.

Land, L. S., and S. P. Dutton, 1979, Reply cementation of sandstones: Journal of Sedimentary Petrology, v. 49, p. 1359–1361.

Larese, R. E., N. L. Haskell, D. R. Prezbindowski, and D. Beju, 1983, Sedimentologic and diagenetic controls on porosity development in selected Jurassic sandstone specimens from the Norwegian and North Seas, Norway—an overview: Proceedings of the North European Margin Symposium, Trondheim, Norway.

Lind, C., and J. Hem, 1975, Effects of organic solutes on chemical reactions of aluminum: United States Geological Survey Water Supply Paper 1827-G, 82 p.

McBride, E. F., 1980, Importance of secondary porosity in sandstones to hydrocarbon exploration: Bulletin of the American Association of Petroleum Geologists, v. 64, p. 742.

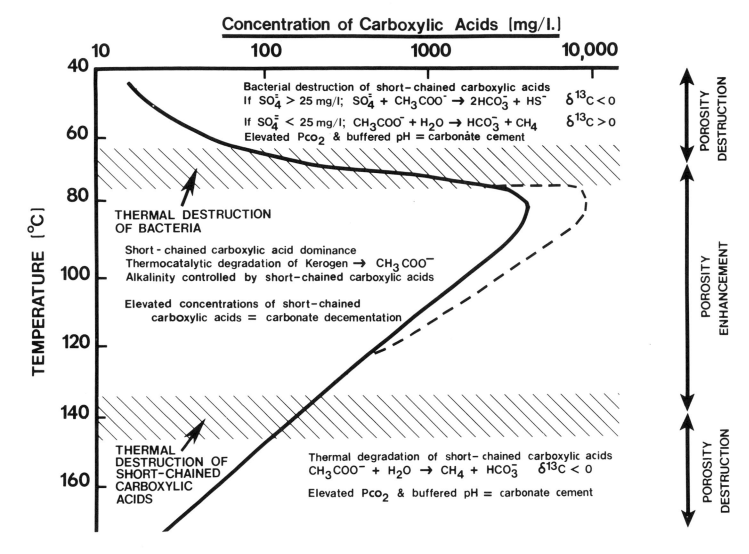

Figure 19—Diagram of the distribution of short chain carboxylic acids in oil field waters (modified from Carothers and Kharaka, 1978). Also shown on the diagram are the areas where bacterial degradation and thermal degradation are prominent. The C^{13} data for the oil field waters has been simplified, but is from Carothers and Kharaka, 1980. For a documentation of early calcite cements (isotopically light) in the zone of bacterial degradation see Larese et al, 1983.

Murphy, R. C., K. Biemann, M. V. Djuricic, and D. Victorovic, 1971, Bulletin of the Chemical Society of Belgrade, v. 36, p. 281.

Petrovic, R., 1976a, Rate control in dissolution of alkali feldspars, I, Study of residual feldspar grains by x-ray photoelectron spectroscopy: Geochimica et Cosmochimica Acta, v. 40 p. 537–548.

———, 1976b, Rate control in feldspar dissolution, II, The protective effect of precipitates: Geochimica et Cosmochimica Acta, v. 40, p. 1509–1521.

Schmidt, V., and D. A. McDonald, 1979, The role of secondary porosity in the course of sandstone diagenesis: Society of Economic Paleontologists and Mineralogists Special Publication 26, p. 175–208.

Seifert, K. E., 1967, Electron microscopy of etched plagioclase feldspar: Journal of the American Ceramic Society, v. 50, p. 660–661.

Tissot, B. P., and D. H. Welte, 1978, Petroleum formation and occurrence: New York, Springer-Verlag, 638 p.

Victorovic, D., 1980, Structure elucidation of kerogen by chemical methods, *in* B. Durand, ed., Kerogen: Paris, Technip, p. 301–338.

Origin of Secondary Porosity and Cement Distribution in a Sandstone/Shale Sequence from the Frio Formation (Oligocene)

G. K. Moncure
Conoco Inc.
Houston, Texas

R. W. Lahann
Conoco Inc.
Ponca City, Oklahoma

R. M. Siebert
Conoco Inc.
Ponca City, Oklahoma

INTRODUCTION

The diagenetic history of the Frio Formation (Oligocene) sandstones from the General Crude Oil/Department of Energy test wells at Pleasant Bayou, Brazoria County, Texas, has been outlined in a number of papers (Kaiser and Richmann, 1981; Milliken et al, 1981; Land and Milliken, 1981; Loucks et al, 1980a, 1980b). In a common approach to diagenetic analysis, these studies utilize data obtained from a series of samples over a large depth interval of a thick sedimentary sequence. Such results provide a depth–temperature–pressure framework for interpretation of diagenetic events. We have completed a very detailed investigation, which examines the distribution and origins of cements and secondary porosity within a small interval of one of these wells. We analyzed closely spaced core samples from a sandstone/shale interval from the GCO/-DOE No. 1 Test Well (Fig. 1). Specifically, this more detailed study was designed to investigate the movement of material from a shale into a sandstone during progressive diagenesis which results in cementation and secondary porosity. Many investigations have proposed that various cements found in sandstones were derived from adjacent shales (Johnson, 1920; Fothergill, 1955; Siever, 1962; Fuchtbauer, 1967, 1974; Weaver and Beck, 1971; Land and Dutton, 1978; Curtis, 1978; Boles and Franks, 1979; Krystinik, 1981). However, until now this mass transport and subsequent

ABSTRACT. Petrographic, SEM, and chemical analyses of closely spaced samples from a core of sandstone and shale (Oligocene Frio Formation, Brazoria County, Texas) reveal a mechanism for secondary porosity development. Maturation of organic and inorganic materials in the shale produced a solvent solution which, upon expulsion, resulted in zoned reservoir quality in the adjacent sandstone. Framework grain dissolution (secondary porosity) originated at the sandstone/shale contact zone (near the solvent source). Aluminum in this zone was not conserved by the process but instead was removed by mobile, shale-derived organic complexers. The production of these complexers (ligands) appears to be essential to the process of framework grain dissolution. Aluminum removal elevated the silica activity and resulted in precipitation of authigenic quartz cement.

Secondary porosity was developed to a lesser extent farther away from the shale. Imported aluminum from the contact zone and a failure to complex aluminum adequately resulted in kaolinite precipitation. This sink for silica prohibited quartz precipitation.

This general process of framework grain dissolution is probably common in sandstone/shale sequences. In summary, secondary porosity development is accentuated by: (1) high initial permeability, (2) increased relative thickness of shale to sandstone, (3) increased organic content in the shale, and (4) abundant soluble grains (potential secondary pores).

precipitation was not documented. This study provides documentation for the interval of investigation as well as corroborative evidence for the theory of framework grain dissolution advanced by Siebert et al, in this volume. We will use this theory as a framework for interpreting some of the observed diagenetic effects in our samples.

In this report we differentiate between framework grain dissolution (FGD) and framework grain alteration (FGA). The former process is the dissolution of grains and removal of much if not all of that material (presumably in solution) from the immediate vicinity. The FGD process has the greatest potential for creation of secondary

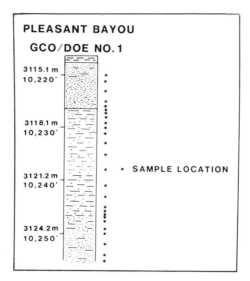

PLEASANT BAYOU

GCO/DOE NO. 1

3115.1 m
10,220'

3118.1 m
10,230'

3121.2 m
10,240'

3124.2m
10,250'

• SAMPLE LOCATION

Figure 1—Section of core upon which
this study is based.

porosity. The latter case (FGA) is a
process in which much of the material
from the dissolving grain precipitates in
the secondary pores and surrounding
intergranular pores as authigenic minerals (for example, clays, zeolites, carbonates). This process produces little
new porosity and tends to reduce overall permeability because the intergranular pore networks become clogged by
cement.

Framework grain dissolution (FGD)
is the more difficult of the two processes to explain. A chemically open
hydrologic system must be maintained
for the duration of dissolution. Aluminum (from feldspar and rock fragment
dissolution) is probably the least
mobile major element in the system
because of its very low solubility (that
is, saturation with respect to clay minerals is reached at very low aluminum
activity). As a result, the normally low
aluminum solubility (less than 10
mg/liter at 100°C) is a major limitation
to its transport and removal.

The precipitation of authigenic aluminosilicates (that is, framework grain
alteration) can be avoided in two ways:
(1) a high fluid flux can promote low
aluminum activity; unfortunately, the
water volume requirements for significant (about 5% rock volume) FGD are
very high (see Siebert et al, this volume)
and (2) complexing of aluminum could
maintain low aluminum (Al^{+3}) activities while allowing high total aluminum
concentrations in solution, thus reducing water volume requirements to reasonable levels.

GEOLOGIC SETTING

The core samples examined in this
study are from the General Crude Oil/
Department of Energy (GCO/DOE)
Pleasant Bayou Geothermal Test Well
No. 1, Brazoria County, Texas. The
study interval (10,221–10,257 ft or
3115.4–3126.3 m) is near the top of the
geopressured zone (Freed, 1980) and
has a temperature of about 100°C. The
samples are also near the top of the oil
generation window and have an average vitrinite reflectance value of about
0.6. We lack data for detailed environmental interpretations, but deposition
was a part of the Frio (Oligocene) fluvial deltaic system (Tyler, personal
communication). The presence of
marine microfossils in the core indicates a marine environment.

The core interval contains a smectitic
shale interlayered between two fine- to
very fine-grained feldspathic litharenites. The lower sandstone fines upward
gradationally into shale; detrital matrix
clay content is very high. Consequently,
porosities and permeabilities are low
and movement of fluid from shale to
sand was retarded. FGD in the lower
sandstone is negligible. The upper
sandstone, in contrast, rests on the
shale with a rather sharp contact; only
a few burrows and/or small shale rip-
up clasts indicate minor grain mixing at
the contact. Detrital matrix clay content is low. Porosities and permeabilities were apparently high enough to
permit passage of expelled shale fluids

during diagenesis. FGD is significant in
the upper sandstone.

SHALE TREND

Inorganic

The shale over the interval of interest 10,226.5–10,245 ft (3117–3122.7 m)
has an average composition of 56%
total clay mineral, 15% plagioclase,
23% quartz, and 7% calcite as determined by X-ray diffraction. These
results are in good agreement with
those reported by Freed (1980). The
clay fraction is dominated (perhaps
80–90%) by a mixed-layer montmorillonite–illite of about 50% expandability. A minor amount of low expandability detrital mica is present, along
with a variable amount of chlorite.

In order to obtain a qualitative measure of the amount of chlorite relative
to mixed-layer clay minerals, oriented
clay mounts were heated at 300°C and
the area of the 7-Å X-ray diffraction
peak (chlorite 002) was compared with
the area of the 10-Å peak (total of detrital mica and collapsed mixed-layer
clay) and this ratio is plotted against
depth in Figure 2; clearly the relative
chlorite abundance increases toward
the sandstone/shale contact.

Organic

Organic geochemical analyses were
also performed on the ten shale samples: total organic carbon (TOC), total
extractables and extractable hydrocarbons, kerogen type, and vitrinite reflectance were determined. Organic carbon
content averaged about 0.25%, a rather
low value in terms of source rock
potential. Total extractables and
extractable hydrocarbons tended to be
relatively high at depths that had
"high" organic carbon (see Fig. 2). In
addition to the correlation of total
organic carbon and extractables, there
is an overall upward increase in the
amount of extractable organics. The
association of increased extractables
with total organic carbon suggests in
situ hydrocarbon generation and the
upward increase in extractables suggests local vertical migration toward
the sandstone above.

Figure 2 shows the depth profile for
resins and asphaltenes (RASP) and
percent extractables, relative to TOC.

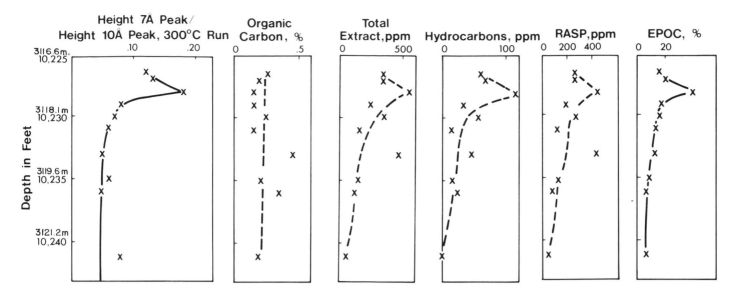

Figure 2—Clay mineral and organic geochemical variations within the shale from depth 10,226.51 to 10,242 ft (3117–3122 m). RASP and EPOC refer to resins and asphaltenes and extractable percentage of organic carbon, respectively. Chlorite variation is shown by the ratio chlorite 002 (7-Å peak) /mica 001 (10-Å peak).

The RASP appears to increase upward although not so rapidly as do hydrocarbons; this is probably due to less rapid migration of organics. The close parallel between the percent extractables curve and the relative chlorite enrichment curve suggests migration of inorganic constituents as a cause for the chlorite distribution.

SANDSTONE TRENDS

Secondary Porosity

Figure 3a and 3b illustrate the changes in the amount of leached feldspar and rock fragment porosity with depth. Average values from two sets of thin-section point-count data (different operators) are shown in this and other figures. The amount of FGD in feldspar grains decreases away from the sandstone/shale contact. A trend in the FGD owing to removal of rock fragments is not apparent and may result from variations in the amount and chemical stability of original rock fragments (predominantly volcanic rock fragments with minor metamorphics). Added together, the total FGD

decreases away from the sandstone/shale contact (Fig. 3b) implying a shale source for the solvent fluid. This trend could be misleading if, for example, the total amount of FGD simply reflected availability of soluble grains. Considering this possibility, Figure 4 illustrates the change with depth in total FGD normalized with respect to total original (potentially) soluble grains. This diagram confirms that the degree of leaching systematically decreases away from the sandstone/shale contact. We believe the FGD trend results from proximity to fresh shale-sourced organic acids.

Chlorite

The distribution of chlorite within the sandstone is shown in Figure 5. A sharp decrease in chlorite content with increasing distance from the sandstone/shale contact is evident. Chlorite in the contact sandstone is present as thin isopachous rims (Fig. 6). From X-ray diffraction and electron microprobe analyses, the chlorite is an iron-rich variety (see Table 1). Elsewhere in the sandstone, "chlorite" forms even thinner, discontinuous rims. The mineralogy and composition of the thin rims farther from the contact zone was difficult to verify. We feel that the textural and chemical evidence suggests that the material that precipitated as chlorite originated in the adjacent shale.

Calcite

Calcite is volumetrically minor in these rocks. It occurs both as randomly distributed pore-filling cement and as small patches clustered near foram and echinoderm fragments. The abundance of carbonate cement does not vary systematically with depth. We obtained compositional analyses of the carbonate cements by electron microprobe. The results of these analyses are plotted in Figure 7 according to petrographic type. Problems with this comparative analysis arise from: (1) difficulty in recognizing cement associated with fossil fragments, owing to two-dimensional sampling of a thin section, and (2) possible compositional zoning "averaged" by a broad electron beam during analysis. The data in Figure 7 indicate that, at all sample depths, calcite cements that are associated with fossil fragments contain essentially no magnesium. The remaining points, which contain larger amounts of magnesium, are probably of a different generation. These remaining points show increasing iron and magnesium content toward the sandstone/shale contact.

In summary, we feel that at least two generations of calcite cement exist; one physically associated with and probably derived from carbonate fossil fragments by compaction (pressure solution) and another showing iron and magnesium enrichment toward the sandstone/shale contact. There is no indication of decementation.

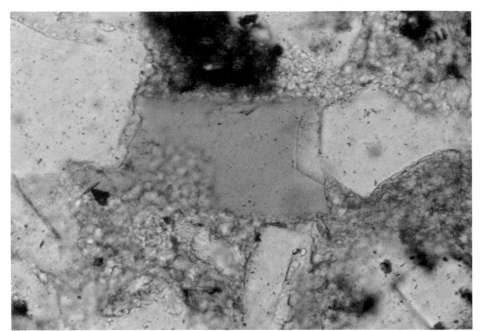

Figure 6 — Photomicrograph showing thin, isopachous chlorite rims in the contact sandstone (3098.9 m). Here the rim surrounds a secondary pore. Note the quartz overgrowth inside the pore and over the chlorite. Figure width = .81 mm.

TABLE 1

Formula Stoichiometry, Assuming 28 O
$(Mg, Al, Fe)_{12} [(Si, Al)_8 O_{20}](OH)_{16}$

Analysis	Ca	Na	K	Mn	Fe[1]	Mg	Al	Si	Ti
1	0.01	0.04	0.01	0.01	5.05	2.97	5.54	5.81	0.01
2	0.02	0.03	0.01	0.02	5.06	2.79	5.35	6.04	—
3	0.01	0.10	0.01	0.01	5.62	2.45	5.37	5.70	0.01
4	0.01	0.01	0.01	0.01	5.15	2.77	5.38	5.99	—
5	0.02	0.05	—	0.01	5.60	2.29	5.26	6.08	0.01

[1]All Fe reported as Fe^{+2H}

Table 1—Chlorite compositions from depth 10,226.5 ft (3117 m). All compositions are based on microprobe analyses.

A common feature in this sandstone (as well as sandstone of other formations, see Siebert et al, this volume) is the presence of dissolution pores adjacent to undissolved calcite foraminifera tests, echinoderm spines, etc. (Fig. 8). This association imposes restrictions on the chemical nature of the solvent solution.

Kaolinite

Between the sandstone/shale contact and 0.3 m into the sandstone, authigenic kaolinite occurs as a patchy primary pore-filling cement and as a minor replacement of rock fragments. Within this zone, kaolinite does not occupy secondary porosity after feldspar. Beyond 0.3 m from the contact, authigenic kaolinite occupies sites of dissolved grains and is also a primary pore-filling cement. We interpret this as indicating that the kaolinite in and about the dissolved grain sites was due to FGA.

The distribution of kaolinite within the sandstone is illustrated in Figure 9. The data plotted are averages from point counts by two operators. Clearly, the sandstone at the sandstone/shale contact contains the least amount of kaolinite, less than 1%. The abundance increases to about 8% at 0.5 m from the contact and decreases upsection from that point. The distribution of kaolinite is not a simple function of shale proximity but apparently reflects a combination of FGD, transport and reprecipitation and local FGA.

Quartz Overgrowths

Quartz overgrowths exist in small to negligible amounts but are distributed systematically with respect to depth (Fig. 10). The plot demonstrates that

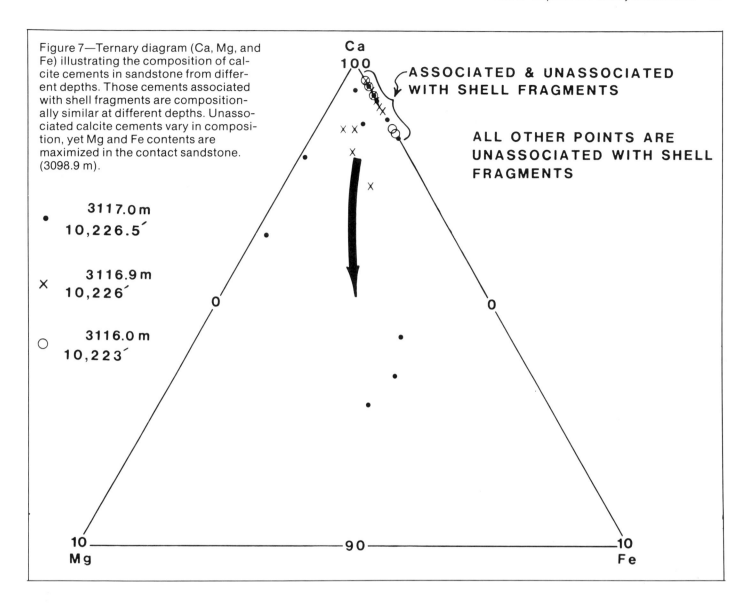

Figure 7—Ternary diagram (Ca, Mg, and Fe) illustrating the composition of calcite cements in sandstone from different depths. Those cements associated with shell fragments are compositionally similar at different depths. Unassociated calcite cements vary in composition, yet Mg and Fe contents are maximized in the contact sandstone. (3098.9 m).

ASSOCIATED & UNASSOCIATED WITH SHELL FRAGMENTS

ALL OTHER POINTS ARE UNASSOCIATED WITH SHELL FRAGMENTS

3117.0 m
10,226.5´

3116.9 m
10,226´

3116.0 m
10,223´

the frequency of occurrence of authigenic quartz increases sharply at the sandstone/shale contact (these data are from 1000 point counts). The overgrowths in the contact sandstone appear to be selectively distributed according to porosity and permeability variations. They are most heavily concentrated in areas where: (1) diagenetic modification is the least apparent and original porosity and permeability was high, or (2) in areas where framework grain dissolution is high (little or no kaolinite is present) and porosity and permeability was accentuated.

DIAGENETIC HISTORY

The foregoing discussion provides evidence demonstrating that fluids that were modified by diagenesis in the shale passed into the overlying sandstone and exerted significant influence on its diagenesis. In this discussion we review the paragenetic events and use these in conjunction with mass balance calculations to account for the distribution of authigenic products.

The transport and precipitation of chlorite in the upper shale zone, as discussed previously, invaded the narrow margin of sandstone at the contact. This chlorite formation resulted from expulsion of chlorite components from the shale. Chlorite rims around secondary pores (Fig. 6) and a total lack of chlorite inside these pores or on partially dissolved grain surfaces (Fig. 11) indicate that chlorite cementation took place prior to significant framework

grain dissolution. This hypothesis is consistent with the observation by Land and Dutton (1978) of very early chlorite grain coatings. It is likely that one or more generations of calcite cement were deposited from the Fe- and Mg-rich waters expelled during chlorite precipitation. This supposition is based on the increasing Fe and Mg content in some calcite cements toward the sandstone/shale contact.

Framework grain dissolution/alteration reached its greatest intensity following chlorite precipitation and probably during maximum conversion of smectite to illite in the shale. Precipitation of quartz and kaolinite took place during and after FGD/FGA. Complexation of aluminum in the shale by organic complexing agents was main-

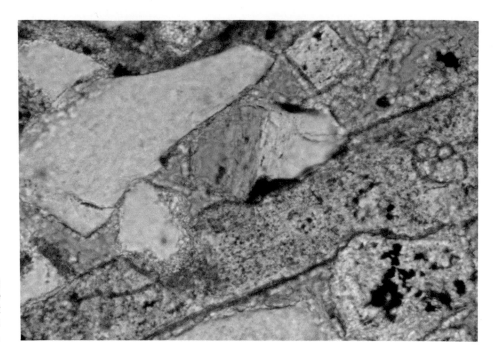

Figure 8—Photomicrograph illustrating a secondary pore (presumably from the dissolution of a feldspar grain) adjacent to an echinoderm spine (faint pink color). Figure width = .81 mm.

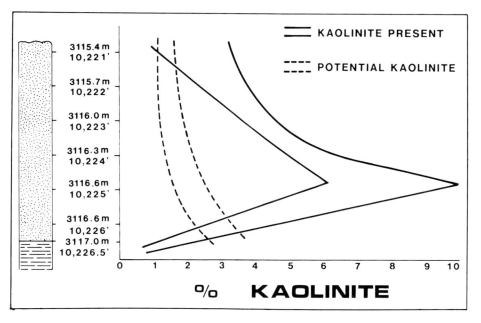

Figure 9—Plot showing the distribution of kaolinite cements (depicted as a band—from two sets of point-count data). Potential kaolinite is calculated on the assumption that the system is closed with respect to Al, that is, all available Al precipitates as kaolinite. The discrepancy indicates that this assumption is false.

tained at low concentrations because aluminum was buffered to low levels by the reaction of smectite to illite, an aluminum-consuming process (Siebert et al, this volume). When this shale solution was expelled into the contact sandstone, framework grains began to dissolve and liberate aluminum and silica among other ions. The Al^{+3} produced by the dissolving grains was organically complexed. This action prevented (initially) supersaturation with respect to kaolinite, thus prohibiting its precipitation and permitting increased concentration of aqueous silica. As a result, quartz overgrowths developed, especially in the contact sandstone.

Farther away from the sandstone/shale contact FGD was less extensive (Fig. 3). This may be explained by exhaustion of organic complexing agents (because of low organic content of the shale) by aluminum uptake as fluids passed through the contact sandstone. The remaining acid content of the shale water (as aqueous CO_2, HCO_3, etc.) was consumed in converting feldspars and rock fragments to kaolinite with the concomitant precipitation of calcite (see Siebert et al, this volume). Precipitation of kaolinite away from the contact decreased aqueous silica levels and quartz overgrowths (as a result) are not as abundant. The presence of undissolved, delicate car-

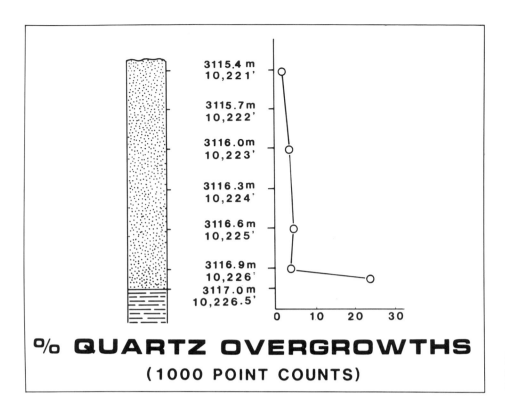

Figure 10—Distribution of quartz overgrowths in the sandstone (from point-count data).

bonate skeletal grains adjacent to leached feldspar pores is significant. Obviously, solvent solutions were saturated or supersaturated with respect to calcite during FGD. This observation is consistent with the presence of calcite in the shale and argues for an aluminum complexing model rather than a simple CO_2 (that is, acid) dissolution mechanism for FGD.

An in situ (sandstone) source of silica for quartz and kaolinite cement was tested by mass balance calculations (Fig. 12). The solid points represent the potential quartz cement assuming all silica from FGD precipitated as quartz overgrowths. The data source for this calculation is the averaged point count of dissolved grains plotted in Figure 3. The open circles represent the amount of silica cement possible after the subtraction of the silica consumed in precipitating the measured (observed) amounts of kaolinite. The resulting theoretical quartz cement trend corresponds well to observed occurrence of quartz overgrowths (Fig. 10), although the calculation predicts somewhat more quartz overgrowths in the contact sandstone than were observed. Based on this calculation, we do not invoke an

external source of silica. However, this deficiency in authigenic quartz does imply, without statistical certainty, that silica may have been exported from the contact sandstone, perhaps by an organic complex.

A similar calculation was performed to demonstrate aluminum mobility within the sandstone. The values for observed kaolinite abundance were compared with calculated values assuming all aluminum from FGD precipitated as kaolinite (Fig. 9). Based on this calculation, Zone 1, the contact zone, is deficient with respect to kaolinite precipitation. Zone 2 contains excess kaolinite suggesting that aluminum from Zone 1 was transported to and precipitated in Zone 2. Zone 3, farthest from the shale, contains about the same amount of kaolinite as would be expected from FGD and may represent a relatively closed system where kaolinite precipitation was controlled by in situ sources of aluminum and silica.

DISCUSSION

The evidence indicates that shale diagenesis exerts considerable influence

on dissolution or alteration of framework grains and the distribution of authigenic cements in sandstones. Furthermore, we propose that the general paragenesis and mechanisms described in this investigation (and Siebert et al, this volume) are very common in lithologically similar sandstone/shale sequences. Qualitatively, we expect FGD to increase in importance over FGA as the: (1) shale becomes thicker relative to sandstone, and (2) the organic content of the shale increases. These increase the amount and effectiveness (that is, more complexing agents) of the FGD solvent and allow a given volume of shale to produce FGD over more sandstone.

Some major requirements for significant framework grain dissolution suggested by this study are as follows: (1) high permeability relative to other fluid expulsion pathways in the shale. A shale encased by sandstones having significantly different permeabilities, as in our example, will transmit the bulk of the fluid to the more permeable sandstone. Initially, the environment of deposition controls the relative permeabilities of the possible expulsion pathways. For example, sorting and the

Figure 11 — SEM photomicrograph shows chlorite forming the original border to a largely dissolved feldspar grain. Left side figure width = .12 mm; right side enlargement of boxed area on left = .03 mm width.

presence or absence of matrix clay controlled the relative permeability of the two boundary sandstones in our examples. The environment of deposition can also control the sand's composition, which in turn controls the type and amount of early diagenesis, which can lead to large differential permeabilities. For example, compaction of ductile grains or carbonate cements derived from aragonite shell fragments can produce large permeability declines. Also, the abundance of soluble grains (feldspars and rock fragments) determines the theoretical upper limit to the amount of framework grain dissolution. This limit can be approached when the total of soluble grains is low (less than about 5%), when the ratio of available solvent to sand is high, and when the solvent is undersaturated with respect to the most stable soluble grain (for example, albite). In addition, the shales must be thick enough and sufficiently rich in organics to produce enough solvent.

Assuming these requirements are met, framework grain dissolution will proceed until: (1) the solvent capacity of the shale system is exhausted, or (2) all soluble grains with which the solvent is undersaturated have dissolved. Albi-

tization, which is a common alteration process in feldspathic sandstones, occurs in part along with framework grain dissolution. In this case, the solvent is undersaturated with respect to a dissolving calcic plagioclase host yet saturated or slightly supersaturated with respect to albite. The end result of albitization is the formation of a relatively stable product thereby removing that material from the pool of soluble grains. In this study the effects are minor in that albitization is just starting (petrographic and microprobe analyses). Removal of aluminum by ligands during albitization would severely modify stability relations calculated by Boles (1979).

ACKNOWLEDGMENTS

The authors with to acknowledge, with appreciation, the stimulating consultations with R. C. Surdam, J. N. Shearer, R. Parker, R. McLimons, and N. Tyler. We also wish to thank L. Blubaugh, C. Robertson, and M. Bourne for their laboratory assistance. The authors wish to thank W. J. Hoover and B. N. Fukui for their figure designs.

Also, the authors wish to express great appreciation to the Texas Bureau of Economic Geology for kindly providing some samples for this study, and Conoco, Inc. for permitting release of this work.

Finally, we express our appreciation to M. M. Nelson for typing the manuscript, over and over again.

SELECTED REFERENCES

Boles, J. R., 1979, Active albitization of plagioclase in Gulf Coast Tertiary sandstones: Geological Society of America Abstracts with Programs, v. 11, p. 391.

Boles, J. R., and S. G. Franks, 1979, Clay diagenesis in Wilcox sandstones of southwest Texas: implications of smectite diagenesis on sandstone cementation: Journal of Sedimentary Petrology, v. 49, p. 55–70.

Curtis, C. D., 1978, Possible links between sandstone diagenesis and depth-related geochemical reactions occurring in enclosing mudstones: Quarterly Journal of the Geological Society of London, v. 135, p. 107–117.

Fothergill, C. A., 1955, The cementation of oil reservoir sands and its origin: Proceedings of the 4th World Petroleum Congress, p. 301–314.

Freed, R. L., 1980, Shale mineralogy of the No. 1 Pleasant Bayou geothermal test well: a progress report, *in* M. H. Dorfman and W. L. Fisher, eds., Proceedings of the Fourth United States Gulf Coast Geopressured-Geothermal Energy Con-

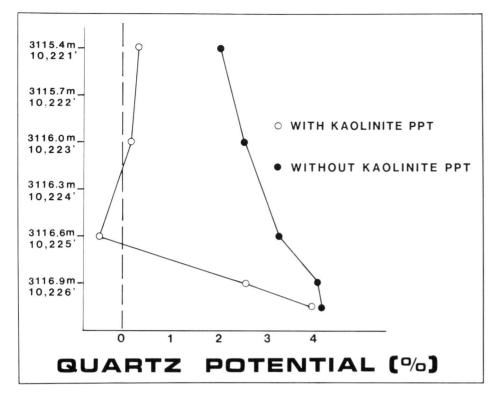

○ WITH KAOLINITE PPT

● WITHOUT KAOLINITE PPT

QUARTZ POTENTIAL (%)

Figure 12—Silica balance diagram. The plot of solid circles indicates the potential volume percent of cement if all the available silica from framework grain dissolution were to precipitate as quartz overgrowths. Subtracted from these values is the amount of silica consumed by precipitation of kaolinite (open circles). Compare with Figure 10.

ference: Austin, University of Texas, p. 153–165.

Fuchtbauer, H., 1967, Influence of different types of diagenesis on sandstone porosity: Proceedings of the 7th World Petroleum Congress, p. 354–369.

———, 1974, Sediments and sedimentary rocks, II: New York, Halsted Press, 464 p.

Johnson, R. H., 1920, The cementation process in sandstones: Bulletin of the American Association of Petroleum Geologists, v. 4, p. 33–35.

Kaiser, W. R., and D. L. Richmann, 1981, Predicting diagenetic history and reservoir quality in the Frio Formation of Brazoria County, Texas and Pleasant Bayou test wells, *in* Proceedings of the Fifth Louisiana Geological Survey United States Gulf Coast Geopressured-Geothermal Energy Conference: Baton Rouge, Louisiana State University and United States Department of Energy, p.

67–74.

Krystinik, L. F., 1981, Pore-filling cements: products of shale dewatering in the Upper Miocene Stevens Sandstone, Elk Hills, Kern County, CA, (abs.): Rocky Mountain Association of Geologists.

Land, L. S., and K. L. Milliken, 1981, Feldspar diagenesis in the Frio Formation, Brazoria County, Texas Gulf Coast: Geology, v. 9, p. 314–318.

Land, L. S., and S. P. Dutton, 1978, Cementation of a Pennsylvanian deltaic sandstone: isotopic data: Journal of Sedimentary Petrology, v. 48, p. 1167–1176.

Loucks, R. G., D. L. Richmann, and K. L. Milliken, 1980a, Factors controlling porosity and permeability in geopressured Frio sandstone reservoirs, General Crude Oil/Department of Energy Pleasant Bayou test wells, Brazoria County, Texas, *in* M. H. Dorfman and W. L. Fisher, eds., Proceedings of the Fourth United States Gulf Coast Geopressured-Geothermal

Energy Conference: Austin, University of Texas, p. 46–82.

———, 1980b, Factors controlling reservoir quality in Tertiary sandstones and their significance to geopressured-geothermal production: United States Department of Energy, Division of Geothermal Energy, DE-AC08-79ET27111, p. 188.

Milliken, K. L., L. S. Land, and R. G. Loucks, 1981, History of burial diagenesis determined from isotopic geochemistry, Frio Formation, Brazoria County, Texas: Bulletin of the American Association of Petroleum Geologists, v. 65, no. 8.

Siever, R., 1962, Silica solubility, 0–200°C. and the diagenesis of siliceous sediments: Journal of Geology, v. 79, p. 127–151.

Weaver, C. E., and K. C. Beck, 1971, Clay water diagenesis during burial: how mud becomes gneiss: Geological Society of America Special Paper 143, p. 96.

A Theory of Framework Grain Dissolution in Sandstones

R. M. Siebert
Conoco Inc.
Ponca City, Oklahoma

G. K. Moncure
Conoco Inc.
Houston, Texas

R. W. Lahann
Conoco Inc.
Ponca City, Oklahoma

INTRODUCTION

Framework grain dissolution (FGD) is the phenomenon by which feldspar and rock fragments are dissolved and the components removed in solution from the sandstone. This creates additional porosity (secondary porosity) in the sandstone. The FGD phenomenon is considered distinct from *framework grain alteration*, in which the mass of the dissolved grains is precipitated as authigenic minerals (for example, clay, zeolite) in the surrounding pore system. We also exclude the process by which carbonates (for example, calcite) chemically replace sand grains or fill porosity and are subsequently dissolved to produce secondary porosity (that is, decementation or decarbonization).

One of the earliest published works in which FGD porosity was recognized as abundant and significant to reservoir porosity enhancement was that by Heald and Larese (1973). They reported up to 4.4% feldspar dissolution porosity in the Mt. Simon Sandstone (Cambrian of Ohio) and also noted that the waters effecting the dissolution were not "acidic" because most of the sandstones were high in carbonate. Other workers have subsequently reported on the importance of FGD to reservoir properties. Prominent among these are Loucks et al (1977) and Hayes et al (1976). Noteworthy among those authors who have published on the decementation type of secondary porosity are Schmidt et al (1977), McBride (1977), Stanton (1977), Dutton (1977), and Lindquist (1977). Most of these authors considered FGD po-

ABSTRACT. Framework grain dissolution (FGD) involving feldspars and rock fragments was found to be significant to reservoir properties in sandstones with more than 10% soluble grains. FGD porosity ranged up to approximately 70% and averages about 30% of the visible porosity in a study of some reservoir sandstones. FGD does not appreciably increase reservoir permeability. However, the amount of FGD porosity developed was found to be a function of the sandstone's initial permeability.

We propose that clay and organic maturation in shales produce the necessary water, acid, and complexing agents for FGD. The FGD solvent is expelled into the sandstones where feldspars and rock fragments are dissolved, and the resulting aqueous aluminum is complexed for transport out of the sandstone.

rosity to be very minor volumetrically or, at the most, subordinate to the decementation type of secondary porosity.

Our purpose here is to present observations on FGD and to propose a theory for the FGD process based on these observations.

OBSERVATIONS ON FRAMEWORK GRAIN DISSOLUTION

Petrographic observations indicate considerable variation in the FGD process and the resulting textures. Feldspars and rock fragments can dissolve completely, leaving an obvious grain-size pore in the sandstone that shows little evidence of the original grain. Often the process leaves behind an insoluble rind of an authigenic mineral or a small amount of insoluble debris. The most common result of FGD is

incomplete dissolution of the grain. FGD is commonly compositionally selective as in the removal of a more soluble calcic core of a zoned plagioclase or one of the components in a microperthite. FGD in rock fragments often leaves a frothlike aggregate of authigenic clays or the less soluble rock fragment components. The porosity of these rock fragment remnants or ghosts is usually very high, but is typically not very effective porosity for hydrocarbon storage because of the small pore sizes. Some examples of these features are presented in Figures 1 to 8.

We and others (Hayes, 1979, for example) have observed dissolution of sanadine, orthoclase, microcline, and the plagioclases with the possible exception of the more albitic end-members. There does appear to be a general hierarchy of feldspar solubility in which the more calcic plagioclases

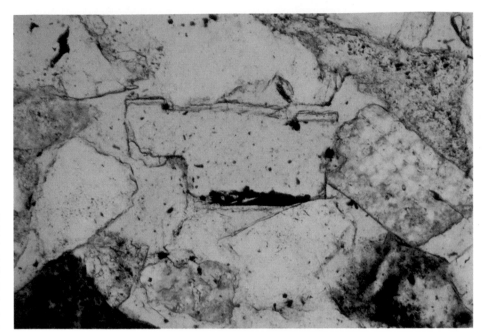

Figure 1—Photomicrograph of FGD pore (after feldspar), which is outlined by an insoluble potassium-feldspar overgrowth. Note lack of compaction effects on overgrowth. Black material at bottom of overgrowth may be insoluble debris left after FGD. Cretaceous Shannon Formation, Wyoming (4677.5 ft). Field width = 0.89 mm.

are attacked first. Almost completely dissolved plagioclase grains are found very commonly next to other plagioclase grains (presumably less calcic), which show no dissolution or surface etching.

Our observations indicate that the bulk of FGD occurs upon moderate to deep burial, rather than near the surface or in a surface-weathering environment. The dissolution remnants of feldspars are typically delicate structures, and they seldom show mechanical disruption or crushing as one might expect during the early stages of compaction. The presence of rims of authigenic minerals around dissolution pores (see Figs. 1 and 2) also supports a "deeper" burial environment for the process. As the rims are delicate and are seldom crushed, the dissolution of the core grain must have occurred after significant compaction (that is, burial depth) had occurred.

The relationship between pore-filling cements and dissolution pores provides additional evidence for the timing of FGD. The intergranular pore filling of calcite and kaolinite in Figures 4 and 5 does not extend into the FGD pore, indicating that the dissolution occurred after the cementation. In other instances, cements (such as calcite and quartz overgrowths) do invade dissolution pores. These cements are considered

diagenetically late (that is, deep burial) cements. FGD apparently operates after early cementation and before or contemporaneously with late cementation events.

A final but important observation is the relatively common coexistence of carbonate fossils, carbonate rock fragments, or early calcite cements with FGD porosity. The example in Figure 6 shows a dissolution pore (after feldspar) next to a partially crushed foram test. This relationship indicates that the solution that dissolved the feldspar grain was not corrosive to the carbonate mineral (probably calcite). This relationship places constraints on the chemistry of the solvent solution.

ABUNDANCE OF FRAMEWORK GRAIN DISSOLUTION POROSITY

We have found that framework grain dissolution (FGD) porosity constitutes a very significant fraction of total porosity in most feldspathic litharenite samples examined. In a thin section study[1] of 71 samples from 7 formations

---

[1]The sandstone samples were pressure impregnated with clear, blue-dyed epoxy, and the thin sections were chemically stained to color the feldspars. We feel that this is the minimum treatment required to make reliable identification of FGD porosity.

(mostly hydrocarbon-bearing reservoirs from the U.S. Gulf Coast Tertiary and the Cretaceous of the Rocky Mountains), the FGD porosity was found to range from 0 to 69% of the total visible porosity and to average 28.6% (standard deviation = 10.5) of the overall porosity. FGD porosity in terms of volume percent of the rock averaged 4.4% (standard deviation = 1.9) and ranged from 0 to 9.6%. These data, along with averages and ranges for the individual formations, are presented in Table 1. Visual estimates and point counts of numerous samples from many other formations indicate that these results are quite representative for sandstones containing more than about 10 to 15% feldspar and rock fragments. Our observations indicate that framework grain dissolution is essentially ubiquitous in sandstones that are in or somewhat above the "hydrocarbon generation window" or have received fluids from such rocks. Consequently, FGD is observed over a very large range of burial depths in a given formation.

An unexpected result of this study is that the total amount of FGD is apparently independent of the total amount of feldspar and rock fragments (that is, the total potentially soluble grains) in the samples. The average volume percent of potentially soluble grains for

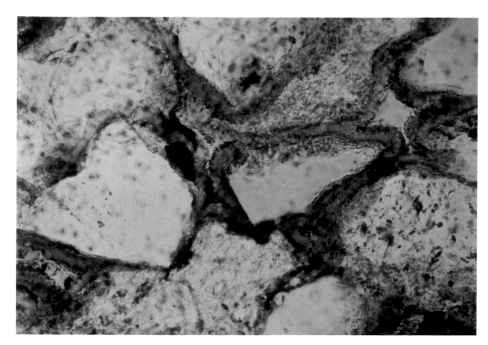

Figure 2—Photomicrograph of dissolved grain outlined by an isopachous coat of authigenic chlorite. FGD is post-chlorite precipitation. Note the apparent lack of compaction of the chlorite coat. Oligocene Frio Formation, South Texas (5518 ft). Field width = 0.35 mm.

each formation has been included in Table 1, and there is no correlation between these and the volume percent of dissolved grains. This result suggests some inherent upper limit to the FGD process, probably the ultimate capacity of the sediments to produce FGD solvent.

FATE OF THE DISSOLVED MATERIAL

Our data indicate that most, if not all, of the dissolved material resulting from FGD is removed from the vicinity and, thus, new additional porosity results. This removal is indicated by lack of sufficient volumes of authigenic minerals to account for the dissolved mass. The following chemical reaction is considered representative of the case where no aluminum or silica is removed from the formation.

$$(2NaAlSi_3O_8 \cdot CaAl_2Si_2O_8) + 4H^+(aq)$$
$$\text{(andesine plagioclase)} \qquad (1)$$

$$+ 2H_2O \rightarrow 2Al_2Si_2O_5(OH)_4 + 4SiO_2$$
$$\text{(authigenic} \quad \text{(quartz}$$
$$\text{kaolinite)} \quad \text{overgrowth)}$$

$$+ 2Na^+(aq) + Ca^{+2}(aq)$$

Using the appropriate molar volumes for equation 1, we calculate that one volume of dissolved plagioclase could produce 0.66 volumes of kaolinite and 0.30 volumes of quartz. Since the authigenic clay contents of most of the samples studied are dominated by kaolinite, the measured amounts of authigenic clay can be compared with the amounts expected if no aluminum and silica were transported out of the sandstone. We assume for this calculation that the amount of aluminum derived from one volume of rock fragments is about the same as for feldspar. The results presented in Table 2 show the actual clay content to be much lower than expected if no aluminum were removed from the sandstone. In fact, the differences between the measured and calculated values may be even larger than indicated since we have probably underestimated the microporosity content of kaolinite patches during point counting and thus overestimated the amount of kaolinite.

The mass balance for authigenic quartz is more complex. The calculated authigenic quartz volumes based on equation 1 are presented as Case 1 in Table 2. The values are only slightly larger than the measured values in the BNU and ZJ wells. However, the calculated value for well FYX is much

smaller than the actual amount of authigenic quartz, indicating sources for authigenic quartz other than FGD. These sources may also be providing authigenic quartz in wells BNU and ZJ. We note that when we calculated the amount of authigenic quartz from FGD on the basis of no silica precipitation in clay (Case 2 in Table 2), the measured authigenic quartz in wells BNU and ZJ was much less than the calculated values.

We conclude from the above lines of evidence that little of the aluminum from FGD is precipitated in the sandstones as authigenic minerals. Our observations show this to be generally true of many other samples examined. Moreover, the possibility exists that much of the authigenic minerals in these sandstones is due to an earlier diagenetic event (for example, framework grain alteration) rather than associated with the FGD event.

RESERVOIR PROPERTIES AND FGD

Besides the obvious increase in total reservoir porosity and, possibly, lower irreducible water saturation per unit porosity (for example, see Schmidt et al, 1977), we have concluded that the

TABLE 1

Formation	Number of Samples	Average[1] Soluble Grains	% of Total Porosity[2]		% of Rock	
			Average	Range	Average	Range
Nodosaria	17	18	27	11–51	4.6	1.8–9.6
Frio	9	59	39	20–69	5.3	1.3–8.1
Vicksburg	16	80	31	17–49	4.9	1.7–8.5
Sussex	13	59	25	16–35	4.1	2.0–5.6
Shannon + Wall Creek	9	56	24	0–34	3.7	0–5.7
Wilcox	7	32	28	24–33	3.8	2.6–5.1
Total	71	—	28.6 ±10.5	0–69	4.4 ±1.9	0–9.6

[1]Vol % average of feldspar, plus rock fragments, in the calculated original composition of the sand.
[2]Values given as vol %. Determined by point counting an average of 565 (±71) points per thin section.

Table 1—Average framework grain dissolution porosity.

TABLE 2

Well Code and Formation	Dissolved[1,7] Grains	Clay[2,7]	Calc.[3] Clay	Quartz[4,7] Overgrowth	Calc.[5] Quartz Case 1	Calc.[6] Quartz Case 2
FYX (Nodosaria)	4.6	1.00	3.04	7.8	1.4	2.8
BNU (Frio)	5.3	0.34	3.50	1.2	1.6	3.2
ZJ (Vicksburg)	3.9	1.77	2.57	0.8	1.2	2.3

[1]Same as FGD porosity. This is an average vol % of feldspar and rock fragments dissolved.
[2]Average vol % from point count. This number is probably too large owing to microporosity in the kaolinite patches.
[3]Calculated kaolinite content assuming that one volume of dissolved grains would yield 0.66 volume of kaolinite (equation 1).
[4]Average vol % of quartz overgrowths by point count.
[5]Calculated vol % of authigenic quartz assuming one volume of dissolved grains would yield 0.3 volume of quartz (equation 1).
[6]Calculated vol % of authigenic quartz assuming that all SiO_2 from dissolved grain is precipitated as authigenic quarz (that is, no SiO_2 is precipitated along with aluminum to produce kaolinite).
[7]Points counts in triplicate for a FYX thin section and in duplicate for a BNU thin section yielded results where the low values were, in the worst case, approximately 55% of the high values. The original data can be obtained from the senior author.

Table 2—Comparison of average volumes of clay and authigenic quartz with calculated equivalent volumes derived from dissolved grains for 3 wells.

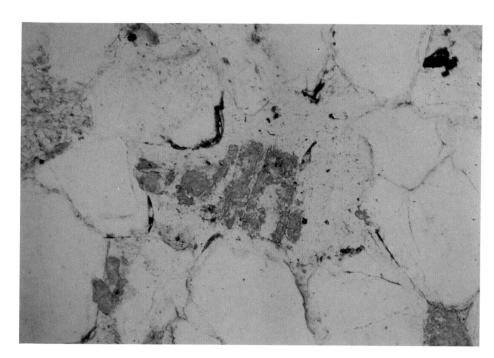

Figure 3—Photomicrograph of partially dissolved plagioclase feldspar in "cage" of welded quartz grains. This type of porosity is expected to survive further burial compaction (that is, porosity reduction) better than primary porosity. Oligocene Nodosaria sandstone, Louisiana (14,556 ft). Field width = 0.89 mm.

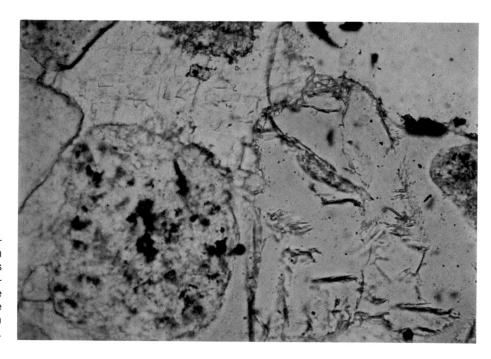

Figure 4—Photomicrograph of dissolved feldspar that is in contact with calcite cement patch (upper left). This calcite has not penetrated the FGD porosity, indicating that the FGD came after calcite cementation. Oligocene Frio, South Texas (3002 ft). Field width = 0.35 mm.

permeability of a sandstone is not greatly increased by addition of moderate amounts (that is, less than 8 vol % of the rock) of FGD porosity. We found a strong, positive correlation between the amount of FGD porosity and the logarithm of permeability. Two examples of this correlation are shown in Figures 9 and 10. This correlation

naturally leads to the question of whether the increasing FGD porosity with increasing permeability is due to (1) large scale enhancement of permeability by FGD, or (2) the sandstone's original permeability controlling the amount of FGD solvent introduced into the sandstone and, thus, the amount of FGD porosity.

Considerations of permeability controlling factors suggest that FGD should not significantly increase permeability. The removal of a few grains (by FGD) enlarges a few pore throats and shortens the flow paths between a few pore throats (that is, fluid no longer has to flow around some grains). However, the total resistance to flow is dom-

Figure 5—Photomicrograph of dissolved feldspar in contact with kaolinite patch (milky area above FGD porosity). Kaolinite has not grown in the FGD porosity, indicating that FGD came after kaolinite precipitation. Oligocene Frio South Texas (3002 ft). Field width = 0.35 mm.

inated by the remaining small pore throats, and the gain in permeability should not be great.

The above proposition was tested experimentally using sand packs composed of lead and zinc metal granules (grain size ≈ 1.0 mm). Two different mixtures of lead and zinc granules were compacted to 20% porosity in plastic sleeves and then water saturated under vacuum in a simple flow apparatus. The time required to flow a fixed volume of water through the packs at constant pressure was taken as a measure of the permeability. The zinc granules were then completely dissolved by flowing dilute sulfuric acid through the packs. After the acid and hydrogen gas (from zinc dissolution) were displaced from the packs, the time required to flow the same volume of water through the packs was measured. This latter time is a measure of the post-FGD permeability. The porosity of the first sand pack (zinc-to-lead ratio of 1 to 20 by volume) was increased from 20 to 24% by FGD, but its permeability increased by only about 10%. The porosity of the second sand pack (zinc-to-lead ratio of 1 to 10 by volume) was increased from 20 to 28%, and its permeability increased by close to 100%. The 4 and 8% porosity increases approximate the average and maximum FGD porosity increases observed

in sandstones. The observed permeability increases are not considered large because we estimate that the permeabilities would have increased by a factor of about 3 and 10 (based on porosity versus permeability trends in "clean" sandstones), respectively, if the experimental porosity increases had been in intergranular porosity.

We conclude that the correlation between total FGD porosity and permeability is not due to the FGD enhancement of permeability. In our artificial FGD experiments using lead-zinc granules, the permeability was increased only by a factor of 2 by increasing the porosity 8 vol %, whereas the permeability varies between 2 and 3 orders of magnitude for an equivalent FGD porosity change (see Figs. 9, 10) in natural samples. We believe that the correlation of increasing FGD porosity with higher permeability results, in some complex manner, from the greater volume of solvent solution that flows through those sandstone layers with higher original permeability.

The major and relevant consequence of the relationship between increasing FGD porosity and increasing permeability is that those sandstone layers with better original reservoir properties (within a given sandstone body) will have their reservoir properties enhanced to the greatest degree.

THEORY OF FRAMEWORK GRAIN DISSOLUTION

Our theory holds that pre-oil generation maturation of organic matter and clays in shales produces the solvent necessary for dissolving sand grains and for transporting much of this dissolved material out of the sandstone. The early maturation of organic matter produces acid (from CO_2) for dissolving aluminosilicates; organic complexing agents (that is, ligands), which hold the dissolved aluminum in solution during transport; and methane, which provides part of the pressure drive for fluid migration. The conversion of smectite (usually montmorillonite) to illite lowers the chemical activity of dissolved aluminum in the shale water and produces some additional free water. The aqueous solvent solution is expelled into adjacent sandstones where the most susceptible aluminosilicate grains are dissolved (in order of their decreasing susceptibility to dissolution). The higher aluminum activities associated with the dissolving feldspars and the availability of complexing agents cause complexing of most of the dissolved aluminum. The spent solvent, along with its load of dissolved material, is ultimately flushed from the sandstone by further dewatering of the shales.

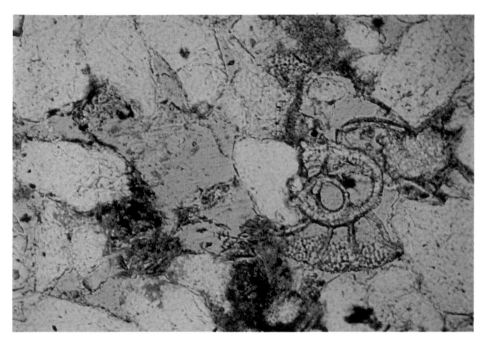

Figure 6—Photomicrograph of dissolved feldspar (left) next to calcite test of a foram. The lack of carbonate dissolution indicates the FGD solvent was at equilibrium with respect to calcite. Oligocene Klumpp "D" sandstone, Louisiana (11,246 ft). Field width = 0.89 mm.

Our theory was developed with the central requirement that it explain or be consistent with the following geochemical considerations and petrographic observations:

(1) There must be water available and a sufficient pressure gradient to move it from the shale into and out of a sandstone.

(2) There must be a source of hydrogen ion for the dissolution reaction.

(3) The aluminum and silica released into solution must be inhibited from precipitating in the sandstone's pores as clays or other authigenic materials.

(4) The dissolved aluminum must be chemically complexed to high concentrations so that it can be removed from the sandstone using the volumes of water available in the sandstone/shale sequence.

(5) The solvent solution must (sometimes at least) be capable of dissolving aluminosilicates but not dissolving coexisting calcite.

Shale Water as Solvent

Shale water is the most logical solvent for FGD in the subsurface because of its availability and proximity to the sandstones. Water is almost continuously available from shale compaction (Dickinson, 1953) and thermal expansion (Barker, 1972) during burial. In addition, bursts of water are produced by (1) the release of absorbed water from smectite clay during the diagenetic conversion to illite (Burst, 1969) and (2) the volume expansion resulting from conversion in shale of solid kerogen to fluid hydrocarbons (especially with the generation of gas) during hydrocarbon generation (Hedburg, 1974). There is no doubt that sufficient pressure gradients exist in the subsurface such that the shale waters are expelled. This is demonstrated by the almost universal decline in shale porosities with increasing depth of burial.

Source of Hydrogen Ion

Geochemical considerations indicate that some aqueous hydrogen ion (that is, acid) is required to take aluminosilicates into solution. This hydrogen ion is provided by carbon dioxide (CO_2) (see Al-Shaieb and Shelton, 1981) generated in the shales from organic matter by thermal maturation processes.

The requirement for hydrogen ion is indicated by the mass action equations for the dissolution process. For the simplified case of pure albite feldspar dissolving in water, an equation describing a possible dissolution pathway is as follows:

$$NaAlSi_3O_8 + 4H^+(aq) \rightarrow Na^+(aq) \quad (2)$$
(albite)

$$+ Al^{+3}(aq) + 3SiO_2(s) + 2H_2O$$
(quartz)

Equation 2 indicates that 4 moles of H^+ are required to produce 1 mole of dissolved aluminum ion (that is, dissolve 1 mole of albite). This equation pertains to albite dissolution at pH values of 3 and lower. An alternative case is the hypothetical dissolution process at very high pH where essentially no hydrogen ion is required (equation 3).

$$NaAlSi_3O_8 + 2H_2O \rightarrow Na^+(aq) \quad (3)$$
(albite)

$$+ Al(OH)_4^-(aq) + 3SiO_2(s)$$
(quartz)

Equations for dissolution paths involving the intermediate hydroxyl complexes [that is, $Al(OH)^{+2}$, $Al(OH)_2^+$, $Al(OH)_3^0$] can also be written, and these show intermediate hydrogen ion consumption. The generally slightly acidic pHs of subsurface solutions indicate that some hydrogen ion (H^+) is consumed during FGD.

Carbon dioxide (CO_2), produced by thermal maturation of organic matter,

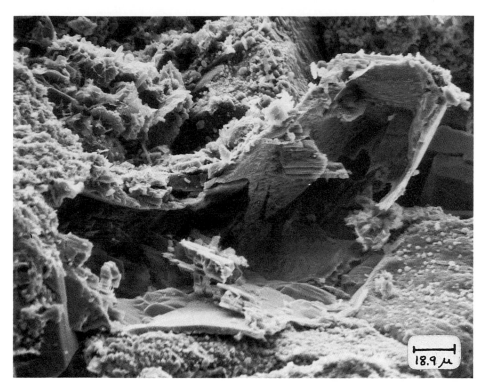

Figure 7—SEM photomicrograph of partially dissolved feldspar. External shell is authigenic feldspar overgrowth. Oligocene Vicksburg Formation, South Texas (5259 ft).

is the main source of acid in the subsurface. This CO_2 is produced dominantly from kerogens beginning with the onset of thermal maturation with the bulk of CO_2 generation occurring at and above the top of the oil generation window. For instance, LaPlante (1974) measured the compositions versus depth of the woody kerogens in several south Louisiana wells and concluded that in over 15,000 ft of burial, the kerogens[2] lost approximately 25% of their original weight as CO_2. This CO_2 dissolves in the pore water of the shale to produce a carbonic acid solution. Much of the hydrogen ion resulting from the dissociation of carbonic acid may immediately react with coproduced, water-soluble organic compounds (for example, carboxylic acid anions) to produce a bicarbonate–organic buffer solution. However, because the solution is a buffer, most of the hydrogen

ion is available for chemical reaction. We also note that while LaPlante ascribed all of the oxygen versus depth decline to CO_2 evolution, a significant amount of this carbon and oxygen may be liberated as water-soluble organic compounds. These compounds may be in part the complexing agents required by our FGD theory. In addition, if part of these compounds are organic acids cleaved from the kerogen in their protonated form, these acid compounds will be an additional primary source of hydrogen ion (see Surdam et al, this volume).

The amount of CO_2 produced by shales is apparently quite sufficient to account for the FGD that is observed. Consider a shale with an original organic matter content expressed as 0.5 wt % *organic carbon* (a typical value for Gulf Coast shales is 0.6 wt % organic carbon). Assuming a reference density (2.3 g/cu cm) for the shale and an original carbon content (70 wt %) for the kerogen, we calculate that 1 liter of shale will ultimately release 4.1 g of CO_2, assuming a 25% conversion figure. This is about 0.09 mole of CO_2, and this amount of CO_2 can account for the dissolution of 4.5 ml or about 0.5 vol % of 1 liter of sandstone (as

feldspar) if the average molar consumption of hydrogen ion per mole of feldspar is 2 to 1. The 2-to-1 ratio is arbitrary, the midpoint between the hydrogen ion consumptions of equations 2 and 3, and is believed to be larger than the actual ratio in the subsurface. Thus about 8 liters of shale could account for the average FGD porosity (that is, 4 vol %) in 1 liter of sandstone. Since we estimate the typical sandstone-to-shale ratio to be greater than 10 to 1 in the typical sedimentary basin, we conclude that there is more than enough CO_2 generation in the subsurface to account for FGD.

Transport of Aluminum

Considerations of the amount of aluminum removed from sandstones versus the amount of water available as a solvent indicate that the aluminum must be chemically complexed to concentrations in excess of 100 mg per liter (total solubility) in order to remove the aluminum. The spent FGD solvent solution must not become excessively supersaturated with respect to clay minerals or precipitation of authigenic clay will result. At the same time, sufficient aluminum must be carried in the available solvent to account for the

[2]A Van Krevelan plot (that is, hydrogen/carbon versus oxygen/carbon ratios) of LaPlante's analytical data show his kerogens to be essentially the same type (that is, Type III or woody kerogen) over the 15,000 ft of burial. This precludes the possibility of the measured oxygen decline with depth being due to a change in kerogen type rather than CO_2 generation.

Figure 8—SEM photomicrograph of FGD in a grain believed to be a microperthite. One of the feldspar components (orthoclase or plagioclase) has been selectively dissolved. Oligocene Nodosaria sandstone, Louisiana (14,563 ft).

TABLE 3

Al Concentration (mg/liter)	0.1	1.0	10.0	100	1000
Volume H$_2$O (liter)	108,000	10,800	1,080	108	10.8
Pore volume at 25% porosity	432,000	43,200	4,320	432	43.2

Table 3—The volume of water required to remove the dissolution products of 4 vol % (as NaAlSi$_3$O$_8$) from 1 liter of sandstones at various levels of aluminum (Al) solubility.

amount of aluminum removed.

We can estimate a crude maximum value for the total aluminum solubility required in the FGD solvent by the following arguments. Consider 1 liter of sandstone with 4 vol % FGD (that is, the average amount) in plagioclase feldspars. The volume of water required to dissolve this volume of feldspar (that is, 40 cu cm) was calculated for various concentrations of total dissolved aluminum and is presented in Table 3.

We can compare this range of water requirements with an estimated water availability. If we assume that 1 liter of a deeply buried shale has yielded 0.5 liter of solvent water (for example, compacting a near-surface Gulf Coast shale with 47% porosity to 20% porosity[3] at 10,000 ft), then at 100 mg per liter total aluminum, 216 liters of shale are required for the dissolution of 4 vol

% of feldspar in 1 liter of sandstone. We estimate that the average shale-to-sand ratio in a shale-prone sedimentary basin like the Gulf Coast is between 10 and 20. Consequently, the maximum aluminum solubility under the conditions specified above would have to be between about 1000 and 2000 mg per liter.

A more realistic aluminum concentration is certainly much lower because of several factors. First, some of the dissolved aluminum may be precipitated as clay locally, particularly early in the burial history when complexing agents are less available. More impor-

---

[3]We note that much of the water filling the porosity in deeply buried shales can be displaced by gas in many formations. This would increase the efficiency of solvent water generation in the upper part of the hydrocarbon generation zone. Water released by the conversion of smectite clays to illites is another source of water above and at the top of the hydrocarbon generation zone.

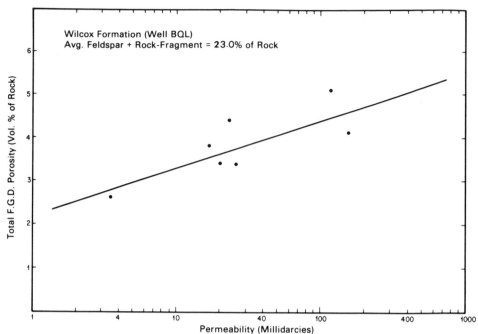

Figure 9—Plot of total FGD porosity (vol % of rock) versus the logarithm of permeability for Wilcox Formation samples (Well BQL).

tantly, reservoir sandstones apparently can receive much more fluid than would be simply predicted by the sand-to-shale ratio. The strong positive correlation of permeability with the amount of FGD indicates that more permeable sandstones receive a much greater proportion of FGD solvent than the low permeability sandstones (for example, see the Frio example in Fig. 10). We note that the 4% average FGD value (Table 1) used above was developed in the much more permeable and usually hydrocarbon-bearing reservoir section of the respective formations. Consequently, the average FGD value, while being representative of hydrocarbon-bearing reservoirs, is much too large for all nonshale rocks (that is, siltstones, impermeable and permeable sandstones) in a sedimentary basin. In other words, the reservoir sandstones have received the FGD solvent from a large volume of shale. This is consistent with the observation that reservoirs usually derive their hydrocarbons from a very large volume of shale. We estimate that the solubility of aluminum in the solvent must be on the order of 100 mg per liter in order to develop FGD porosity.

In order for total dissolved aluminum to reach values in excess of 100 mg per liter, most of the aluminum must be chemically complexed and the nucleation and growth of clays must be inhibited. We estimate that the total aluminum solubility based on equilibrium with kaolinite and common inorganic complexes is less than 10 mg per liter at 100°C. If the required 100 mg per liter concentrations were attained in a complexing agent-free solution, the supersaturation with respect to clay would be excessive and clay precipitation in the sandstone would be expected.

Organic complexing agents are produced in the shales from kerogen by maturation processes that parallel the generation of CO_2 and hydrocarbons. For instance, Carothers and Kharaka (1978) report up to 4900 mg per liter of short chain aliphatic acid anion (that is, acetate, propionate, etc.) in formation waters from 15 oil and gas fields in the San Joaquin Valley, California, and in the Houston and Corpus Christi areas, Texas. Acetate ion accounts for about 90% of these anions. The senior author has estimated the association constant for the aluminum acetate ion-pair to be about 4000 at 100°C. If a subsurface solution contains 5000 mg per liter acetate ion only, the ratio of complexed aluminum to free aluminum is about 300.

The actual types and amounts of soluble complexing agent in the subsurface are poorly known. A literature survey (for example, Collins, 1975) shows that complex mono- and dicarboxylic acids, substituted phenols and pyridines, amino acids, and other complex nitrogen- or oxygen-bearing compounds have been detected or inferred to exist in oil field brines. Many of these types of compounds, the amino acids and other nitrogen-bearing compounds, for example, are extremely strong complexers of aluminum (Sillen and Martell, 1964). Amino acids have been detected in oil field brines but, so far, only in 1 mg per liter concentrations or less (Collins, 1975). However, ammonia has been found in concentrations ranging from 100 to 650 mg per liter in oil field brines from some 41 wells in southwest Louisiana (Dickey et al, 1972), and this ammonia may have been derived from nitrogen-bearing organic compounds that were active in the FGD process earlier in the burial history.

Another possibility for increasing the total dissolved aluminum in the FGD solvent is the complexing of aluminum by aqueous borate species. This complexing is also very strong (Sillen and Martell, 1964) and concentrations of boron ranging from 18 to 75 mg per liter (versus 5 mg per liter in sea water) have been found in oil field brines

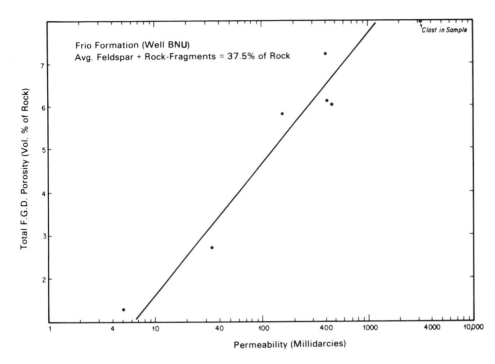

Figure 10—Plot of total FGD porosity (vol % of rock) versus the logarithm of permeability for Frio Formation samples (Well BNU).

(Dickey et al, 1972).

The exact nature of the aluminum species that interact with organic compounds and the nature of the interaction are also poorly known. Complexing agents may interact with free aluminum ion or one of the aluminum hydroxyl species. In addition, we speculate that silica–alumina polymers may exist in solution which are inhibited from crystallization to clays by complexing with organic agents. Irrespective of the nature of the complex, experimental evidence indicates that organic complexing agents (simple mono- and difunctional carboxylic acids) greatly increase the solubility of aluminum from alumina–silicate minerals (see Surdam et al, this volume).

Unfortunately, the actual existence of high aluminum concentrations in subsurface waters has not yet been demonstrated. A few analyses have reported high aluminum concentrations (for example, see Collins, 1975) in oil field brines, but these analyses are suspect because of the difficulty in separating and analyzing for true "aqueous" aluminum. A possible reason for not detecting high aluminum concentrations is the cooling of formation waters when brought to the surface. The lower temperature destabilizes the aluminum complexes and

causes most of the aluminum to precipitate from the solution.

Clay Diagenesis

The role played by the smectite (that is, montmorillonite)-to-illite conversion in shale is to reduce the total aluminum concentration in the shale water so that additional aluminum can be dissolved and complexed when the shale water enters the sandstone. The presence of silt-size feldspar in most shales might be expected to exhaust the ability of the shale water to dissolve aluminosilicates before the solvent enters the adjacent sandstones. However, the smectite-to-illite conversion reaction and related processes operate in most shales to reduce the aluminum activity in the shale water. The general process involves the reaction of feldspars with smectite to form illite plus other minerals (for example, chlorite) (see Hower, et al, 1976; and Moncure et al, this volume). Since the feldspars are dissolving, and relatively stable aluminum-rich clays are forming, the chemical activity of aluminum species is reduced with respect to the situation where feldspar and solution are in equilibrium. The maximum limit of the reduction of aluminum activity can be estimated by comparing the activities of aluminum in equilibrium with micro-

cline and with muscovite at a potassium-to-hydrogen-ion ratio of one. The muscovite is considered representative of a very pure and well-crystallized illite. The ratio of equilibrium aluminum activities for feldspar and mica was calculated at 25°C using the thermodynamic data of Robie et al (1978). The calculation indicates that the aluminum activity can be theoretically reduced by a factor of about 25,000. The real reduction factor is less than this because illite is less stable than muscovite and because of kinetic reasons (that is, the system is not in equilibrium).

The reduced aluminum activity in the shale water means reduced use of the available complexing agents. When the shale water reaches the sandstone, the dissolution of the more unstable or kinetically active feldspars (for example, Ca-rich plagioclases) will dominate the dissolution reactions and tend to keep the activity of aluminum close to the equilibrium values of the least stable feldspars. Consequently, much more aluminum is complexed allowing for aluminum transport out of the sandstone.

We must note that the smectite-to-illite conversion reaction is not (theoretically) absolutely necessary to the generation of solvent and, consequently, to the FGD process. The reaction is necessary only in shales that contain significant amounts of unstable feldspars.

Situations where delicate but unetched calcite fossils coexist with dissolved feldspars are consistent with our theory. Hower et al (1976) reported, in conjunction with their smectite-to-illite diagenesis work, that calcite disappeared from their shales in the depth zone where illite production occurred. This zone corresponds to the zone (near the top of the oil generation window) of significant CO_2 generation. The first CO_2 generated in the shale is (apparently) consumed to dissolve any carbonate present to form a calcium bicarbonate solution. If an excess of calcite is present, the first solutions entering the sandstones will be at equilibrium with calcite and, consequently, will not dissolve any calcite. However, this solution can still dissolve feldspars as illustrated in the following equation:

$$NaAlSi_3O_8 + 2Ca^{+2}(aq) + \qquad (4)$$
(albite)

$$4HCO_3^-(aq) \rightarrow Na^+(aq) +$$

$$Al(OH)_2^+(aq) + 2HCO_3^-(aq) +$$

$$3SiO_2(s) + 2CaCO_3(s)$$
(quartz)   (calcite)

Note that the equation indicates the precipitation of calcite. This is consistent with the common occurrence of late calcite cements in sandstones.

DISCUSSION

The ultimate fate of the dissolved and transported aluminum is problematical. As we seldom see sandstones in more shallow, organically immature zones with excessive or unexplained authigenic clay, we speculate that the transported aluminum is ultimately precipitated within shales at shallower depths or escapes at the surface.

A small-scale example of chemical interactions, including FGD, between shales and sandstones that lends support to the FGD theory is reported in detail in Moncure et al (this volume). To summarize, a thin shale encased in sandstones (Oligocene Frio in the GCO/DOE, Pleasant Bayou Geothermal Test Well No. 1, Brazoria County,

Texas) was found to contain about 0.25 wt % organic carbon, making it a poor source of FGD solvent. Geochemical and mineralogical analysis indicates that thermally generated hydrocarbons and inorganic components had migrated toward and into the upper sand. The fluid expelled from the shale caused significant FGD and chlorite cementation in the first foot of sandstone. Farther away from the shale, the amount of FGD and chlorite cement declines sharply while the amount of kaolinite, apparently derived by framework grain alteration, increases sharply.

Some relatively important factors that are believed to influence the amount of FGD porosity generation and were presented only implicitly in the foregoing discussion are:

(1) Organic Matter Content—The amount of organic matter in a shale will determine the amount of CO_2 and complexing agents that are produced. Obviously, a very lean source rock will produce little FGD porosity along with very few hydrocarbons.

(2) Type of Organic Matter—The dominant type of kerogen (that is, woody versus herbaceous versus algal, etc.) is expected to control the total and relative amounts of CO_2 and complexing agents produced per unit of organic matter.

(3) Sand-to-Shale Ratio—Obviously, if there are more volumes of shale per volume of sand, more solvent solution should be expelled into the sandstone and should result in more FGD porosity.

SUMMARY

We propose that organic and clay maturation in concert are responsible for much framework grain dissolution (secondary porosity). Petrographic observations show that there is often not enough authigenic clay to account for the aluminum removed from the dissolved grains, suggesting that most of the dissolved material was removed from the sandstones. Geochemical considerations indicate that H^+ ions are required for aluminosilicate dissolution and that the aluminum must be complexed to concentrations greater than 100 ppm in order to transport aluminum out of the sandstone using water volumes available in most basins. The

early stages of organic matter maturation generate acid (as CO_2) and organic complexing agents (for example, short chain fatty acids), which can complex aluminum. The aqueous organic complexing agents in the shale complex aluminum at relatively low concentrations because aqueous aluminum activity is depressed by the formation of illite from smectite. The H^+ and complexing agent-bearing solution is expelled into sandstones where the aluminum activity is buffered at higher levels by unstable feldspars, thus allowing higher levels of complexed aluminum. The solution dissolves the feldspars and other aluminosilicate components and complexes much of the resulting aluminum for transport out of the sandstone.

SELECTED REFERENCES

Al-Shaieb, Z., and J. W. Shelton, 1981, Migration of hydrocarbons and secondary porosity in sandstones: Bulletin of the American Association of Petroleum Geologists, v. 65, p. 2433–2436.

Barker, C., 1972, Aquathermal pressuring —role of temperature in development of abnormal-pressure zones: Bulletin of the American Association of Petroleum Geologists, v. 56, p. 2068–2071.

Burst, J. F., 1969, Diagenesis of Gulf Coast clayey sediments and its possible relation to petroleum migration: Bulletin of the American Association of Petroleum Geologists, v. 53, p. 73–93.

Carothers, W. W., and Y. K. Kharaka, 1978, Aliphatic acid anions in oil field waters—implications for the origin of natural gas: Bulletin of the American Association of Petroleum Geologists, v. 62, p. 2441–2453.

Collins, A. G., 1975, Geochemistry of oilfield waters: New York, Elsevier Scientific Publishing Co., 496 p.

Dickey, P. A., A. G. Collins, and I. Fajardo, 1972, Chemical composition of deep formation waters in southwestern Louisiana: Bulletin of the American Association of Petroleum Geologists, v. 56, p. 1530–1533.

Dickinson, G., 1953, Geological aspect of abnormal reservoir pressures in Gulf Coast Louisiana: Bulletin of the American Association of Petroleum Geologists, v. 37, p. 410–432.

Dutton, S. P., 1977, Diagenesis and porosity distribution in deltaic sandstone, Strawn Series (Pennsylvanian), north-central Texas: Gulf Coast Association of Geological Societies Transactions, v. 27, p. 272–277.

Hayes, J. B., J. C. Harms, and T. Wilson, 1976, Contrasts between braided and meandering stream deposits, Beluga and Sterling Formations (Tertiary), Cook Inlet, Alaska, in T. P. Miller, ed., Recent and ancient sedimentary environments in Alaska: Anchorage, Alaska Geological Society, p. J1–J27.

Hayes, J. B., 1979, Sandstone diagenesis—the hole truth, in aspects of diagenesis: Society of Economic Paleontologists and Mineralogists Special Publication 26, p. 127–140.

Heald, M. T., and R. E. Larese, 1973, The significance of the solution of feldspar in porosity development: Journal of Sedimentary Petrology, v. 43, p. 458–460.

Hedburg, H. D., 1974, Relation of methane generation to undercompacted shales, shale diapirs, and mud volcanoes: Bulletin of the American Association of Petroleum Geologists, v. 58, p. 661–673.

Hower, J., E. V. Eslinger, M. E. Hower, and E. A. Perry, 1976, Mechanisms of burial metamorphism of argillaceous sediments, 1, Mineralogical and chemical evidence: Geological Society of America Bulletin, v. 87, p. 725–737.

LaPlante, R. E., 1974, Hydrocarbon generation in Gulf Coast Tertiary sediments: Bulletin of the American Association of Petroleum Geologists, v. 58, p. 1281–1289.

Lindquist, S. J., 1977, Secondary porosity development and subsequent reduction, overpressured Frio Formation sandstone (Oligocene), South Texas: Gulf Coast Association of Geological Societies Transactions, v. 27, p. 99–107.

Loucks, R. G., D. G. Bebout, and W. E. Galloway, 1977, Relationship of porosity formation and preservation to sandstone consolidation history—Gulf Coast Lower Tertiary Frio Formation: Gulf Coast Association of Geological Societies Transactions, v 27, p. 109–119.

McBride, E. F., 1977, Secondary porosity—importance in sandstone reservoirs in Texas: Gulf Coast Association of Geological Societies Transactions, v. 27, p. 121–122.

Robie, R. A., B. S. Hemingway, and J. R. Fisher, 1978, Thermodynamic properties of minerals and related substances at 298.15 k and 1 bar (10^5 pascals) pressure and at higher temperatures: United States Geological Survey Bulletin 1452, p. 456.

Schmidt, V., D. A. McDonald, and R. L. Platt, 1977, Pore geometry and reservoir aspects of secondary porosity in sandstones: Bulletin of Canadian Petroleum Geology, v. 25, p. 271–290.

Sillen, L. G., and A. E. Martell, 1964, Stability constants of metal ion complexes: Special Publications no. 17, London; The Chemical Society, 754. p.

Stanton, G. D., 1977, Secondary porosity in sandstones of the Lower Wilcox (Eocene), Karnes County, Texas: Gulf Coast Association of Geological Societies Transactions, v. 27, p. 197–207.

Diagenesis of Plio–Pleistocene Nonmarine Sandstones, Cagayan Basin, Philippines: Early Development of Secondary Porosity in Volcanic Sandstones

Mark E. Mathisen
Iowa State University
Ames, Iowa

INTRODUCTION

The relatively rapid reduction of volcaniclastic sandstone porosity by compaction and cementation during early burial diagenesis has been well documented in many diagenetic studies (Galloway, 1974, 1979; Burns and Ethridge, 1979; Hayes, 1979; Surdam and Boles, 1979). These studies primarily describe the diagenesis of Tertiary or Mesozoic sands deposited in a marine environment. Several studies (Hayes et al, 1976; Walker et al, 1978) suggest that, in contrast to reducing porosity, early diagenetic processes may actually increase porosity in some shallow nonmarine volcaniclastics. Hayes et al (1976) noted that extensive secondary porosity formed in the Pliocene Sterling Formation of Alaska as a result of dissolution of volcanic glass and heavy minerals. Walker et al (1978) described significant early dissolution of unstable framework grains in Cenozoic desert alluvium, which increased sandstone porosity.

Recent investigations of Plio–Pleistocene volcanic sandstones from the central part of the Cagayan basin (Fig. 1) of Northern Luzon, Philippines (Mathisen, 1981a; Vondra et al, 1981), indicate that the nonmarine sediments contain significant amounts of secondary porosity, which formed at shallow depths. This paper describes the diagenesis of these volcanic sandstones in order to document another example of early secondary porosity development in shallow volcaniclastics. The significance of early secondary porosity development in volcaniclastic sandstones is then discussed with regard to

ABSTRACT. The Plio–Pleistocene nonmarine volcanic sandstones of the Cagayan basin, Philippines, have been significantly altered by early dissolution and cementation processes. The amount and type of alteration vary by formation, depth, and age of the deposit. Plio–Pleistocene fluvial sandstones (litharenites and feldspathic litharenites) buried to depths of 400–900 m, are only slightly compacted, but contain significant amounts of authigenic pore-lining clay and zeolites. Dissolution of plagioclase, heavy minerals, and volcanic rock fragments has occurred in nearly all samples, dissolving up to one-half the framework grains and increasing thin-section porosity to as much as 40%. The overlying Pleistocene sandstones are compositionally different (lithic arkoses and arkoses) and have not been as extensively affected by diagenetic processes. The more extensive alteration of the Plio–Pleistocene sandstones reflects increased diagenetic alteration with burial depth and time as a result of relatively high pore-fluid flow rates in shallow alluvial deposits.

The diagenesis of the Cagayan basin Plio–Pleistocene sandstones indicates that significant secondary porosity can develop in nonmarine volcaniclastics as a result of early silicate dissolution during shallow burial diagenesis. Early dissolution and secondary porosity development have important implications for studies of nonmarine volcaniclastics. Early dissolution processes distort provenance, tectonic setting, and depositional environment interpretations based on the detrital mineralogy of older volcaniclastic sediments. Secondary porosity increases the reservoir quality of volcaniclastics prior to more extensive compaction and cementation. Recognition of similar shallow volcaniclastic reservoirs in the past may have been limited because of low resistivity sand identification problems caused by authigenic smectite.

provenance, depositional environment, and reservoir quality interpretations.

GEOLOGIC SETTING

The Cayagan basin is a north-south trending interarc basin approximately 250 km long and 80 km wide (Fig. 1), which contains 10,000 m of volcaniclas-

tic sediment. The basin began to form in the Late Oligocene–Early Miocene following a reversal of the North Luzon arc polarity and initial uplift of the Cordillera Central volcanic arc to the west (De Boer et al, 1980). Approximately 8000 m of Late Oligocene–Miocene marine sediments, primarily turbidites, were deposited in response to

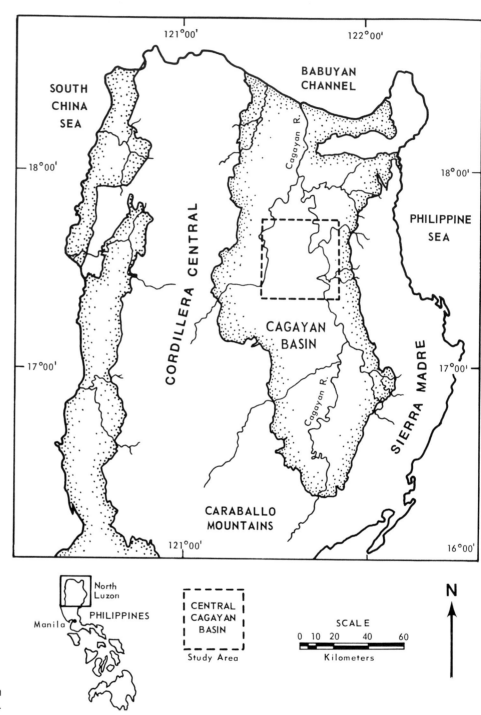

Figure 1—Location map of Cagayan
Basin and study area.

rapid subsidence of the basin floor to
bathyal depths (Durkee and Pederson,
1961; Caagusan, 1980). Most of the
clastic debris was derived from the
active Cordillera Central to the west.
Uplift of the region in the Plio–Pleistocene (Christian, 1964) resulted in the
transition to deltaic and nonmarine
sedimentation (Vondra et al, 1981). In
the Middle to Late Pleistocene, uplift in

the Cordillera Central resulted in décol-
lement and gravity sliding of unstable
uplifted sediments toward the basin,
where large asymmetric to overturned
folds were formed. Erosion of these
folds produced extensive outcrops of
Plio–Pleistocene sediment, which are
being studied to provide geologic documentation of archaeologic sites and
fossil localities (Mathisen and Vondra,

1978; Shutler and Mathisen, 1979;
Vondra et al, 1981).

The Plio–Pleistocene nonmarine vol-
caniclastic sediments have been divided
into two lithostratigraphic units—the
Plio–Pleistocene Ilagan Formation
Upper Member and the Pleistocene
Awidon Mesa Formation (Durkee and
Pederson, 1961; Vondra et al, 1981).
The Ilagan Formation Upper Member

consists of 500 m of fluvial deposits (the Lower Member is formed by 300 m of deltaic deposits). The Awidon Mesa Formation conformably overlies the Ilagan Formation, and consists of 400 m of interbedded fluvial and pyroclastic deposits.

The nonmarine sediments of the Ilagan Formation Upper Member and Awidon Mesa Formation form a coarsening upward volcaniclastic sequence (Mathisen and Vondra, 1983). The Ilagan Formation Upper Member sediments, meandering stream deposits interbedded with tuffs, are overlain by coarser-grained alluvial fan braided stream and pyroclastic deposits of the Awidon Mesa Formation. The changes in depositional environment reflect changes in the volcanic and tectonic history of the adjacent active Cordillera Central (Mathisen and Vondra, 1983).

The Plio–Pleistocene sediments were derived from volcanic, plutonic, and metamorphic rocks exposed in the Cordillera Central, Sierra Madre, and Caraballo volcanic arcs, which border the basin to the west, east, and south. Heavy mineral and petrographic studies (Mathisen, 1981b) indicate that the active Cordillera Central became the dominant source of sediment in the Pleistocene.

PETROGRAPHY

The Plio–Pleistocene volcaniclastic sediments of the central part of the Cagayan basin consist of conglomerates, tuff-breccias and tuffs, sandstones, siltstones, and mudrocks. The composition of 115 outcrop samples was studied using conventional petrographic, X-ray diffraction, and scanning electron microscope techniques. Textural analyses were completed on selected samples from the Awidon Mesa Formation that were not extensively altered. Ilagan Formation samples were unsuitable for grain-size analyses because of extensive diagenetic changes. The petrographic data and methods are described in detail by Mathisen (1981c).

Conglomerates occur in both the Ilagan and Awidon Mesa Formations, but are more abundant and coarser grained in the alluvial fan deposits of the Awidon Mesa Formation. They contain more rock fragments and detrital clay than the sandstones, but have undergone similar diagenetic changes.

Tuff-breccias occur only in the Awidon Mesa Formation. The breccias are composed of dacite or pumice clasts in a feldspathic sand matrix. The pumice appears relatively unaltered in most samples. The feldspars, primarily andesine, are unaltered in the younger deposits, but display minor dissolution features in some of the older deposits.

The mudrocks are primarily claystones and mudstones. Smectite is the most common clay mineral, but kaolinite also occurs in some rocks. The mudrocks contain minor amounts of sand and silt-size bipyramidal quartz, plagioclase, and heavy minerals, which suggests they may have formed, in part, from the alteration of fine-grained volcaniclastic and pyroclastic material. Paleosols characterized by iron enrichment, iron-rich pisolites, or weakly developed blocky structure occur in some mudrocks.

Sandstone Composition

Forty-one alluvial sandstones were analyzed from throughout the study area. The average bulk composition and detrital mineralogy, as determined from counts of 300 points, are recorded in Table 1. The sandstones are predominantly lithic arkoses and feldspathic litharenites, but litharenites and arkoses are also common. The composition varies significantly between formations, as illustrated in QFL triangular plots (Fig. 2). The Upper Member of the Ilagan Formation is composed of feldspathic litharenites and litharenites. Feldspars are more abundant in the Awidon Mesa Formation, resulting in a greater percentage of lithic arkoses and arkoses. The increase in feldspar in the Awidon Mesa Formation is due to increased dacitic volcanism in the Cordillera Central during the Pleistocene.

Framework grains, primarily feldspar and rock fragments, are the dominant component of the sandstones, averaging 58% of each sample (Table 1). Andesine is the principal feldspar. It increases in abundance from the Ilagan Formation (22%) to the Awidon Mesa Formation (48%). K-feldspar also occurs in many samples, but is rare, usually less than 0.5% of the sandstones. The rock fragments are primarily composed of euhedral to subhedral plagioclase laths in a felted microlite groundmass, and are interpreted as microlitic volcanic rock fragments (Dickinson, 1970). The volcanic rock fragments, in contrast to the feldspars, decrease in abundance from the Ilagan Formation (56%) to the Awidon Mesa Formation (31%). Sedimentary, metamorphic, and plutonic igneous rock fragments also occur in the sandstones, but are of minor importance.

Quartz is a constant minor component of the sandstones. It increases in abundance from the Ilagan Formation (3%) to the Awidon Mesa Formation (7%), which contains larger water-clear bipyramidal β quartz crystals up to granule size. Many of the Awidon Mesa quartz crystals are embayed or rounded to varying degrees, reflecting resorption of the crystals by the original magma. In the Ilagan Formation, the β quartz morphology is not as pronounced, and most grains occur as crystal fragments.

Heavy minerals are common in all Cagayan basin nonmarine sandstones. In the Upper Member of the Ilagan Formation, heavy minerals are relatively abundant, averaging 12.5% of each sample. Magnetite, ilmenite, augite, and hypersthene are the dominant heavy minerals, with minor amounts of green-brown hornblende. Heavy minerals are also abundant in the Awidon Mesa Formation, where they constitute 12% of the sandstones. Hornblende is the dominant heavy mineral, averaging 5% of the sandstones. The opaque minerals, magnetite and ilmenite, are also common, forming 4% of the sandstones. Oxyhornblende, augite, and hypersthene occur in some sandstones in variable amounts up to 11%.

The sandstones contain variable amounts of detrital matrix and cement (Table 1). The matrix commonly comprises only a few percent of the sandstones. It is locally abundant, however, forming 18 to 53% of 8 samples in the Awidon Mesa Formation. Compositionally, the matrix may be classified as protomatrix (Dickinson, 1970), unrecrystallized clay with minor amounts of silt. X-ray diffraction analyses indicate that the matrix clays are smectite and/or kaolinite. Three major types of cement, authigenic clays, zeolites, and calcite, occur in the sandstones. Minor amounts of iron oxide cement also occur in some samples. The authigenic clays are primarily smectite, but kaolin-

TABLE 1

Bulk Composition

| | No. of Samples | Framework Grains | Detrital Matrix | Cements | | | Inter-granular Porosity | Intra-granular Porosity |
				Clay	Zeolite	Calcite		
Awidon Mesa Fm.	30	58.7	13.7	7.7	0.5	—	14.5	4.7
Ilagan Fm. Upper Member	11	58.1	1.7	9.9	11.9	—[1]	10.9	6.4

Framework Grain Mineralogy

| | No. of Samples | Quartz | K-feldspar | Plagio-clase | Rock Fragments | | | Amphiboles | | Pyroxenes | | Opaque Minerals |
					Volcanic	Sedi-mentary	Meta-morphic, Plutonic Igneous	Horn-blende	Oxyhorn-blende	Augite	Hyper-sthene	
Awidon Mesa Fm.	30	6.6	—	47.7	31.4	1.3	1.1	5.2	0.4	1.5	0.7	4.2
Ilagan Fm. Upper Member	11	3.4	0.4	22.4	56.2	3.5	2.2	1.4	—	3.5	1.0	6.6

[1]One concretion sample contained 32%.

Table 1—Sandstone bulk composition and detrital mineralogy.

ite is also common. The characteristics of the various cements are described in more detail in the discussion of diagenesis. The cements are more abundant in the Ilagan Formation (22%) than in the Awidon Mesa Formation (8%).

Three types of porosity occur in the sandstones: primary intergranular porosity, secondary intragranular porosity, and microporosity. Intergranular and intragranular porosity values (Table 1) were estimated from point counts. Microporosity occurs in the detrital matrix, authigenic clays, and partially dissolved volcanic rock fragments, but cannot be accurately estimated. Primary intergranular porosity is highest, 14.5%, in the Pleistocene Awidon Mesa Formation, and decreases to an average of 10.9% in the Plio–Pleistocene Ilagan Formation Upper Member. Secondary intragranular porosity formed by partial dissolution of silicate framework grains is

abundant in many Cagayan basin sandstones. It is most abundant in the Plio–Pleistocene Upper Member sandstones, where it averages 6.4%, but is also common (4.7%) in the Awidon Mesa sandstones.

DIAGENESIS

The sandstones of the Ilagan Formation Upper Member and Awidon Mesa Formation display a variety of diagenetic features (Figs. 3, 4, and 5), which reflect alteration by five diagenetic processes. These are: (1) compaction, (2) formation of authigenic clay, (3) dissolution of silicate framework grains, (4) crystallization of authigenic zeolites, and (5) the crystallization of calcite cement.

Compaction

Compaction has occurred in all sandstones, but is not very extensive. Most grains have floating or point contacts, as illustrated in Figures 3 and 5, and have been compacted only by grain

slippage and rotation (Jonas and McBride, 1977). Most volcanic rock fragments have not been deformed. However, slight ductile deformation (McBride, 1979) of rock fragments has occurred in some Ilagan sandstones. The lack of compaction features is consistent with the stratigraphic occurrence of the rocks, which suggests that they have not been buried more than about 900 m, the maximum possible thickness of the Ilagan Formation Upper Member and the Awidon Mesa Formations. The high porosity that remained after compaction, 25–35%, left the sandstones permeable to fluid flow, which led to the precipitation of cements and dissolution of unstable grains.

Authigenic Clay

Authigenic clays have cemented the sandstones of both the Ilagan and Awidon Mesa formations to varying degrees. The clay averages 10% of the Ilagan sandstones, and ranges from 0 to 17% in abundance. In the Awidon

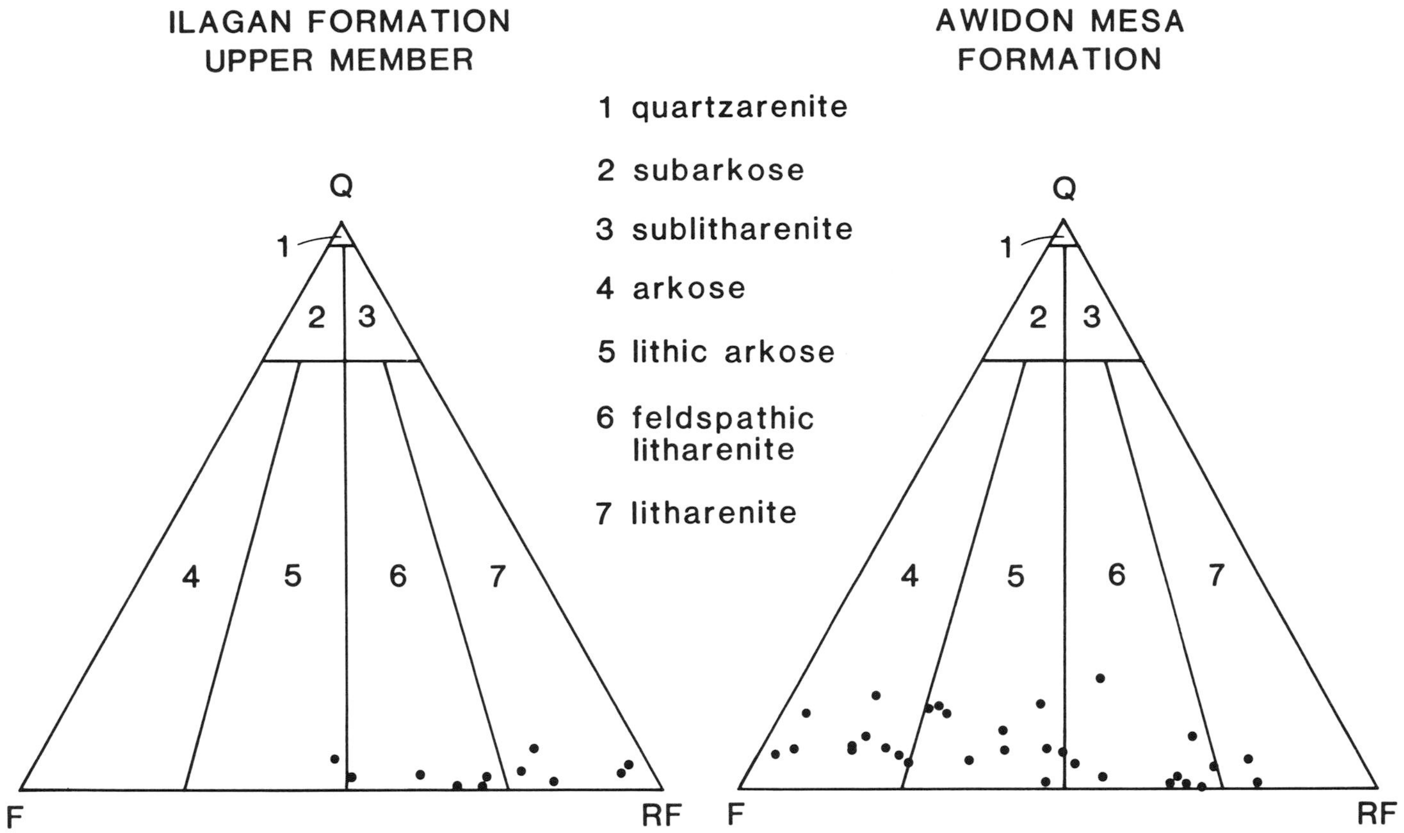

ILAGAN FORMATION
UPPER MEMBER

AWIDON MESA
FORMATION

1 quartzarenite

2 subarkose

3 sublitharenite

4 arkose

5 lithic arkose

6 feldspathic
  litharenite

7 litharenite

Figure 2—Classification (after Folk, 1968) of the Ilagan Formation Upper Member and Awidon Mesa Sandstones.

Mesa sandstones, the clay averages 8%, and ranges from 0 to 26%.

The authigenic origin of the clay is interpreted by the identification of clay rims or clay coats that form rinds around detrital sand grains (Galloway, 1974; Wilson and Pittman, 1977). Smectite commonly forms homogeneous clay rims of uniform distribution and thickness (up to 20 μ) in the Ilagan and Awidon Mesa sandstones, as illustrated in Figure 3b. The smectite rims typically have a honeycomb-like morphology when viewed with the SEM (Fig. 4a). Clay coats, which show preferred orientation parallel to the grain surfaces (Galloway, 1974; Walker et al, 1978), also commonly occur in the Ilagan and Awidon Mesa sandstones. The clay coats (Fig. 3c) are stratified and of variable thickness. Smectite is the dominant clay mineral that forms clay coats and occurs as layers of flat, slightly irregular flakes (Fig. 4b), as illustrated

by Scholle (1979). The clay coats in the Plio–Pleistocene sandstones appear to have been formed by illuviation of colloidal material onto the grains, as described by Galloway (1974) and Walker et al (1978).

X-ray diffraction analyses of authigenic clay minerals from nine sandstones indicate that both smectite and kaolinite have formed authigenically. Smectite is the more common clay cement, while kaolinite, which occurs primarily in the Awidon Mesa Formation, is of minor importance.

The authigenic clay began to form early in the diagenetic history of the sandstones. Significant amounts of clay had coated detrital grains in many samples before significant dissolution of the framework grains occurred, as indicated by the relict clay coats in many samples (Fig. 5a). In numerous samples, the clay continued to accumulate until the pores were almost completely filled.

Dissolution

Most sandstones of the Ilagan Upper Member and Awidon Mesa Formation

contain significant amounts of secondary intragranular porosity or moldic porosity which formed as a result of the early dissolution of heavy minerals, plagioclase feldspars, and volcanic rock fragments. The original form or size of the etched grains is usually indicated by the matrix or authigenic clay rims. Pyroxenes are the only heavy minerals that are extensively etched and nearly always display hacksaw terminations, as described by Edelman and Doeglas (1931). In most sandstones, the grains have been etched continuously or only once during their diagenetic history. In several sandstones, however, multiple clay rims indicate that the dissolution of pyroxene grains took place in two or more steps separated by a period of clay formation, as illustrated in Figure 3d. The feldspars are commonly etched along cleavage traces or in a random pattern (Fig. 3b, c, d). Feldspar dissolution has affected both single grains, which are almost completely dissolved in some samples, and laths in rock fragments. The dissolution of volcanic rock fragments usually is not complete, but produces grains with a distinct

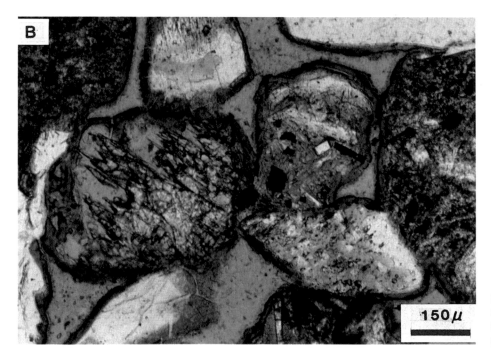

Figure 3—Photomicrographs of diagenetic features observed in Pleistocene Awidon Mesa Formation sandstones. 3a—Relatively unaltered matrix-free fluvial sandstone (plain light) with high primary intergranular porosity that displays only minor compaction and authigenic clay cementation.

Figure 3b—Authigenic smectite clay rims (plain light) of uniform thickness and distribution, and secondary intragranular porosity.

amount of secondary microporosity.

Dissolution has generally been more extensive in the Plio–Pleistocene Ilagan sandstones than the Pleistocene Awidon Mesa sandstones. Secondary intragranular porosity averages 37% of the total porosity in the Ilagan sandstones. In some Ilagan sandstones, the dissolution has been so extensive that most grains have been partially or totally dissolved, as illustrated in Figure 5a. In the Pleistocene Awidon Mesa sandstones, intragranular porosity forms 24.5% of the total porosity. Most grains commonly display evidence of dissolution (Fig. 3c), but are not as extensively dissolved as the Ilagan sandstones.

Figure 3c—Stratified clay coats (plain light) of variable thickness composed of smectite. The secondary intragranular porosity is typical of Pleistocene Awidon Mesa Formation sandstones.

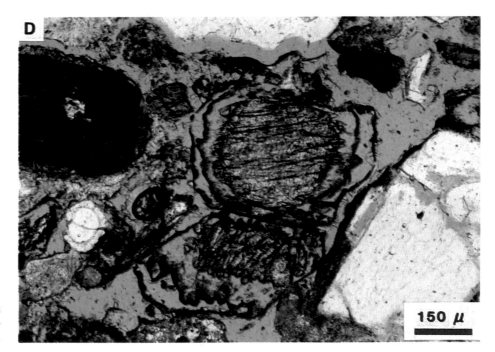

Figure 3d—Multiple smectite clay rims (plain light) document two-step dissolution history of augite grains.

The dissolution of plagioclase and volcanic rock fragments is the result of hydration reactions, which are ubiquitous in volcanogenic terranes (Surdam and Boles, 1979). The hydration reactions are important because they increase the pH of the interstitial solutions and release cations into solution. This increase in pH and salinity has a direct effect on all subsequent diagenetic reactions. Hay (1966) has noted that plagioclase dissolution occurs in saline, alkaline nonmarine environments, and that the rates of plagioclase dissolution are increased with increasing salinity and pH. Silicate dissolution may also occur as a result of the interaction of organic acids formed from

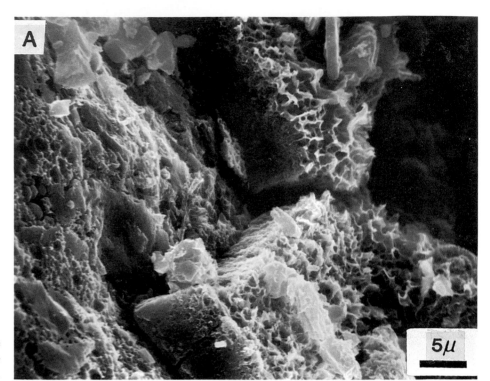

Figure 4—SEM photomicrographs of authigenic cements. 4a—Smectite clay rims overlying grain surface.

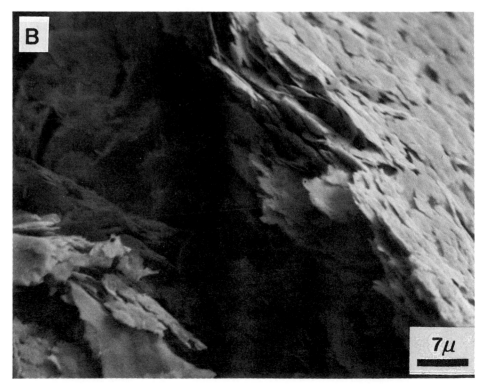

Figure 4b—Stratified smectite clay coats.

Figure 4c—Smectite clay rims and pore-lining clinoptilolite and stilbite crystals.

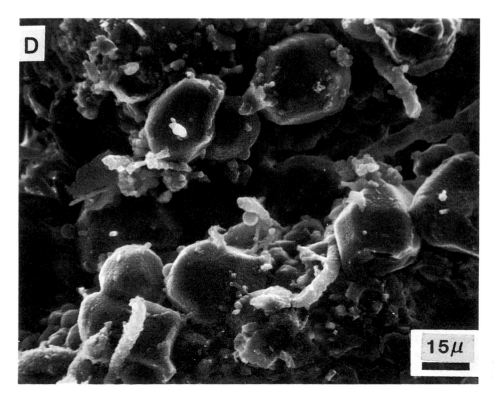

Figure 4d—Pore-lining analcime crystals displaying typical cubo-octahedral form.

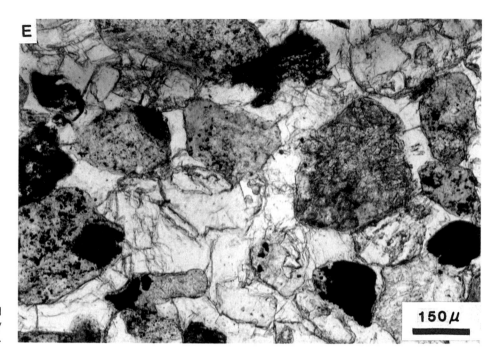

Figure 5e—Pore-filling chabazite and secondary microporosity in partially dissolved rock fragments (plain light).

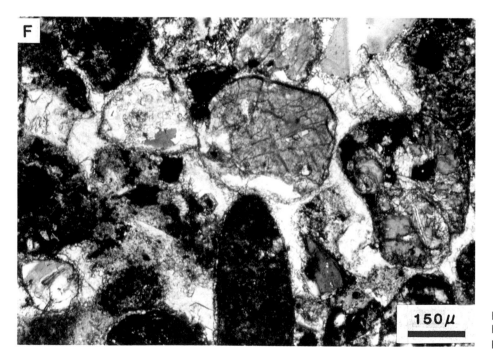

Figure 5f—Late pore-filling calcite filling intergranular and intragranular pores (crossed nicols).

Almost all Ilagan sandstones contain zeolites, which average 16% but form up to 30% of the samples. Zeolites are also abundant in the Ilagan Formation tuffs (Mathisen, 1981c). In the Awidon Mesa Formation, zeolites are rare.

The Ilagan zeolites primarily occur as pore-lining crystals that average 50 μm in length (Fig. 5b, c), but also form pore-filling cements (Fig. 5d, e). Clinoptilolite, stilbite, analcime, and chabazite were identified by X-ray diffraction and crystal morphology.

The distribution of the zeolite mineral species varies stratigraphically in the Ilagan Formation. Clinoptilolite and stilbite are the principal zeolites of the Upper Member, but some crystals of chabazite were also observed. These zeolites usually occur in the same rock associated with authigenic smectite, as indicated by petrographic microscope observations, X-ray diffraction, and SEM analyses (Fig. 4c). In most samples, clinoptilolite and stilbite occur as isolated pore-lining crystals (Fig. 5c), but they also form continuous coatings. In one lithic tuff (Mathisen, 1981c), clinoptilolite and stilbite totally cement the sample, reducing the porosity to 2%. Sandstones at the base of the Upper Member and the Lower Member (Kvale and Vondra, 1982) contain pore-lining analcime (Figs. 4d, 5d) and pore-filling chabazite (Fig. 5e).

Zeolites are common authigenic minerals in volcaniclastic sandstones that form by the interaction of saline, alkaline aqueous solutions, and unstable volcanic detritus (Deffeyes, 1959; Hay, 1966; Surdam and Boles, 1979). Volcanic glass and plagioclase commonly dissolve in saline, alkaline solutions, and contribute silica and aluminum to the pore waters. The zeolites then precipitate from solution. Plagioclase feldspars are abundant in the Ilagan Formation, and have been significantly dissolved in most samples. The zeolite reaction, plagioclase → zeolite, is therefore a major diagenetic reaction that has taken place in the Ilagan sediments. The zeolite reaction, glass → zeolite, may also have taken place, but there is no petrographic evidence that glass shards were present and dissolved from the Ilagan Formation. The association of smectite clay rims with the zeolites may indicate that glass was present. This is because the alteration of glass to smectite is an important factor in providing the chemical environment, particularly an alkaline pH and high activity of silica, suitable to the formation of zeolites (Hay, 1966). The volcanic origin of the Ilagan sediments and the occurrence of volcanic glass in the overlying Awidon Mesa Formation also suggest that glass may have been present in the Ilagan Formation. If the glass was present, dissolution and the reaction, glass → zeolite, has occurred, and no relict shard textures were preserved.

The Ilagan Formation sandstones may be divided into two diagenetic facies, the clinoptilolite–stilbite facies, which occurs throughout the Upper Member, and the analcime–chabazite facies, which occurs at the base of the Upper Member and at greater depth in the Lower Member. Clinoptilolite, stilbite, and chabazite commonly form at shallow depths under low temperatures. Analcime, in contrast, forms by the alteration of low temperature zeolites, such as clinoptilolite, and is stable at higher temperatures and greater burial depths (Hay, 1966). The alteration of low temperature zeolites to analcime with increased temperature and depth of burial has been attributed to: (1) burial metamorphic reactions as a result of increased temperature and pressure (Aoyagi and Kazama, 1980); (2) increases in salinity and pH (Sheppard and Gude, 1969); (3) kinetic factors where analcime forms later than zeolites such as clinoptilolite (Sheppard and Gude, 1969); and (4) various combinations of these factors (Hay, 1966; Coombs and Whetten, 1967). The Ilagan Formation zeolite facies have been subjected to different burial depths, temperatures, and pressures, as well as different reaction times. These factors have definitely contributed to the differentiation of the analcime–chabazite facies. Increases in salinity and pH may also have been important factors, but these cannot be assessed at this time.

Calcite

Sparry calcite is a localized cement that forms concretions in the Ilagan Formation. The calcite primarily fills pores that are lined with authigenic clay and zeolites, but also partially replaces some plagioclase grains. The formation of calcite cement is a common diagenetic reaction in volcaniclastic sediments (Galloway, 1974). Surdam and Boles (1979) note that the initial hydration reactions of volcaniclastic rocks are sources of Ca^{+2} and that early diagenetic changes in organic matter are one of the major sources of HCO_3^-. They suggest that the amount and distribution of organic matter in sediments is probably the limiting factor relative to carbonization reactions. The Ilagan concretions, therefore, indicate that a significant amount of organic matter was deposited with the sediments. Calcite concretions were then formed after decay of the organic matter and the formation of HCO_3^-.

Diagenetic Sequence

The Ilagan and Awidon Mesa sandstones display diagenetic features indicative of both mechanical and chemical diagenesis. The distribution of these features with depth is summarized in Figure 6. Compaction is the main mechanical event, and began shortly after deposition with a limited amount of grain slippage and rotation. Chemical diagenesis also began shortly after deposition, and has had the greatest effect on the sandstones. Authigenic clay rims and coats began to form first at shallow depths, several hundred meters, and continued to form with burial to depths of approximately 900 m. Dissolution of pyroxenes, plagioclase, and volcanic rock fragments also began to occur at shallow depths. The relict clay rims and coats that indicate the original grain form suggest that significant amounts of authigenic clay had formed before the grains were extensively dissolved. The relict rims and coats also indicate that significant amounts of additional compaction did not occur with depth. Extensive dissolution of framework grains occurred at relatively shallow depths and has also been extensive at greater depths where grains were dissolved after zeolite pore-filling cement has formed, as illustrated in Figure 5d. Pore-lining zeolites of the clinoptilolite–stilbite facies began to form in sandstones buried several hundred meters, and are abundant in samples to a depth of approximately 900 m. At this depth, the analcime–chabazite zeolite facies begins and extends to greater depths, reflecting a greater depth of burial, longer reaction time, and possible geochemical variations of pore fluids. The last diagenetic event in some sandstones was the precipitation of pore-filling sparry calcite. The calcite has formed concretions at depths of 400 to 900 m, cementing sandstones that already contained pore-lining clays, zeolites, and significant amounts of secondary intragranular porosity, as seen in Figure 5f.

The diagenetic sequence just described, which is based on outcrop samples and maximum possible burial depths, is interpreted to reflect diagenetic change with increasing depth of burial and age of the deposit in contrast to alteration by weathering processes. As illustrated in Figure 5c and d, authigenic clay and extensive secondary intragranular porosity formed prior to zeolite cementation. The growth of zeolites on clay rims and in intragranular pores indicates that the authigenic clay and secondary pores existed at the greater depths where zeolite cementation occurred. Alteration during weathering may have occurred before

Figure 6—Diagenesis of the Ilagan and Awidon Mesa sandstones. (The vertical bars indicate the interpreted depth range of major diagenetic events.)

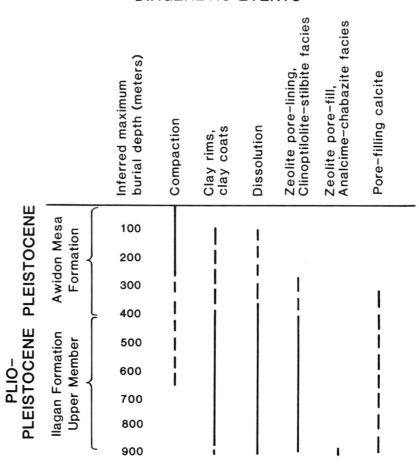

deposition and/or after folding and erosion of the anticlines. Alteration probably occurred during weathering in the source area, but this cannot be distinguished from the extensive diagenetic alteration. The weathering would have affected all samples equally because the climate did not change significantly in Northern Luzon during the Plio–Pleistocene (Vondra et al, 1981). The increase in the alteration of the Ilagan Formation sandstones in contrast to the Awidon Mesa Formation sandstones must, therefore, reflect diagenetic alteration. Alteration during weathering of the sampled outcrops may also have occurred, but cannot be differentiated from the diagenetic alteration. Both the Ilagan and Awidon Mesa formations were weathered for the same period of time following erosion of the anticlines. Alteration of the samples during outcrop weathering should be similar for both formations if the alteration occurred during weathering. The more extensive alteration of the Ilagan Formation cannot be attributed to weathering, and must, therefore, reflect increased diagenetic alteration with burial depth and time. Diagenetic interpretations of outcrop samples have also been made by Galloway (1974) and Hayes et al (1976), who did not notice any significant alteration of the outcrop samples by weathering when compared with core samples.

The diagenetic features and sequence of the Ilagan and Awidon Mesa sandstones reflect diagenetic stages that are common during the burial diagenesis phase of volcaniclastic diagenesis (Galloway, 1974). These stages essentially consist of compaction and the formation of authigenic clay rims and coats and then phyllosilicate and/or zeolite pore-filling cement, as described by Galloway (1974) and Walker et al (1978). Calcite pore-filling cements commonly form first in some volcaniclastic sandstones (Galloway, 1974),

but did not occur early in the Ilagan and Awidon Mesa formations. This probably reflects the fluvial environment of the Ilagan Formation Upper Member and Awidon Mesa sandstones, which contained less Ca^{+2} and HCO_3^- than volcaniclastic rocks deposited in marine environments. Significant amounts of calcite have formed concretions, however, during the later stages of diagenesis. The lack of complex replacement and alteration features indicates that the Ilagan and Awidon Mesa sandstones have not been subjected to the advanced burial metamorphic phase defined by Galloway (1974).

IMPLICATIONS FOR PROVENANCE, DEPOSITIONAL ENVIRONMENT, AND RESERVOIR QUALITY INTERPRETATIONS

Dissolution of framework grains and the development of secondary porosity is commonly thought to be a late-stage

diagenetic event occurring primarily at greater depths (Schmidt et al, 1977; Schmidt and McDonald, 1979; Pittman, 1979). The early dissolution that has been documented in the literature is commonly thought to be relatively minor, localized in extent, or to involve primarily unstable volcanic glass (Hayes et al, 1976). The Cagayan basin Plio–Pleistocene sandstones indicate that significant amounts of secondary porosity can develop in nonmarine volcanic sandstones during shallow burial diagenesis as a result of silicate framework grain dissolution. This early development of secondary porosity has important implications for provenance, depositional environment, and reservoir quality interpretations of volcanic sandstones.

Provenance
The mineralogy of Cenozoic volcaniclastic sandstones is commonly used to interpret provenance and other factors

controlling the detrital mineralogy such as tectonic setting (Hayes et al, 1979). In many cases, workers are aware of significant diagenetic changes in volcaniclastics and attempt to compensate for them in the final interpretation. In some studies, however, the significance of diagenetic alteration is not discussed (Schwab, 1981; Dickinson, 1982). Numerous authors (Hay, 1957; Walker et al, 1978) have cautioned that the diagenetic overprint needs to be assessed to determine the validity of interpretations based on detrital mineralogy. Walker et al (1978) have emphasized this fact by suggesting that ancient first-cycle alluvium rarely has the same mineralogy, texture, or chemical composition of the original deposit.

The extensive early dissolution of the Plio–Pleistocene Cagayan basin sandstones indicates the potential magnitude of these diagenetic changes. As illustrated in Figure 5a, approximately half of the original silicate framework grains, primarily plagioclase, rock fragments, and heavy minerals, have been dissolved. As noted in the previous section, only 34% of the rock represents the original detritus. With additional cementation and compaction typical of more deeply buried rocks, the original composition could not be accurately interpreted. Instead of being classified as a well-sorted quartz-deficient litharenite or arkose, the sample would contain relatively high percentages of quartz as a result of the dissolution of the more unstable components. This suggests that volcaniclastics may be easily enriched in quartz, as noted by Walker et al (1978), and that percentages of quartz tabulated for many volcaniclastic sediments (Schwab, 1981; Dickinson, 1982) may not accurately reflect the composition of the original sediment.

Depositional Environment

The early diagenesis of volcaniclastic sandstones may distort environmental interpretations by modification of sandstone composition and texture. Numerous studies have documented the distortion of grain-size data by the diagenetic alteration of sandstones (Mousinho de Meis and Amadon, 1974; Galloway, 1974; Wilson and Pittman, 1977). The abundant authigenic clay developed during early diagenesis may be deformed following framework grain dissolution and compaction to the extent that it may be confused with detrital matrix. Partially dissolved unstable framework grains may cause volcaniclastic sandstones to appear more poorly sorted after compaction. Thus, a well-sorted fluvial sand may be rapidly altered to a more poorly sorted matrix-rich rock, which would suggest an entirely different depositional environment.

Reservoir Quality

Most studies of volcaniclastic diagenesis suggest that volcanic sandstones have poor reservoir quality owing to the relatively rapid formation of pore-filling cements (Galloway, 1979; Davies et al, 1979; Surdam and Boles, 1979). This is certainly true for older, more deeply buried sandstones (Surdam and Boles, 1979), and appears to be a controlling factor for younger marine deposits (Galloway, 1979; Burns and Ethridge, 1979). Nonmarine deposits, however, at shallow burial depths, may contain significant amounts of porosity owing to unstable grain dissolution caused by relatively rapid rates of pore-fluid flow. Hayes et al (1976) have documented abnormally high porosities, 30–40%, in the fluvial Sterling Formation of Alaska, which were formed by the dissolution of volcanic glass. Walker et al (1978) suggest that significant amounts of early secondary porosity may form in first-cycle desert alluvium. The extensive dissolution of many Cagayan basin Plio–Pleistocene fluvial volcanic sandstones provides another example of early secondary porosity development at shallow burial depths.

Secondary porosity in the Cagayan basin sandstones enhances reservoir quality by offsetting the normal porosity reduction with depth by development of intragranular or moldic pores. Average thin-section porosity values decrease from 19.2% in the Pleistocene sandstones to 17.3% in the Plio–Pleistocene sands buried up to approximately 900 m. This porosity decrease with depth is small owing to significant increases in secondary porosity with depth. Secondary porosity in the Cagayan sandstones increases from an average of 24.5% of the total porosity in the Pleistocene sandstones to 37% in the Plio–Pleistocene sandstones. In addition to conserving porosity in most sandstones, dissolution processes have enhanced porosity considerably in some samples by dissolving up to one-half of the framework grains, thereby producing 40% thin-section porosity, as illustrated in Figure 5a. The dominant porosity type produced by dissolution processes, intragranular or moldic porosity, also significantly increases reservoir quality. Morgan and Gordon (1970) determined that reservoir quality is higher in rocks with moldic porosity due to high porosities, permeabilities, and flow rates.

Authigenic clays may affect reservoir quality depending on the type of clay and its distribution, as described by Wilson and Pittman (1977). The authigenic smectite clay rims and coats typical of the Cagayan basin sandstones would probably not offset the increases in reservoir quality produced by dissolution processes. The clays may actually bind water and result in production of water-free hydrocarbons from low resistivity sands, as described by Almon and Schultz (1979). The low resistivity sand phenomenon suggests that hydrocarbon zones in smectite-rich volcaniclastic sandstones with early secondary porosity may have been overlooked in the past when only conventional exploration log analysis methods were used.

SUMMARY

The Plio–Pleistocene fluvial deposits of the Cagayan basin have been extensively altered by diagenetic processes at shallow burial depths. Authigenic clay rims and clay coats began to form soon after deposition and initial compaction. Dissolution processes accelerated by relatively rapid pore-fluid flow began dissolving unstable detrital components, primarily rock fragments, heavy minerals, and plagioclase. At greater depths of burial, 400–900 m, extensive dissolution removed as many as one-half of the original silicate framework grains. Pore-lining zeolites, clinoptilolite, and stilbite began to form at these depths, followed by pore-filling chabazite and analcime at depths of 900 m.

The diagenesis of the Cagayan basin sandstones is an example of the type and magnitude of early diagenetic alteration that may occur in other nonmarine volcaniclastic sandstones. Early framework grain dissolution processes can modify the detrital mineralogy so that accurate interpretations of provenance, tectonic history, and depositional environments would be difficult, if not impossible. Dissolution processes can also produce significant amounts of early secondary porosity, which may enhance reservoir quality at shallow depths, despite the formation of numerous cements. Authigenic smectite may bind formation water, resulting in water-free hydrocarbon production. With deeper burial and additional cementation, the early dissolution porosity and corresponding reservoir quality would be reduced to the much lower values recorded in most volcanic sandstone diagenesis studies.

ACKNOWLEDGMENTS

This study was funded by National Science Foundation Grant Int-7901802 to Carl F. Vondra and by the Philippine National Museum. Special thanks are extended to Carl Vondra for guidance and encouragement and to the staff of the Philippine National Museum Geology Division for assistance with the field work. I would also like to thank Joshua Cocker, Thane McCulloh, and Ganapathy Shanmugam for reviewing the initial manuscript.

SELECTED REFERENCES

Almon, W. R., and A. L. Schultz, 1979, Electric log detection of diagenetically altered reservoirs and diagenetic traps: Gulf Coast Association of Geological Societies Transactions, v. 29, p. 1–10.

Aoyagi, K., and T. Kazama, 1980, Transformational changes of clay minerals, zeolites, and silica minerals during diagenesis: Sedimentology, v. 27, p. 179–188.

Blatt, H., 1979, Diagenetic processes in sandstones, in P. A. Scholle and P. R. Schluger, eds., Aspects of diagenesis: Society of Economic Paleontologists and Mineralogists Special Publication 26, p. 141–152.

Burns, L. K., and F. G. Ethridge, 1979, Petrology and diagenetic effects of lithic sandstones: Paleocene and Eocene Umpqua Formation, southwest Oregon, in P. A. Scholle and P. R. Schluger, eds., Aspects of diagenesis: Society of Economic Paleontologists and Mineralogists Special Publication 26, p. 307–317.

Caagusan, N. L., 1977, Source material compaction history and hydrocarbon occurrence in the Cagayan Valley basin, Luzon, Philippines, in SEAPEX Program, Offshore Southeast Asia Conference: Southeast Asia Petroleum Exploration Society, Singapore, 19 p.

———, 1980, Stratigraphy and evolution of the Cagayan Valley basin, Luzon, Philippines: Geology and Paleontology of Southeast Asia, v. 21, p. 163–182.

Christian, L. B., 1964, Post-Oligocene tectonic history of the Cagayan basin, Philippines: The Philippine Geologist, v. 18, p. 114–147.

Coombs, D. S., and J. T. Whetten, 1967, Composition of analcime from sedimentary and burial metamorphic rocks: Geological Society of America Bulletin, v. 78, p. 269–282.

Davids, D. K., W. R. Almon, S. B. Bonis, and B. E. Hunter, 1979, Deposition and diagenesis of Tertiary-Holocene volcaniclastics, Guatemala, in P. A. Scholle and P. R. Schluger, eds., Aspects of diagenesis: Society of Economic Paleontologists and Mineralogists Special Publication 26, p. 281–306.

De Boer, J., L. A. Odom, P. C. Ragland, F. G. Snider, and N. R. Tilford, 1980, The Bataan orogene: eastward subduction, tectonic rotations, and volcanism in the western Pacific (Philippines): Tectonophysics, v. 67, p. 251–282.

Deffeyes, K. S., 1959, Zeolites in sedimentary rocks: Journal of Sedimentary Petrology, v. 29, p. 602–609.

Dickinson, W. R., 1970, Interpreting detrital modes of graywacke and arkose: Journal of Sedimentary Petrology, v. 40, p. 695–707.

———, 1982, Compositions of sandstones in circum-Pacific subduction complexes and forearc basins: Bulletin of the American Association of Petroleum Geologists, v. 66, p. 121–137.

Durkee, E. F., and S. L. Pederson, 1961, Geology of Northern Luzon, Philippines: Bulletin of the American Association of Petroleum Geologists, v. 45, p. 137–168.

Edelman, C. H., and D. J. Doeglas, 1931, Reliktstrukturen detritischer pyroxene und amphibole: Mineralogische und petrographische mitteilungen, v. 42, p. 482–490.

Folk, R. L., 1968, Petrology of sedimentary rocks, Austin, Hemphill Publishing Co., 170 p.

Galloway, W. E., 1974, Deposition and diagenetic alteration of sandstone in northeast Pacific arc-related basins: implications for graywacke genesis: Geological Society of America Bulletin, v. 85, p. 379–390.

———, 1979, Diagenetic control of reservoir quality in arc-derived sandstones: implications for petroleum exploration, in P. A. Scholle and P. R. Schluger, eds., Aspects of diagenesis: Society of Economic Paleontologists and Mineralogists Special Publication 26, p. 251–262.

Hay, R. L., 1957, Mineral alteration in rocks of Middle Eocene age, Absoroka Range, Wyoming: Journal of Sediment-ary Petrology, v. 27, p. 32–40.

———, 1966, Zeolites and zeolitic reactions in sedimentary rocks: Geological Society of America Special Paper 85.

Hayes, J. B., 1979, Sandstone diagenesis—the hole truth, in P. A. Scholle and P. R. Schluger, eds., Aspects of diagenesis: Society of Economic Paleontologists and Mineralogists Special Publication 26, p. 127–140.

Hayes, J. B., J. C. Harms, and T. Wilson, 1976, Contrasts between braided and meandering stream deposits, Beluga and Sterling Formations (Tertiary), Cook Inlet, Alaska, in T. P. Miller, ed., Recent and ancient sedimentary environments in Alaska: Anchorage, Alaska Geological Society, p. J1–J27.

Huang, W. H., and W. D. Keller, 1972, Organic acids as agents of chemical weathering of silicate minerals: Nature, v. 239, p. 149–151.

Jonas, E. C., and E. F. McBride, 1977, Diagenesis of sandstone and shale: application to exploration for hydrocarbons: Continuing Education Program Publication 1, Austin, University of Texas, 120 p.

Kvale, E. P., and C. F. Vondra, 1982, Description and diagenetic history of Mio-Pliocene transitional marine facies of northcentral Cagayan Valley, Luzon, Philippines: Geological Society of America Abstract, v. 14, p. 538.

Mathisen, M. E., 1981a, Diagenesis of the Upper Cenozoic volcaniclastic sandstones of the Cagayan basin, Northern Luzon, Philippines: Geological Society of America Abstract, v. 13, p. 288.

———, 1981b, Provenance of the Plio-Pleistocene volcaniclastic sediments of the central Cagayan basin, Northern Luzon, Philippines: Geological Society of America Abstract, v. 13, p. 309.

———, 1981c, Plio-Pleistocene geology of the central Cagayan Valley, Northern Luzon, Philippines: Unpublished Ph.D. dissertation, Ames, IA, Iowa State University, 209 p.

Mathisen, M. E., and C. F. Vondra, 1978, Pleistocene geology fauna and early man in the Cagayan Valley, Northern Luzon, Philippines: Geological Society of America Abstract, v. 10, p. 451.

———, 1983, The fluvial and pyroclastic deposits of the Cagayan basin, Northern Luzon, Philippines—an example of nonmarine volcaniclastic sedimentation in an interarc basin: Sedimentology, v. 30, p. 369–392.

McBride, E. F., 1979, Ductile deformation porosity loss during compaction: Oil and Gas Journal, v. 77, p. 92–94.

Morgan, J. T., and D. T. Gordon, 1970, Influence of pore geometry on water-oil relative permeability: Journal of Petroleum Technology, October, p. 1199–1208.

Mousinho de Meis, M. R., and E. S.

Amadon, 1974, Note on weathered arkosic beds: Journal of Sedimentary Petrology, v. 44, p. 727–737.

Pittman, E. D., 1979, Recent advances in sandstone diagenesis: Annual Review of Earth and Planetary Sciences, v. 7, p. 39–62.

Schmidt, V., and D. A. McDonald, 1979, The role of secondary porosity in the course of sandstone diagenesis, *in* P. A. Scholle and P. R. Schluger, eds., Aspects of diagenesis: Society of Economic Paleontologists and Mineralogists Special Publication 26, p. 175–207.

Schmidt, V., D. A. McDonald, and R. L. Platt, 1977, Pore geometry and reservoir aspects of secondary porosity in sandstones: Bulletin of Canadian Petroleum Geology, v. 25, p. 271–290.

Scholle, P. A., 1979, Constituents, textures, cements, and porosities of sandstones and associated rocks: American Association of Petroleum Geologists Memoir 28, 201 p.

Schwab, F. L., 1981, Evolution of the western continental margin, French–Italian Alps: sandstone mineralogy as an index of plate tectonic setting: Journal of Geology, v. 89, p. 349–368.

Sheppard, R. A., and A. J. Gude, 1969, Diagenesis of tuffs in the Barstow Formation, Mud Hills, San Bernardino County, California: United States Geological Survey Professional Paper 634, 35 p.

Shutler, R., and M. Mathisen, 1979, Pleistocene studies in the Cagayan Valley of Northern Luzon, Philippines: Journal of the Hong Kong Archeological Society, v. 8, p. 105–114.

Surdam, R. C., and J. R. Boles, 1979, Diagenesis of volcanic sandstones, *in* P. A. Scholle and P. R. Schluger, eds., Aspects of diagenesis: Society of Economic Paleontologists and Mineralogists Special Publication 26, p. 227–242.

Surdam, R. C., S. Boese, and L. J. Crossey, 1982, Role of organic and inorganic reactions in development of secondary porosity in sandstones: American Association of Petroleum Geologists Abstract, v. 66, p. 635.

Vondra, C. F., M. E. Mathisen, D. R. Burggraf, and E. P. Kvale, 1981, Plio–Pleistocene paleoenvironments of Northern Luzon, Philippines, *in* G. R. Rapp and C. F. Vondra, eds., Hominid sites: their paleoenvironmental settings: American Association for the Advancement of Science Symposium, v. 63, p. 255–310.

Walker, T. R., B. Waugh, and A. J. Crone, 1978, Diagenesis in first-cycle desert alluvium of Cenozoic age, southwestern United States and northwestern Mexico: Geological Society of America Bulletin, v. 89, p. 19–32.

Wilson, M. D., and E. D. Pittman, 1977, Authigenic clays in sandstone: recognition and influence on reservoir properties and paleoenvironmental analysis: Journal of Sedimentary Petrology, v. 47, p. 3–31.

Predicting Reservoir Quality and Diagenetic History in the Frio Formation (Oligocene) of Texas*

W. R. Kaiser
Bureau of Economic Geology
The University of Texas
Austin, Texas

INTRODUCTION

Diagenetic studies of Gulf Coast Tertiary rocks (Lindquist, 1977; Bebout et al, 1978; Loucks et al, 1979a, 1981) have produced much data on the regional variations in detrital and authigenic mineralogy. They have progressed from broad regional assessment of reservoir quality, stressing sandstone petrography, to detailed comparison of sandstone mineralogy and elemental composition of calcite from selected areas (Loucks et al, 1979b; Richmann et al, 1980). These studies have advanced our understanding of the occurrence and quality of deep subsurface reservoirs. Factors responsible for the variation in reservoir quality were identified through comparison of detrital and authigenic mineralogies and petrophysical properties. Loucks et al (1981) established a diagenetic sequence relating diagenesis to depth of burial.

The role of brine chemistry in diagenesis is poorly understood and heretofore has not been an integral part of diagenetic studies. Logically, then, the next step in the effort to better understand diagenesis and to predict reservoir quality is to investigate the role of pore fluids (formation waters) or aspects of water–rock interaction. The objective here is to relate water–rock interaction to sandstone diagenesis, using equilibrium thermodynamics, or solution–mineral equilibria, to test the relative stability of authigenic and detrital minerals with respect to forma-

ABSTRACT. Principles of equilibrium thermodynamics were applied and found useful in evaluating aspects of Frio diagenesis. Solution-mineral equilibria as predictors of reservoir quality and diagenetic history were tested by comparing formation waters from regions of good and poor reservoir quality, the upper and lower Texas coast, respectively. Comparison among waters from these regions was made on activity diagrams of 16 diagenetic reactions such as calcite = ferroan calcite, kaolinite = chlorite, and Ca-montmorillonite = Na-montmorillonite. Relative position of tested waters, with respect to the stability field of authigenic minerals occluding permeability and porosity, was related to reservoir quality. Solution–mineral equilibria are indicators of reservoir quality; equilibria in hydropressured waters best reflect reservoir quality. Activity indices favoring chlorite and ferroan calcite stability and large $\log([Ca^{+2}]^{.16}/[Na^+]^{.33})$ ratios are the best indicators of reservoir quality in deep Frio sandstones. Change in ionic strength, analytical molality, and activity indices is correlated with geopressuring. Variation in mole ratios and activity indices with depth is largest between 8000 and 11,000 ft (2440 and 3355 m), the transition zone between the hydropressured and geopressured intervals. The variation is attributed to more active water–rock interaction, or diagenesis, in the transition zone. Diagenesis in the Frio Formation is a function of temperature, pH, activity, and pressure. Predictions made from solution–mineral equilibria add new insight into relative mineral stabilities and in situ pH and are consistent with the diagenetic sequence developed from petrographic data. Calcite equilibrium favors precipitation of calcite early in the burial history. Two stages of chlorite formation are postulated, one early in the hydropressured interval at the expense of smectite grain coats and another late in the geopressured interval at the expense of kaolinite cement. Chlorite and illite are the stable layer silicates in deep Frio sandstones. Albitization of feldspar is initiated in the hydropressured interval at temperatures less than 100°C.

tion waters. Solution–mineral equilibria as predictors of reservoir quality and diagenetic history are evaluated by comparing, on activity diagrams of diagenetic reactions, waters from regions of good and poor reservoir quality.

Analyzed formation waters came from 51 Frio oil and gas fields located in 13 coastal counties, stretching across Texas from the Rio Grande to the Sabine River, divided here into upper, middle, and lower coast (south Texas) (Fig. 1). They are primarily Na⁺-Cl⁻ waters having total dissolved solids content ranging from 6903 to 248,454

*Publication authorized by Director, Bureau of Economic Geology, The University of Texas at Austin.

TABLE 3

Mineral	$\Delta H°_f$	$\Delta G°_f$	Heat Capacity Power Function
Quartz	−217.65	−204.65	$11.22 + 8.20 \times 10^{-3}T - 2.70 \times 10^5 T^{-2}$
Calcite	−288.77	−270.10	$24.98 + 5.24 \times 10^{-3}T - 6.20 \times 10^5 T^{-2}$
Ferroan calcite	−283.29	−264.72	$24.98 + 6.32 \times 10^{-3}T - 5.89 \times 10^5 T^{-2}$
Kaolinite	−982.22	−907.70	$72.77 + 29.20 \times 10^{-3}T - 21.52 \times 10^5 T^{-2}$
Albite	−939.68	−886.31	$61.70 + 13.90 \times 10^{-3}T - 15.01 \times 10^5 T^{-2}$
Chlorite	−1916.96	−1782.98	$173.40 + 40.71 \times 10^{-3}T - 37.47 \times 10^5 T^{-2}$
Illite	−1393.34	−1313.13	$92.39 + 28.20 \times 10^{-3}T - 24.00 \times 10^5 T^{-2}$
Na-montmorillonite	−1349.10	−1255.99	$93.75 + 33.16 \times 10^{-3}T - 23.80 \times 10^5 T^{-2}$
Plagioclase (An30)	−960.11	−907.07	$62.18 + 14.17 \times 10^{-3}T - 15.14 \times 10^5 T^{-2}$
Microcline	−949.19	−895.37	$76.62 + 4.31 \times 10^{-3}T - 29.94 \times 10^5 T^{-2}$
Ca-montmorillonite	−1352.15	−1257.17	$87.17 + 29.35 \times 10^{-3}T - 22.47 \times 10^5 T^{-2}$
$H_4SiO_4^0$	Estimated from the reaction SiO_2 (quartz) $+ 2H_2O = H_4SiO_4^0$.		
H_2O	Data at 1 bar from Robie et al (1978) and at 600 bars from Fisher and Zen (1971).		

Table 3—Thermodynamic functions at 25°C in kcal/mole.

Heat capacity was calculated using a structural algorithm after Helgeson et al (1978). The assumption is that the standard heat capacity of reaction ($C°_{p,r}$) among oxides and silicates of the same or similar structural class equals zero at 25°C and 1 bar ($\Delta C°_{p,r} = 0$). To estimate Maier-Kelley heat capacity power functions end-member values for layer silicates were chosen from Helgeson et al (1978). Kaolinite's power function is well established. In most cases the structural algorithm affords estimates of heat capacity at higher temperatures that are within 2% of experimentally determined values.

Standard enthalpies of formation were estimated using a graphic method of Helgeson (personal communication, 1971) described in Eugster and Chou (1973). Enthalpies ($\Delta H°_f$) of layer silicates divided by the number of tetrahedrally coordinated Si and Al ions ($Si_{tet} + Al_{tet}$) plotted against $\Sigma \Delta H°_f$ of octahedral, exchange, and hydroxyl ions (as aqueous species) divided by ($Si_{tet} + Al_{tet}$) yields a straight line having a slope of approximately 1. Plotting of estimated and experimentally determined values also yields a line slope of

approximately 1, indicating close agreement between values. The method provides acceptable estimates and in many cases generates values within the limits of experimental uncertainty.

Free energies of formation were estimated using the method of Tardy and Garrels (1974), on the assumption that oxide and hydroxide components of layer silicates have fixed free energies of formation within the silicate. The general rule is that all components are treated as oxides with the exception of Mg, which is treated as $Mg(OH)_2$ in all silicates. There is close agreement between estimated and experimentally determined values.

Activities of ionic species were calculated by computer using an updated version of SOLMNEQ (solution–mineral equilibrium computations) (Kharaka and Barnes, 1973). SOLMNEQ uses the extended Debye-Hückel equation, designed for aqueous solutions having ionic strengths of up to 3.0 in which more than 80% of the solute is Na^+ and Cl^-, to estimate activity coefficients. These chemical constraints correspond to the bulk of Frio waters. The range of temperatures covered is 25 to 300°C, at pressures of the liquid–vapor equilibrium for water; best results are achieved at temperatures of up to

200°C. The equilibrium distribution of inorganic species commonly present in natural waters is computed from chemical analyses, temperature, and pH. Values of ion activity coefficients have the smallest uncertainty (less than an order of magnitude), whereas $\log K_r$, $\log K_f$, or gas fugacities may be uncertain to a few orders of magnitude.

Stability Relations

Reaction Pairs

Calcite equilibrium, chloritization, kaolinization, cation exchange, and albitization reactions were written to test for relative mineral stability in Frio Formation waters. Twelve key reaction pairs are given in Table 4 and $\log K_r$ in Table 5. The Na-montmorillonite and K-spar analogs of equations 3, 6, 7, and 10 are not shown, but they were considered in this study and yielded similar results. Reactions were written conserving Al, as suggested by Garrels and Christ (1965).

Equation 1 describes the relative stability of calcite and ferroan calcite and helps explain variations in calcite composition with depth. In effect, calcite governs reservoir quality in the Frio Formation, because it is the major inhibitor of porosity and permeability.

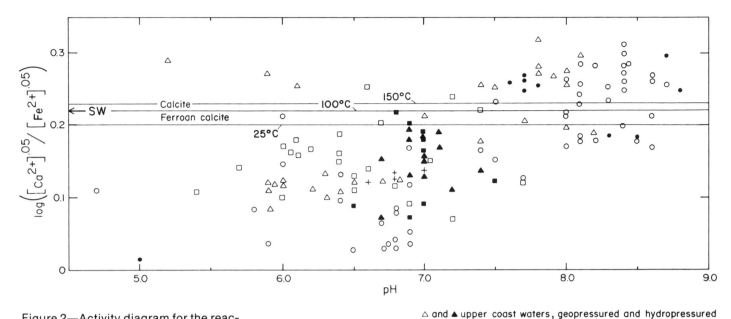

Figure 2—Activity diagram for the reaction calcite = ferroan calcite (equation 1, Table 4). Open symbols designate geopressured waters, solid hydropressured waters, crosses East White Point Field hydropressured waters. On the ordinate, for reference, SW = log $([Ca^{+2}]^{.05}/[Fe^{+2}]^{.05})$ ratio in sea water. Wellhead pH values on the abscissa. Relative stability of calcite and ferroan calcite independent of pH.

△ and ▲ upper coast waters, geopressured and hydropressured
○ and ● middle coast waters, geopressured and hydropressured
□ and ■ lower coast waters, geopressured and hydropressured

Equations 2, 3, and 4 are chloritization reactions. In equation 2, the relative stability of kaolinite and chlorite is tested, whereas equations 3 and 4 are important to the diagenesis of clay grain coats or the relative stability of Ca-montmorillonite and illite. Equations 5, 6, 7, and 8 are kaolinization reactions; equation 5, kaolinization of plagioclase, is perhaps the most important among these as a generator of voids and as a source of Al. Chloritization frees H^+, whereas kaolinization consumes H^+, perhaps playing a role in pH control by acting as buffers. Both sets of reactions free $H_4SiO_4^0$ as well as cations (K^+, Na^+, and Ca^{+2}) important in diagenesis. Chloritization consumes Fe^{+2} and Mg^{+2}, whereas kaolinization frees Fe^{+2} and Mg^{+2}.

Equation 9 represents the cation-exchange reaction between Ca-montmorillonite and Na-montmorillonite. Equation 10 represents the smectite/illite transformation whose role in diagenesis is moderately well known and will not be discussed here (Hower et al,

1976; Boles and Franks, 1979). Finally, two albitization reactions (nos. 11 and 12) are considered. Note that albitization of plagioclase and K-spar frees Ca^{+2} and K^+ and consumes Na^+ and $H_4SiO_4^0$.

Activity Diagrams
Frio waters were thermodynamically tested by plotting activity indices (log activity ratios and products) for each water on activity diagrams of 16 reaction pairs (Table 4). All waters were plotted assuming a water activity of 1. Nine diagrams are presented (Figs. 2, 3, 5, 6, 7, 8, 9, 10, 11). Stability relations are shown at subsurface temperatures and 1 bar pressure to represent hydropressured conditions and 600 bars to represent geopressured conditions.

The pressure dependence of ionic and mineral equilibria over the range of interest in the Frio is slight and considered to be negligible here. Disassociation equilibria of aqueous species are affected little by pressure at temperatures of less than 300°C (Helgeson, 1969; Wolery, 1979) whereas mineral equilibria are affected even less at high pressure and temperature (Helgeson et al, 1978). The effect of pressure on the reaction pairs evaluated in Table 4 is manifest in a more positive free energy

of water (Fisher and Zen, 1971), expanding the stability field of the product mineral (for example, Figs. 7, 8). The effect on ions and minerals was ignored in calculating log K_r at 600 bars pressure.

Relative position of tested waters on the activity diagrams, with respect to the stability field of authigenic minerals occluding permeability and porosity, was related to mineralogical trends or reservoir quality. Waters from regions of good and poor reservoir quality, the upper and lower Texas coast, respectively, were compared. Sandstone reservoirs having permeabilities in excess of 20 md and abundant deep secondary porosity exist on the upper Texas coast, whereas permeability in sandstones of the lower coast or south Texas is less than 1 md.

All activity diagrams display geographic clustering of data points. They also show clusters of points parallel to the slopes of theoretical stability boundaries (Figs. 7, 8), parallel and oblique (Figs. 3, 5, 6), primarily oblique (Fig. 10), and mainly perpendicular (Fig. 9). The significance of these trends is uncertain. Points parallel to stability boundaries probably indicated that both phases coexist whereas oblique and perpendicular trends suggest one phase is dissolving or exchanging

TABLE 4[1]

1. $CaCO_3 + 0.05Fe^{+2} = Ca_{.95}Fe_{.05}CO_3 + 0.05Ca^{+2}$
2. $1.4Kao + 2.3Mg^{+2} + 2.3Fe^{+2} + 6.2H_2O = Chl + 0.2H_4SiO_4^0 + 9.2H^+$
3. $1.8Ca\text{-}mont + 1.85Mg^{+2} + 1.85Fe^{+2} + 14.8H_2O = Chl + 0.29Ca^{+2} + 4.6H_4SiO_4^0 + 6.8H^+$
4. $Illite + 1.64Mg^{+2} + 1.89Fe^{+2} + 8.24H_2O = 0.82Chl + 0.6K^+ + 1.37H_4SiO_4^0 + 6.46H^+$
5. $Plagio + 1.3H^+ + 3.45H_2O = 0.65Kao + 0.3Ca^{+2} + 0.7Na^+ + 1.4H_4SiO_4^0$
6. $2Ab + 2H^+ + 9H_2O = Kao + 2Na^+ + 4H_4SiO_4^0$
7. $Ca\text{-}mont + 1.32H^+ + 4.78H_2O = 0.78Kao + 0.25Mg^{+2} + 0.25Fe^{+2} + 0.16Ca^{+2} + 2.44H_4SiO_4^0$
8. $Illite + 1.1H^+ + 3.15H_2O = 1.15Kao + 0.6^+ + 1.2H_4SiO_4^0 + 0.25Mg^{+2}$
9. $Ca\text{-}mont + 0.33Na^+ = Na\text{-}mont + 0.16Ca^{+2}$
10. $Ca\text{-}mont + 0.41K^+ + 0.57H^+ + 2.64H_2O = 0.68Illite + 0.08Mg^{+2} + 0.25Fe^{+2} + 0.16Ca^{+2} + 1.62H_4SiO_4^0$
11. $Plagio (An30) + 0.6Na^+ + 1.2H_4SiO_4^0 = 1.3Ab + 0.3Ca^{+2} + 2.4H_2O$
12. $K\text{-}spar + Na^+ = Ab + K^+$

[1]Reactions written conserving Al in solid phase.

Table 4—Reaction pairs.

Figure 3—Activity diagram of the reaction kaolinite = chlorite (equation 2). SW = log activity product in sea water.

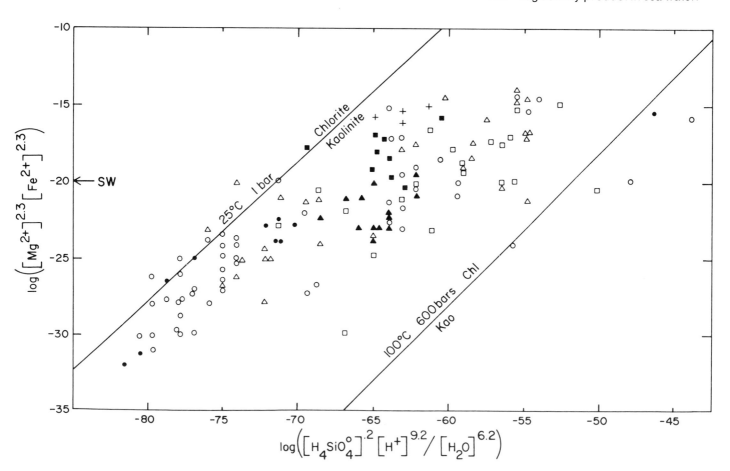

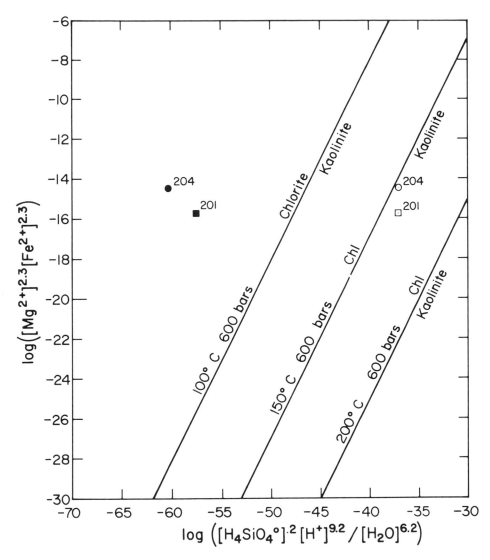

Figure 4—Activity diagram of the reaction kaolinite = chlorite. Upper Texas coast, Brazoria County, Pleasant Bayou geothermal test well waters at wellhead pH (solid symbols) and in situ pH (open) (from Kaiser and Richmann, 1981).

cations at the expense of the other.

Calcite equilibrium—Calcite stability is the key to reservoir quality in the Frio Formation; its stability or precipitation is favored by higher pH, temperature, and activity of ferrous iron, or $[Fe^{+2}]$. Therefore, the inverse of these are indicators of calcite instability or dissolution—in effect, good reservoir quality. Inferences about these indicators are drawn from the calcite/ferroan calcite activity diagram, geothermal gradients, and available petrographic data.

Middle and upper coast waters at subsurface temperatures plot in the calcite and ferroan calcite stability fields. Almost all lower coast or south Texas waters fall in the ferroan calcite field (Fig. 2). All south Texas and upper coast hydropressured waters fall in the ferroan calcite field. These two areas are on opposite ends of the reservoir quality spectrum because of the presence and absence, respectively, of late-stage ferroan calcite cement.

Among the factors governing calcite stability (pH, $[Fe^{+2}]$, $[Ca^{+2}]$, alkalinity, and temperature), pH is a critical but difficult parameter to estimate at in situ conditions. Using a mass-balance approach, Kharaka et al (1979) calculated an in situ pH of 4.1 for two upper coast geopressured waters (Pleasant Bayou geothermal test well, Brazoria County). At an equilibrium pH of 4, a concentration of total dissolved carbonate species of approximately 60,000 mg/liter is required for calcite precipitation in Pleasant Bayou waters; a concentration that seems unrealistically high for natural waters. The point is

that high pH favors calcite precipitation or stability. The absence of late-stage, porosity-occluding ferroan calcite in upper coast Frio sandstones is attributed to low in situ pH, whereas its presence in south Texas is attributed to higher pH and secondarily to higher temperature and $[Fe^{+2}]$. Overall, calcite becomes more stable at higher temperature, and in the presence of Fe^{+2}, ferroan calcite is more stable than nonferroan calcite.

Early prediction was that south Texas hydropressured waters would plot in the ferroan calcite stability field (Kaiser and Richmann, 1981; Kaiser, 1982); that is, in the stability field of the mineral mainly responsible for occluding permeability and porosity. Indeed, the position of these south Texas waters confirms that prediction (Fig. 2). All

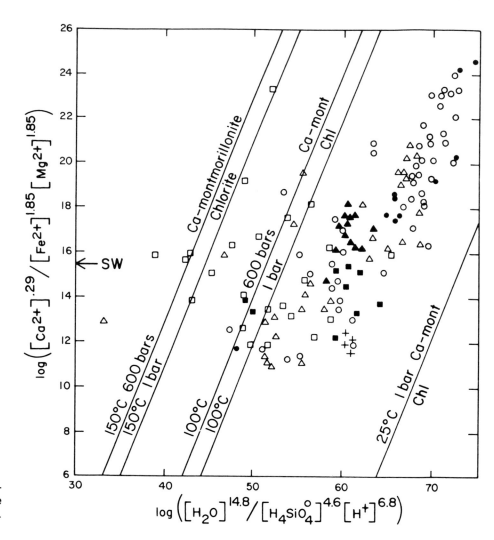

Figure 5—Activity diagram of the reaction Ca-montmorillonite = chlorite (equation 3).

but three middle coast hydropressured waters plot in the calcite stability field, suggesting from large log $([Ca^{+2}]^{.05}/[Fe^{+2}]^{.05})$ ratios good reservoir quality in the middle coast deep Frio. On the other hand, if high wellhead pHs actually reflect high in situ pH in middle coast waters then one might predict late-stage, nonferroan calcite cement and poor reservoir quality. Hydropressured waters of the East White Point Field, San Patricio County (crosses, Fig. 2) further illustrate, in the absence of in situ pH data, the use of log $[Ca^{+2}]^{.05}/[Fe^{+2}]^{.05})$ ratio as a quality indicator. Poor reservoir quality is indicated at East White Point from small log $([Ca^{+2}]^{.05}/[Fe^{+2}]^{.05})$ ratios. Note that the log ratio is independent of pH (equation 1, Table 4) and knowledge of pH is not needed.

Chloritization—Three activity diagrams are presented to evaluate chlorite stability (Figs. 3, 5, 6). The relative stability of chlorite and kaolinite is discussed first. Equation 3 (Table 4) is evaluated with reference to the fate of Ca-montmorillonite grain coats or cutans, coatings on framework grains formed during pedogenesis (Brewer, 1964), and leads to consideration of chlorite/illite stability (equation 4).

All Frio waters plot in the chlorite stability field under geopressured conditions represented by 100°C and 600 bars pressure, which both contradicts and supports petrographic evidence (Fig. 3). Chlorite is generally absent in upper coast sandstones and abundant in south Texas sandstones. Factors controlling chlorite stability relative to kaolinite are pH, temperature, and log $([Mg^{+2}][Fe^{+2}])$ product. Note that log $[H^+]$ is multiplied by 9.2. The role of pH in two upper coast waters is illustrated in Figure 4. At a wellhead pH of 6.35

they plot in the chlorite stability field whereas at an in situ pH of 4 they are shifted into the kaolinite field. At pH values of 4 to 6, not unreasonable in situ values for subsurface waters, all tested upper coast waters plot in the kaolinite stability field, consistent with petrographic evidence. The general absence of chlorite in upper coast sandstones is explained by pHs and temperatures too low for chlorite stability. For example, Pleasant Bayou waters would fall in the chlorite field if temperatures rose above 150°C to approximately 175°C (Fig. 4).

Apparently, chloritization can be initiated at low temperatures (50 to 75°C) if pH is high, approximately 8. Note in particular that at these temperatures and pH, hydropressured waters of the middle coast fall in the chlorite stability field (Fig. 3). However, at assumed in situ pH values of 5 to 7,

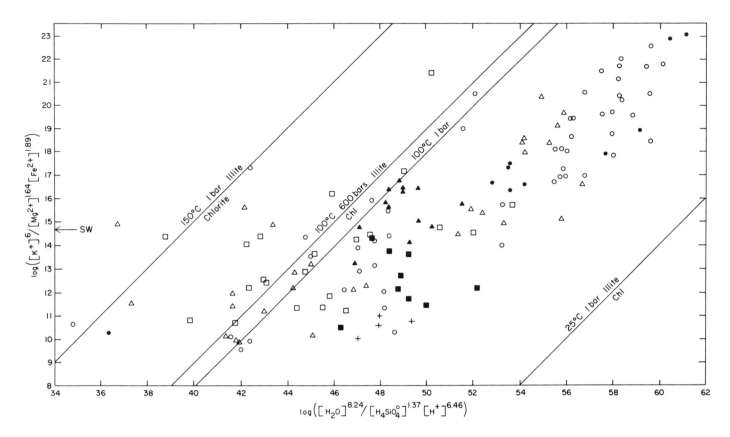

Figure 6—Activity diagram of the reaction illite = chlorite (equation 4).

TABLE 5

T°C	P bar(s)	Reaction Pairs of Table 4											
		1	2	3	4	5	6	7	8	9	10	11	12
25	1	0.20	−51.7	−57.6	−46.4	4.7	0.32	−3.2	−3.7	−0.49	−0.45	5.1	−3.03
50	1	0.21	−45.1	−50.0	−41.0	4.3	0.47	−2.8	−3.9	−0.37	0.11	4.7	−2.66
75	1	0.22	−39.3	−43.6	−36.3	3.9	0.39	−2.5	−4.1	−0.26	0.50	4.3	−2.35
100	1	0.22	−33.9	−38.0	−32.2	3.6	0.20	−2.4	−4.4	−0.18	0.80	3.8	−2.07
	600	0.22	−32.2	−35.7	−30.8	4.2	1.76	−0.8	−4.0	−0.18	1.35	2.9	−2.07
125	1	0.23	−29.5	−33.1	−28.5	3.2	−0.06	−2.3	−4.7	−0.11	1.04	3.6	−1.84
	600	0.23	−27.2	−30.8	−27.2	4.0	1.37	0.5	−4.2	−0.11	1.43	2.4	−1.84
150	1	0.23	−25.2	−28.8	−25.2	2.9	−0.46	−2.4	−5.0	−0.04	1.17	3.4	−1.65
	600	0.23	−22.8	−27.1	−23.9	3.7	1.08	0.2	−4.4	−0.04	1.60	1.8	−1.65
200	1	0.24	−17.8	−21.6	−19.4	2.3	−1.37	−2.6	−5.5	0.05	1.27	2.9	−1.35
	600	0.24	−15.1	−19.4	−18.2	3.1	−0.05	1.3	−5.0	0.05	1.71	0.5	−1.35

Table 5—Log K of reaction (log K_r).

Figure 7—Activity diagram of the reaction plagioclase (An30) = kaolinite (equation 5).

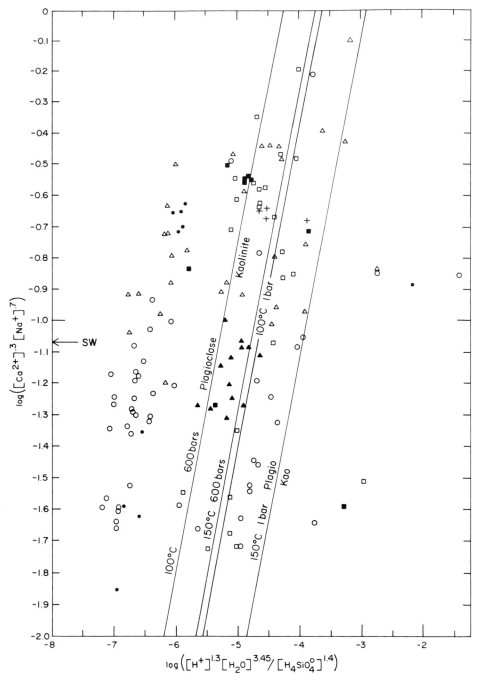

these waters are shifted right, deep into the kaolinite field at 50 to 75°C. At the same pH values, south Texas hydropressured waters remain in the chlorite stability field because of larger $\log([Mg^{+2}]^{2.3}[Fe^{+2}]^{2.3})$ products, in excess of -20 or that for sea water. Clearly, higher temperature (higher geothermal gradient), pH, $[Mg^{+2}]$ and $[Fe^{+2}]$, and volcanic-rich detritus promote chlorite formation in south Texas (Kaiser, 1982).

Prior prediction was that south Texas hydropressured waters would plot in the chlorite stability field (Kaiser and Richmann, 1981; Kaiser, 1982), serving as a direct indicator of poor reservoir quality, because chlorite is an important inhibitor of permeability and porosity in the deep Frio of south Texas. Indeed, this turned out to be the case. Consequently, poor reservoir quality is predicted in the deep Frio of the East White Point Field (Fig. 3) as was also predicted from small log $([Ca^{+2}]^{.05}/[Fe^{+2}]^{.05})$ ratios. Insofar as kaolinite is an indicator of acidic waters and possible undersaturation with respect to calcite, then hydropressured waters clustering in or toward the kaolinite stability field might be an indirect indicator of deep secondary porosity.

Ca-montmorillonite is stable with respect to chlorite at 25°C in near-surface waters. At temperatures between 50 and 75°C chloritization is promoted in hydropressured waters by high pH and small log $([Ca^{+2}]^{.29}/[Fe^{+2}]^{1.85}[Mg^{+2}]^{1.85})$ ratios of less than 15.4 or that for sea water (Fig. 5). Again, because chlorite is favored in south Texas hydropressured waters, it is postulated that montmorillonite grain coats underwent chloritization to chlorite rims in south Texas. Clay coats may also undergo transformation to mixed layer smectite/illite (S/I). Thermodynamic data developed in this study (Table 5) indicate the S/I transformation begins, in the presence of K^+, between 50 and 75°C. All Frio waters with respect to the montmorillonite/illite reaction pair plot deeply into the illite stability field. Illite in turn becomes unstable relative to chlorite as diagenesis proceeds as shown in Fig. 6. Illite is stable at low temperature and pressure; however, as temperature and pressure increase, the chlorite stability field expands to include most of the Frio waters (Fig. 6). Thus, chlorite, among the layer silicates reported petrographically, is the layer silicate ultimately stable in deep Frio sandstones. Low pH favors illite stability, as illustrated by middle coast waters. Even at the rather high in situ pH of 7, they are shifted left into the illite stability field at 100°C. Again, south Texas hydropressured waters plot in the chlorite field, separated from upper and middle coast waters at the log activity ratio for sea water. Petrographic and SEM analysis of deep Frio sandstones in south Texas shows chlorite is abundant relative to illite, which is consistent with solution–mineral equilibria data.

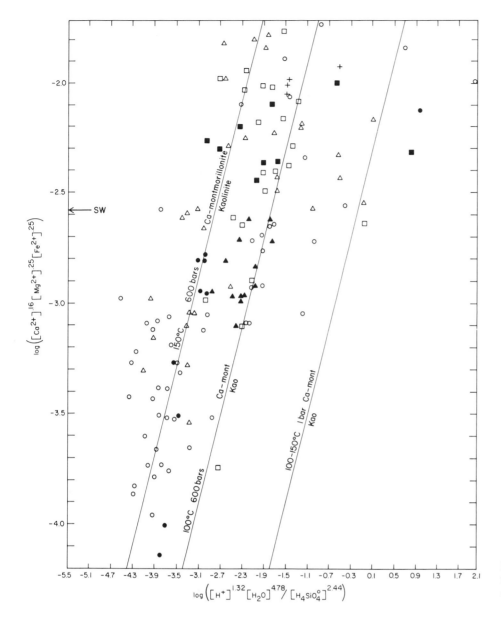

Figure 8—Activity diagram of the reaction Ca-montmorillonite = kaolinite (equation 7).

Kaolinization—Reactions of major importance in the Frio are numbers 5 and 7 (Table 4). Petrographic evidence is unequivocal that kaolinite has replaced plagioclase; however, there is no clear evidence that kaolinization of albite and K-spar occurred during Frio diagenesis. Conversion of montmorillonite to kaolinite is probably most important early in the shallow diagenetic environment, perhaps in low-pH meteoric water and later at the top of geopressure in low-pH compactional water. Relative to illite (equation 8), kaolinite is highly unstable in Frio waters at subsurface temperatures. Activity indices fall deep into the illite

stability field, indicating that illite is stable relative to kaolinite in the deep Frio.

In hydropressured waters, at 100°C and 1 bar pressure, kaolinite and plagioclase coexist. Under geopressured conditions and lowered pH the kaolinite stability field expands (Fig. 7). Petrographically, kaolinite begins forming as a grain replacement of plagioclase, consuming the whole grain as diagenesis proceeds, and eventually growing outward to fill surrounding pore space (Loucks et al, 1981). This sequence is consistent with geochemical calculations indicating an expanding kaolinite field. In passing from hydro-

pressured to geopressured conditions, the kaolinite field expands to its maximum extent relative to plagioclase. Because most dissolved grains are either plagioclase or volcanic rock fragments, the effect of kaolinization on reservoir quality should be most evident at the top of the geopressured zone. Thereafter as temperature increases kaolinite's stability field decreases in size. The absence of K-spar and albite alteration to kaolinite is predictable from thermodynamics. K-spar is stable with respect to kaolinite, and albite's stability field is large in Frio waters; therefore, K-spar and albite are less subject to kaolinization than is

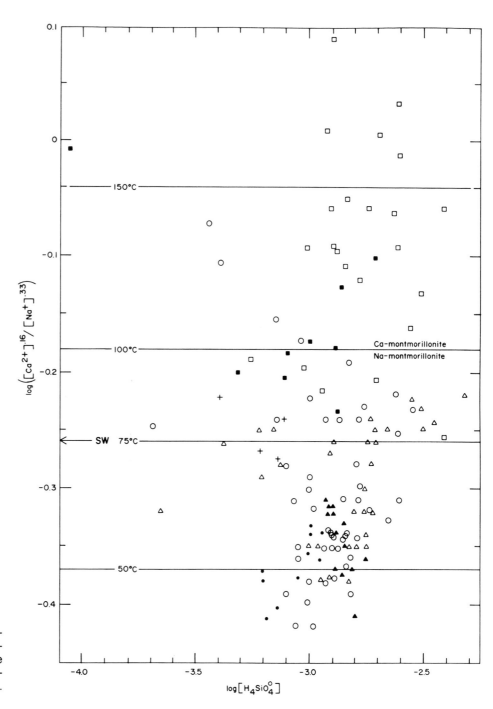

Figure 9—Activity diagram of the reaction Ca-montmorillonite = Na-montmorillonite (equation 9). Relative stability of montmorillonite independent of log $[H_4SiO_4^0]$.

plagioclase.

Ca-montmorillonite is stable with respect to kaolinite in high-pH, silica-rich Frio waters up to approximately 100°C. Figure 8 shows that as temperature and pressure increase the kaolinite stability field expands and suggests that Ca-montmorillonite could coexist with kaolinite in geopressured waters at pH greater than 6 or 7. Hydropressured waters of the upper and middle coast

favor kaolinite stability and indirectly indicate superior reservoir quality; they once again are separated from south Texas waters at the activity index for sea water (SW on Fig. 8). Kaolinization of montmorillonite is favored over chloritization in upper and middle coast waters because, as discussed above, temperature, $[Mg^{+2}]$ and $[Fe^{+2}]$, and in situ pH are lower.

Cation exchange—Relative abun-

dance of Ca- or Na-montmorillonite is controlled by temperature and the log $([Ca^{+2}]^{16}/[Na^+]^{33})$ ratio; as temperature increases the Na-montmorillonite field expands to include, at approximately 100°C, most hydropressured waters (Fig. 9). In most geopressured waters, Na-montmorillonite and illite or Na-montmorillonite and chlorite would likely coexist in upper/middle coast and south Texas sandstones, respec-

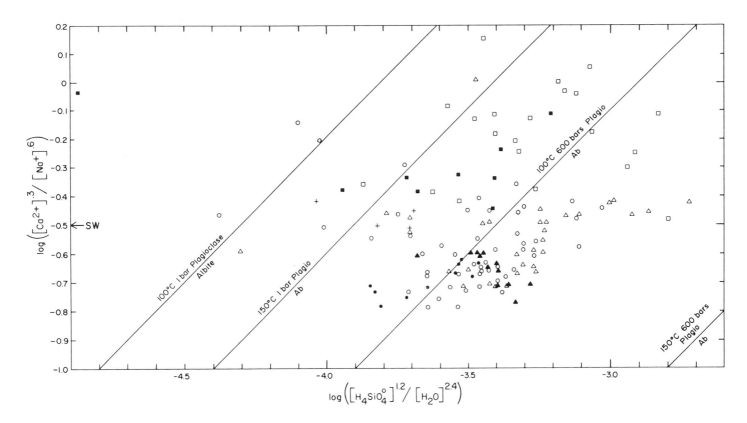

Figure 10—Activity diagram of the reaction plagioclase (An30) = albite (equation 11).

tively. Geographic clustering of data points on the montmorillonite activity diagram is excellent. All south Texas waters have large $log([Ca^{+2}]^{.16}/[Na^+]^{.33})$ ratios, whereas the smallest ratios occur in middle and lower coast waters. Small ratios are postulated to be indicators of good reservoir quality at depth and large ratios poor quality. Recall that south Texas waters also have small log $([Ca^{+2}]^{.05}/[Fe^{+2}]^{.05})$ ratios and large log $([Mg^{+2}]^{2.3}[Fe^{+2}]^{2.3})$ products.

Albitization—Two reactions were considered: albitization of plagioclase and K-spar (equations 11 and 12, Table 4). Under hydropressured conditions, represented by 1 bar pressure and temperatures of 100°C or less, all Frio waters relative to the plagioclase/albite reaction pair plot well into the albite stability field (Fig. 10). This implies that albitization is initiated in the hydropressured interval at less than 100°C, probably between 50 and 75°C. At these temperatures the albite stability field expands to fill most of Figure

10. Under geopressured conditions, represented by 600 bars pressure, the plagioclase stability field expands because of higher temperature and pressure. In effect, albitization becomes more difficult with depth. At 125°C almost all Frio waters plot in the plagioclase field, suggesting that albitization is essentially completed at approximately 125°C and certainly before 150°C. Figure 10 indicates a temperature of approximately 100°C in south Texas. Indeed, microprobe analysis shows that albitization is completed earlier in south Texas than on the upper coast. The depth to complete albitization is considerably shallower (about 8350 ft; 2546 m) in Hidalgo County than in Brazoria County (about 12,200 ft; 3720 m).

On the basis of thermodynamic calculations, K-spar is stable relative to albite but less so at increased temperatures, whereas albite is stable relative to plagioclase. This means that plagioclase is albitized before K-spar in the Frio as shown from petrographic data (Kaiser and Richmann, 1981). Albitization or Na-metasomatism of K-spar (microcline) requires a $log([K^+]/[Na^+])$ ratio appropriate for the temperature.

As temperature and $log([K^+]/[Na^+])$ ratio increase, the albite stability field expands to include all but one Frio hydropressured water at 100°C (Fig. 11). Hydropressured waters cluster in the albite stability field between boundaries at 50 and 85°C suggesting that albitization of K-spar was initiated between 50 and 85°C in Frio Formation waters. In effect, lower $[Na^+]$ is required for albite stability at higher temperatures; K-spar is favored in near-surface waters at lower temperatures. K-spar and albite cements or overgrowths would be stable in Frio low-temperature, hydropressured waters; they are minor, volumetrically insignificant, near-surface cements in Frio sandstones.

Geopressuring

Variation in ionic strength, analytical molality, and activity indices versus depth, shows a correlation with geopressuring of uncertain geochemical significance. The correlation between ionic strength (I) and the pressure regime is obvious: hydropressured waters decrease or show no strong trend in I with depth to approximately 10,000 ft (3050 m) or the top of geo-

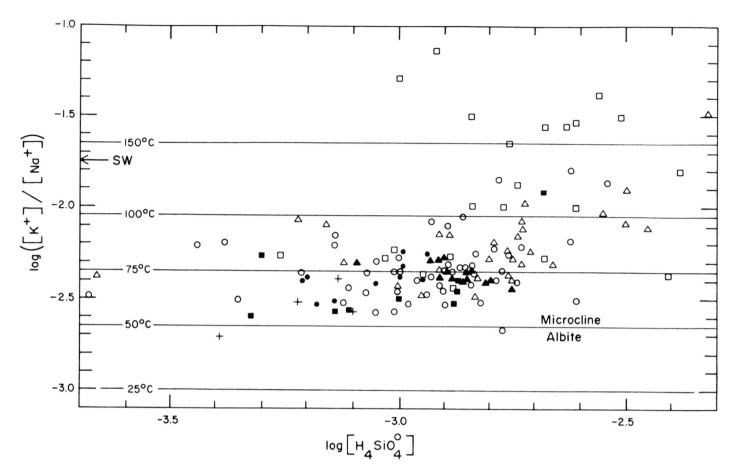

Figure 11—Activity diagram of the reaction microcline = albite (equation 12). Relative stability of albite and microcline independent of log [H₄SiO₄].

pressure (Fig. 12). In geopressured waters, I uniformly increases with depth to approximately 14,000 ft (4270 m), then apparently begins to decrease.

Several mole ratios (Ca^{+2}/Na^+, Cl^-/Ca^{+2}, Cl^-/Mg^{+2}, Cl^-/K^+, Mg^{+2}/Ca^{+2}, K^+/Na^+, Cl^-/Na^+, Ca^{+2}/Fe^{+2}, and Sr^{+2}/Ca^{+2}) were plotted against depth to investigate the relationship of molality to geopressuring. Among these, the Ca^{+2}/Na^+ and Cl^-/Ca^{+2} ratios show the best correlation with the pressure regime. The Ca^{+2}/Na^+ ratio ranges widely in the hydropressured interval, from 0.002 to 1.7, and shows no distinct trend or perhaps a decrease to the top of geopressure. Here there is a slight shift to lower ratios and then a uniform increase in values in excess of that for sea water (0.022) (Fig. 13), reflecting the effects of Ca^{+2}-generating diagenetic reactions

(Table 4). Variation in the Cl^-/Ca^{+2} ratio is largest (about 5 to 510) between 8000 and 11,000 ft (2440 and 3355 m), the transition zone between the hydropressured and geopressured intervals (Fig. 14). The smallest variation (5 to 20) is in waters deep in the hydropressured or geopressured interval. A similar correlation with geopressuring is shown by the Cl^-/Mg^{+2} and Cl^-/K^+ ratios. The variation observed is attributed to more active water–rock interaction, or diagenesis in the transition zone. In other words, less active diagenesis or perhaps fluid flow in the hydropressured and geopressured intervals is reflected in less variation among mole ratios. In the former interval, diagenesis is just beginning; whereas, in the latter interval, formation waters are probably equilibrating with minerals.

The Mg^{+2}/Ca^{+2} ratio uniformly decreases from the top of geopressure downward, reflecting loss of Mg^{+2} into authigenic chlorite, calcite, and possibly dolomite and gain of Ca^{+2}. Varia-

tion among hydropressured waters is large, ranging from 0.002 to 1.0 (sea water = 5.6). The K^+/Na^+ ratio shows no correlation with pressure; it shows a gradual increase with depth and may be controlled mainly by temperature-dependent water–rock interaction, such as albitization and breakdown of detrital illite. The latter source of K^+ is suggested by the fact that the K^+/Na^+ ratio continues to increase below the depth of K-spar albitization. Most Cl^-/Na^+ ratios are approximately 1.0; however, ratios in south Texas waters exceed 1.0, ranging from greater than 1.0 to 4.4. Clearly, not all Frio waters can be attributed to halite dissolution.

Mineral transformations were correlated with geopressuring by plotting activity indices, derived from the diagenetic reaction pairs (Table 4), against depth. The intent was to relate specific reactions to geopressuring, high-resistivity shale (cap rock) at the top of geopressure, and ultimately brine evolution. One of these reactions is the Ca–Na–montmorillonite

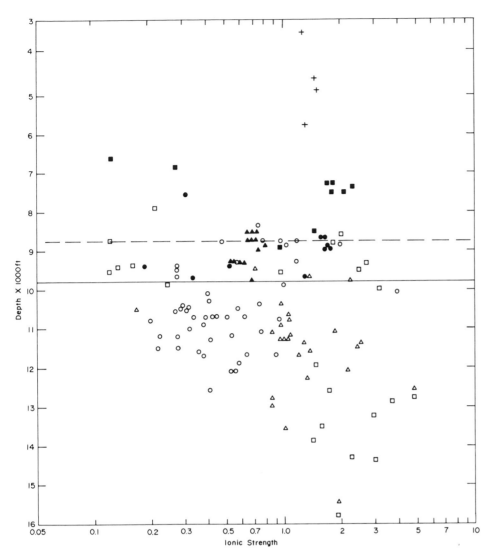

Figure 12—Ionic strength versus depth, I = 0.7 in sea water. I in Frio waters ranges from 0.12 to 4.8. Solid line designates approximate depth to deepest hydropressured waters in upper and middle coast, and dashed line likewise in lower coast or south Texas. Open symbols designate geopressured waters, solid hydropressured waters.

exchange. From the top of geopressure downward, roughly coincident with the 100°C isotherm, the log $([Ca^{+2}]^{.16}/[Na^+]^{.33})$ ratio uniformly increases; whereas the variation for hydropressured waters is large (Fig. 15). Alternatively, it could be argued that this ratio decreases with depth in south Texas geopressured waters, where Ca^{+2}-rich waters possibly favor Ca-montmorillonite stability and Na^+ release (Fig. 9). Besides the exchange reaction, exchanging one Ca^{+2} for two Na^+ ions, other Ca^{+2}-generating (and Na^+-consuming) reactions are probably reflected in the log $([Ca^{+2}]^{.16}/[Na^+]^{.33})$ ratio. Among these are Ca-montmorillonite = chlorite, Ca-montmorillonite = kaolinite, plagioclase = kaolinite, calcite = ferroan calcite, smectite =

illite, and albitization of plagioclase. The latter is an important source of Ca^{+2}; however, total albitization of detrital plagioclase would probably release less than 50% of the Ca^{+2} necessary for diagenesis in the Frio Formation (Kaiser and Richmann, 1981). Clearly, additional sources of Ca^{+2} must be sought in chloritization, kaolinization, calcite, and illitization reactions.

Plots of activity indices versus depth for the clay-mineral reactions discussed here display distinctive funnel-shaped patterns of data distribution (Fig. 16). Variation in the activity index is largest between 8000 and 11,000 ft (2440 and 3355 m), the hydropressure–geopressure transition zone. Considerably less variation is displayed by distinctly hy-

dropressured or geopressured waters. This implies that most of the diagenesis occurs in the transition zone. Activity indices converge on the top of geopressure and get larger or smaller, signaling favorability of one clay mineral over another. In Figure 16, the index gets larger, in effect converging on Ca-montmorillonite favorability. Chlorite either is unstable in the transition zone or is forming freeing Ca^{+2} and consuming Fe^{+2} and Mg^{+2}. With respect to montmorillonite, illite and kaolinite stability are favored in the transition zone, whereas montmorillonite, illite, and kaolinite are favored with respect to chlorite. Chlorite stability is favored in distinctly hydropressured and geopressured waters, suggesting two stages of chlorite formation (Fig. 16).

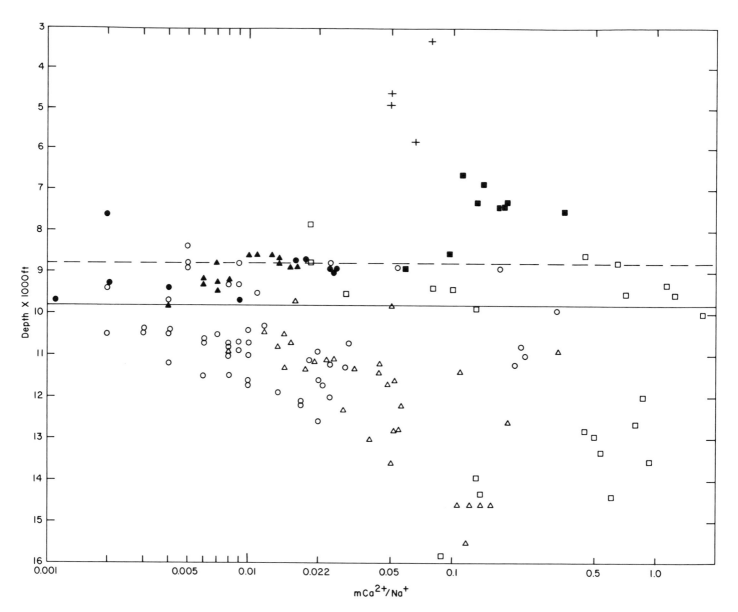

Figure 13—Ca^{+2}/Na$^+$ mole ratio versus depth, ratio = 0.022 in sea water.

DIAGENETIC HISTORY

Major diagenetic events in the Frio Formation have been established petrographically by Loucks et al (1981) (Fig. 17). Within this diagenetic framework and in view of limited ability to evaluate reaction kinetics solution–mineral equilibria can be used to sharpen our perception of calcite equilibrium, relative mineral stabilities, and timing of diagenetic events. For example, some petrographically unanswered questions concern episodic precipitation and leaching of authigenic calcite, timing of albitization, and stages of chlorite formation.

Calcite Equilibrium

Multiple generation of authigenic calcite is the rule in the Frio Formation. Major episodes of calcite precipitation occurred in several stages throughout the burial history. Leaching of calcite is also episodic and is responsible for considerable secondary porosity in Frio sandstones. These episodes are reflected in the presence of several distinctive calcite compositions at various depths (Loucks et al, 1981). Calcite sequencing is predictable from calcite equilibrium and the trend of log $([Ca^{+2}]^{.05}/[Fe^{+2}]^{.05})$ ratio and Ca^{+2}/Fe^{+2} mole ratio with depth. To form nonferroan calcite, the virtual absence of Fe^{+2} is required; waters having a [Ca^{+2}]/[Fe^{+2}] ratio of less than 10,000 will first yield ferroan calcite. Nonferroan calcite is stable in surface sea water (Fig. 2) and probably would be favored in sulfide-rich hydropressured waters before initiation of clay mineral transformations and release of Fe^{+2}. Ferroan calcite might be expected in sulfide-poor geopressured waters subject to

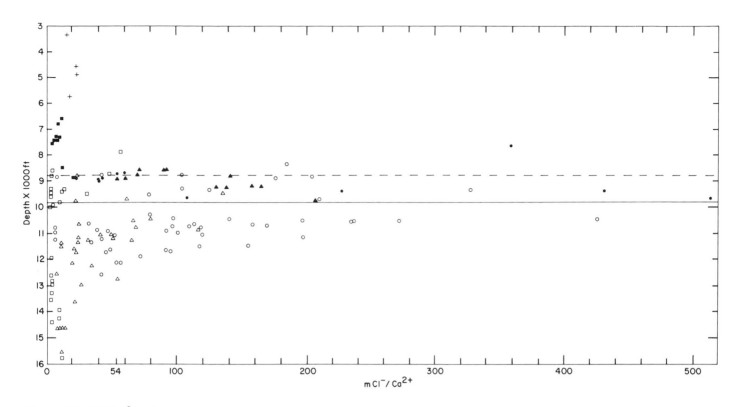

Figure 14—Cl⁻/Ca^{+2} mole ratio versus depth, ratio = 54 in sea water.

deeper burial and higher temperatures. Ferroan calcite can also form at shallow depths in response to circulation of meteoric ground water. Indeed, in anaerobic fresh water, $[Fe^{+2}]$ is commonly high relative to $[Ca^{+2}]$ and siderite is stable. In addition, pure calcite might reappear at depth, even in the presence of some Fe^{+2} in very Ca^{+2}- and sulfide-rich waters, such as those that might be discharging from the underlying Mesozoic strata, as suggested by Galloway (1982). Timing and depth of calcite precipitation are strongly affected by pH. Because pH generally decreases with depth in elisian (compactional) and abyssal (thermobaric) waters, calcite precipitation should be expected early in diagenesis and at shallow depth.

Frio log($[Ca^{+2}]^{.05}/[Fe^{+2}]^{.05}$) ratios range widely, from 0.01 to 0.32, plotting deep in the ferroan calcite and calcite stability fields, respectively. Reflected, perhaps, is constant reequilibration with changing water chemistry or, in other words, multiple generation

of calcite. Similarly, the Ca^{+2}/Fe^{+2} mole ratio displays a tremendous range of values from 2 to 600,000. No obvious correlation is seen with geopressuring. Late-stage ferroan calcite in south Texas is consistent with the trend of log ($[Ca^{+2}]^{.05}/[Fe^{+2}]^{.05}$) ratio with depth; in these Frio waters it decreases. Likewise, a weak trend of decrease with depth is evident in the Ca^{+2}/Fe^{+2} mole ratio and also indicates that ferroan calcite is favored at depth.

Timing of Albitization

Albitization follows formation of quartz overgrowths (Fig. 17), which precipitate at approximately 60 to 80°C (Land, this volume). On the basis of microprobe data, regional geothermal gradients, and isotopic data, albitization occurs nearly contemporaneous with major kaolinite precipitation (Kaiser and Richmann, 1981). The last major authigenic mineral to form on the upper coast is kaolinite cement and it follows and/or coincides with major calcite leaching, a low pH event (Loucks et al, 1981). Kaolinite cement is inferred to follow albitization because kaolinite's temperature of for-

mation, approximately 100°C as calculated from isotopic data (Loucks et al, 1981), is higher than that for initiation of albitization, which is estimated thermodynamically at between 50 and 85°C.

Chlorite Formation

Two stages of chlorite formation are proposed from geochemical calculations and confirmed by petrographic data (Fig. 17). In the first stage, chlorite forms at the expense of Ca-montmorillonite (smectite) grain coats to yield chlorite rims under hydropressured conditions at less than 75°C. Chlorite will appear early in those sediments or rocks that are rich in Fe^{+2} and Mg^{+2} (Velde, 1972), not unlike the volcanic-rich Frio sediments of south Texas. Certainly, chlorite, in particular 7-Å varieties, would be favored in the Fe^{+2}- and Mg^{+2}-rich waters of south Texas. A second stage is postulated to form at the expense of kaolinite cement to yield chlorite cement under geopressured conditions at greater than 100°C, accounting for the absence of kaolinite in the deep Frio of south Texas.

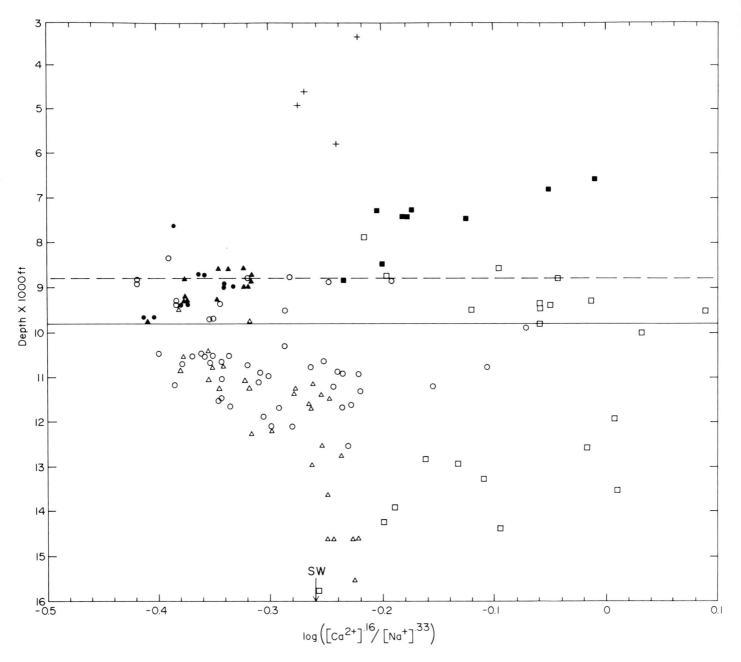

Figure 15—Log([Ca^{+2}]$^{.16}$/[Na$^+$]$^{.33}$) ratio, derived from the reaction Ca-montmorillonite = Na-montmorillonite, versus depth. SW = log activity ratio in sea water.

CONCLUSIONS

(1) Diagenesis in the Frio Formation is a function of temperature, pH, activity, and pressure. Most diagenesis occurs in the transition zone between the hydropressured and geopressured interval at relatively low temperatures.

(2) Solution–mineral equilibria in hydropressured waters best reflect reservoir quality because most diagenesis occurs in the transition zone. Activity indices favoring chlorite and ferroan calcite stability and large log([Ca^{+2}]$^{.16}$/[Na$^+$]$^{.33}$) ratios are the best indicators of reservoir quality in deep Frio sandstone reservoirs.

(3) Predictions about Frio diagenesis based on equilibrium thermodynamics are consistent with those established petrographically or from isotopic data,

suggesting that estimated equilibrium constants (log K$_f$ and K$_r$) are sufficiently accurate to describe diagenetic reactions in the Frio Formation.

(4) Water-chemistry data complement petrographic data, adding insight on diagenetic sequence, in situ pH, and relative mineral stabilities. Calcite equilibrium favors precipitation of calcite early in the burial history. In situ pH values of geopressured waters are inferred to be 2 or 3 pH units lower than those measured at the wellhead.

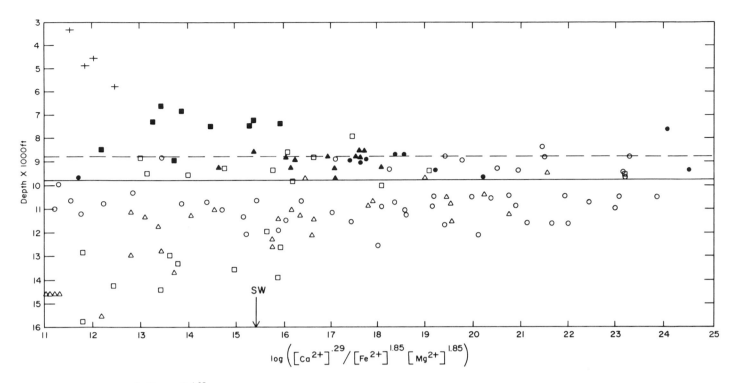

Figure 16—Log($[Ca^{+2}]^{.29}/[Fe^{+2}]^{1.85}$ $[Mg^{+2}]^{1.85}$) ratio, derived from the reaction Ca-montmorillonite = chlorite, versus depth. Larger log ratios favor montmorillonite stability.

Low temperature and pH explain the general absence of chlorite in the upper Texas coast. Chlorite and illite are the stable layer silicates in deep Frio sandstones. The smectite/illite transformation is not the only clay-mineral transformation of importance in diagenesis.

(5) Predicting the course of Gulf Coast burial diagenesis will require application of the principles of solution–mineral equilibria to data on bulk composition, geothermal gradient, and geopressuring.

ACKNOWLEDGMENTS

This research was funded in part by the U.S. Department of Energy, Division of Geothermal Energy, under Contract No. DE-AC08-79ET27111. B. Dan Legett assisted in the calculation of thermodynamic functions; he computerized the calculation of log K_f and log K_r values. Numerous companies provided access to their properties for water sampling. Special thanks go to the personnel of Exxon's Kingsville office. James F. O'Connell collected Frio water samples, compiled water-chemistry data, and modified computer routines for plotting activity indices and mole ratios. Chemical analyses were done under the direction of Clara L. Ho, chemist-in-charge, Bureau of Economic Geology, Mineral Studies Laboratory. Dan F. Scranton, chief cartographer, supervised drafting of the illustrations.

SELECTED REFERENCES

Bebout, D. G., R. G. Loucks, and A. R. Gregory, 1978, Frio sandstone reservoirs in the deep subsurface along the Texas Gulf Coast—their potential for the production of geopressured geothermal energy: Austin, University of Texas, Bureau of Economic Geology Report of Investigations 91, 92 p.

Boles, J. R., and S. G. Franks, 1979, Clay diagenesis in Wilcox sandstones of southwest Texas: implications of smectite diagenesis on sandstone cementation: Journal of Sedimentary Petrology, v. 49, no. 1, p. 55–70.

Brewer, R., 1964, Fabric and mineral analysis of soils, Pedological features, I, Cutans: New York, John Wiley and Sons, p. 205–233.

Carroll, D., 1970, Clay minerals: a guide to their X-ray identification: Geological Society of America Special Paper 126, 80 p.

Eugster, H. P., and I. M. Chou, 1973, The depositional environments of Precambrian banded iron-formations: Economic Geology, v. 68, p. 1144–1168.

Fisher, J. A., and E.-A. Zen, 1971, Thermochemical calculations from hydrothermal phase equilibrium data and the free energy of H_2O: American Journal of Science, v. 270, p. 297–314.

Galloway, W. E., 1982, Epigenetic zonation and fluid flow history of uranium-bearing fluvial aquifer systems, south Texas uranium province: Austin, University of Texas, Bureau of Economic Geology Report of Investigations 119, 31 p.

Galloway, W. E., and W. R. Kaiser, 1980, Catahoula Formation of the Texas coastal plain: origin, geochemical evolution, and characteristics of uranium deposits: Austin, University of Texas, Bureau of Economic Geology Report of Investigations 100, 81 p.

Garrels, R. M., and C. L. Christ, 1965, Solutions, minerals, and equilibria, New York, Harper and Row, p. 306–351.

Helgeson, H. C., 1969, Thermodynamics of hydrothermal systems at elevated temperatures and pressures: American Journal of Science, v. 267, p. 729–804.

Helgeson, H. C., J. M. Delaney, H. W. Nesbitt, and D. K. Bird, 1978, Summary and critique of the thermodynamic properties of rock forming minerals: American Journal of Science, v. 278-A, 229 p.

Hower, J., E. V. Eslinger, M. E. Hower, and E. A. Perry, 1976, Mechanism of burial metamorphism of argillaceous sediments; I, Mineralogical and chemical evidence:

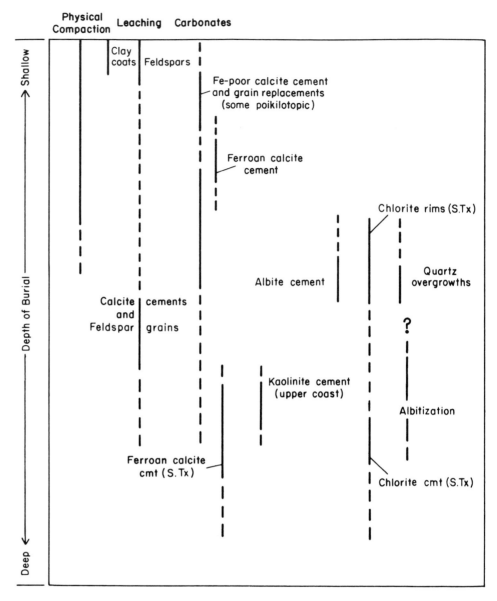

Figure 17—Major diagenetic events in the Frio Formation (modified from Loucks et al, 1981).

Geological Society of America Bulletin, v. 87, no. 5, p. 725–737.

Kaiser, W. R., 1982, Predicting diagenetic history and reservoir quality in the Frio Formation (Oligocene) of Texas (abs.): Eleventh International Congress on Sedimentology, Hamilton, ON, McMaster University, Abstracts of Papers, p. 119–120.

Kaiser, W. R., and D. L. Richmann, 1981, Predicting diagenetic history and reservoir quality in the Frio Formation of Brazoria County, Texas and Pleasant Bayou test wells, in D. G. Bebout and A. L. Bachman, eds., Proceedings of the Fifth United States Gulf Coast Geopressured-Geothermal Energy Conference: Baton Rouge, Louisiana State University, p. 67–74.

Kerrick, Y. K., and L. S. Darken, 1975, Sta-

tistical thermodynamic models for ideal oxide and silicate solid solutions with applications to plagioclase: Geochimica et Cosmochimica Acta, v. 39, p. 1431–1442.

Kharaka, Y. K., and I. Barnes, 1973, SOLMNEQ: Solution–mineral equilibrium computations: Springfield, VA, National Technical Information Service Technical Report PB-215899, 81 p.

Kharaka, Y. K., E. Callender, and W. W. Carothers, 1977, Geochemistry of geopressured geothermal waters from the Texas Gulf Coast, in J. Meriwether, ed., Proceedings of the Third United States Gulf Coast Geopressured-Geothermal Energy Conference: Lafayette, LA, University of Southwestern Louisiana Center for Energy Studies, v. 1, p. GI 121–165.

Kharaka, Y. K., M. S. Lico, V. A. Wright,

and W. W. Carothers, 1979, Geochemistry of formation waters from Pleasant Bayou No. 2 well and adjacent areas in coastal Texas, in M. H. Dorfman and W. L. Fisher, eds., Proceedings of the Fourth United States Gulf Coast Geopressured-Geothermal Energy Conference: Austin, University of Texas, Center for Energy Studies, v. 1, p. 168–193.

Lindquist, S. J., 1977, Secondary porosity development and subsequent reduction, overpressured Frio Formation sandstone (Oligocene), South Texas: Gulf Coast Association of Geological Societies Transactions, v. 27, p. 99–107.

Loucks, R. G., M. M. Dodge, and W. E. Galloway, 1979a, Sandstone consolidation analysis to delineate areas of high-quality reservoirs suitable for production of geopressured geothermal energy along the Texas Gulf Coast: Austin, University of Texas, Bureau of Economic Geology Contract Report no. EG-77-5-05-5554 for United States Department of Energy, 97 p.

Loucks, R. G., D. L. Richmann, and K. L. Milliken, 1979b, Factors controlling porosity and permeability in geopressured Frio sandstone reservoirs, General Crude Oil/Department of Energy Pleasant Bayou test wells, Brazoria County, Texas, *in* M. H. Dorfman and W. L. Fisher, eds., Proceedings of the Fourth United States Gulf Coast Geopressured-Geothermal Energy Conference: Austin, University of Texas, Center for Energy Studies, v. 1, p. 46–82.

————, 1981, Factors controlling reservoir quality in Tertiary sandstones and their significance to geopressured geothermal production: Austin, University of Texas, Bureau of Economic Geology Report of Investigations 111, 41 p.

Morton, R. A., C. M. Garrett, J. S. Posey, J. H. Han, and L. A. Jirik, 1981, Salinity variations and chemical compositions of waters in the Frio Formation, Texas Gulf Coast: Austin, University of Texas, Bureau of Economic Geology Contract Report no. DOE/ET/27111-5 for United States Department of Energy, 96 p.

Richmann, D. L., K. L. Milliken, R. G. Loucks, and M. M. Dodge, 1980, Mineralogy diagenesis, and porosity in Vicksburg sandstones, McAllen Ranch Field, Hidalgo County, Texas: Gulf Coast Association of Geological Societies Transactions, v. 30, p. 473–481.

Robie, R. A., B. S. Hemingway, and J. R. Fisher, 1978, Thermodynamic properties of minerals and related substances at 298.15K and 1 bar (10^5 pascals) pressure and at higher temperature: United States Geological Survey Bulletin 1452, 456 p.

Saxena, S. K., 1973, Thermodynamics of rock-forming crystalline solutions: New York, Springer-Verlag, 188 p.

Tardy, Y., and R. M. Garrels, 1974, A method of estimating the Gibbs energies of formation of layer silicates: Geochimica et Cosmochimica Acta. v. 38, p. 1101–1116.

Velde, B., 1972, Phase equilibria for dioctahedral expandable phases in sediments and sedimentary rocks: Association International pour l'Etude des Argiles (A.I.P.E.A.), International Clay Conference, Madrid, preprints, v. 1, p. 285–300.

Worley, T. J., 1979, Calculation of chemical equilibrium between aqueous solutions and minerals: EQ3/6 software package: Livermore, CA, University of California Lawrence Livermore Laboratory, UCRL-52658, 41 p.

Wood, B. J., and D. G. Fraser, 1976, Elementary thermodynamics for geologists, Multicomponent solids and fluids: Oxford University Press, p. 81–126.

Secondary Porosity Reactions in the Stevens Sandstone, San Joaquin Valley, California

James R. Boles
University of California
Santa Barbara, California

INTRODUCTION

Secondary porosity or the subsurface leaching of sandstone components has been increasingly recognized as an important reservoir-forming process (Hayes, 1979; Schmidt et al, 1977). Of critical interest to the petroleum geologist is the ability to predict where such leach zones occur and their geometry. Predictive models must be based on a detailed understanding of such zones in natural settings, including knowledge of what phases have been dissolved, and their relationship to avenues of mass transfer. From a thermodynamic viewpoint, the problem is simply one of predicting mineral stability (that is, precipitation versus dissolution) as a function of burial P–T conditions and pore-water chemistry. Surprisingly little is known in detail about phase compositions, by-products, and geometries of the leach zones, even though they are apparently common in the subsurface. We know even less about the mass transfer mechanisms, so critical to the process, but recent studies suggest convection processes may be involved (Wood and Hewitt, 1982).

In this paper, leach zones from the North Coles Levee Field, San Joaquin Valley, California, are described in some detail (Fig. 1). The geologic setting of this field is an example of an active burial setting, in that there are no significant unconformities indicating uplift and subaerial exposure. Thus, present-day burial conditions are believed to represent the maximum P–T conditions (for example, see Webb, 1981). Further, present-day pore-water compositions suggest that secondary porosity development could be active today. The field affords an unusually good opportunity to study

ABSTRACT. Secondary porosity in Miocene Stevens sandstones of the North Coles Levee Field results from dissolution of calcite, ferroan dolomite, and calcic plagioclase (An30). Kaolinite is a leach product of plagioclase, and mass balance calculations indicate that alumina is conserved on a thin-section scale. Iron released from dissolution of Fe-carbonate is possibly conserved in late-stage pyrite.

Detrital K-feldspar and early formed albite fracture filling in plagioclase are unaffected by the leach fluids, suggesting that these components are stable with respect to the leach fluids, whereas the anorthite component is unstable.

Thermodynamic considerations indicate present-day pore waters at 100° C, 260 bars fluid pressures, are nearly at equilibrium for the reaction:

$$CaCO_3 + CaAl_2Si_2O_8 + H_2O + 3H^+ = 2Ca^{+2} + HCO_3^- + Al_2Si_2O_5(OH)_4$$

and for a similar reaction involving dolomite.

Compaction following or perhaps contemporaneous with leaching has resulted in at least two diagenetic events. One involves albitization of plagioclase at stressed grain contacts with quartz, possibly as a result of higher silica activities stabilizing albite at these contacts. The other involves crushing of detrital biotite resulting in crystallization of carbonate adjacent to it. This phenomenon is due to depletion of H^+ in the pore water adjacent to newly formed mica surfaces as $K^+ \gtrless H^+$ exchange occurs between pore water and mica.

leach zones in detail because of the closeness of well spacings, abundance of core material, and supporting data on the field. To date, approximately 200 m of core from 5 wells in the field have been studied. Some of the preliminary results are presented here.

NORTH COLES LEVEE FIELD

The North Coles Levee (NCL) Field was discovered in 1938, and produces from the Upper Miocene Monterey Formation within Stevens sandstones in 12 to 70 m thick zones at 2440–2940 m (Hardoin, 1962). The field has undergone a water-flood secondary recovery program since 1954, but uncontaminated formation waters are still obtainable from some areas in the field.

Structurally, the NCL Field is on the western end of the Bakersfield arch and is a simple east-west trending, east-plunging anticline. Faulting is not generally recognized in the field.

The NCL Field is at essentially hydrostatic pressures since discovery with fluid pressures of 283 bars recorded at 2590 m (A. Tinnemier,

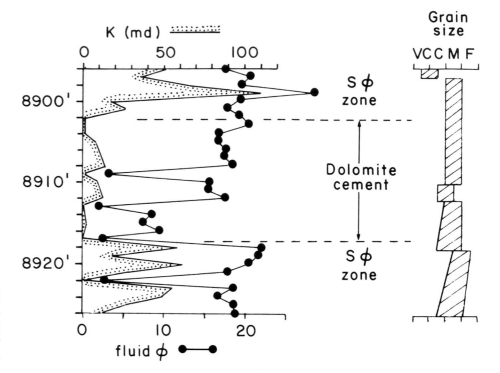

Figure 4—Dolomite cement remnant in secondary porosity zone. Note how cement distribution is not generally controlled by grain size, and marked differences in porosity and permeability occur over small vertical distances.

been interpreted as secondary pore spaces. It also fills fractures in detrital grains and, in some cases, pyrite-filled fractures cross-cut kaolinite. Thus, pyrite is a very late-stage phase in these rocks.

SECONDARY POROSITY ZONES

Dissolution of carbonate cements, both dolomite and calcite, and plagioclase feldspar is the cause of secondary porosity in the Stevens sands of the NCL Field. Leach zones are best defined by relatively high permeability (Fig. 4). Fluid porosities provide a less reliable discrimination between zones. Leach zones in the cores are dark colored with noticeably less induration than the hard light gray cemented zones. Samples have been studied from each zone as well as at their contacts. The interfaces between leached and unleached zones are sharp, on a scale less than 1 cm and are usually subparallel to bedding.

In one well, a single thin section was made across such a sharply defined leach zone (Figs. 4, 5) allowing comparison of the two zones over a few millimeters distance. This sample suggests that the carbonate cement, in this case dolomite, and plagioclase feldspar are leached by the same dissolution event. Plagioclase grains are virtually untouched in the carbonate-cemented area but are extensively leached in the dissolution zone. The cemented rock contains 30% dolomite and 3–4% total porosity. The leach zone contains 20–22% measured fluid porosity within 20 cm of the cemented rock. Approximately one-half to two-thirds of the total porosity is due to carbonate dissolution and one-half to one-third is due to plagioclase dissolution. Calcic plagioclase is preferentially leached along twin planes and fractures, but albite-fracture fillings remain as delicate residual bridges crossing the leached feldspar pore pseudomorphs (Fig. 3). The importance of this observation is that it demonstrates that albite was stable with respect to leach fluids, but calcic plagioclase was not. Detrital K-feldspar is virtually untouched in both zones.

The detailed study of the reaction interface also clearly shows the origin of the interstitial kaolinite. Kaolinite is rarely found *within* a leached plagioclase, but is found within a few micrometers of such grains. Mass balance calculations indicate that aluminum is preserved on the scale of this thin section. Point-count analyses indicate the leach zone within 2 cm of the cement zone contains 4.3 vol % removed plagioclase and 4.3 vol % kaolinite (2200 counts). The kaolinite percent estimate is a maximum value, as it does not account for microporosity in the clusters of kaolinite booklets indicated by the uptake of blue epoxy. The mean composition (microprobe) of the plagioclase is Ab70An30, which could produce a volume of kaolinite 64% of the volume of dissolved plagioclase. This correlates well with that observed in thin section (plagioclase/kaolinite = 1/1), considering the microporosity of the kaolinite.

SECONDARY POROSITY REACTIONS

The above petrographic observations allow the following conclusions to be drawn regarding secondary porosity reactions:

Reactants	Products
calcite	kaolinite
dolomite (±Fe)	albite (apparently stable but not necessarily precipitated)
plagioclase	pyrite

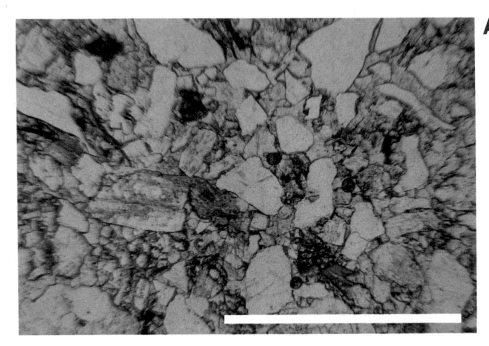

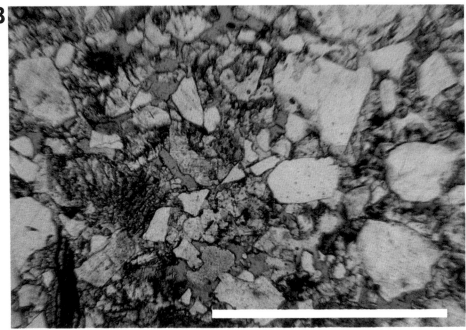

Figure 5—Photomicrograph comparing cemented (**A**) and leached (**B**) zones at 8918.3 ft. Photomicrographs taken 1.5 cm apart. Porosity filled with blue epoxy. See Figure 4 for location of sample. Bar length = 2 mm.

As for sources and sinks of components, aluminum is the only component that can be demonstrated, on a thin-section scale, to be conserved between reactants and products. Iron, released from dissolution of ferrous dolomite and ankerite, may be conserved in pyrite, but this has yet to be proven. Regarding the sink for iron, it is interesting that a *chlorite* leach product does not result from dissolution of ferrous dolomite and plagioclase. Apparently, the leach fluids are stable with respect to kaolinite $\pm$ pyrite + Mg^{+2} rather than an Fe–Mg chlorite. This observation places an additional chemical constraint on the nature of the leaching pore fluids.

The major components removed during the leaching include calcium, magnesium, and possibly silica and iron (if pyrite is not present).

The simplest type of leach reactions that can be written are:

(1) for calcite + anorthite removal

$$CaCO_3 + CaAl_2Si_2O_8 + H_2O + 3H^+ =$$
$$\quad Cc \qquad \quad Anor$$

$$2Ca^{+2} + HCO_3^- + Al_2Si_2O_5(OH)_4$$
$$\qquad\qquad\qquad\qquad Kaol$$

(2) for dolomite + anorthite removal

$$CaMg(CO_3)_2 + CaAl_2Si_2O_8 + H_2O +$$
$$\quad Dol \qquad\qquad Anor$$

$$4H^+ = 2Ca^{+2} + Mg^{+2} + 2HCO_3^- +$$

$$Al_2Si_2O_5(OH)_4$$
$$\quad Kaol$$

Alternative reactions involve considering the plagioclase composition observed in Stevens sandstones (that is, $0.7NaAlSi_3O_8 \cdot 0.3CaAl_2Si_2O_8$). For reaction 1 this alternative reaction is:

(1a) $CaCO_3 + Na_{.7}Ca_{.3}Al_{1.3}Si_{2.7}O_8 +$ $3.45\ H_2O + 2.3\ H^+ = 1.3Ca^{+2} + HCO_3^-$ $+ .65Al_2Si_2O_5(OH)_4 + .7Na^+ +$ $1.4H_4SiO_4$

Note that reaction 1a produces less kaolinite and also releases sodium and silica, compared with reaction 1. Reaction 1a, or a similar reaction involving dolomite, is a viable reaction in Stevens sandstones. However, reaction 1a is misleading, in that the leaching pore

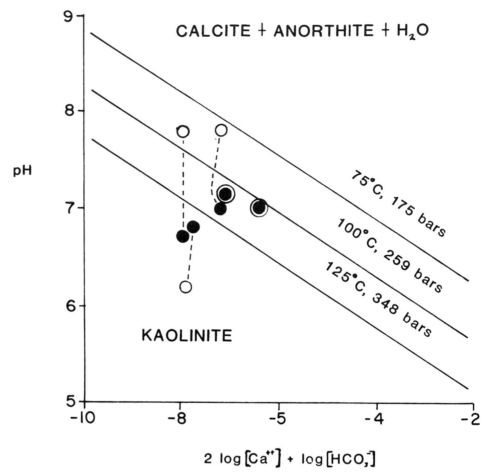

Figure 6—Equilibrium relations for reaction:
$CaCO_3 + CaAl_2Si_2O_8 + H_2O + 3H^+ = 2Ca^{+2} + HCO_3^- + Al_2Si_2O_5(OH)_4$
Anorthite activity assumed to be 0.40. (●) water analyses from NCL field at 100° C producing depth, corrected for CO_2 degassing. (○) uncorrected analyses.

waters are stable with respect to pure albite (see petrographic observations), and therefore, were not evaluated thermodynamically.

Figures 6 and 7 show the equilibrium activity diagram for reactions 1 and 2 at between 75 and 125°C, assuming appropriate hydrostatic fluid pressures for each temperature. These diagrams were constructed using the methods described by Garrels and Christ (1967) and thermodynamically evaluated by C. Bruton, ARCO Geoscience Group, using the SUPCRT[1] computer program. Detrital plagioclase from the NCL field contains 30 mole % anorthite component and thus would have an activity of 0.4 (Saxena and Ribbe, 1972). This is accounted for in Figures 6 and 7 and has the effect of lowering the equilibrium boundaries by 0.1 pH units. Also plotted on Figures 6 and 7

[1]Program written under direction of H. Helgeson (University of California, Berkeley) by P. Kirkham, J. Walther, J. Delany, and G. Flowers. Last update January 1978.

are pore-water analyses from the North Coles Levee Field, which were first corrected for possible CO_2 degassing and its effect on surface pH. Dr. Bruton, ARCO Geoscience Group, corrected for degassing by assuming that the P_{CO_2} in the gas phase of the reservoir (known from gas analyses) is in equilibrium with the pore water. Using equilibrium constants for the carbonic acid system and knowing the total dissolved carbonate species (from the HCO_3^- in the water analyses) and assuming the pH is controlled by the carbonate system, the subsurface pH was calculated. Some calculated pHs are *lower* than measured surface pH values by as much as 1.0 pH unit (see Figs. 6, 7). Activities of ionic species were then calculated using the EQ3 computer program (Woolery, 1979). In general, both figures show that present-day waters in the leach zones of the field (≃100°C temperature) are nearly on the equilibrium boundary or in the dissolution stability field (that is, kaolinite), indicating that

the leach process could be taking place today. Several waters on Figure 7 plot in the dolomite stability field in large part because of their high $a_{Mg^{+2}}$ contribution. Whether or not this is real is uncertain from present data. Currently, attempts are being made to obtain higher quality water analyses from the area.

Another important factor that could displace these water analyses toward the kaolinite (dissolution) field is the HCO_3^- value. Conventional oil field water analyses do not account for organic acid anions included within titration measured alkalinity (HCO_3^-) values. Carothers and Kharaka (1978) report that the HCO_3^- in pore-water analyses of San Joaquin oil fields contain up to 100% organic acid anions. If this were true at North Coles Levee, the water analyses would plot to the left to an unknown extent. This factor is presently being assessed.

The absence of chlorite as a by-product of leaching, and the presence of

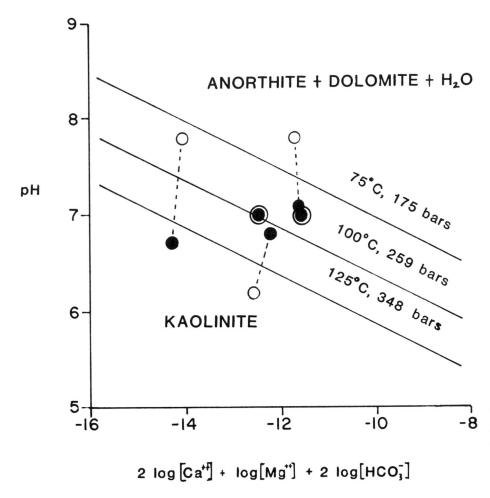

Figure 7—Equilibrium relations for reaction $CaMg(CO_3)_2 + CaAl_2Si_2O_8 + H_2O + 4H^+ = 2Ca^{+2} + Mg^{+2} + 2HCO_3^- + Al_2Si_2O_5(OH)_4$.
Dolomite assumed disordered structural state, anorthite activity = 0.40. (●) water analyses from NCL field at 100° C producing depth, corrected for CO_2 degassing. (○) uncorrected analyses.

kaolinite suggests the following reaction:

$$Mg_5Al_2Si_3O_{10}(OH)_8 + 10H^+ = Al_2Si_2O_5(OH)_4 + 5Mg^{+2} + SiO_2 + 7H_2O$$

Figure 8 shows that most Stevens pore waters favor a Mg-chlorite rather than kaolinite, in contrast to the observed petrography. Perhaps a more realistic reaction involves Fe,Mg chlorite + sulfur + H^+ = kaolinite + pyrite. This reaction cannot be properly assessed at this time owing to the lack of thermodynamic data on Fe–Mg chlorites and an oxidation state of sulfur species in the pore waters.

THE AFTERMATH OF SECONDARY POROSITY

One of the most interesting aspects of leaching in the San Joaquin NCL Field is the resulting late-stage compaction and its effect on diagenesis. Since this reservoir has apparently always been at hydrostatic pressures, grain–grain contact stresses are high relative to fluid pressures. When the rock was extensively cemented, the detrital grain contacts were protected to some degree by carbonate. After leaching of the carbonate at depths, presumably around 2.5 km burial, late-stage fracturing or deformation of detrital grains commences. One observed result is the onset of albitization at contacts between plagioclase and resistant grains, usually quartz (see Fig. 2). The albitization is localized as a reaction "halo" around the stressed contact rather than being concentrated along any associated fracturing. A possible explanation for this phenomenon is that higher silica activities result from the increased stress at the plagioclase–quartz contacts. High silica activities would stabilize albite with respect to calcic plagioclase (see Boles and Coombs, 1977; Boles, 1982).

Another result of removal of carbonate cement is late-stage deformation of detrital biotite. Commonly associated between the cleavage flakes of this crushed biotite is carbonate, ferrous dolomite in one case (Fig. 9). The explanation for this mineral association lies in the fact that newly formed biotite surfaces release K^+ to the pore water in exchange for H^+ (see Garrels and Howard, 1959). Thus, because the mica is a very effective H^+ sink, an increase in pH around the mica results and carbonate is stabilized.

Garrels and Howard (1959) show that the equilibrium constant for the reaction $H^+ + K\text{-mica} = H\text{-mica} + K^+$ is 10^{7-8} at 25°C, and $10^{6.8}$ at 65°C. Pore waters in NCL rocks have $\frac{[K^+]}{[H^+]}$ activity ratios of about 10^{4-5}, indicating the hydrogenated surface is stable. We have recently completed some experiments and theoretical calculations that verify the significance of biotite as an H^+ sink (Boles and Johnson, in press). The process becomes particularly significant where biotite is abundant,

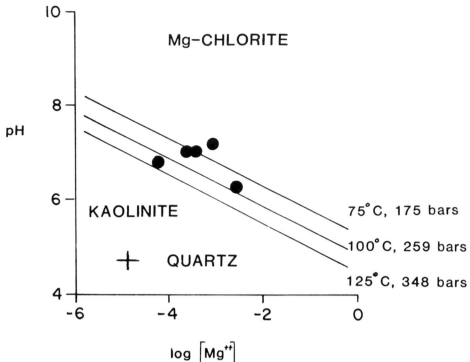

Figure 8—Equilibrium relations for reaction:
$$Mg_5Al_2Si_3O_{10}(OH)_8 + 10H^+ =$$
$$Al_2Si_2O_5(OH)_4 + 5Mg^{+2} + SiO_2 + 7H_2O.$$
($\bullet$) water analyses from NCL field at 100° C corrected for CO_2 degassing.

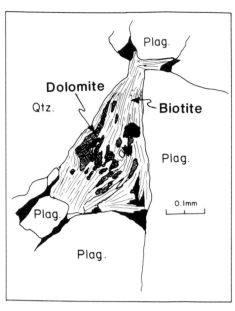

Figure 9—Dolomite in crushed detrital biotite. Mineral association is due to H⁺ affinity of newly formed mica surfaces during late-stage post-leaching compaction.

where significant compaction occurs, and where biotite/pore-water volumes are high.

ACKNOWLEDGMENTS

The author gratefully acknowledges the ARCO Geoscience Group, Dallas, Texas, for financial assistance on this project. Dr. Carol Bruton, ARCO Geoscience, assisted in the thermodynamic calculations and refinement of water analyses. Dr. Steve Franks provided useful discussion during the course of the study. The ARCO Bakersfield District Office, including Dave Woltz, District Geologist, Art Tinnemier, District Engineer, and Bill Bazeley (ARCO, Denver), were most helpful in providing core material and relevant data. Partial funding of this project in its later stages was provided by NSF Grant EAR82-12932.

SELECTED REFERENCES

Boles, J. R., 1982, Active albitization of plagioclase Gulf Coast Tertiary: American Journal of Science, v. 282, p. 165–180.

Boles, J. R., and D. S. Coombs, 1977, Zeolite facies alteration of sandstones in the Southland Syncline, New Zealand: American Journal of Science, v. 277, p. 982–1012.

Boles, J. R., and K. S. Johnson, in press, Influence of mica surfaces on pore water pH: Chemical Geology.

Carothers, W. S., and Y. K. Kharaka, 1978, Aliphatic acid anions in oil-field waters—implications for origin of natural gas: Bulletin of the American Association of Petroleum Geologists, v. 62, p. 2441–2453.

Garrels, R. M., and C. L. Christ, 1965, Solutions, minerals, and equilibria: San Francisco, Freeman, Cooper, and Co., 450 p.

Garrels, R. M., and P. Howard, 1959, Reactions of feldspar and mica with water at low temperature and pressure: Clay and clay minerals 2, Proceedings of the 6th National Conference, p. 68–88.

Hardoin, J. L., 1962, North Coles Levee Oil Field, in Summary of operations, California Oil Fields: California Division of Oil and Gas, 48th Annual Report, p. 53–61.

Hayes, J. B., 1979, Sandstone diagenesis—the hole truth: in P. A. Scholle and P. R. Schluger, eds., Aspects of diagenesis, Society of Economic Paleontologists and Mineralogists Special Publication 26, p. 127–139.

Saxena, S. K., and P. H. Ribbe, 1972, Activity-composition relations in feldspar: Contributions to Mineralogy and Petrology, v. 37, p. 131–138.

Schmidt, V., D. A. McDonald, and R. L. Platt, 1977, Pore geometry and reservoir aspects of secondary porosity in sandstones: Bulletin of Canadian Petroleum Geology, v. 25, p. 271–290.

Walker, R. G., and E. Mutti, 1973, Turbidite facies and facies associations: Anaheim, CA, Society of Economic Paleontologists and Mineralogists Pacific Section Short Course Notes, p. 119–157.

Webb, G. W., 1981, Stevens and earlier Miocene turbidite sandstones, southern San Joaquin Valley, California: Bulletin of the American Association of Petroleum Geologists, v. 65, p. 438–465.

Wood, J. R., and T. A. Hewett, 1982, Fluid convection and mass transfer in porous sandstones—A theoretical model: Geochimica et Cosmochimica Acta, v. 46, p. 1707–1713.

Woolery, T. J., 1979, Calculation of chemical equilibrium between aqueous solution and minerals: EQ3/6 software package: Livermore, CA, University of California Lawrence Livermore Lab UCRL-52658, 41 p.

Secondary Porosity in Laumontite-Bearing Sandstones

Laura J. Crossey
B. Ronald Frost
Ronald C. Surdam
University of Wyoming
Laramie, Wyoming

INTRODUCTION

The development of laumontite cements as a result of progressive diagenesis in some sedimentary rocks (particularly those containing volcanogenic material) has been well documented by many workers (Coombs, 1954; Boles and Coombs, 1977; Galloway, 1974; McCulloh et al, 1978). As these cements are highly destructive to reservoir potential, laumontite-bearing sedimentary rocks are commonly considered economic basement in the search for hydrocarbons. Galloway (1974) has defined several stages of progressive diagenesis (see Fig. 1). The constriction of pore throats by the development of authigenic clay rims (Stage 2), marks the entrance to the permeability basement. The porosity basement is entered during the development of authigenic cements such as calcite and laumontite (Stage 3). To evaluate the validity of the use of laumontite as a marker of economic basement, we have examined the stability of laumontite using both thermodynamic data and the experimental dissolution behavior of natural laumontites in organic acid solutions.

LOW-TEMPERATURE STABILITY RELATIONS OF LAUMONTITE

It has long been recognized that the stability of zeolite minerals is more strongly controlled by the composition of associated fluids than by temperature (Zen, 1961). $T-X_{CO_2}$ topologies calculated by Thompson (1971) and by Ivanov and Gurevich (1975) show that at low temperatures (where zoisite and

ABSTRACT. Thermodynamic calculations indicate that, at diagenetic temperatures, laumontite is stable only in the presence of fluids of high pH and low P_{CO_2}. This is supported by experimental dissolution studies that suggest that laumontite is soluble in the presence of carboxylic acids. As both CO_2 and carboxylic acids are produced prior to and during hydrocarbon generation, laumontite is unlikely to form in sandstones plumbed to source rocks during the maturation process. It is more likely that early formed laumontite cements will be destroyed and secondary porosity created by the processes associated with maturing kerogen. Thus potential reservoir rocks may be found in or below laumontite-bearing sandstones. Laumontite in hydrocarbon environments would most likely have formed late relative to hydrocarbon maturation. The sedimentary basins of California may demonstrate such a late-stage, hydrothermal origin for laumontite. The concept of laumontite as an economic basement for hydrocarbon exploration must be carefully evaluated on a case-by-case basis.

epidote are unlikely to be stable; see Frost, 1980), laumontite is restricted to conditions where X_{CO_2} is less than 0.01. At diagenetic temperatures (75–150°C) two of the major equilibria controlling laumontite stability are:

laumontite + CO_2 = calcite + kaolinite + 2 quartz + $3H_2O$
(see reaction 8, Table 1)

laumontite + $2H^+$ = kaolinite + 2 quartz + $3H_2O$ + Ca^{+2}
(see reaction 9, Table 1)

The equilibrium conditions for these two reactions can be evaluated by using thermochemical data for the solution of laumontite, kaolinite, calcite, and quartz in chloride brines given by

Frantz et al (1981) combined with expressions for the dissociation of aqueous HCl^0 (Helgeson, 1969) and $CaCl_2^0$ (Frantz and Marshall, 1983) (see Table 1).

There are several potential sources of error inherent in the data presented in Table 1 and the diagrams subsequently derived. First, there are uncertainties associated with the experimental determination of ΔG values. Second, in the evaluations here, log K is assumed to be a linear function of temperature (that is, $\Delta H/dt = 0$). Finally, in many cases high temperature data have been extrapolated to relatively low diagenetic temperatures. Because of the lack of experimental data in the pressure and temperature range of consideration, the magnitude of the errors can-

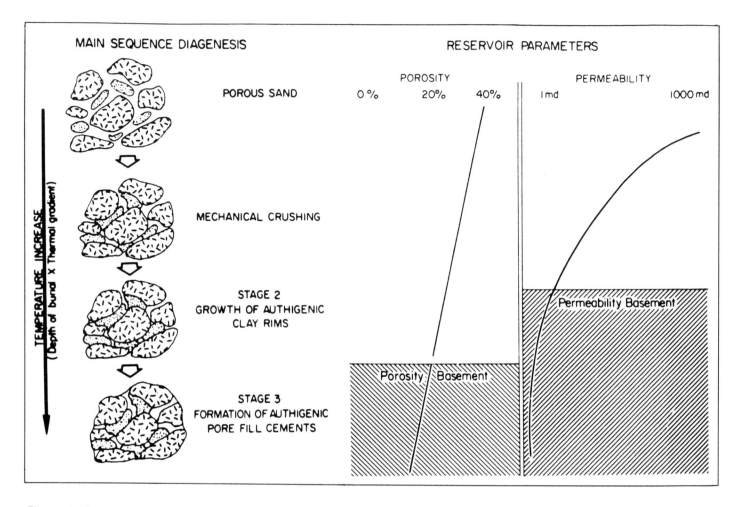

Figure 1—Sequential diagenetic stages in arc-derived sandstones and relative positions of porosity and permeability basement, assuming economic reservoirs require 20% porosity and 10 md permeability. From Galloway. Copyright © 1974 The Geological Society of America.

not be estimated, and the diagrams presented here should not be used to estimate fluid compositions quantitatively. However, the topologies of the phase diagrams will not change regardless of the magnitude of the errors; even exceptionally large errors would not alter the conclusion that laumontite is stable only in solutions of high pH and low P_{CO_2}.

From data in Table 1, the following equilibrium expression for reaction 8 at 1 kb can be derived:

$$\log K_{(8)} = \frac{3017}{T} - 3.666$$

In the range of diagenetic temperatures, the CO_2 content of the fluid is likely to

be very low, and it is thus assumed that f_{H_2O} for fluid in equilibrium with laumontite, kaolinite, calcite, and quartz is that of pure H_2O at the temperature and pressure of interest (Burnham et al, 1969). Figure 2 shows the equilibrium curve for reaction 8 as calculated in this manner, and it indicates that at diagenetic temperatures even extremely small amounts of CO_2 in the fluid can lead to the breakdown of laumontite. Because there is no convenient way to determine the fugacity coefficients for CO_2 in H_2O–CO_2 mixtures at high pressures and sub-solvus temperatures, the CO_2 variable in Figure 2 is expressed as f_{CO_2} rather than P_{CO_2} or X_{CO_2}. The existence of the CO_2–H_2O solvus at the temperature of interest (Takenouchi and Kennedy, 1964) indicates that γCO_2 will be greater than 1.00. Thus, the values given for f_{CO_2} will indicate the maximum P_{CO_2} possible, with the actual P_{CO_2} being perhaps an order of magnitude lower than the fugacity. Figure 3 shows the stability

limits of laumontite where calcite is not stable and the controlling equilibrium is reaction 9. Under these conditions, laumontite will break down to kaolinite and quartz except in solutions that are very alkaline. An increase in temperature will contract the laumontite stability field only slightly.

As can be noted from Figures 2 and 3, the independent variables controlling laumontite stability at a fixed total pressure are: T, $a_{Ca^{+2}}$, pH, and f_{CO_2}. As mentioned above, f_{H_2O} is assumed to be that for pure H_2O at the temperature and pressure of interest. The combined effects of pH, f_{CO_2}, and $a_{Ca^{+2}}$ are shown on isobaric, isothermal pH–log ΣCO_3 diagrams where log $a_{Ca^{+2}}$ and log f_{CO_2} are contoured (Fig. 4a, b). The independent CO_2 variable is expressed in these figures as ΣCO_3 ($a_{H_2CO_2} + a_{HCO_3^-} + a_{CO_3^{-2}}$), as f_{CO_2} will vary with pH. On these diagrams (Fig. 4a, b) reaction 8, which limits the stability of laumontite with respect to kaolinite and calcite, follows an f_{CO_2} isopleth. This reaction is

TABLE 1

Number	Reaction	Log K 100°C	Log K 150°C	Reference
(1)	$Lm + 2HCl^0 + 3H_2O = CaCl_2{}^0 + Al_2O_3 + 4H_4SiO_4$	-0.64	-0.69	1
(2)	$Kao + 2H_2O = Al_2O_3 + 2H_4SiO_4$	-10.76	-8.71	1
(3)	$Q + 2H_2O = H_4SiO_4$	-2.30	-2.09	1
(4)	$Cc + 2HCl_0 = CaCl_2{}^0 + CO_2 + H_2O$	10.29	8.74	1
(5)	$HCl^0 = H^+ + Cl^-$	3.0	1.4	2
(6)	$CaCl_2{}^0 = Ca^{+2} + 2Cl^-$	6.54	6.06	3
(7)	$3Kao + 2K^+ = 2\,Mus + 2H^+ + 3H_2O^{1,2}$	-6.7	-3.9	4
(8)	$Lm + CO_2 = Cc + Kao + 2Q + 2H_2O$	4.43	3.46	a
(9)	$Lm + 2H^+ = Kao + 2Q + 3H_2O + Ca^{+2}$	15.25	15.46	b
(10)	$Cc + 2H^+ = Ca^{+2} + CO_2 + H_2O$	10.8	12.0	c
(11)	$Lm + 4H^+ + 2K^+ = 3Ca^{+2} + 2\,Mus + 6Q + 2H_2O$	39.05	42.49	d
(12)	$CO_2 + H_2O = H_2CO_3{}^2$	-1.97	-2.07	2
(13)	$H_2CO_3 = H^+ + HCO_3{}^-$	-6.45	-6.73	2
(14)	$HCO_3{}^- = H^+ + CO_3{}^{-2}$	-10.16	-10.29	2
(15)	$H_2O = H^+ + OH^-$	-12.26	-11.64	2
(16)	$CO_2(g) + H_2O = H^+ + HCO_3{}^-$	-8.42	-8.80	e
(17)	$CO_2(g) + H_2O = 2H^+ + CO_3{}^{-2}$	-18.58	-19.09	f
(18)	$Cc + 2H^+ = Ca^{+2} + H_2CO_3$	8.83	9.93	g
(19)	$Cc + H^+ = Ca^{+2} + HCO_3{}^-$	2.38	3.20	h
(20)	$Cc = Ca^{+2} + CO_3{}^-$	-7.78	-7.09	i
(21)	$Lm + 3CO_2 = Cc + Kao + 2Q + 2H_2CO_3$	0.49	-0.68	j
(22)	$Lm + 3CO_2 = Cc + Kao + 2Q + 2HCO_3{}^- + 2H^+$	-12.41	-14.14	k
(23)	$Lm + 3CO_2 = Cc + Kao + 2Q + 4H^+ + 2CO_3{}^{-2}$	-32.73	-34.72	l

[1] Lm = Laumontite, Kao = Kaolinite, Q = Quartz, Cc = Calcite, Mus = Muscovite
[2] $H_2CO_3 = H_2CO_3 + CO_2$ (aq)

1. Frantz et al (1981)
2. Estimated from data in Helgeson (1969)
3. Frantz and Marshall (1983)
4. Calculated from data in Montoya and Hemley (1975)
a. $\log K_8 = \log K_1 - K_2 - \log 2 \log K_3 - \log K_4$; (Frantz, 1981)

   ($\log K_8 = \dfrac{3017}{T} - 3.666$ at 1 kb)

b. $\log K_9 = \log K_1 - \log K_2 - 2 \log K_3 - 2 \log K_5 + \log K_6$
c. $\log K_{10} = \log K_4 - 2 \log K_5 + \log K_6$
d. $\log K_{11} = 3 \log K_9 + \log K_7$
e. $\log K_{16} = \log K_{12} + \log K_{13}$
f. $\log K_{17} = \log K_{12} + \log K_{14}$
g. $\log K_{18} = \log K_{10} + \log K_{12}$
h. $\log K_{19} = \log K_{10} + \log K_{16}$
i. $\log K_{20} = \log K_{10} + \log K_{17}$
j. $\log K_{21} = \log K_8 + 2 \log K_{12}$
k. $\log K_{22} = \log K_8 + 2 \log K_{16}$
l. $\log K_{23} = \log K_8 + 2 \log K_{17}$

Table 1—Equilibrium expressions for reactions controlling laumontite stability at 1 kb.

Figure 2—Calculated equilibrium curve for reaction: $Lm + CO_2 = Cc + Kao + 2Q + 2H_2O$ (reaction 8, Table 1) as a function of $\log f_{CO_2}$ and temperature.

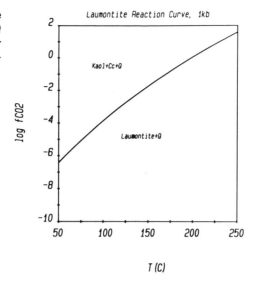

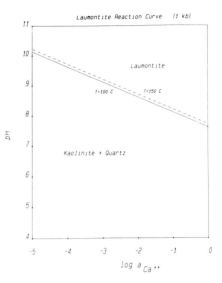

Figure 3—Calculated equilibrium curve for reaction $Lm + 2H^+ = Kao + 2Q + 3H_2O + Ca^{+2}$ (reaction 9, Table 1) as a function of pH and $\log a_{Ca^{+2}}$.

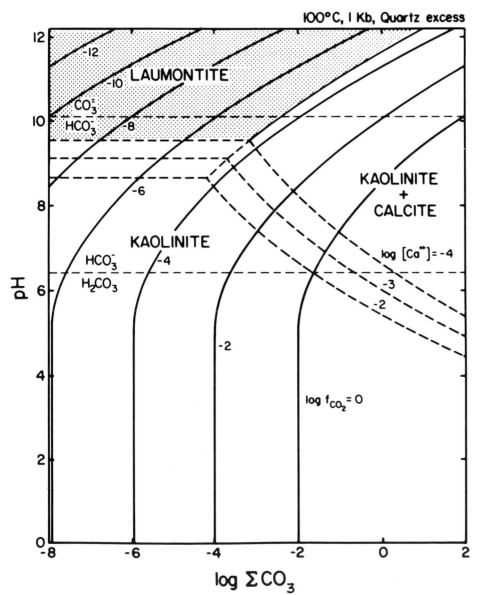

Figure 4a—Stability diagram for laumontite at 100°C as defined by $Lm + CO_2 = Cc + Kao + 2Q + 2H_2O$ and $Lm + 2H^+ = Kao + 2Q + 3H_2O + Ca^{+2}$ (reactions 8 and 9, Table 1). Solid lines represent f_{CO_2} isopleths, heavy dashed lines represent contours of constant Ca^{+2} activity, and light dashed lines separate the pH regions where different carbonate species are dominant.

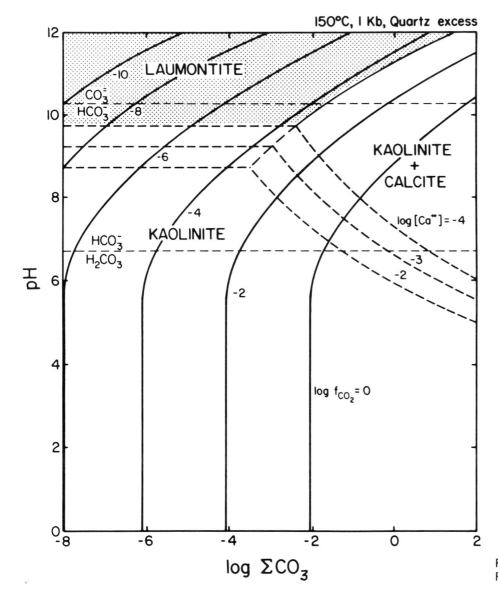

Figure 4b—Stability diagram, same as Fig. 4a, drawn for T = 150°C.

independent of $a_{Ca^{+2}}$, and (though written as independent of pH) depends on pH to the extent that f_{CO_2} varies with pH at a given ΣCO_3 value. Reaction 9, which limits the stability of laumontite with respect to kaolinite and quartz, is independent of ΣCO_3; but the position of the equilibrium curve will shift to lower pH as $a_{Ca^{+2}}$ increases. Comparison of Figures 4a and 4b shows that, as temperature increases, laumontite will be stable at a higher f_{CO_2} but will be restricted to more alkaline solutions.

Another series of diagrams (Figs. 5, 6a, b) have been constructed applying to the condition where, because of low ΣCO_3 and the presence of K$^+$, laumontite will become unstable with respect

to a K-mica phase plus quartz. Thermodynamic data for muscovite have been used to describe the K-mica phase in these diagrams, although at these relatively low temperatures phengite or illite may be more stable. Thus the diagrams predict an upper limit to the stability of laumontite with respect to a K-mica phase.

Figure 5 shows that even with respect to a K-mica phase plus quartz, under conditions of low ΣCO_3, laumontite is stable only in alkaline solutions (reaction 11). In this instance, the effect of temperature is more pronounced for the stability of K-mica with respect to laumontite than for kaolinite. At 150°C the laumontite field is displaced

upward by one pH unit (for a given $a_{Ca^{+2}}$ and a_{K^+}) compared to its position at 100°C. Figures 6a and 6b show the combined effects of pH, a_{K^+} and $a_{Ca^{+2}}$ on isobaric, isothermal diagrams (again, $a_{Ca^{+2}}$ is contoured). As in Figures 4a and 5b, reaction 9 represents the $a_{Ca^{+2}}$-dependent laumontite stability boundary. In the K$^+$-bearing system, as a_{K^+} increases, reaction 11, which is dependent on pH, a_{K^+} and $a_{Ca^{+2}}$, becomes the stability boundary of laumontite. In this system, the stability fields for laumontite and kaolinite contract considerably with respect to a K-mica phase as temperature increases (Fig. 6a, b).

The above diagrams clearly show

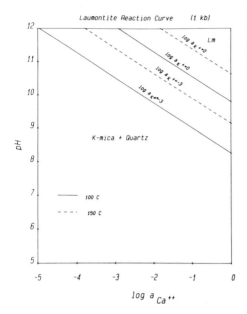

Figure 5—Equilibrium curve for reaction Lm + 4H$^+$ + 2K$^+$ = 3Ca^{+2} + 2Mus + 6Q + 2H$_2$O (reaction 11, Table 1) as a function of pH and log a$_{Ca+2}$ at 100°C and 150°C (dashed lines).

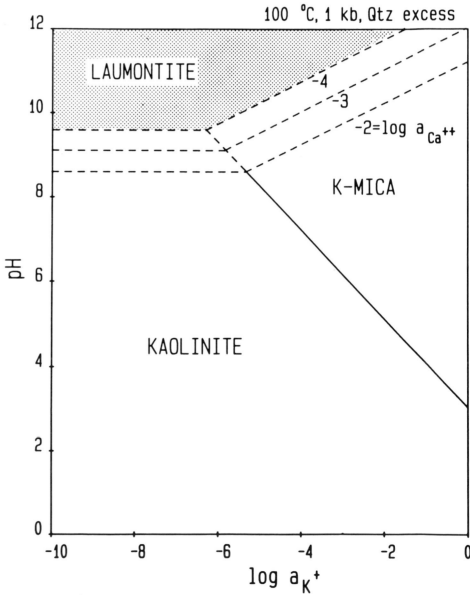

Figure 6a—Stability diagram for laumontite at 100°C as defined by Lm + 4H$^+$ + 2K$^+$ = 3Ca^{+2} + 2Mus + 6Q + 2H$_2$O and Lm + 2H$^+$ = Kao + 2Q + 3H$_2$O + 2Ca^{+2}, and 3Kao + 2K$^+$ = 2Mus + 2H$^+$ + 3H$_2$O (reactions 11, 9, and 7, respectively, Table 1).

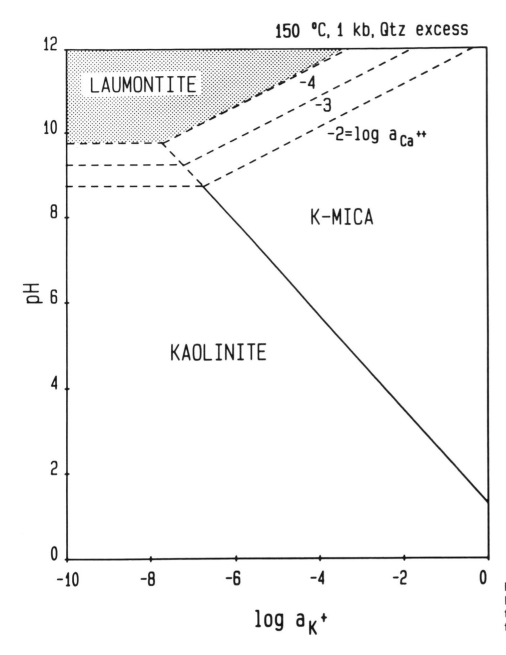

150 °C, 1 kb, Qtz excess

LAUMONTITE

-4

-3

$-2 = \log a_{Ca^{++}}$

K-MICA

KAOLINITE

pH

$\log a_{K^+}$

Figure 6b—Stability diagram, same as Fig. 6a, drawn for 150° C. The laumontite field is smaller at the higher temperature.

that laumontite can be destabilized by one or more of the following changes in fluid composition: (1) increasing the acidity, (2) increasing the CO_2 content, (3) decreasing the $a_{Ca^{+2}}$, and (4) increasing the a_{K^+}. The destruction and replacement of laumontite as described above are important to the production of secondary porosity because of the large molar volume of laumontite (see Table 2). Reaction 8 is accompanied by a 12% decrease in the volume of the solids, while reaction 9 produces a volume decrease of 30%. Sample calculations assuming an initial content of

15% laumontite cement show that net increases in porosity of 2 and 5% result from reaction 8 and 9, respectively (Table 3).

CO_2 and bicarbonate are produced from organic matter disseminated in sediments (see Fig. 7; Curtis, 1978). Immediately after burial, sulfate reduction occurs, followed by fermentation as depth increases. As temperature and depth continue to increase, other mechanisms predominate. Carothers and Kharaka (1978) have described the decarboxylation of carboxylic acids (acetic acid → CO_2 + methane). Figure

8 schematically shows the relationship between CO_2 production and hydrocarbon generation; in both humic and sapropelic source materials CO_2 is generated in greatest abundance just prior to hydrocarbon generation and migration. As CO_2 is produced throughout the diagenesis of organic-rich sediments, early laumontite may be associated with sediments of low organic content (such as the Triassic of Southland New Zealand, Boles and Coombs, 1977; Landis and Bishop, 1972). Laumontite in sands associated with shales rich enough in organic mat-

TABLE 2

Mineral	Composition	(100° C) Molar Volume (cu cm)	Abbreviation
Calcite	$CaCO_3$	36.934	Cc
Kaolinite	$Al_2Si_2O_5(OH)_4$	99.52	Kao
Laumontite	$CaAl_2Si_4O_{12} \cdot 4H_2O$	207.25	Lm
Quartz	SiO_2	22.688	Q
Muscovite	$KAl_2(Si_3Al)O_{10}(OH)_2$	140.71	K-mica

Table 2—Mineral compositions.

TABLE 3

Reaction (see Table 1)	Molar Volume Change (100° C)	Porosity Change Based on 15% Lm Cement
(8)	$\dfrac{Lm - (Kao + 2Q + Cc)}{Lm} = -12\%$	+2%
(9)	$\dfrac{Lm - (Kao + 2Q)}{Lm} = -30\%$	+5%
(11)	$\dfrac{Lm - (2K + mica + 6Q)}{Lm} = +34\%$	-5%

Table 3—Molar volume change of solids and resultant porosity change.

ter to be source rocks would most likely be late formed with respect to hydrocarbon maturation.

As mentioned above, short chain aliphatic acids such as acetic, oxalic, and propionic acids are a by-product of organic matter maturation. Surdam et al (this volume) have described oil field brines containing up to 10,000 ppm acetic acid. These acids, like CO_2, are produced throughout burial, peaking just prior to hydrocarbon maturation (see Fig. 9), and may contribute 50 to 100% of the alkalinity in oil field formation waters (Carothers and Kharaka, 1978). Surdam et al (this volume) have shown that carboxylic acids can increase the solubility of aluminosili-

cates by Al^{+3} complexation (see Fig. 10).

To evaluate the stability of laumontite in short chain organic acid solvents, natural samples of laumontite were tested in oxalic acid solutions ranging in pH from 5 to 9. Aliquots of 50 mg of ground laumontite (purified by heavy liquid centrifugation and verified with X-ray diffraction analysis) were placed in Teflon vials and 5 ml buffered solution added (Na-acetate barbitol buffers were used). Samples were analyzed after 2 weeks of constant agitation at 100°C. Samples were run both with and without oxalic acid (1000 ppm) to ascertain any pH effect. Total dissolved silica values were used as a relative

measure of stability. Ground andesine was tested in identical solutions for comparative purposes. Some large pieces of laumontite were also included for SEM textural observations.

An SEM photomicrograph of laumontite prior to dissolution is contrasted with an SEM photomicrograph of post-dissolution laumontite in Figure 11; post-dissolution laumontite exhibits rounded edges and enlarged cleavage cracks, as well as etched textures (dissolution pits). A general trend of increasing stability with increasing pH was observed for both laumontite and andesine in the absence of oxalic acid (see Fig. 12). In the presence of 1000 ppm oxalic acid both andesine

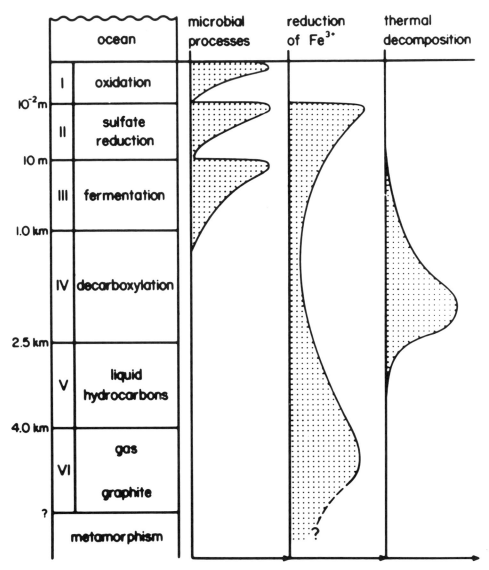

Figure 7—CO_2 production within different diagenetic zones as a result of different processes. From Curtis. Copyright © 1978 Blackwell Scientific Publications Ltd. Used with permission.

and laumontite are less stable (have higher dissovled silica values) than in the solutions buffered at the same pH without the carboxylic acid. Between pH 8 and 9 the curves for laumontite and andesine cross, with laumontite having the lowest dissolved silica at the higher pH. These results agree well with the theoretical prediction of the lower pH limit of laumontite stability discussed in the previous section (Fig. 6a). This indicates that the thermodynamic uncertainties associated with the calculation cannot introduce enough error to alter the conclusion that laumontite is unstable in fluids of low (or moderate) pH or high P_{CO_2}.

CONCLUSIONS

Theoretical and experimental evidence combined indicate that: (1) laumontite is restricted to fluids of high pH and low P_{CO_2}; and (2) laumontite is unstable in the presence of carboxylic acids. Because both CO_2 and carboxylic acids are generated prior to hydrocarbon maturation in organic-rich (source) rocks, early formed laumontite cements in fluid communication with source rocks is not feasible; any early formed laumontite would become destabilized. It is likely that most observed laumontite cements are formed late relative to hydrocarbon

maturation and are related to hydrothermal events. At temperatures in excess of 150°C, carboxylic acids would have thermally degraded (Carothers and Kharaka, 1978) and the dilute waters would have an elevated pH owing to hydration of aluminosilicates. For example, laumontite cements in California plagioclase-bearing Cretaceous and Eocene sands may be related to thermal events associated with the migration of the Mendocino triple point. During the Miocene the triple point migrated northward beneath many of the west coast basins (Zandt and Furlong, 1982) producing a thermal pulse that certainly influenced

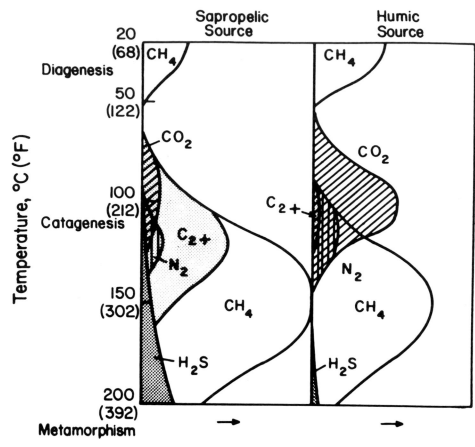

Figure 8—Relative yield of gas from organic matter in fine-grained sediments. C_{2+} represents hydrocarbon heavier than CH_4 in gas phase. N_2 is generated initially as NH_3. Modified from PETROLEUM GEOCHEMISTRY AND GEOLOGY by John M. Hunt. W. H. Freeman and Company. Copyright © 1979. Used with permission.

Relative Yield of Gas from Organic Matter in Fine-Grained Sediments

the maturation of hydrocarbons in those basins and may also have produced late, hydrothermal laumontite. Close examination may reveal the spatial distribution of the laumontite cements to be fault controlled.

Some examples of production beneath laumontite-bearing sands are: the Temblor Formation of the Kettleman North Dome in the San Joaquin Valley, California (Merino, 1975a), and the Vicksburg formation of the McAllen Ranch field in Hidalgo County, Texas (T. Dunn, 1983, personal communication).

Theoretical and experimental work demonstrate that the presence of laumontite does not preclude the possibility of porosity at greater depth. In fact, in some cases laumontite cement may form a diagenetic seal. Early formed

laumontite would not survive exposure to fluids expulsed by maturing source rocks. The most likely occurrence of laumontite is late formed, probably in response to a more alkaline thermal event. The implication of this work to exploration is that laumontite should not be used indiscriminately as an economic basement indicator.

ACKNOWLEDGMENTS

The writers thank Amoco Production Company for its support. This work was originally presented at the AAPG annual meeting in Calgary, 1982. We appreciate helpful discussion with T. Dunn, M. Lewan, J. Boles, H. Eugster, and J. Drever. We thank T. Dunn and R. Stewart for sending us sample material. Particular thanks go to Benita Chavez for her patience with us while preparing the manuscript.

SELECTED REFERENCES

Boles, J. R., and D. S. Coombs, 1977, Zeolite facies alteration of sandstones in the Southland Syncline, New Zealand: American Journal of Science, v. 277, p. 982–1012.

Burnham, C. W., J. R. Holloway, and N. F. Davis, 1969, Thermodynamic properties of water to 1,000°C: Geology Society of America Special Paper 132, 96 p.

Carothers, W. W., and Y. K. Kharaka, 1978, Aliphatic acid anions in oil-field waters—implication for origin of natural gas: Bulletin of the American Association of Petroleum Geologists, v. 62, p. 2441–2453.

Coombs, D. S., 1954, The nature and alteration of some Triassic sediments from Southland, New Zealand, Royal Society of New Zealand Transactions: v. 82, p. 65–109.

Curtis, C. D., 1978, Possible links between sandstone diagenesis and depth related

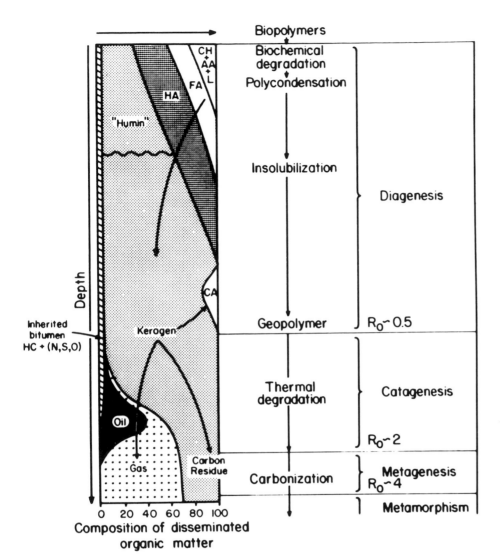

Biopolymers

Biochemical degradation

Polycondensation

Insolubilization

Geopolymer $R_O \sim 0.5$

Thermal degradation   Catagenesis

$R_O \sim 2$

Carbonization   Metagenesis
$R_O \sim 4$

Metamorphism

Diagenesis

Composition of disseminated organic matter

Figure 9—Composition of organic matter in different temperature regimes. (**CH**) Carbohydrates, (**AA**) amino acids, (**FA**) fulvic acids, (**HA**) humic acids, (**L**) lipids, (**HC**) hydrocarbons, (**N,S,O**) nonhydrocarbon (N, S, O compounds), (**CA**) carboxylic acids. From Surdam et al, this volume, modified to include carboxylic acids from Tissot and Welte. Copyright © 1978 Springer-Verlag New York. Used with permission.

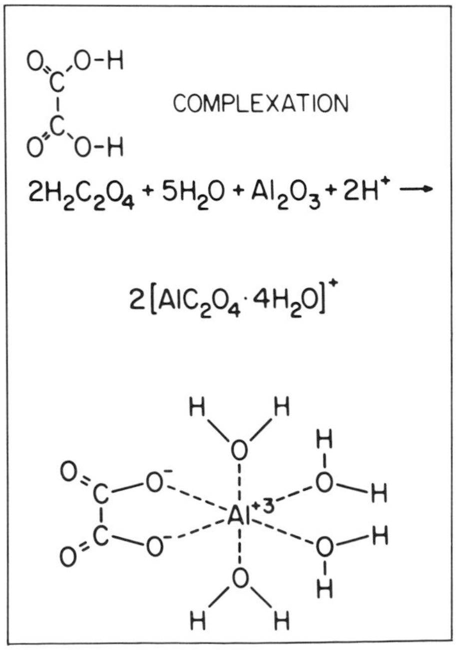

COMPLEXATION

$$2H_2C_2O_4 + 5H_2O + Al_2O_3 + 2H^+ \longrightarrow$$

$$2[AlC_2O_4 \cdot 4H_2O]^+$$

Figure 10—Schematic diagram of complexation of Al^{+3} by oxalic acid.

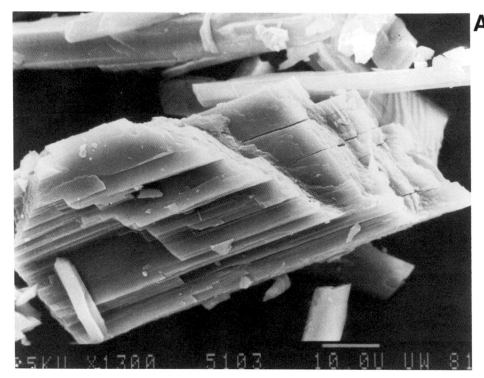

A

Figure 11—(**A**) SEM photomicrograph of laumontite fragments prior to dissolution experiment (1300×), (**B**) after 2 weeks at 100°C in 1000 ppm oxalic acid solution (1600×). Note dissolution pits and rounding of fragment edges.

geochemical reactions occurring in enclosing mudstones: Quarterly Journal of the Geology Society of London, v. 135, p. 107–117.

Frantz, J. D., and W. L. Marshall, 1983, Electrical conductances and ionization constants of calcium chloride and magnesium chloride in aqueous solutions at temperatures to 600°C and pressures to 4000 bars: American Journal of Science, v. 282, p. 1666–1693.

Frantz, J. D., R. K. Popp, N. Z. Boctor, 1981, Mineral–solution equilibria, V, Solubilities of rock forming minerals in supercritical fluids: Geochimica et Cosmochimica Acta, v. 45, p. 69–78.

Frost, B. R., 1980, Observations on the boundary between zeolite facies and prehnite–pumpellyite facies: Contributions to Mineralogy and Petrology, v. 73, p. 365–373.

Galloway, W. E., 1974, Deposition and diagenetic alteration of sandstone in Northeast Pacific arc-related basins: implications for graywacke genesis: Geological Society of America Bulletin, v. 85, p. 379–390.

Helgeson, H. C., 1969, Thermodynamics of hydrothermal systems at elevated temperatures: American Journal of Science, v. 267, p. 729–804.

Helgeson, H. C., J. M. Delany, W. H. Nesbett, and D. K. Bird, 1978, Summary and critique of the thermodynamic properties of rock-forming minerals: American

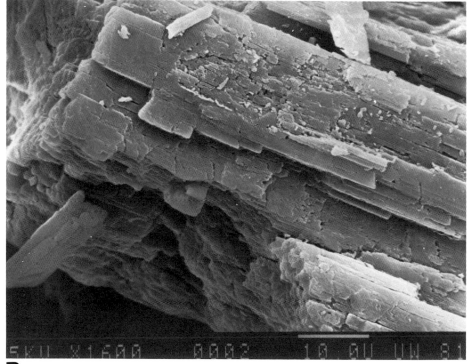

B

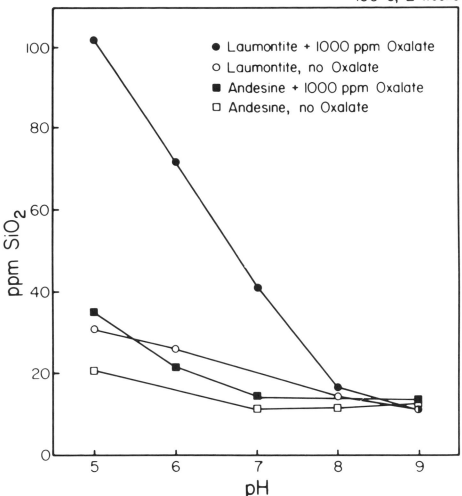

Figure 12—Total dissolved silica plotted as a function of the pH of experimental solutions after 2 weeks at 100° C. Open symbols represent laumontite and andesine in buffered solution with no complexing oxalic acid; closed symbols indicate initial oxalic acid concentration of 1000 ppm.

Journal of Science, v. 298A, p. 1–229.

Hunt, J. M., 1979, Petroleum geochemistry and geology: San Francisco, W. H. Freeman and Co.

Ivanov, I. P., and L. P. Gurevich, 1975, Experimental study of $T-X_{CO_2}$ boundaries of metamorphic zeolite facies: Contributions to Mineralogy and Petrology, v. 53, p. 55–60.

Landis, C. A., and D. G. Bishop, 1972, Plate tectonics and regional metamorphic relations in the southern part of the New Zealand geosyncline: Geological Society of America Bulletin, v. 83, p. 2267–2284.

Marshall, W. L., 1972, Predictions of the geochemical behavior of aqueous electrolytes at high temperatures and pressures: Chemical Geology, v. 10, p. 9–68.

McCulloh, T. H., and R. J. Steward, 1979, Subsurface laumontite crystallization and porosity destruction in Neogene sedimentary basins: Geology Society of America Abstracts with Programs, v. 11, p. 475.

McCulloh, T. H., V. A. Frizzell, R. J. Stewart, and I. Barnes, 1981, Precipitation of laumontite with quartz, thenardite, and gypsum at Sespe Hot Springs, Western Transverse Ranges, California: Clays and Clay Minerals, v. 29, p. 353–364.

Merino, E, 1975a, Diagenesis in Tertiary sandstones from the Kettleman North Dome, California, I, Diagenetic mineralogy: Journal of Sedimentary Petrology, v. 45, p. 320–336.

———, 1975b, Diagenesis in Tertiary sandstones from the Kettleman North Dome, California, II, Interstitial solutions: distribution of aqueous species at 100°C and chemical relation to the diagenetic mineralogy: Geochimica et Cosmochimica Acta, v. 39, p. 1629–1645.

Montoya, J. W., and J. J. Hemley, 1975, Activity relations and stabilities in alkali feldspar and mica alteration reactions: Economic Geology, v. 70, p. 577–594.

Stalder, P. J., 1979, Organic and inorganic metamorphism in the Taveyannaz sandstone of the Swiss Alps and equivalent sandstones in France and Italy: Journal of Sedimentary Petrology, v. 49, p. 463–482.

Takenouchi, S., and G. C. Kennedy, 1964, The binary system H_2O-CO_2 at high temperatures and pressures: American Journal of Science, v. 262, p. 1055–1074.

Thompson, A. B., 1971, P_{CO_2} in low-grade metamorphism: zeolite, carbonate, clay mineral, prehnite relations in the system $CaO-Al_2O_3-SiO_2-CO_2-H_2O$: Contributions to Mineralogy and Petrology, v. 33, p. 145–161.

Tissot, B. P., and D. H. Welte, 1978, Petroleum formation and occurrence: New York, Springer-Verlag.

Zandt, G., and K. Furlong, 1982, Evolution and thickness of the lithosphere beneath coastal California: Geology, v. 10, p. 376–381.

Zen, E., 1961, The zeolite facies: an interpretation: American Journal of Science, v. 259, p. 401–409.

Diagenetic Mineralogy and Controls on Albitization and Laumontite Formation in Paleogene Arkoses, Santa Ynez Mountains, California

Kenneth P. Helmold
Cities Service Oil & Gas Corp.
Tulsa, Oklahoma

Peter C. van de Kamp
GeoResources Associates
Napa, California

INTRODUCTION

Early diagenetic studies of sandstones were primarily concerned with cementation and resultant pore-space reduction in mineralogically mature quartzose sandstones (Waldschmidt, 1941; Heald, 1950; Siever, 1959). The limited diagenetic products obtained in this relatively simple chemical system led to the investigation of more complex sandstones, particuarly those rich in chemically unstable volcaniclastic and pyroclastic debris (Coombs, 1954; Dickinson, 1962; Brown and Thayer, 1963). Detailed diagenetic studies of arkosic sandstones are not nearly as numerous as those of other compositional types because they neither belong to a simple chemical system, nor contain abundant labile rock fragments. Galloway (1974) proposed a generalized diagenetic model for arkoses of the Queen Charlotte Basin that is also applicable to other northeast Pacific arc-related basins containing immature sandstones. Heald and Baker (1977) studied primary porosity reduction and occurrence of secondary porosity in the feldspathic Mt. Simon and Rose Run Sandstones, and Stanley and Benson (1979) documented low-temperature diagenetic products in nonmarine vitric and arkosic sandstones of the High Plains sequence. Early diagenesis of feldspathic continental red beds has been extensively studied by Walker (1976) and Walker et al (1978) with particular emphasis given to the role of diagenesis in controlling the ultimate texture and composition of these sandstones.

This study documents the nature,

ABSTRACT. The stratigraphic and lateral distributions of authigenic minerals in feldspar-rich Paleogene sandstones of the Santa Ynez Mountains, California, are important in determining their reservoir quality. The sandstones were deposited in an east-west elongate basin during two regressive episodes. Deep-water turbidites were overlain by shallow-water traction deposits and eventually by continental fluvial deposits as the basin was progressively filled from the east. Modal analyses document a common provenance for all the Paleogene sandstones consisting primarily of acidic to intermediate plutonic rocks, with minor volcanic, metamorphic, and sedimentary components. The average detrital mode of 27 sandstones is $Q_{37}F_{54}L_9$, and the average partial mode including only the monocrystalline mineral grains is $Qm_{39}P_{40}K_{21}$.

Textural relationships and the stratigraphic distribution of diagenetic minerals delineate the paragenetic sequence: (1) syndepositional to very early pyrite; (2) early concretionary calcite cement; (3) incipient dissolution of detrital heavy minerals and feldspars; (4) clay pore linings and pore fillings; (5) formation of sphene and anatase; (6) incipient albitization of detrital plagioclase; (7) quartz, plagioclase, and K-feldspar overgrowths; (8) dissolution of feldspar creating secondary porosity; (9) local precipitation of pore-filling kaolinite; (10) laumontite cementation and replacement of plagioclase; (11) barite cementation and replacement of detrital grains; and (12) late-stage calcite replacement of detrital grains and earlier cements.

Organic metamorphism, as expressed by vitrinite reflectance (R_0), provides a means to correlate mineral diagenesis in the sandstones with the thermal history of the Santa Ynez basin. In the eastern end of the basin (Wheeler Gorge) incipient albitization is first recognized at 0.5% R_0 corresponding to a paleotemperature of 110°C (4572 m burial depth), with complete albitization first occurring at a reflectance of 0.90% R_0 corresponding to a paleotemperature of 165°C (5425 m burial depth). The first occurrence of laumontite is in the turbidite beds of the basal Matilija Formation (5669 m burial depth) at approximately 1.0% R_0 reflectance (173°C). Further to the west, at Point Conception (Gerber No. 1 well), the first occurrence of laumontite is at an estimated burial depth of only 2515 m, corresponding to approximately 0.5% R_0 and a paleotemperature of 110°C. In this well, incipient albitization begins at 0.35% R_0 (77°C), with complete albitization occurring at roughly the same burial depth (2515 m) and reflectance (0.5% R_0) as the first occurrence of laumontite. *(continued)*

The top of the laumontite zone occurs at greater burial depths and paleotemperatures in the eastern portion of the Santa Ynez basin than in the west. Laumontite distribution appears to be controlled by pore-fluid chemistry and post-compaction permeability variations, which are responsible for creating differences in fluid pressure between petrologically similar sandstones. "Dynamic" overpressuring may have occurred in the turbidite facies of the Juncal and lower Matilija Formations, whereby pore fluids enriched in Na^+ from the dewatering of smectite-rich shales permeated into the turbidite sandstones at a faster rate than they were expelled. Under these conditions, a continuous supply of Na^+ would have been delivered to the sandstones to allow albitization of calcium-bearing plagioclase, which in turn supplied Ca^{+2} necessary for the formation of laumontite.

The authigenic minerals in the lower Paleogene sandstones of the Santa Ynez Mountains render them ineffective as reservoirs. Better reservoir prospects occur in the upper Paleogene and Neogene sandstones, particularly in the western part of the basin where they have not been subjected to deep burial, and secondary porosity is well developed.

occurrence, and paragenetic sequence of authigenic minerals in Paleogene marine arkoses of the Santa Ynez Mountains, California, and elucidates the relationship between albitization, zeolite cementation, and burial history. The Paleogene strata of the Santa Ynez Mountains provide an excellent setting for this study owing to their great stratigraphic thickness, homoclinal structure, good exposure, and the extensive previous work on the stratigraphy and sedimentation of the region.

REGIONAL GEOLOGY

The Santa Ynez Mountains consist of east-west trending strike ridges that extend for more than 120 km from east of Ojai to Point Conception (Fig. 1). The Paleogene strata crop out in a steep southerly dipping homocline, which is truncated to the north by the Santa Ynez Fault and is overturned north of Santa Barbara and Ojai forming the Montecito and Matilija Overturns (Kerr and Schenck, 1928). The homocline is interrupted in the vicinity of San Marcos Pass by a series of northwest-southeast trending folds, which form a structural saddle in the Santa Ynez Range. Movement along the Santa Ynez Fault has been predominantly vertical (Dibblee, 1950, 1966;

Page et al, 1951) with the southern Santa Ynez block upthrown. Associated strike-slip movement is suggested to be left-lateral (Page et al, 1951; Dibblee, 1966; Dickinson, 1969a) with estimated displacement of at least several kilometers (Dibblee, 1966; McCulloh, 1981), The structure and geologic history are discussed in detail by Page et al (1951), Dibblee (1950, 1966), and Dickinson (1969a).

South of the Santa Ynez Fault, the Tertiary strata are underlain disconformably by the Upper Cretaceous Jalama Formation (van de Kamp et al, 1974) and, in places, by Paleocene reefoid algal beds of the Sierra Blanca Limestone (Dickinson, 1969a). The Paleogene marine clastic sequence consists of turbidite deposits of the Juncal Formation, which includes a middle sandstone member informally named the Camino Cielo Sandstone, transitional deep- to shallow-marine Matilija Sandstone (Link and Welton, 1982), turbidite and basinal siltstones and shales of the Cozy Dell Formation, and shallow marine sandstones of the "Coldwater" Formation (Fig. 2). All formation contacts within the sequence are transitional and conformable (van de Kamp et al, 1974). The marine sequence is overlain by the Oligocene continental red beds of the Sespe For-

mation, which is in turn unconformably overlain by the transgressive Upper Oligocene Vaqueros Sandstone (Ingle, 1980). The Tertiary section is capped by deep marine siltstones and shales of the Miocene Rincon and Monterey Formations. Pliocene strata are thin or altogether absent owing to the initial rise of the Santa Ynez Mountains during that time (Dibblee, 1966).

Correlation of the Paleogene clastic sequence across the structural complexities at San Marcos Pass has caused much controversy. Recent studies (Stauffer, 1967; van de Kamp et al, 1974) have followed the suggestion of Bailey (1952) and have recorrelated the Paleogene clastic sequence west of San Marcos Pass with the type sections to the east. The correlations of Stauffer (1967) and van de Kamp et al (1974) have been adopted for this study.

Generalized facies models for the Paleogene strata (van de Kamp et al, 1974; Link, 1975; Link and Welton, 1982) consist of a deltaic complex prograding across a narrow shelf and feeding a submarine fan system located on the lower slope and basin floor. Continued basin filling resulted in deepwater turbidites to be overlain by shallow-water traction deposits and eventually by continental fluvial deposits. Two such regressive events are recorded in the Paleogene strata (Fig. 2). The first episode is represented by the transition from the turbidite and basin plain deposits of the Juncal and basal Matilija Formations to the shallow marine deposits of the middle and upper Matilija Formation (van de Kamp et al, 1974; Link and Welton, 1982). The second episode is represented by a similar transition from the basin deposits of the Cozy Dell Formation to the nonmarine deposits of the Sespe Formation (Fig. 2). Protracted westward progradation of the delta and submarine fan complexes resulted in filling of the basin by time transgressive units that decrease in age to the west. Detailed descriptions of the stratigraphy and sedimentology can be found in Dibblee (1950, 1966), Stauffer (1965, 1967), O'Brien (1973), and van de Kamp et al (1974).

This study is confined to the Paleogene marine clastic sequence (Juncal to "Coldwater" Formations) in the seaward dipping homocline south of the

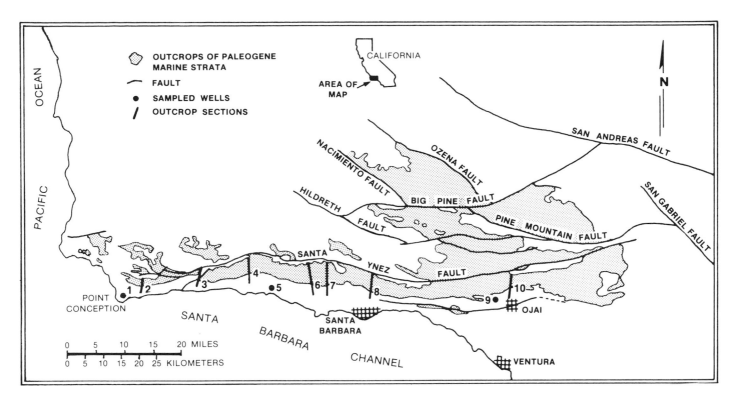

Figure 1 Index map of western Transverse Ranges showing the distribution of Paleogene marine strata and the location of outcrop sections and cores studied (modified from Nilsen and Clarke, 1975). KEY: (**1**) Standard Oil, Gerber No. 1, (**2**) Gato Canyon, (**3**) Gaviota Pass, (**4**) Refugio Pass, (**5**) Union Oil, Gila Land No. 1, (**6**) Tecolote Tunnel, (**7**) San Marcos Pass, (**8**) Gibraltar Road, (**9**) Honolulu-Sunray Oil, Dunshee No. 1, (**10**) Wheeler Gorge.

Santa Ynez Fault because of the conformable nature of the strata, lack of major faults cutting the section, and relative ease of estimating former burial depths. A few Cretaceous sandstones were also examined to determine whether trends in detrital and diagenetic mineralogy exist throughout the complete stratigraphic section.

METHODS OF INVESTIGATION

Approximately 200 thin sections of fine- to medium-grained sandstones, stained for both plagioclase and K-feldspar (Laniz et al, 1964), were analyzed for diagenetic texture and mineralogy using optical microscopy. Samples containing abundant authigenic clay, silicate overgrowths, or other diagenetic features were studied with the scanning electron microscope (SEM). Energy dispersive X-ray (EDX) analyses were conducted on sufficient samples of authigenic clay to confirm identifications based on crystal morphology (Wilson and Pittman, 1977).

The clay fraction of the sandstones (both <5-μm and <2-μm fractions) was separated by sedimentation and analyzed using standard X-ray diffraction techniques (Gibbs, 1968; Jackson, 1969). The <5-μm fraction was analyzed because authigenic clays in sandstones are frequently coarser than 2 μm. The composition of mixed-layer clay was determined from the <2-μm fraction because it contains less quartz and feldspar admixtures. The unit cell dimensions of laumontite were determined using powder patterns obtained with a Debye-Scherrer camera. Microprobe analyses were conducted on albite and K-feldspar overgrowths, laumontite cement, and detrital plagioclase grains. Microanalyses of authigenic clays proved unsatisfactory owing to volatilization of several components.

Techniques of sampling, specimen preparation, and phytoclast selection for vitrinite reflectance analysis are those of Bostick (1979) and Bostick and Alpern (1977). Most analyses were conducted on siltstones and shales that are interbedded with the Paleogene sandstones, but a few were performed on sandstones where interbedded shales are absent.

DETRITAL MODES

The Paleogene sandstones are composed of quartz (mainly monocrystalline grains), plagioclase, and K-feldspar with minor sedimentary rock fragments (argillite, shale/slate, and siltstone), volcanic rock fragments (felsitic, microlitic/lathwork, and hypabyssal), metamorphic rock fragments (schist), chert, and mica, and accessory opaque and heavy minerals. The average detrital mode of 27 sandstones from Gibraltar Road is $Q_{37}F_{54}L_9$. The average partial mode including only the monocrystalline mineral grains is $Qm_{38}P_{40}K_{21}$. The dominance of quartz and feldspar (with total quartz $<75\%$) over lithic fragments classifies these sandstones as arkoses and lithic arkoses according to the scheme of Folk (1974; Fig. 3). Plagioclase generally exceeds K-feldspar with a mean P/F ratio of 0.6. The high plagioclase content is unusual compared with the "typical" arkose in

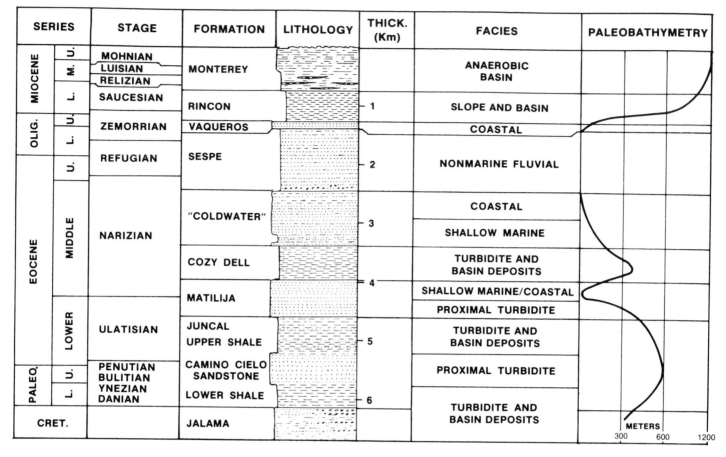

SERIES			STAGE	FORMATION	LITHOLOGY	THICK. (Km)	FACIES	PALEOBATHYMETRY
MIOCENE	U.		MOHNIAN	MONTEREY			ANAEROBIC BASIN	
	M.		LUISIAN					
			RELIZIAN					
	L.		SAUCESIAN	RINCON		1	SLOPE AND BASIN	
OLIG.	U.		ZEMORRIAN	VAQUEROS			COASTAL	
	L.							
EOCENE	U.		REFUGIAN	SESPE		2	NONMARINE FLUVIAL	
	MIDDLE		NARIZIAN	"COLDWATER"		3	COASTAL	
							SHALLOW MARINE	
				COZY DELL		4	TURBIDITE AND BASIN DEPOSITS	
				MATILIJA			SHALLOW MARINE/COASTAL	
	LOWER		ULATISIAN				PROXIMAL TURBIDITE	
				JUNCAL UPPER SHALE		5	TURBIDITE AND BASIN DEPOSITS	
PALEO.	U.		PENUTIAN BULITIAN	CAMINO CIELO SANDSTONE			PROXIMAL TURBIDITE	
	L.		YNEZIAN DANIAN	LOWER SHALE		6	TURBIDITE AND BASIN DEPOSITS	
CRET.				JALAMA				METERS 300 600 1200

Figure 2—Generalized stratigraphic column, facies, and paleobathymetry of the Tertiary strata south of the Santa Ynez Fault, in the vicinity of Santa Barbara (modified from Dibblee, copyright © 1966 California Division of Mines and Geology; van de Kamp et al, copyright © 1974 Blackwell Scientific Publications Ltd.; Ingle, 1980). West of San Marcos Pass the Juncal Formation is underlain conformably by the Anita Formation.

which K-feldspar is the dominant feldspar (Pettijohn et al, 1972), and is probably due to provenance in that these sandstones were derived from plutonic rocks in which plagioclase exceeds K-feldspar (van de Kamp et al, 1976).

Detrital matrix rarely exceeds 6%; thus, these sandstones are classified as arenites (Dott, 1964). The presence of iron oxide stain on authigenic clays, particularly in outcrop samples, can easily result in an overestimation of the matrix content. The arkoses range from fine-grained siltstones to coarse-grained sandstones. Framework grains vary from angular to subrounded and are variably sorted with the majority of the sands being only moderately sorted. In general, the turbidite sandstones of the Juncal and basal Matilija Formations tend to be more poorly sorted than the overlying sandstones.

The detrital modes of the Juncal, Matilija, Cozy Dell, and "Coldwater" Formations are all strikingly similar as indicated by the tight clustering of data points in both the QFL and QmPK diagrams (Fig. 3). One petrologic difference between the Camino Cielo Sandstone Member of the Juncal Formation and the overlying sandstones not apparent from the QFL and QmPK diagrams is the higher percentage of detrital micas (3–13%; average 7% versus 0–4%; average 2%) in the Camino Cielo Sandstone (Stauffer, 1967). The similarity of detrital modes for all sandstones of the Paleogene clastic sequence indicates they share a common provenance consisting primarily of plutonic rocks with minor volcanic, metamorphic and sedimentary components. The source terrane was an area of extensive tectonism and high relief resulting in rapid erosion and only minor chemical weathering of the sediment (van de Kamp et al, 1976).

COMPACTION OF SANDSTONES

Maxwell (1964) demonstrated that well-sorted quartz sands are compacted from initial porosities of 35 to 40% to 13 to 20% with up to 30,000 psi total pressure. Such pressures exist at 9144-m depth (30,000 ft) in the crust. Porosity measurements have shown that most sandstones have considerably lower porosities at much shallower depths. Point-count and porosity data for numerous sandstones show that compaction in nature generally reduces porosity to 15 to 25% in well-sorted sandstones buried 1524 m (5000 ft) or more. Beyond this point, cements significantly reduce porosity. This has been documented in studies on sandstones we have examined in numerous basins.

Compaction effects in the Santa

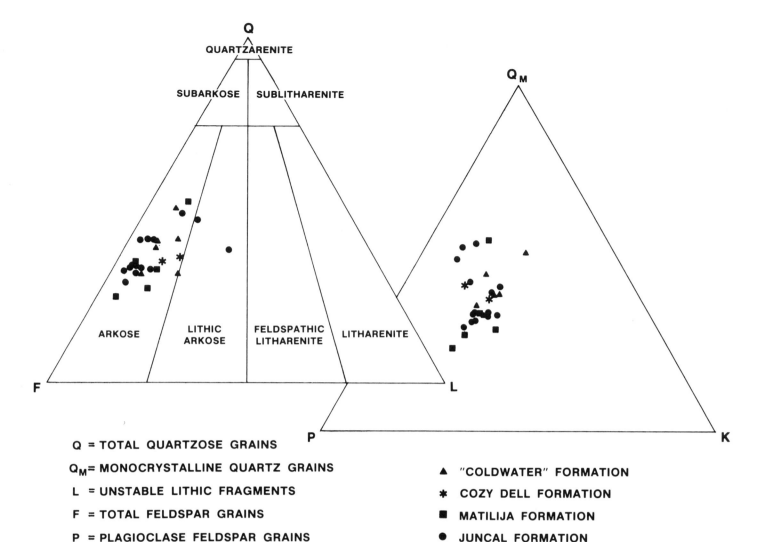

Q = TOTAL QUARTZOSE GRAINS

Q_M = MONOCRYSTALLINE QUARTZ GRAINS

L = UNSTABLE LITHIC FRAGMENTS

F = TOTAL FELDSPAR GRAINS

P = PLAGIOCLASE FELDSPAR GRAINS

K = POTASSIUM FELDSPAR GRAINS

▲ "COLDWATER" FORMATION

✱ COZY DELL FORMATION

■ MATILIJA FORMATION

● JUNCAL FORMATION

Figure 3—Triangular diagrams showing similarity in detrital modes for 27 Paleogene sandstones from Gibraltar Road. QFL diagram shows composition of entire population of framework grains (classification scheme of Folk, 1974). Q_mPK diagram shows composition of just monocrystalline mineral grains.

Ynez sandstones are registered in the detrital micas, which are flat to slightly bent in the least-buried rocks. Micas evidence increasing bending and crushing in progressively deeper-buried sandstones (Fig. 4a). Feldspar grains may be bent or crushed (Fig. 4b, c) and quartz grains may be strained or fractured (Fig 4d) in the deeply buried sandstones. Volcanic and siltstone fragments are plastically deformed in

deeply buried sandstones. Sutured and interpenetrant grain boundaries resulting from pressure solution are common in the deeper-buried rocks (Fig. 4e); in the very deeply buried sandstones (>4572 m) the rocks have interlocking mosaic textures owing to extreme pressure solution. In many cases, quartz and feldspar overgrowths and cements contribute to porosity reduction. In the deeper-buried sandstones, the combined effects of compaction and cementation have resulted in a very tight, interlocking, grain-supported fabric with only minimal porosity preserved (Link and Welton, 1982; Fig. 4f).

In the rocks studied, for a given vertical section, compaction and pressure solution are least in the "Coldwater" Formation and greatest in the lower

Juncal and Cretaceous sandstones. Laterally, depth of burial of the entire section increases from west to east and this is very clearly reflected in the decreased porosity and permeability in any given formation from west to east (Table 1, Fig. 18).

AUTHIGENIC MINERALS IN SANDSTONES

Recognition and Paragenesis of Authigenic Minerals

Mineral textures chosen as diagnostic of an authigenic origin are similar to those of Merino (1975): overgrowths, especially on quartz, plagioclase, and K-feldspar grains; cement; euhedral crystals, especially of quartz, albite, sphene, and anatase; sutured bounda-

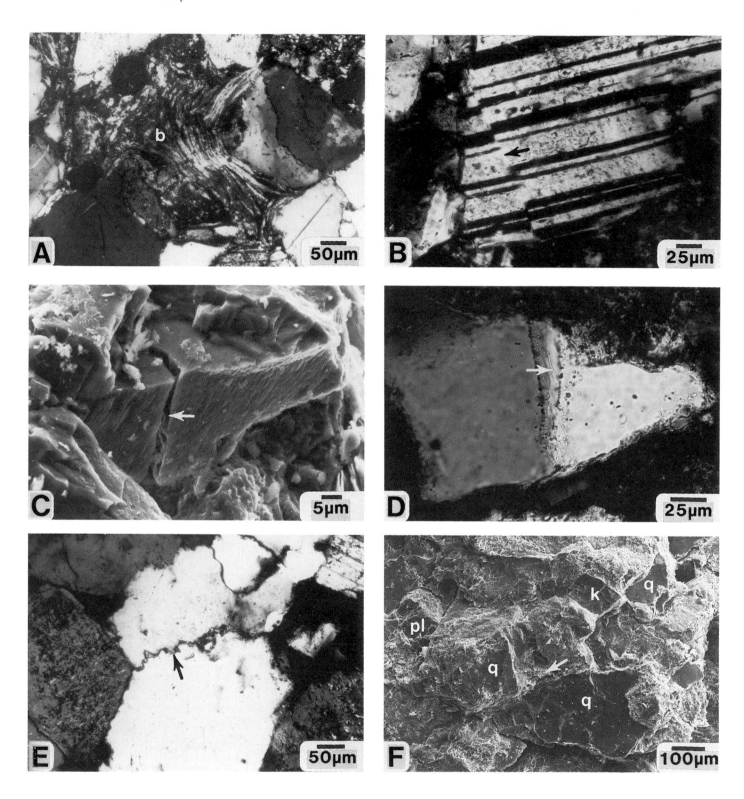

ries; intergrowths of euhedral quartz and clay minerals with delicate fabrics; pore linings and pore fillings of clay minerals; grain fractures healed in situ by quartz and albite; and replacement of detrital grains by calcite and laumontite. Albitization of Ca-bearing plagioclase (oligoclase and andesine) is documented as a diagenetic effect, whereas the occurrence of sericite and saussurite as alterations of plagioclase is not accepted as proof of their diagenetic origin.

The textural relationships between authigenic minerals observed in thin sections and SEM photomicrographs, and their distribution through the stratigraphic sections (Helmold, 1980) delineate the paragenetic sequence (Fig. 5): (1) syndepositional to very early pyrite; (2) early concretionary calcite

Figure 4a—Biotite (**b**) that has been strongly deformed between adjacent framework grains. Thin-section photomicrograph, crossed nicols. Matilija Formation, Gibraltar Road.

Figure 4b—Twin lamellae in plagioclase grain offset along fracture (arrow). Thin-section photomicrograph, crossed nicols. Camino Cielo Sandstone, Standard Oil, Gerber No. 1.

Figure 4c—Fracture (arrow) in plagioclase grain along which there has been slight rotation. SEM photomicrograph. Camino Cielo Sandstone, Gibraltar Road.

Figure 4d—Authigenic quartz with medial suture (arrow) filling fracture in detrital quartz grain. The difference in optic orientation between halves of grain suggests that rotation accompanied fracturing. Thin-section photomicrograph, crossed nicols. Camino Cielo Sandstone, Standard Oil, Gerber No. 1.

Figure 4e—Sutured contact (arrow) is the result of pressure solution between adjacent quartz grains. Thin-section photomicrograph, crossed nicols. Matilija Formation, Gibraltar Road.

Figure 4f—Very tight packing of quartz (**q**), plagioclase (**pl**), and K-feldspar (**k**) grains with only minor interstitial authigenic clay (arrow). SEM photomicrograph. Matilija Formation, Wheeler Gorge.

---

cement; (3) incipient dissolution of detrital heavy minerals and feldspars; (4) clay pore linings and pore fillings; (5) formation of sphene and anatase; (6) incipient albitization of detrital plagioclase; (7) quartz, plagioclase, and K-feldspar overgrowths; (8) dissolution of feldspar creating secondary porosity; (9) local precipitation of pore-filling kaolinite; (10) laumontite cementation and replacement of plagioclase; (11) barite cementation and replacement of detrital grains; and (12) late-stage calcite replacement of detrital grains and earlier cements.

Description of Authigenic Minerals

Pyrite

Authigenic pyrite occurs in minor amounts in sandstones throughout the Santa Ynez Range, most commonly replacing woody material, but it also occurs in the interior of foram tests, as small framboids throughout the rock, and as euhedral crystals along the cleavages of biotite grains.

Pyrite associated with fossil tests is of early origin, forming in local reducing environments associated with decaying organic matter. Glover (1963) and Merino (1975) interpreted pyrite replacing woody material to have an early origin. Other pyrite may also have formed early but textural evidence is inconclusive. In outcrop samples, almost all pyrite has altered to hematite as a result of surficial weathering.

Clay Minerals

Four species of authigenic clay minerals occur in the Paleogene sandstones including mixed-layer chlorite/smectite, chlorite, mixed-layer illite/smectite, and kaolinite. Most samples contain more than one clay species, usually with either one of the mixed-layer clays or chlorite being the dominant variety. In addition, a minor amount of illite is recognized in the X-ray diffraction patterns of most samples. It is interpreted to be a product of deuteric or hydrothermal alteration of plagioclase in the source area and is, therefore, detrital.

Mixed-layer chlorite/smectite (C/S) is one of the most common authigenic clays in the sandstones throughout the Santa Ynez Range, varying from a minor constituent in chlorite-rich sandstones (Wheeler Gorge section) to the dominant, and in some cases, sole, authigenic clay in others (Gibraltar Road section). In moderately compacted sandstones, C/S occurs as continuous linings of otherwise unfilled pores, completely coating framework grains except at points of contact (Fig. 6a–d). It also occurs as bridges between detrital grains (Fig. 6e) and as molds of chemically unstable grains (Fig. 6f). In the more highly compacted sandstones, C/S occurs as discontinuous rims and patches between framework grains, and commonly separates them from overgrowths, indicating its prior formation.

Both randomly interstratified and ordered C/S (corrensite) have been identified by X-ray diffraction analyses. The two are differentiated by the presence of a 29-Å superlattice peak in the ordered C/S that shifts to 31 Å upon glycol treatment (Reynolds, 1980; Hower, 1981; Fig. 7). Elemental compositions, as determined by EDX analyses, are similar for both types of C/S, and are characterized by Fe, Mg, and Ca in addition to Si and Al. The Fe/Mg and Fe/Ca ratios are invariably much greater than unity, but are difficult to quantify precisely, particularly in outcrop samples, where clays may be profusely coated by iron oxides related to surficial weathering. The crystal morphology of these clays, as revealed by SEM, commonly resembles the cellular or crinkly structures characteristic of smectite (Wilson and Pittman, 1977). Observable variations in degree of crenulation are probably due to differences in hydration state. The ordered C/S has a crystal morphology very similar to that of corrensite from the Guadalupe Formation (Permian age) of west Texas (Tompkins, 1981). Authigenic corrensite has also been reported from volcaniclastic sandstones of Japan (Iijima and Utada, 1971) and the disturbed belt of Montana (Hoffman, 1976), where it is interpreted to be a decomposition product of ferromagnesium minerals.

A decrease in percentage of expandable layers in mixed-layer clays with increasing depth of burial (that is, increasing temperature and pressure) has been documented for both shales and sandstones (Perry and Hower, 1970; Hower et al, 1976; Boles and Franks, 1979). While these studies document the variation in expandability of mixed-layer illite/smectite (I/S), similar variations in expandability of C/S may be expected in rocks where the latter is the dominant species. The proportion of smectite layers in C/S of the Paleogene sandstones was calculated using the methods of Reynolds (1980) and Hower (1981). Samples from the Gibraltar Road section show a gradual decrease in expandability with increasing depth of burial (Fig. 8). C/S from the Sespe and upper portion of the "Coldwater" Formations are highly expandable, containing greater than 75% smectite interlayers. Samples from the lower "Coldwater," Cozy Dell, Matilija, and upper Juncal Formations have highly variable expandabilities, but generally the C/S contains between 25 to 75 % smectite interlayers. C/S from the Camino Cielo Sandstone Member of the Juncal Formation and the Cretaceous Jalama Formation have

TABLE 1

	Standard Oil Gerber No. 1				Tecolote Tunnel				Gibraltar Road				Wheeler Gorge			
	Porosity (%)		Permeability (md)		Porosity (%)		Permeability (md)		Porosity (%)		Permeability (md)		Porosity (%)		Permeability (md)	
Formation	Average	Range	Average	Range	Average	Range	Average	Range	Average	Range	Average	Range	Average	Range	Average	Range
Sespe					18.8	4.5–28.2	371	0.1–2540								
"Coldwater"	22.1	2.7–34.0	130	0–820	13.7	3.6–32.9	146	0–1600					11.0	9.6–12.3	0.22	0.17–0.27
Cozy Dell					7.7	6.6–9.3	0.3	0–0.5					5.5		0.04	
Matilija	16.1	9.6–26.4	2.2	0.5–13	8.3	2.4–12.3	3.3	0–28	9.4	8.0–10.7	1.49	0.41–2.32	3.9	2.6–5.5	0.002	0.002–0.002
Juncal	14.4	1.6–21.3	0.9	0–10	3.8	0.8–7.2	<0.1	0–1.8					1.5	1.3–1.7	0.003	0.002–0.003
Jalama					2.6	0.4–6.7	0						2.5	1.9–3.1	0.085	0.005–0.165

Table 1—Summary of porosity and permeability data.

low expandabilities and, except for one sample, contain less than 25% smectite layers. The two Cretaceous samples have very few, if any, smectite interlayers, and are therefore essentially discrete chlorite. Ordered C/S (corrensite) is interspersed among randomly interstratified C/S throughout most of the Paleogene section, but is apparently absent in the Camino Cielo and Cretaceous sandstones (Fig. 8).

Sandstones from the Wheeler Gorge section do not show the same gradual decrease in expandability of C/S clay with depth. Instead, there is an abrupt change between the upper portion of the "Coldwater" Formation, where C/S has greater than 75% smectite interlayers, and the remainder of the Paleogene and Cretaceous section, in which discrete chlorite is the dominant authigenic clay (Fig. 8). Ordered C/S is restricted to the uppermost part of the Paleogene section where expandabilities are highest.

Chlorite is the dominant authigenic clay in the Paleogene and Cretaceous sandstones of the Wheeler Gorge section and is also a common authigenic constituent of the less deeply buried sandstones to the west. It is differentiated from C/S in X-ray diffraction

patterns by 14-Å [$(001)_{14}$] and 7-Å [$(002)_{14}$] basal reflections that do not shift upon glycolation (Hower, 1981; Fig. 7). Using this criterion some of the clays reported as discrete chlorite may actually contain a few percent of interstratified smectite. Chlorite is petrographically similar to C/S, occurring as thin discontinuous rims on framework grains and occasionally separating them from overgrowths (Fig. 9a, b). This textural relationship suggests that chlorite (or its partially expandable precursor) formed fairly early in the diagenetic history of the sandstones, and certainly prior to the main episode of quartz and feldspar cementation.

EDX analyses indicate that Fe and Mg are the major cations, in addition to Si and Al. Compositionally, chlorite is distinguished from C/S by the lack of Ca, but with that single exception, the elemental compositions of the two clays, as determined by EDX analyses, are similar. Chlorite most commonly formed as individual euhedral platelets (Fig. 9c, d) that are usually, but not always, arranged in honeycomb pattern. Although superficially resembling C/S, chlorite may be distinguished by the presence of discrete platelets that cannot be resolved in C/S (Wilson and Pittman, 1977).

Although not totally substantiated by the data presented in this paper, it is hypothesized that much, if not all, of

the authigenic chlorite in the sandstones, particularly that in the Lower Eocene and Cretaceous strata, was originally a C/S phase. The presence of chlorite beneath quartz overgrowths in these sandstones clearly indicates its early origin, and the decrease in expandability of C/S with increasing depth shown for the Gibraltar Road section (Fig. 8) suggests a diagenetic conversion of C/S to chlorite. As previously noted, Boles and Franks (1979) documented a similar decrease in expandability of I/S clay with depth for the Eocene Wilcox sandstones of the Gulf Coast.

Mixed-layer illite/smectite (I/S) is the dominant authigenic clay in the upper portion of the Paleogene section studied in the Gerber No. 1 well, but it is rare or absent in the Lower Eocene strata where chlorite is the dominant clay mineral. It is uncommon in all of the more eastern Paleogene sandstones. It occurs as pore linings (Fig 9e) and is commonly intergrown with pore-filling kaolinite (Fig. 9f) present above a well depth of 1347 m.

Both randomly interstratified and ordered I/S (allevardite) have been identified by X-ray diffraction analyses. They are most easily differentiated by the "allevardite" ordering condition, which introduces a 24-Å reflection that expands to 27 Å upon glycolation (Reynolds, 1980; Hower, 1981). Ran-

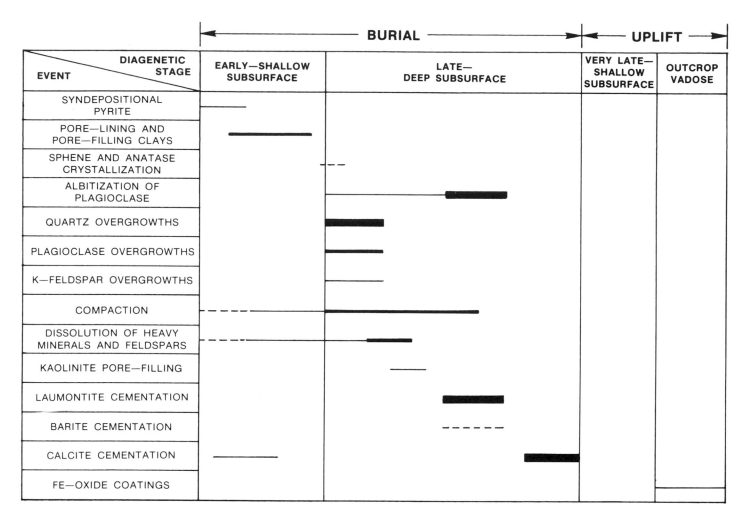

EVENT / DIAGENETIC STAGE	BURIAL			UPLIFT	
	EARLY—SHALLOW SUBSURFACE	LATE— DEEP SUBSURFACE		VERY LATE— SHALLOW SUBSURFACE	OUTCROP VADOSE
SYNDEPOSITIONAL PYRITE	───				
PORE—LINING AND PORE—FILLING CLAYS	━━━━━				
SPHENE AND ANATASE CRYSTALLIZATION		── ──			
ALBITIZATION OF PLAGIOCLASE		───────━━━━━			
QUARTZ OVERGROWTHS		━━━━━			
PLAGIOCLASE OVERGROWTHS		━━━━			
K—FELDSPAR OVERGROWTHS		────			
COMPACTION	─ ─ ─ ─ ────────────				
DISSOLUTION OF HEAVY MINERALS AND FELDSPARS	─ ─ ─ ━━━━━				
KAOLINITE PORE—FILLING		────			
LAUMONTITE CEMENTATION		━━━━━			
BARITE CEMENTATION		─ ─ ─ ─ ─			
CALCITE CEMENTATION	────────		━━━━		
FE—OXIDE COATINGS					

Figure 5—General diagenetic stages and events in the Paleogene sandstones. Time increases towards the right; depth increases towards the right until uplift. Stage and abundance of diagenetic components are indicated by length and width of bars, respectively.

domly interstratified I/S is by far the dominant species, with ordered I/S being confined to a zone in the upper portion of the Camino Cielo Sandstone. EDX analyses indicate the I/S contains K, Ca, and Fe (in addition to Si and Al) as the primary cations and occasional minor Na. A few analyses indicate minor Mg, which may actually be derived from intermixed chlorite. The I/S crystals vary in morphology from well-developed laths that project away from the substrate to a more sheetlike appearance (Wilson and Pittman, 1977).

Kaolinite has been detected only in the Gerber No. 1 well above a depth of 1347 m (estimated maximum burial depth of 2566 m) corresponding to the uppermost portion of the Camino Cielo Sandstone. It occurs as well-developed booklets that partially to completely fill intergranular pores. Kaolinite is occasionally intergrown with I/S (Fig. 9f). It occurs over the same depth interval in which there is substantial secondary porosity created by the dissolution of both plagioclase and K-feldspar. Although kaolinite was not observed to replace detrital feldspars directly, the necessary Si and Al for kaolinite precipitation was probably derived indirectly from the dissolved feldspars.

Sphene

Numerous authigenic crystals of sphene have been recognized in SEM photomicrographs (Fig. 10a) of several of the Paleogene sandstones that also

contain larger detrital sphene as an accessory. The crystals are 10 to 25 μm long and all have euhedral terminations. They are identified on the basis of an EDX elemental composition of subequal amounts of Si, Ti, and Ca with trace to minor amounts of Fe and Al. The detected Fe and Al may represent substitutions for Ti (Deer et al, 1962), but in view of the expected chemical purity resulting from the low formation temperature, it probably originated from adjacent iron- and aluminum-bearing phases. Confirming X-ray diffraction data were not obtained because of the scarcity of the crystals (authigenic sphene has not been recognized in thin section).

Most of the authigenic sphene occurs as individual crystals that have overgrown an earlier generation of authigenic clay (Fig. 10a); however, a single detrital rutile grain appears to be partially replaced by sphene that may be of

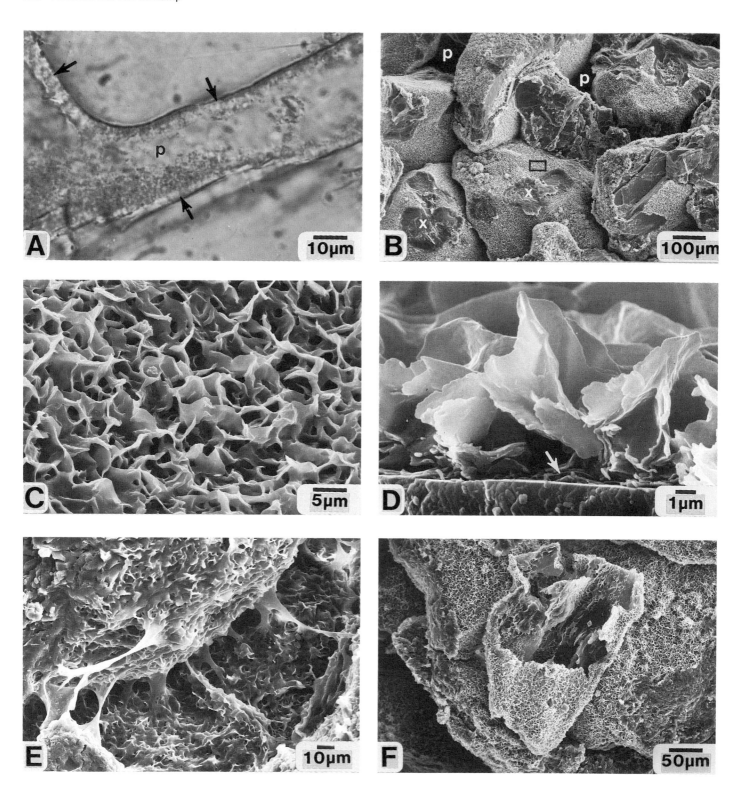

authigenic origin. Although the sphene definitely postdates the formation of authigenic clays, the timing of its genesis relative to other authigenic components remains in doubt. If sphene precipitated shortly after the pore-lining clays, the required Ca and Ti may have been derived from the early dissolution of chemically unstable heavy minerals such as epidote and rutile. A concise review of the reported occurrences of authigenic sphene in sandstones is given by Merino (1975).

Anatase

Authigenic crystals of TiO_2, tentatively identified by SEM as anatase, are present in the Paleogene sandstones in trace amounts. They are 10 to 30 μm long and occur both as isolated crystals partially engulfing authigenic clay and

Figure 6a—Authigenic ordered chlorite/smectite (corrensite, arrows) lining detrital quartz grains. Remaining pore space (**p**) is filled by epoxy. Thin-section photomicrograph, plane light. "Coldwater" Formation, Gibraltar Road.

Figure 6b—Authigenic ordered chlorite/smectite (corrensite) coating detrital grains. Note intergranular pores (**p**) and the absence of coatings at grain contacts (**x**). SEM photomicrograph. "Coldwater" Formation, Gibraltar Road.

Figure 6c—Enlarged view of outlined area in Fig. 6b, showing characteristic cellular morphology of chlorite/smectite. SEM photomicrograph. "Coldwater" Formation, Gibraltar Road.

Figure 6d—Cross section of chlorite/smectite coating on feldspar grain. Initial clay platelets are oriented parallel to grain surface (arrow), while later platelets are perpendicular to surface. SEM photomicrograph. "Coldwater" Formation, Gibraltar Road.

Figure 6e—Authigenic chlorite/smectite bridging pore between two extensively coated detrital grains. SEM photomicrograph. "Coldwater" Formation, Wheeler Gorge.

Figure 6f—Chlorite/smectite mold formed by the intrastratal dissolution of a detrital grain of unknown composition. SEM photomicrograph. "Coldwater" Formation, Gibraltar Road.

---

as aggregates of crystals that have replaced framework grains (Fig. 10b), presumably titaniferous heavy minerals. Like sphene, the exact timing of anatase genesis is problematical, but because of similarities in their chemistry and modes of occurrence, they are believed to be coeval.

Quartz

The major occurrence of secondary quartz is as overgrowths on detrital quartz grains (Fig. 10c). Overgrowths are present throughout the stratigraphic section amounting to as much as 2–3% of the whole rock, but commonly are less abundant. They range up to 0.05 mm thick and are separated from the detrital hosts by "dust" seams, rims of small hematite crystals, or authigenic clay coats (Fig. 9a). In other

cases host grains are differentiated from overgrowths by the presence of acicular inclusions or deformation lamellae that are not present in the secondary quartz. In the "Coldwater" Formation some overgrowths lack any separation from the detrital grains and are recognized solely by the presence of euhedral rhombohedral and prismatic crystal faces. In the Matilija and Juncal Formations where effects of pressure solution are apparent, all traces of a secondary origin for similar overgrowths may be destroyed. The presence of euhedral overgrowths in laumontite cemented sandstones indicates they formed prior to laumontite cementation (Fig. 10c).

Examination with SEM reveals that large well-developed overgrowths form in a manner similar to that described by Waugh (1970) and Pittman (1972). In addition, SEM reveals small authigenic quartz crystals (Fig. 10d) not observed with the petrographic microscope. They range up to 30 μm in length and have euhedral rhombohedral and prism faces attesting to a secondary origin. These quartz crystals are usually associated with authigenic clay and in some cases they are intimately intergrown.

Secondary quartz occurs as fracture fillings in detrital quartz grains (Fig. 4d). Fractures produced in quartz clasts during the early stages of compaction are subsequently healed by secondary quartz. Fracture boundaries are often delineated by "dust" seams or lines of hematite crystals similar to those associated with overgrowths. Medial sutures in the fracture fillings indicate that quartz grew inward from the walls of the fracture and are conclusive evidence for a post-depositional origin. Where grain rotation accompanied fracturing a marked optical discontinuity is observed at the medial suture (Fig. 4d).

K-Feldspar

Authigenic K-feldspar in the Paleogene sandstones occurs as overgrowths on orthoclase, microcline, and perthite grains, with which they usually show a marked optical discontinuity (Fig. 10e). Overgrowths range up to 0.03 mm thick and never exceed 1% of the whole rock. Most are small euhedral projections on the detrital host (Fig. 10f), but a few completely encase the detrital core (Fig. 10e). They are most abundant in the

western part of the basin, especially in the Gerber No. 1 well, where they have been etched and in some cases partially dissolved to create secondary porosity.

Microanalyses show that the overgrowths are nearly pure K-feldspar with an average composition of $Ab_{0.6}An_{0.0}Or_{99.4}$ (Table 2). Kastner (1971) reported similar compositions for authigenic microcline from carbonate rocks. In contrast, the average composition of the detrital K-feldspar hosts is $Ab_{8.8}An_{0.1}Or_{91.1}$ (Table 2). The order of magnitude difference in sodium content between overgrowths and hosts accounts for the observed optical discontinuity. Several K-feldspar grains have rims with very low sodium concentrations, which are interpreted to be the product of weathering during erosion and subsequent transport.

Albite

Authigenic albite occurs in the Paleogene sandstones as (1) fracture fillings in detrital plagioclase grains, (2) overgrowths on framework plagioclase, and (3) a replacement (albitization) of detrital oligoclase and andesine.

Minor amounts of secondary albite are present throughout the Paleogene sections as water-clear fillings in fractured plagioclase grains (Fig. 11a) similar to those originally described by Merino (1975). They range in length up to 0.22 mm, but are generally smaller. Medial sutures similar to those associated with fractured quartz grains are always present. Where grain rotation accompanied fracturing, twin lamellae may be offset along the medial suture with the grain fragments exhibiting optical discontinuity. Microanalyses show that the average composition of the fracture-filling albite is $Ab_{97.0}An_{1.8}Or_{1.2}$ (Table 2). This agrees with Kastner (1971) and Kastner and Siever (1979) who determined that authigenic albite in carbonate rocks is a very pure Na-form distinct from the wide compositional range of detrital plagioclase. Microanalyses of the fractured plagioclase grains usually show a composition very similar to the secondary albite, as indicated by their near optical continuity. The high sodium content of the detrital plagioclase is due to in situ albitization.

Clear albite overgrowths on detrital plagioclase grains (Fig. 11b–d) are

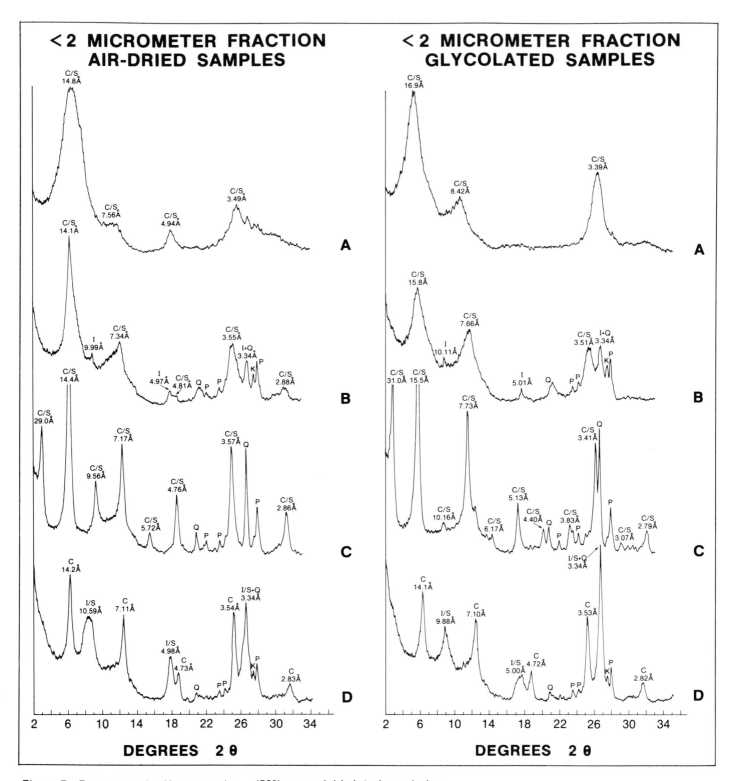

Figure 7—Representative X-ray powder diffraction patterns of the <2 μm fraction of four sandstones from Gibraltar Road. C/S = chlorite/smectite, I/S = illite/smectite, C = chlorite, I = illite, Q = quartz, P = plagioclase, K = K-feldspar. (**A**) Randomly interstratified chlorite/smectite (88% expandable interlayers); "Coldwater" Formation. (**B**) Randomly interstratified chlorite/smectite (50% expandable interlayers) plus minor illite, quartz, and feldspar; Matilija Formation. (**C**) Ordered chlorite/smectite (corrensite; 42% expandable interlayers) plus quartz and feldspar; Juncal Formation. (**D**) Chlorite plus randomly interstratified illite/smectite (20% expandable interlayers) and minor quartz and feldspar; Jalama Formation.

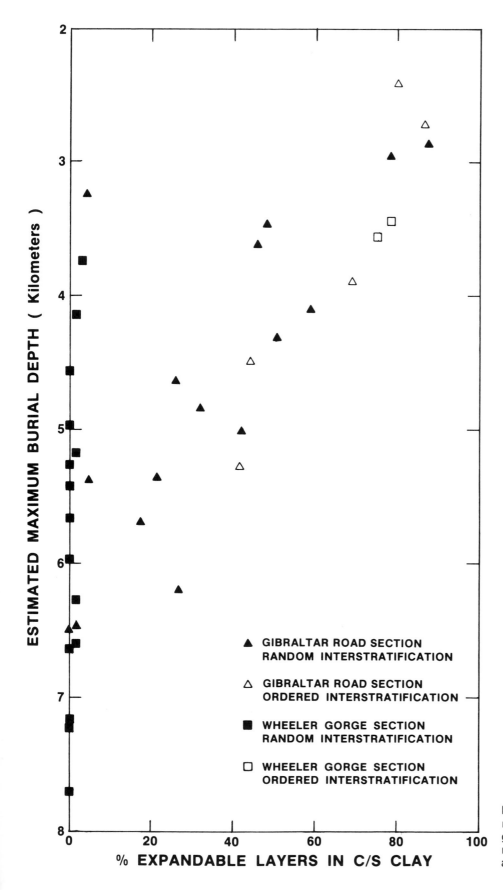

Figure 8—Variation in expandability of mixed-layer chlorite/smectite in Paleogene sandstones with estimated maximum burial depth for Gibraltar Road and Wheeler Gorge sections.

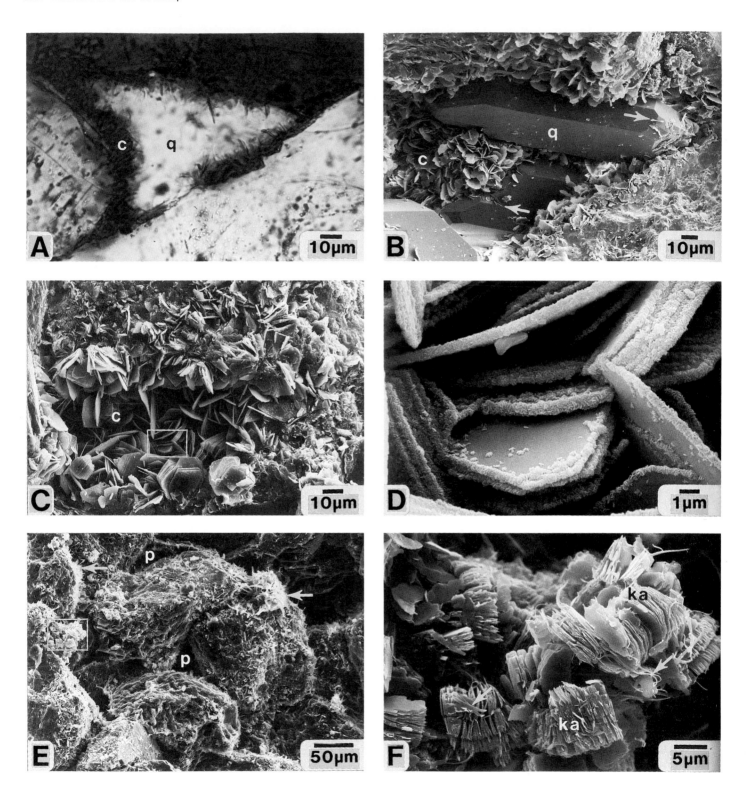

present in the Paleogene sandstones in amounts rarely exceeding 1 to 2% of the whole rock. They range up to 0.03 mm thick and almost always have euhedral faces. Microanalyses show that the average composition of the

albite overgrowths is $Ab_{98.7}An_{0.8}Or_{0.5}$ (Table 2). This agrees with analyses of Merino (1975) of albite overgrowths from the Temblor Formation of Kettleman North Dome, California. Compromise boundaries between albite, K-

feldspar, and quartz overgrowths suggest they are coeval (Fig. 11e).

Both detrital oligoclase and diagenetically albitized plagioclase grains act as hosts for albite overgrowths. Oligoclase hosts show little evidence of alteration

Figure 9a—Triangular pore occluded by initial lining of authigenic chlorite (**c**) and later syntaxial quartz overgrowth (**q**). Note that chlorite precipitated normal to substrate. Thin-section photomicrograph, crossed nicols. Matilija Formation, Wheeler Gorge.

Figure 9b—Pore-lining chlorite (**c**) beneath later quartz overgrowths (**q**). Note that several chlorite platelets are partially engulfed by the overgrowths (arrows). SEM photomicrograph. Matilija Formation, Wheeler Gorge.

Figure 9c—Dense growth of chlorite (**c**) that has partially occluded intergranular pore. SEM photomicrograph. Camino Cielo Sandstone, Gibraltar Road.

Figure 9d—Enlarged view of outlined area in Fig. 9c, showing iron oxide coating along edges of chlorite platelets. The oxides are the result of surficial weathering as evidenced by their absence in core samples. SEM photomicrograph. Camino Cielo Sandstone, Gibraltar Road.

Figure 9e—Randomly interstratified illite/smectite (arrows) coating framework grains, and kaolinite booklets (outlined area) partially occluding pores. Note abundant intergranular porosity (**p**). SEM photomicrograph. "Coldwater" Formation, Standard Oil, Gerber No. 1.

Figure 9f—Enlarged view of outlined area in Fig. 9e, showing intergrown kaolinite (**ka**) and illite/smectite (arrows). SEM photomicrograph. "Coldwater" Formation, Standard Oil, Gerber No. 1.

and are not optically continuous with overgrowths (Fig. 11c). Microanalyses show they have a range in composition with an average composition of $Ab_{78.1}An_{20.9}Or_{1.0}$ (Table 2). Albitized plagioclase hosts are murky brown in plane light owing to extensive alteration and are optically continuous with overgrowths (Fig. 11d). They invariably are similar in composition to the overgrowths, with an average composition of $Ab_{97.6}An_{1.7}Or_{0.7}$ (Table 2). The degree of optical continuity between overgrowth and host is directly related to the amount of $CaAl_2Si_2O_8$ in the detrital grain. Albite overgrowths are

optically continuous with albitized plagioclase hosts, but are markedly optically discontinuous with calcium-bearing plagioclase hosts.

Detrital plagioclase grains with clear albite rims that resemble albite overgrowths except for the lack of euhedral crystal faces are present in a few samples (Fig. 11f). Microanalyses show the albite rims have an average composition of $Ab_{90.8}An_{8.1}Or_{1.1}$ (Table 2). Based on their impure albitic chemistry and detrital nature, they are thought to be the rims of zoned plagioclase grains with selectively albitized cores.

Diagenetic albite occurs in the Paleogene and Cretaceous strata as replacement of calcium-bearing plagioclase grains. Albitized plagioclase is characterized by a murky brown appearance resulting from numerous "dusty" inclusions (Boles, 1982). Albitized grains from the Camino Cielo Sandstone are partially replaced by laumontite, which presumably formed from calcium released during albitization, and silicon and aluminum ions in pore fluids. The albitization reaction is discussed below in more detail.

Laumontite

Laumontite occurs only in the lower part of the Paleogene section where it is common in the Camino Cielo Sandstone Member of the Juncal Formation throughout the entire Santa Ynez Range, and it occurs locally in the basal Matilija Formation. Where available for sampling (Wheeler Gorge and Tecolote Tunnel), Cretaceous sandstones of the Jalama Formation also contain laumontite. Laumontite varies in abundance from trace amounts to 15% of the whole rock. Several distinct modes of occurrence are recognized (Helmold, 1979): (1) patches interstitial to framework grains (cement); (2) alterations within detrital plagioclase grains; (3) displacive patches along the cleavage of detrital biotite; and (4) fracture-filling veins up to 2 mm wide.

Laumontite cement is optically clear, usually with well-developed cleavage and undulatory extinction, and where extensive it occludes all visible porosity (Fig. 12a). It may fill pores that are partially lined by authigenic clay or silicate overgrowths, suggesting its later genesis. Cement may be optically continuous with laumontite replacing detrital plagioclase, and where replacement is

pervasive, distinction between the two modes is especially difficult.

Laumontite replacement of detrital plagioclase varies from incipient alteration within a grain (Fig. 12b, c) to complete replacement, and is the most common mode of occurrence of laumontite. Where replacement is extensive, the laumontite is usually murky brown and contains inclusions of sericite and epidote inherited from the plagioclase precursor.

Laumontite commonly occurs as displacive patches along biotite cleavage in sandstones that are pervasively zeolitized. This varies from slight doming of the cleavage to complete disaggregation of biotite into single cleavage flakes (Fig. 12d). The similarity in chemistry between displacive laumontite and that of other modes of occurrence (Table 3) suggests that it is not altering directly from biotite, but is merely prying apart the biotite as a result of the force of crystallization.

Laumontite veins occur in the Camino Cielo Sandstone throughout the entire Santa Ynez Range and are most evident in core samples, presumably owing to the lack of outcrop weathering. They appear as powdery white veins 1 to 2 mm wide and are up to tens of centimeters long. Many veins consist solely of prismatic laumontite, whereas others contain scattered inclusions of detrital grains (Fig. 12e). Laumontite crystallization occurred in open fractures in the rock and is therefore considered a late-stage diagenetic event.

Laumontite crystallization postdates the formation of all other authigenic minerals except calcite. Laumontite replacement of detrital grains, its sharp boundary relationships with silicate overgrowths and authigenic clay rims, and the presence of laumontite veins in hand samples all support its late origin. In addition to textural evidence, the restriction of laumontite to the deepest portion of the stratigraphic section also suggests a late origin.

Unit cell parameters, which are diagnostic of hydration state, were determined for laumontite separates via X-ray diffraction analysis. A total of 32 reflections was processed to derive cell constants according to the procedure of Evans et al (1963). The calculated a and b cell dimensions and β angle, which are most sensitive to hydration state,

TABLE 2

	K-feldspar Overgrowths	K-feldspar Grains	Fracture-Filling Albite	Albite Overgrowths	Albitized Plagioclase Grains	Oligoclase Grains	Detrital Albite Rims	Partially Dissolved Albite Grains	Partially Dissolved Oligoclase Grains	Partially Dissolved Andesine Grains
Number of spots analyzed										
	24	26	5	16	7	5	2	27	5	10
SiO_2	64.46 ± 0.51	64.37 ± 0.49	67.83 ± 0.40	68.15 ± 0.27	67.77 ± 0.56	62.40 ± 0.66	66.59 ± 0.01	68.43 ± 0.57	64.58 ± 1.24	60.31 ± 0.55
TiO_2	—	—	0.01 ± 0.01	0.01 ± 0.01	0.00 ± 0.00	0.00 ± 0.00	0.00 ± 0.00	—	—	—
Al_2O_3	18.55 ± 0.15	18.49 ± 0.14	19.71 ± 0.51	19.48 ± 0.10	19.63 ± 0.15	23.16 ± 0.50	20.28 ± 0.16	19.71 ± 0.38	21.85 ± 0.98	24.93 ± 0.53
Fe_2O_3	0.03 ± 0.03	0.08 ± 0.04	0.06 ± 0.02	0.03 ± 0.03	0.02 ± 0.01	0.11 ± 0.05	0.08 ± 0.11	0.05 ± 0.04	0.09 ± 0.04	0.14 ± 0.04
MgO	—	—	0.00 ± 0.00	0.00 ± 0.00	0.00 ± 0.00	0.00 ± 0.00	0.00 ± 0.00	—	—	—
CaO	0.01 ± 0.02	0.03 ± 0.03	0.38 ± 0.27	0.17 ± 0.09	0.37 ± 0.14	4.29 ± 0.46	1.74 ± 0.48	0.31 ± 0.35	3.36 ± 0.30	6.33 ± 0.70
Na_2O	0.06 ± 0.01	0.98 ± 0.30	11.25 ± 0.32	11.54 ± 0.09	11.44 ± 0.11	8.87 ± 0.33	10.72 ± 0.05	11.45 ± 0.44	9.46 ± 0.16	7.83 ± 0.32
K_2O	14.74 ± 0.34	15.40 ± 0.45	0.20 ± 0.18	0.09 ± 0.02	0.13 ± 0.04	0.18 ± 0.05	0.20 ± 0.12	0.12 ± 0.10	0.30 ± 0.03	0.24 ± 0.10
Total	97.85 ± 0.44	99.35 ± 0.60	99.44 ± 0.52	99.47 ± 0.38	99.36 ± 0.55	99.01 ± 0.55	99.61 ± 0.46	100.07 ± 0.76	99.64 ± 0.48	99.78 ± 0.40
Feldspar composition (mole %)										
Ab	0.57 ± 0.12	8.82 ± 2.65	97.03 ± 1.50	98.70 ± 0.41	97.55 ± 0.85	78.10 ± 1.89	90.78 ± 1.43	97.82 ± 1.80	82.18 ± 1.35	68.21 ± 3.08
An	0.06 ± 0.12	0.13 ± 0.13	1.82 ± 1.32	0.79 ± 0.40	1.73 ± 0.64	20.88 ± 2.19	8.13 ± 2.11	1.48 ± 1.60	16.12 ± 1.45	30.43 ± 3.24
Or	99.37 ± 0.19	91.05 ± 2.71	1.15 ± 1.02	0.51 ± 0.10	0.72 ± 0.25	1.02 ± 0.30	1.09 ± 0.69	0.70 ± 0.68	1.70 ± 0.18	1.36 ± 0.56

Table 2—Electron microprobe analyses of authigenic and detrital K-feldspar and plagioclase.

TABLE 3

	Laumontite Cement and Laumontite Replacing Plagioclase	Laumontite Displacing Biotite Cleavage	Laumontite from East Taiwan Ophiolite (Liou et al, 1977)	Laumontite and Leonhardite, Worldwide Occurrences (Coombs, 1952)[1]
Number of spots analyzed				
	7	2	3[2]	14[2]
SiO_2	52.43 ± 1.14	52.15 ± 1.73	51.69 ± 0.55	51.38 ± 1.32
Al_2O_3	21.62 ± 0.45	21.76 ± 1.01	21.01 ± 0.43	21.73 ± 0.98
Fe_2O_3	0.01 ± 0.01	0.09 ± 0.04	0.82 ± 1.12	0.69 ± 0.73
CaO	11.30 ± 0.41	11.25 ± 0.14	11.70 ± 0.83	11.06 ± 1.45
Na_2O	0.12 ± 0.10	0.05 ± 0.04	0.09 ± 0.07	0.67 ± 0.75
K_2O	0.15 ± 0.12	0.07 ± 0.03	0.39 ± 0.19	0.73 ± 1.02
Anhydrous total	85.63 ± 1.00	85.37 ± 2.78	85.42 ± 2.27	86.16 ± 1.14

[1] Analyses via wet chemical methods.
[2] Not all samples were analyzed for all elements, therefore anhydrous total does not equal sum of the oxides.

Table 3—Electron microprobe analyses of laumontite.

TABLE 4

	Leonhardite from Camino Cielo Sandstone[1]	Leonhardite (Madsen and Murata, 1970)	Leonhardite (Lapham, 1963)	Leonhardite (Coombs, 1952)	Laumontite (Coombs, 1952)
a (Å)	14.74 ± .02	14.72 ± .02	14.75 ± .05	14.75 ± .03	14.90 ± .05
b (Å)	13.09 ± .03	13.07 ± .01	13.08 ± .02	13.10 ± .02	13.17 ± .02
c (Å)	7.57 ± .01	7.56 ± .01	7.57 ± .05	7.55 ± .01	7.55 ± .05
β	111.9° ± .2°	111.9° ± .2°	112°	112.0° ± .2°	111.5° ± .5°
Vol (Å³)	1355 ± 3	1349			

[1]Cell values derived using least-squares refinement procedure of Evans et al (1963), based on 32 reflections.

Table 4—Unit-cell data for laumontite and leonhardite.

are similar to published values for leonhardite, the partially dehydrated variety of laumontite (Table 4). The name laumontite as used in this paper, therefore, does not connote a particular hydration state.

Microanalyses show that the laumontite is a pure calcium–aluminum hydrous silicate with only minor amounts of sodium, potassium, and iron (Table 3). Composition does not vary with mode of occurrence (that is, cement, replacement, or displacive patches), suggesting there is no selective partitioning of ions on the basis of texture. Laumontite from the Camino Cielo Sandstone is similar in composition to that of the East Taiwan Ophiolite (Liou et al, 1977) and several occurrences studied by Coombs (1952; Table 3).

Barite

Barite occurs in one sandstone sample of the upper shale member of the Juncal Formation from Gibraltar Road. It is texturally similar to laumontite, occurring both as cement and replacement of detrital grains (Fig. 12f). Plagioclase and K-feldspar are the most likely precursors of barite, although the near complete replacement of framework grains precludes their positive identification. Hawkins (1978) described a similar replacement of K-feldspar and kaolinite, and the

corrosion of quartz grains by barite from Carboniferous sandstones of the North Sea. Barite genesis postdates the formation of quartz and albite overgrowths as evidenced by sharp boundary relationships, and is thought to be roughly coeval with laumontitization.

Calcite

Calcite is a common cement throughout the entire Santa Ynez Range, comprising up to 40% of some sandstones. Where poorly developed, it fills pore spaces and replaces plagioclase. In rocks with abundant calcite cement, all silicates are partially or completely replaced and no porosity is visible in thin section. Two generations are distinguished based on their distribution and texture: early localized, concretionary calcite cement, and late, widely distributed, replacement calcite.

Early calcite cement occurs locally as round concretions ranging in size from a few centimeters to 1 m in diameter and has a poikilitic texture with crystals up to 2 cm across. The cement crystallized before significant compaction, and before the formation of clay coats and silicate overgrowths, thereby preventing further diagenetic modifications of the affected sandstones.

Late-stage calcite is not confined to concretionary zones, being widely distributed throughout the stratigraphic section, and is more abundant than the early cement. The individual crystals are less than 0.1 mm in diameter and

are equidimensional giving the calcite a granular texture. They are often murky, containing visible impurities resulting from the incomplete replacement of detrital grains and matrix (Fig. 13b). This late-stage calcite originated after the sandstone had undergone initial compaction and further reduction of pore space by growth of authigenic minerals. An important criterion for the recognition of late-stage calcite in thoroughly cemented sandstones is the presence of additional cements and the alteration or deformation of labile and ductile framework grains (Galloway, 1974). The presence of overgrowths on detrital quartz (Fig. 13a) and feldspar, localized barite cement, laumontite cement and replacement of detrital plagioclase, and deformed detrital biotite and ductile lithic fragments in calcite-cemented sandstones all point to late crystallization of calcite. The replacement of laumontite by calcite has been observed in a few instances also suggesting late calcite genesis. Abundant twinning of late-stage calcite is possibly the result of deformation associated with uplift of the Santa Ynez Range, and therefore suggests the calcite was emplaced prior to uplift.

SECONDARY POROSITY

Secondary porosity created by the dissolution of feldspar occurs throughout the Santa Ynez Range, and is best

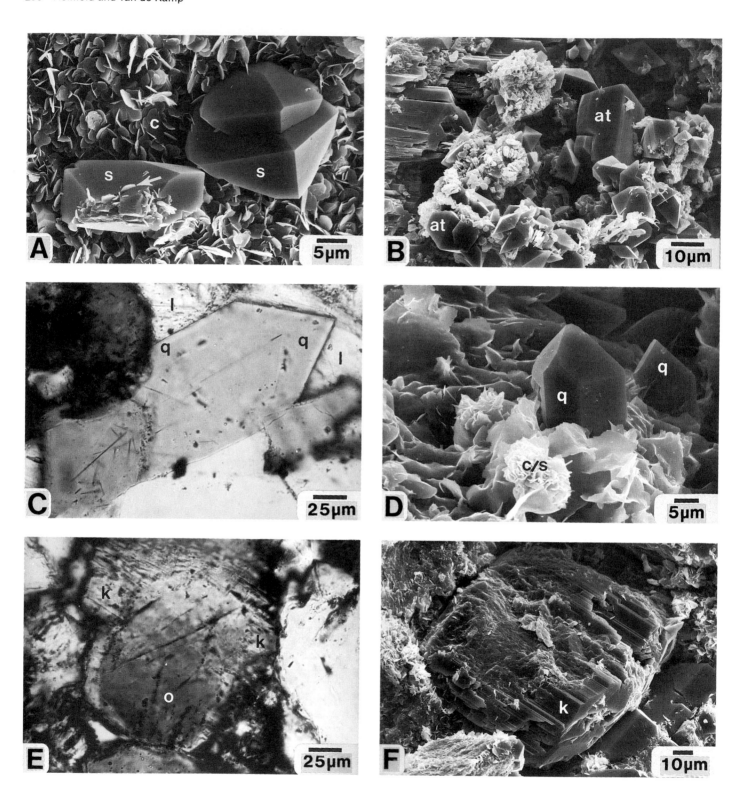

developed in the western end of the basin, especially in the Gerber No. 1 well above a depth of 1372 m (maximum burial depth of 2591 m). Dissolution commonly proceeds parallel to cleavages, which act as conduits for reactive fluids, resulting in a reticulate structure in partially dissolved grains (Fig. 13c). Dissolution affects both plagioclase and K-feldspar, but large dissolution voids formed by the complete dissolution of the feldspar core occur only in plagioclase. Microanalyses show that the remnants of dissolved plagioclase vary in composition from albite ($Ab_{97.8}An_{1.5}Or_{0.7}$) to andesine ($Ab_{68.2}An_{30.4}Or_{1.4}$; Table 2). Dissolution may transect cleavage as evidenced by

Figure 10a—Euhedral crystals of authigenic sphere (**s**) that have overgrown an earlier generation of chlorite (**c**). Note chlorite platelets partially engulfed by sphere (arrow). SEM photomicrograph. Camino Cielo Sandstone, Standard Oil, Gerber No. 1.

Figure 10b—Euhedral crystals of authigenic anatase (**at**) that have replaced a detrital grain, presumably a titaniferous heavy mineral. SEM photomicrograph. Cozy Dell Formation, Standard Oil, Gerber No. 1.

Figure 10c—Quartz overgrowth (**q**) with euhedral rhombohedral and prism faces in sharp contact with laumontite cement (**l**). This relationship suggests quartz overgrowths formed prior to laumontite. Thin-section photomicrograph, crossed nicols. Camino Cielo Sandstone, Standard Oil, Gerber No. 1.

Figure 10d—Authigenic quartz crystals (**q**) on substrate of randomly interstratified chlorite/smectite (**c/s**). SEM photomicrograph. Matilija Formation, Gibraltar Road.

Figure 10e—K-feldspar overgrowth (**k**) almost completely surrounding orthoclase host (**o**). Optical discontinuity is the result of the difference in sodium content between overgrowth (<0.1 wt % Na_2O) and core (approx. 1.0 wt % Na_2O). Thin-section photomicrograph, crossed nicols. "Coldwater" Formation, Standard Oil, Gerber No. 1.

Figure 10f—Euhedral K-feldspar overgrowth (**k**) on detrital K-feldspar host. SEM photomicrograph. Camino Cielo Sandstone, Standard Oil, Gerber No. 1.

---

plagioclase grains with circular dissolution voids (Fig. 13d, e) similar to those described by Heald and Larese (1973) in K-feldspars. Feldspar dissolution undoubtedly occurred in situ, as grains with delicate reticulate structures and hollow cores could not withstand transport and subsequent compaction. A few feldspar overgrowths are present in the dissolution voids of partially dissolved feldspars (Fig. 13f), but the scarcity of this feature suggests the bulk of feldspar dissolution postdated the main episode of overgrowth formation. Par-

tial dissolution of K-feldspar overgrowths, particularly in the Gerber No. 1 well, further indicates that the main period of dissolution followed overgrowth formation.

ALBITIZATION OF PLAGIOCLASE

As noted above, diagenetic albite occurs in the Paleogene sandstones as a replacement of calcium-bearing grains. This was first detected by optical methods, and subsequently the stratigraphic distribution of albitized plagioclase was determined by means of microprobe analyses of framework plagioclase for sodium, calcium, and potassium (Helmold, 1980). Calculations reveal that if silicon and aluminum are present in stoichiometric proportions, the average total weight percent oxides of 1051 analyses is 100.1 ± 2.1. This indicates that the partial analyses are an accurate measure of plagioclase composition. Sandstones chosen for analysis contain no, or in a few cases, only minor calcite cement. The analyses (Fig. 14, Table 5) indicate that there is an increase in albite component and concurrent decrease in anorthite component with increasing burial depth. The orthoclase component decreases only slightly, or not at all, from shallow to deep samples. Standard deviations of both the average albite and anorthite contents decrease with increasing burial depth, indicating the composition of the plagioclase population is more homogeneous at greater burial depths.

Variations from the general trend (Table 5) are believed to result from localized zones of low permeability, which retard the albitization reaction, rather than differences in detrital composition. In addition, many of the deviations may represent statistical variation, which can be eliminated by analyzing a larger number of grains per sample.

Although detrital K-feldspar grains were not systematically included in the analyses, sufficient random analyses were conducted on K-feldspars to indicate they have not been affected by albitization. The lack of visible alteration of K-feldspars, similar to that displayed by albitized plagioclase, also

suggests they have not undergone albitization.

The albitized sandstones can be divided into two zones based on the average Ab content of framework plagioclase. A zone of incipient albitization separates the overlying unalbitized sandstones from the completely albitized ones below. Framework plagioclase in this zone has an average composition of Ab_{95} to Ab_{98}, and is distinctly more albitic than plagioclase in the overlying sandstones, which ranges in average composition from Ab_{84} to Ab_{94} (Table 5). Below this zone of incipient albitization framework plagioclase has an average composition of Ab_{98} or greater, signifying the complete albitization of detrital calcium-bearing plagioclase. This is the main zone of albitization. Incipient albitization is apparent in the sandstones overlying the laumontite zone, but wholesale albitization occurs only in those sandstones that have been extensively zeolitized. This relationship suggests that the precipitation of laumontite and massive albitization are coeval, as is also suggested by the chemically complementary nature of the reactions (Coombs, 1954; Merino, 1975).

During the initial stages of albitization, the conversion of the anorthite component of oligoclase and andesine proceeds as an equal volume replacement reaction with the receipt of Na^+ ions from solution and the release of Ca^{+2} ions into solution (Merino, 1975; Boles, 1982):

$$(V_m = 100.16) \quad (1)$$
$$0.8NaAlSi_3O_8 \cdot 0.2CaAl_2Si_2O_8 +$$
Oligoclase

$$0.203H_4SiO_4 + 0.201Na^+ + 0.796H^+ =$$

$$(V_m = 100.07)$$
$$1.001NaAlSi_3O_8 + 0.2Ca^{+2} + 0.199Al^{+3}$$
Albite

$$+ 0.804H_2O$$

At greater burial depths where incipient laumontization has initiated, calcium-bearing plagioclase is partly albitized and partially replaced by laumontite according to the reaction (Boles and Coombs, 1977):

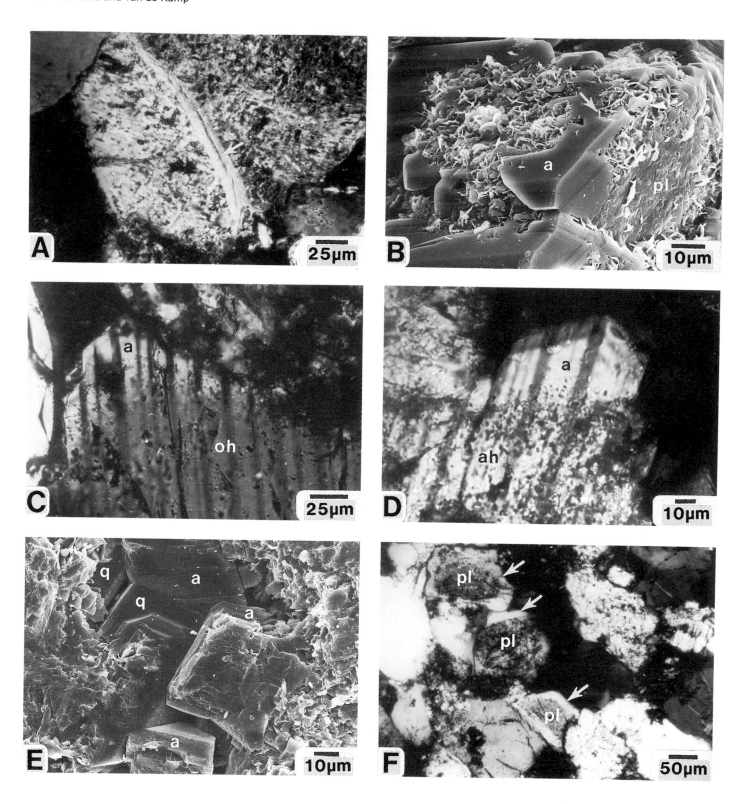

Figure 11a—Clear fracture-filling albite in albitized plagioclase grain. Medial suture (arrow) indicates that albite grew inward from walls of fracture. Thin-section photomicrograph, crossed nicols. "Coldwater" Formation, Wheeler Gorge.

Figure 11b—Albite overgrowth (**a**) on detrital plagioclase grain (**pl**) that is coated with authigenic clay. Note clay is partially engulfed by overgrowth (arrow). SEM photomicrograph. "Coldwater" Formation, Gibraltar Road.

Figure 11c—Clear albite overgrowth (**a**) on relatively unaltered oligoclase host (**oh**). Lack of optical continuity results from the difference in chemistry. Thin-section photomicrograph, crossed nicols. "Coldwater" Formation, Gibraltar Road.

Figure 11d—Clear albite overgrowth (**a**) on murky albitized plagioclase host (**ah**). Optical continuity is an indication of similar chemistry, which is confirmed by microanalyses. Thin-section photomicrograph, crossed nicols. "Coldwater" Formation, Gibraltar Road.

Figure 11e—Quartz (**q**) and albite (**a**) overgrowths occluding intergranular pore. Compromise boundaries between overgrowths suggest that authigenic quartz and albite are coeval. SEM photomicrograph. Matilija Formation, Wheeler Gorge.

Figure 11f—Rims of albitic plagioclase (arrows) surrounding albitized plagioclase cores (**pl**). They probably are the rims of zoned plagioclase grains with selectively albitized cores. Thin-section photomicrograph, crossed nicols. Juncal Formation, Gibraltar Road.

$$NaAlSi_3O_8 \cdot CaAl_2Si_2O_8 + \qquad (2)$$
$$\text{Plagioclase}$$

$$2SiO_2 + 4H_2O = NaAlSi_3O_8 +$$
$$\text{Quartz} \qquad \text{Albite}$$

$$CaAl_2Si_4O_{12} \cdot 4H_2O$$
$$\text{Laumontite}$$

Boles and Coombs (1977) calculated that for a given volume of plagioclase altered, an equal volume of albite–laumontite intergrowth plus additional laumontite will be produced to replace quartz, fill pores, and displace biotite cleavages.

Boles and Coombs (1977) have proposed an additional reaction, especially for large volumes of sandstones in which albitization has been complete, that necessitates the supply of Na^+ ions from solution and the release of Ca^{+2} ions into solution:

$$NaAlSi_3O_8 \cdot CaAl_2Si_2O_8 + 3SiO_2 + \qquad (3)$$
$$\text{Plagioclase} \qquad \text{Quartz}$$

$$2H_2O + Na^+ =$$

$$2NaAlSi_3O_8 + 0.5CaAl_2Si_4O_{12} \cdot 4H_2O +$$
$$\text{Albite} \qquad \text{Laumontite}$$

$$0.5Ca^{+2}$$

Both reactions 1 and 3 require substantial amounts of sodium and silica to proceed. While connate water could supply some of the necessary Na^+ ions, the sodium concentration in sea water is insufficient for it to be the only source unless numerous pore volumes move through the rock. The conversion of smectite to illite in the interbedded shales is the most likely source for large quantities of Na^+ ions as shown by the general reaction (modified from Hower et al, 1976):

$$smectite + Al^{+3} + K^+ = \qquad (4)$$
$$illite + Si^{+4} + Na^+ + Ca^{+2} + Mg^{+2} +$$
$$Fe^{+2, +3} + H_2O$$

The Na^+ ions released are not taken up in the formation of authigenic chlorite in the shales and, therefore, are free to migrate to the sandstones, assuming the shales do not act as a completely closed system with respect to sodium. Silica is also required in reactions 1, 2, and 3. Boles and Coombs (1977) assume local replacement of quartz as the primary source, but as shown by Towe (1962), Hower et al (1976), and Boles and Franks (1979), large quantities of Si^{+4} are released by the conversion of smectite to illite, which if free to migrate into sandstones, may form secondary quartz overgrowths or take part in albitization. The virtual disappearance of the smectite component of mixed-layered clays in the shales at the base of the Paleogene sequence (Helmold and Snider, 1981), the same interval over which albitization and laumontization are pronounced, lends credence to smectite layers as a possible source of Na^+ and Si^{+4} ions.

Land and Milliken (1981) and Boles (1982) documented similar trends in plagioclase compositions from the Frio and Wilcox Formations of the Gulf Coast Tertiary. In addition, Boles (1982) noted the presence of authigenic calcite within many albitized grains, and speculated that it formed from Ca^{+2} released during albitization in combination with carbonate ions from pore fluids. The precipitation of calcite as a by-product of the albitization reaction differs from the cases of the Santa Ynez Basin and other California Tertiary basins (Castaño and Sparks, 1974) where laumontite precipitates concurrently with albitization. This difference may result from variations in the activity (a) of CO_2 relative to that of H_2O between the two geographic areas (Zen, 1961; Albee and Zen, 1969). Where a_{CO_2}/a_{H_2O} is high, as may be the case in the Gulf Coast Tertiary, calcite will precipitate as a by-product of albitization according to the reaction:

$$NaAlSi_3O_8 \cdot CaAl_2Si_2O_8 + 4SiO_2 + \qquad (5)$$
$$\text{Plagioclase} \qquad \text{Quartz}$$

$$2Na^+ + CO_3^{-2} = 3NaAlSi_3O_8 + CaCO_3$$
$$\text{Albite} \qquad \text{Calcite}$$

On the other hand, where a_{CO_2}/a_{H_2O} is low, laumontite precipitation will accompany albitization according to reaction 2 or 3.

ORIGIN OF CALCITE

The abundant secondary calcite in the Paleogene sandstones requires a sizeable source of Ca^{+2} ions. In the deeper parts of the basin where laumontite is abundant, it may react with newly acquired CO_2 to form calcite and quartz (Helmold, 1980):

$$CaAl_2Si_4O_{12} \cdot 4H_2O + CO_2 + 6H^+ \qquad (6)$$
$$\text{Laumontite}$$

$$= CaCO_3 + 4SiO_2 + 7H_2O + 2Al^{+3}$$
$$\text{Calcite} \quad \text{Quartz}$$

This reaction is suggested by the calcite replacement of laumontite observed in a few thin sections. In addition, Ca^{+2} ions released during albitization and laumontitization (reaction 3) could remain in solution until an increase in a_{CO_2} initiates calcite precipitation. These, however, are plausible sources

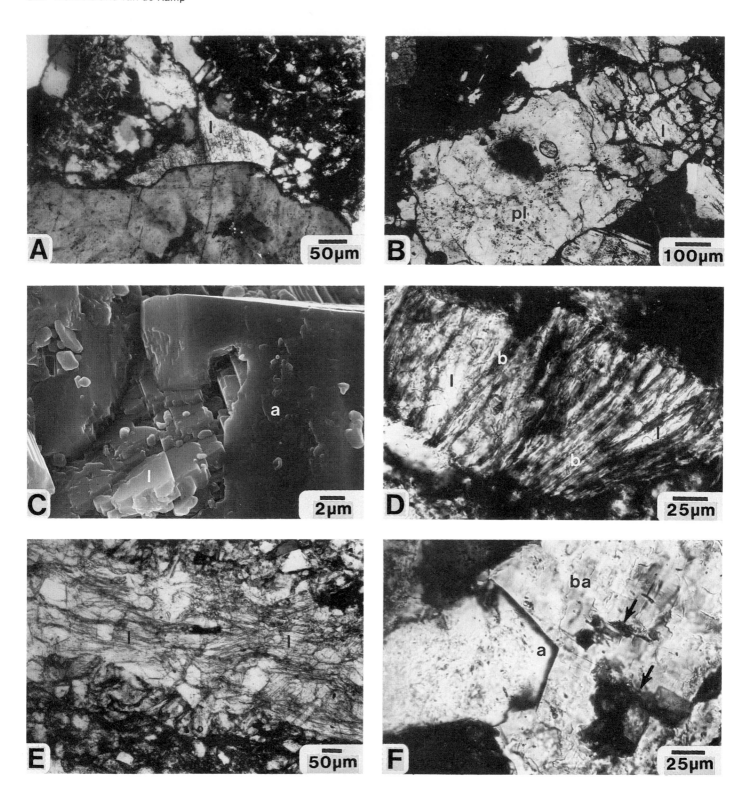

of Ca^{+2} only for the calcite in the Juncal Formation and locally in the basal Matilija Formation, since the highly indurated nature of these sandstones makes it doubtful that Ca^{+2} ions released could migrate any significant distance upsection. In the younger, less deeply buried sandstones dissolution of pyroboles and epidote releases Ca^{+2} ions into solution, which if not flushed from the system may later be incorporated in late-stage calcite upon an increase in a$_{CO_2}$. It is doubtful, however, that this could release enough Ca^{+2} ions to account for the amount of calcite in the sandstones, especially in those where calcite amounts to up to 40% of the whole rock. This raises the

Figure 12a—Laumontite cement (**l**) filling intergranular pore. Note characteristic undulatory extinction. Thin-section photomicrograph, crossed nicols. Camino Cielo Sandstone, Gibraltar Road.

Figure 12b—Detrital plagioclase grain (**pl**) altering to patches of laumontite (**l**). Microanalyses show that the plagioclase is nearly pure albite. Thin-section photomicrograph, crossed nicols. Camino Cielo Sandstone, Gibraltar Road.

Figure 12c—Albitized plagioclase grain with albite overgrowth (**a**) that is partially replaced by laumontite (**l**). SEM photomicrograph. Camino Cielo Sandstone, Standard Oil, Gerber No. 1.

Figure 12d—Authigenic laumontite (**l**) that has completely displaced cleavages of detrital biotite (**b**). Thin-section photomicrograph, crossed nicols. Camino Cielo Sandstone, Standard Oil, Gerber No. 1.

Figure 12e—Laumontite vein (**l**) in siltstone. Note several detrital grains within vein. Thin-section photomicrograph, plane light. Matilija Formation, Honolulu-Sunray Oil, Dunshee No. 1.

Figure 12f—Detrital grain of unknown composition (arrows) almost completely replaced by barite (**ba**), which abuts albite overgrowth (**a**). Euhedral nature of overgrowth suggests that it formed prior to barite replacement. Thin-section photomicrograph, crossed nicols. Juncal Formation, Gibraltar Road.

---

possibility of introduction of calcium into the system from interstitial pore water, with the ultimate source unknown.

With the possible exception of calcium required for the formation of late-stage calcite, it is believed that all components necessary for the crystallization of authigenic minerals in the sandstones were derived from the redistribution of ions within the sandstones or were transported over short distances from the interbedded shales.

PALEOTEMPERATURES AND BURIAL HISTORY

In order to estimate temperatures at which the major diagenetic reactions proceeded, paleotemperatures for the Paleogene sandstones were determined from levels of organic metamorphism (LOM) in the interbedded shales. Measurements of vitrinite reflectance (R_0) were utilized in this study as indicators of LOM. Details of the use of organic metamorphism in geological investigations are discussed by Dow (1977) and Bostick (1979). For all three sections studied, there is a well-defined increase in vitrinite reflectance with increasing burial depth (Fig. 15). The reflectance gradients for the Wheeler Gorge and Gibraltar Road sections are similar (0.33 and 0.34 log % R_0/km, respectively) and are lower than that for the Gerber No. 1 well (0.53 log % R_0/km). The differences are probably the result of variations in sediment thickness and, hence, sedimentation rates, along the east-west axis of the basin.

Knowledge of the burial history of a sedimentary basin is required to convert LOM into useful paleotemperature data. In particular, the effective burial time (t_{eff}), which is defined as the length of time during which the strata were within 15°C of the maximum burial temperature (Hood and Castaño, 1974), is an essential parameter to evaluate. Burial history curves (Fig. 16) were constructed based on the data of Ingle (1980), Dibblee (1950, 1966), Redwine (1952), Dickinson and Lowe (1966), Dickinson (1969a, 1969b), Lowe (1969), van de Kamp and Harper (1969), Curran et al (1971), Nagle and Parker (1971), Edwards (1972), and Givens (1974). These curves show that the effective burial time (dotted lines in Fig. 16) of the Paleogene sandstones is approximately 3 million years.

Paleotemperatures were calculated from the vitrinite reflectance data using the time–temperature–reflectance relationship of Hood and Castaño (1974), as modified by Bostick et al (1978) assuming a 3-m.y. effective burial time. Paleotemperatures for samples with high maturation levels ($R_0 > 0.5\%$) are considered reliable, whereas those for samples with lower maturation levels are questionable because of the uncertainty in the time–temperature–reflectance relationship (Bostick et al, 1978). As expected, there is a linear increase in paleotemperature with increasing burial depth over the depth interval studied (Fig. 17). Using these data, the maximum paleotemperatures at formation tops (Table 6), and at the top of the albitized and laumontite zones are estimated (Table 7).

CORRELATION OF MINERAL DIAGENESIS WITH ORGANIC METAMORPHISM

Correlations of albitization and laumontitization with LOM are considered together because the reactions are chemically complementary and because they both have proceeded at various temperatures along the basin axis. In the Wheeler Gorge section incipient albitization is first recognized at 0.5% R_0 corresponding to a paleotemperature of 110°C (4572-m burial depth), with complete albitization first occurring at 0.90% R_0 corresponding to a paleotemperature of 165°C (5425-m burial depth; Table 7). The first occurrence of laumontite is in the turbidite beds of the basal Matilija Formation (5669-m burial depth) at approximately 1.0% R_0 (173°C; Helmold, 1979). In the Gibraltar Road section, correlations between diagenetic mineral zones, LOM, and paleotemperatures are similar to those of the Wheeler Gorge section owing to the similarity in burial depths (Table 7). To the west, the estimated burial depth at which the top of the laumontite zone occurs gradually decreases to approximately 4572 m in the vicinity of Tecolote Tunnel area, as the basin begins to shallow. Further to the west, at Point Conception (Gerber No. 1 well), the first occurrence of laumontite is at an estimated burial depth of only 2515 m, corresponding to approximately 0.5% R_0 and a paleotemperature of 110°C. In this well, incipient albitization begins at 0.35% R_0 (77°C), with complete albitization occurring at roughly the same burial depth (2515 m) and LOM (0.5% R_0) as the first occurrence of laumontite.

CONTROLS ON LAUMONTITE FORMATION

The top of the laumontite zone occurs at greater burial depths and paleotemperatures in the eastern portion of the Santa Ynez Basin than in the west. It is, therefore, concluded that temperature and burial depth (pressure) alone do not play a critical role in the stratigraphic position of laumontite formation. Throughout the study area

TABLE 5

Formation	Estimated Maximum Burial Depth (m)	Number of Grains Analyzed	Average Composition (Mole %)			Standard Deviation			Lowest Ab Content (Mole %)	Highest Ab Content (Mole %)
			Ab	An	Or	Ab	An	Or		
Standard Oil, Gerber No. 1										
Alegria[1]	1975	30	89.8	9.2	1.0	± 9.2	± 9.1	±0.5	73.5	99.2
"Coldwater"	1812	30	87.2	11.5	1.3	±10.4	±10.1	±0.6	68.7	98.8
"Coldwater"	1847	30	84.2	14.9	0.9	±11.1	±10.9	±0.4	64.9	99.3
"Coldwater"	1935	30	86.5	12.4	1.1	±11.7	±11.4	±0.5	68.7	99.4
"Coldwater"	2014	30	85.6	13.3	1.1	±10.4	±10.2	±0.5	71.9	99.2
Cozy Dell	2152	30	96.3	3.0	0.7	± 4.5	± 4.2	±0.4	83.0	99.5
Matilija	2273	30	96.2	3.2	0.6	± 4.9	± 4.7	±0.3	81.6	99.2
Matilija	2370	30	91.7	7.5	0.8	± 7.1	± 6.7	±0.4	78.1	99.2
Camino Cielo Ss.	2515	30	97.8	1.6	0.6	± 2.0	± 1.8	±0.4	88.9	99.1
Camino Cielo Ss.	2960	30	97.3	2.1	0.6	± 2.4	± 2.2	±0.4	91.1	99.2
Camino Cielo Ss.	3050	30	98.6	0.9	0.5	± 1.1	± 1.0	±0.2	94.0	99.4
Camino Cielo Ss.	3251	30	97.8	1.7	0.5	± 2.3	± 2.2	±0.2	88.1	99.2
Gibraltar Road Section										
"Coldwater"	2957	30	91.7	7.3	1.0	± 7.4	± 7.1	±0.5	77.9	99.3
"Coldwater"	3399	27	93.7	5.4	0.9	± 7.3	± 7.1	±0.4	76.6	99.3
Cozy Dell	3901	30	92.0	6.8	1.2	± 6.7	± 6.4	±0.7	78.4	99.3
Matilija	4313	27	98.0	1.3	0.7	± 1.5	± 1.2	±0.3	94.2	99.1
Upper Juncal	5014	35	95.3	3.6	1.1	± 4.3	± 4.0	±0.6	84.4	99.3
Camino Cielo Ss.	5364	26	97.2	2.0	0.8	± 2.7	± 2.4	±0.6	88.6	98.8
Camino Cielo Ss.	5974	26	97.4	1.5	1.1	± 1.9	± 1.5	±0.9	92.9	99.1
Wheeler Gorge Section										
"Coldwater"	3444	34	89.5	9.7	0.8	± 8.8	± 8.6	±0.5	73.7	98.7
"Coldwater"	3566	33	90.2	8.5	1.3	± 8.0	± 8.0	±1.8	73.4	99.2
"Coldwater"	3749	30	89.1	9.3	1.6	± 9.4	± 9.1	±2.0	69.3	98.9
Cozy Dell	4161	30	92.2	5.9	1.9	± 8.1	± 7.3	±2.3	72.4	99.3
Cozy Dell	4572	33	95.2	3.8	1.0	± 6.5	± 6.3	±1.0	72.1	99.4
Matilija	5182	33	93.1	5.7	1.2	± 6.3	± 6.2	±1.3	77.5	98.7
Matilija	5304	34	96.6	2.6	0.8	± 4.6	± 4.5	±0.3	82.8	99.3
Matilija	5425	34	98.2	1.1	0.7	± 1.2	± 0.9	±0.6	95.4	99.7
Matilija	5547	30	97.4	1.9	0.7	± 2.1	± 2.1	±0.4	87.5	98.6
Matilija	5669	36	94.7	4.6	0.7	± 5.6	± 5.4	±0.5	81.2	99.1
Upper Juncal	5791	32	94.2	4.7	1.1	± 5.8	± 5.6	±0.8	79.6	99.2
Upper Juncal	6264	31	97.0	2.0	1.0	± 4.2	± 3.9	±1.2	81.6	99.5
Camino Cielo Ss.	6614	36	94.1	4.9	1.0	± 5.4	± 5.2	±1.1	81.8	98.9
Lower Juncal	6828	34	98.1	1.1	0.8	± 1.8	± 1.4	±0.6	92.2	99.6
Lower Juncal	7163	30	97.8	1.3	0.9	± 1.5	± 1.2	±0.7	94.3	99.6

[1]Marine equivalent of the Sespe Formation.

Table 5—Electron microprobe analyses of plagioclase in Paleogene sandstones, Santa Ynez Mountains.

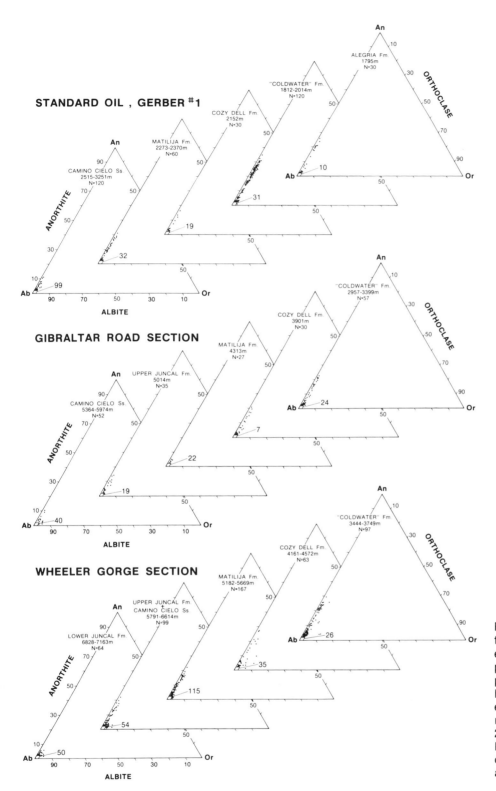

Figure 14—Modal plagioclase composition of Paleogene sandstones based on electron microprobe analyses. Each point represents a single analysis of a plagioclase grain selected at random. K-feldspar grains were systematically excluded from the analyses. The number of grains containing less than 2% An or Or is shown at the Ab apex. Depths are estimated maximum burial depths. N equals number of grains analyzed.

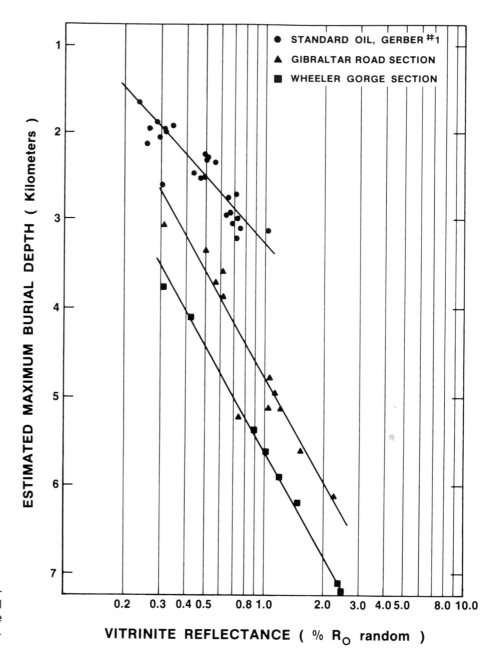

Figure 15—Variation in vitrinite reflectance with estimated maximum burial depth. Depth axis is linear; reflectance axis is logarithmic.

overburden calculations assuming an average rock density of 2.55 ($P_T = 250$ bars/km). All data points representing T–P conditions found in the Camino Cielo Sandstone and Cretaceous sandstones ($P_f = P_T$) lie to the left of the curve for $P_f/P_T = 1$, in the stability field of laumontite (open symbols in Fig. 19). For the overlying sandstones where fluid pressure is presumed to have been hydrostatic ($P_f = 0.5P_T$), points representing encountered temperatures and pressures lie to the right of the curve for $P_f/P_T = 0.50$, in the field of anorthite + 2 quartz + $4H_2O$ (closed symbols in

Fig. 19). The diagram indicates that for temperatures less than 75°C and pressures less than 1 kb, laumontite should be the stable phase. In this case it is reasonable to argue that owing to the low temperature and pressure, slow reaction kinetics enable anorthite + quartz to remain as metastable phases.

MODEL FOR LAUMONTITE FORMATION

Variations in pore-fluid chemistry and in fluid pressure between turbidite and shallow marine facies are the two

most plausible controls on the distribution of laumontite. One possible model to explain the variation in fluid pressure between petrologically similar sandstones entails the formation of laumontite in a "hard" overpressured zone in the eastern Santa Ynez Mountains, a zone where diagenetic reactions proceeded slowly because of the low rate of pore-fluid throughput (Yoder, 1955; Hay, 1966). "Dynamic" overpressuring may have occurred in the turbidite facies of the Juncal and lower Matilija Formations during the time of laumontite formation (Fig. 20). In this

MILLIONS OF YEARS BEFORE PRESENT

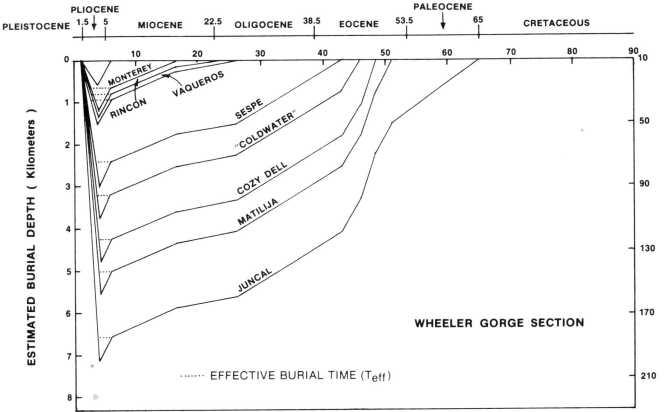

Figure 16—Burial history diagram for Tertiary strata of the Wheeler Gorge section.

situation, pore fluids enriched in Na^+ from the dewatering of smectite-rich shales are envisioned to have permeated into the turbidite sandstones at a faster rate than they were expelled, thereby creating excess fluid pressure in the sandstones. A certain amount of leakage at the updip ends of the sandstone bodies would have allowed for continued fluid throughput from the surrounding shales, but would have been insufficient to allow the reestablishment of hydrostatic pressures. Under these conditions a continuous supply of Na^+ would have been delivered to the sandstones to allow albitization of calcium-bearing plagioclase, which in turn supplied Ca^{+2} necessary for the formation of laumontite. Pressolved quartz grains indicate that pressure solution in these sandstones occurred before the onset of geopressuring. The absence of laumontite in the turbidite sandstones of the overlying Cozy Dell Formation suggests that hydrostatic conditions prevailed in those turbidites. This may be a result of higher permeabilities in these less-deeply buried sandstones.

The overlying more permeable sandstones of the Matilija and "Coldwater" Formations were deposited in a shallow marine environment and are more laterally continuous. Once pore fluids derived from shale dewatering enter these more porous and permeable sandstones, they were free to migrate updip and along strike and, consequently, remained under hydrostatic pressure (Fig. 20). Under these conditions the anorthite component in the detrital plagioclase remained the stable phase and did not alter to laumontite (Fig. 19).

The two assumptions involved in this model are that low permeability of the turbidite sandstones encased in a shale lithosome was capable of elevating P_f close to P_T, and that there is a dramatic shift to higher temperatures in the equilibrium curve for the reaction laumontite = anorthite + 2 quartz + $4H_2O$ as fluid pressure is raised from hydrostatic to lithostatic conditions. In addition, it should be emphasized that this model does not take into account variations in pore-fluid chemistry between the turbidite sandstones and overlying shallow marine strata. Such variations could also exert a control on the distribution of laumontite. The relative importance of geopressure versus pore-fluid chemistry in the formation of laumontite remains unknown.

APPLICATIONS TO HYDROCARBON EXPLORATION

Porosity and permeability are the two critical parameters that control the reservoir quality of a sandstone. The three processes that most effectively reduce porosity and permeability in the arkosic sandstones of the Santa Ynez Mountains (and presumably in equivalent Paleogene strata in the Santa Barbara Basin) are (1) compaction, (2) formation of laumontite concurrent with albitization, and (3) formation of late-stage calcite. While the precipita-

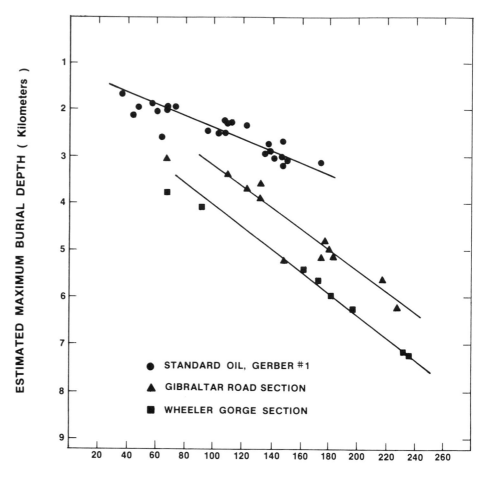

Figure 17—Variation in paleotemperature with estimated maximum burial depth. Paleotemperatures are estimated from vitrinite reflectance data.

TABLE 6

Formation	Estimated Maximum Burial Depth (m)			Vitrinite Reflectance (% R_0)			Maximum Paleotemperature (°C)		
	Gerber No. 1	Gibraltar Road	Wheeler Gorge	Gerber No. 1	Gibraltar Road	Wheeler Gorge	Gerber No. 1	Gibraltar Road	Wheeler Gorge
"Coldwater"	1798	2774	3048	0.26	—	—	48	—	—
Cozy Dell	2088	3749	3810	0.33	0.54	0.38	72	118	85
Matilija	2256	4298	4907	0.39	0.74	0.64	88	148	135
Juncal	2454	4846	5669	0.47	1.00	1.00	103	173	173
Camino Cielo Sandstone	2515	5334	6553	0.50	1.30	1.65	110	190	205
Jalama	—	6370	7193	—	2.40	2.36	—	236	233

Table 6—Estimated maximum burial depths, vitrinite reflectance, and maximum paleotemperatures for formation tops.

TABLE 7

Diagenetic Zone	Estimated Maximum Burial Depth (m)			Vitrinite Reflectance (% R_0)			Maximum Paleotemperature (°C)		
	Gerber No. 1	Gibraltar Road	Wheeler Gorge	Gerber No. 1	Gibraltar Road	Wheeler Gorge	Gerber No. 1	Gibraltar Road	Wheeler Gorge
Top of incipient albitized zone (>75% Ab)	2152	—	4572	0.35	—	0.50	77	—	110
Top of albitized zone (>98% Ab)	2515	4313	5425	0.50	0.75	0.90	110	150	165
Top of laumontite zone	2515	5669	5669	0.50	1.30	1.00	110	190	173

Table 7—Estimated maximum burial depths, vitrinite reflectance, and maximum paleotemperatures for top of albitized and laumontite zones.

tion of pore-lining and pore-filling clays, and quartz and feldspar cementation do have an adverse effect on reservoir quality, hydrocarbons are currently being produced from upper Paleogene and Neogene strata where these are the principal diagenetic products (Curran et al, 1971).

In the past few years with the refinement of the concept of secondary porosity, the presence of calcite cement is no longer perceived as the termination of a sandstone's potential as a hydrocarbon reservoir. The onset of organic maturation, and the resulting liberation of substantial quantities of CO_2 and organic acids into pore fluids, create a mechanism capable of producing secondary porosity in carbonate cemented rocks ahead of migrating hydrocarbons (Schmidt and McDonald, 1979; Surdam et al, 1982). Keeping this concept in mind, the presence of extensive calcite cement in outcrop samples of the Paleogene sandstones does not necessarily imply that equiva-

lent strata in the subsurface will be equally cemented. The extra duration of burial to which subsurface equivalents have been subjected may have allowed for the creation of secondary dissolution porosity in carbonate cements, especially if cementation preceded the main episode of hydrocarbon maturation and migration.

It is widely accepted among petroleum geologists that the presence of laumontite cement in a sandstone signals the attainment of economic basement (McCulloh et al, 1978; Galloway, 1979). This is based on the pervasive nature of laumontite cement and its insolubility in carbonic acid (H_2CO_3). Recent work based on theoretical considerations, however, has shown that under certain conditions laumontite may be unstable in the diagenetic environment (Frost et al, 1982). In the presence of pore fluids with high CO_2 activity, laumontite may break down to calcite, clay, and quartz:

$$CaAl_2Si_4O_{12} \cdot 4H_2O + CO_2 = CaCO_3 \quad (7)$$
$$\text{Laumontite} \qquad\qquad \text{Calcite}$$

$$+ \; Al_2Si_2O_5(OH)_4 + 2SiO_2 + 2H_2O$$
$$\text{Kaolinite} \qquad \text{Quartz}$$

Alternatively, with the addition of hydrogen ions (that is, a decrease in pH), laumontite may break down to clay and quartz:

$$CaAl_2Si_4O_{12} \cdot 4H_2O + 2H^+ = \quad (8)$$
$$\text{Laumontite}$$

$$Al_2Si_2O_5(OH)_4 + 2SiO_2 + Ca^{+2} + 3H_2O$$
$$\text{Kaolinite} \qquad \text{Quartz}$$

Therefore, either an increase in a_{CO_2} or a decrease in pH can result in the destruction of laumontite to yield more stable products. Frost et al (1982) have calculated that the decomposition of laumontite by reaction 7 results in a volume decrease of 10%, whereas laumontite destruction via reaction 8 leads to a volume decrease of 28%. A smaller volume decrease is expected from reaction 7 because an additional crystalline product (calcite) is precipitated as a cement. In either case, the decreased volume of the products may result in the creation of secondary porosity. If laumontite decomposition proceeds according to reaction 7, a later decrease in pH of the pore waters will result in the dissolution of the neoformed calcite, thereby creating addi-

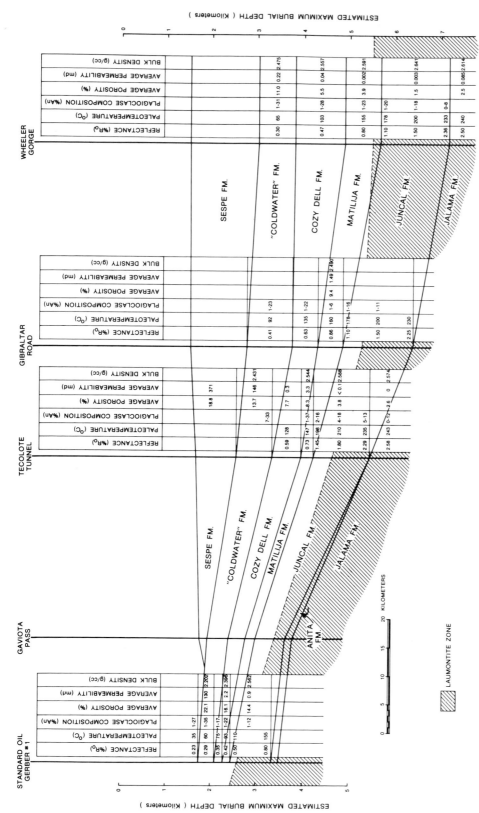

Figure 18—Summary of vitrinite reflectance, paleotemperature, plagioclase composition, and petrophysical properties of Paleogene and Cretaceous sandstones.

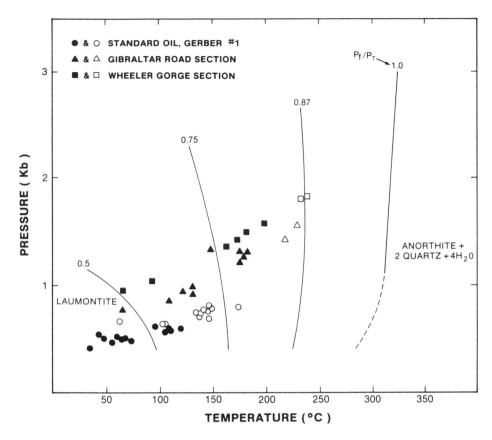

Figure 19—Equilibrium curves for the reaction laumontite = anorthite + 2 quartz + 4H$_2$O at various P$_f$/P$_T$ ratios. Experimentally derived curve for P$_f$/P$_T$ = 1 is from Thompson (1970), curves for P$_f$/P$_T$ = 0.50, 0.75, and 0.87 are schematic after the hypothetical reaction of Boles and Coombs (1977). Closed symbols = P-T conditions for "Coldwater," Cozy Dell, and Matilija sandstones. Open symbols = P-T conditions for Camino Cielo and Cretaceous sandstones. See discussion in text.

tional secondary porosity.

In spite of the considerations of Frost et al (1982), it must be pointed out that no visible secondary porosity was observed in the laumontite-cemented sandstones of the Santa Ynez Mountains and that only minor replacement of laumontite by calcite was noted. This may be a result of the late origin of laumontite, in that the main episode of cementation did not occur until the sandstones were beyond the major decarboxylation stage of organic maturation. In this diagenetic environment decreasing CO$_2$ activity associated with the waning stages of decarboxylation would not favor laumontite dissolution. At least in this particular basin it would be prudent to consider the top of the laumontite zone as economic basement.

In the search for new reservoirs in the Santa Barbara Basin, particular attention should be paid to the upper Paleogene sandstones (that is, Vaqueros, Sespe, "Coldwater," and Cozy Dell Formations) and their lateral equivalents. Not only do these sandstones retain high primary (depositional) po-

rosities, but they have a high potential for secondary porosity. As noted above, the upper Paleogene sandstones from the Gerber No. 1 well have a substantial amount of secondary porosity created through the subsurface leaching of plagioclase and K-feldspars. It is expected that equivalent sandstones, particularly those to the south and west of Point Conception, in the offshore regions, should have sufficient porosity (both primary and secondary) to be effective hydrocarbon reservoirs.

CONCLUSIONS

The similarity of detrital modes of the Paleogene sandstones of the Santa Ynez Mountains indicates they share a common provenance consisting primarily of plutonic rocks of acidic to intermediate composition, with minor volcanic, metamorphic and sedimentary components. Depth-dependent variations in diagenetic mineralogy are, therefore, not a result of a varying provenance through time.

A paragenetic sequence has been documented for the diagenetic miner-

alogy, with the major events consisting of early precipitation of pore-lining and pore-filling clays, formation of overgrowths on quartz and feldspar, dissolution of plagioclase and K-feldspar, concurrent albitization of detrital plagioclase and formation of laumontite, and precipitation of late-stage calcite. Authigenic clays in the sandstones include mixed-layer chlorite/smectite, chlorite, mixed-layer illite/smectite, and kaolinite, usually with either one of the mixed-layer clays or chlorite being the dominant variety. Authigenic feldspar is chemically pure albite and K-feldspar, and is distinct from detrital feldspars, which have a varied chemistry. Laumontite is a pure calcium–aluminum hydrous silicate that has an invariant composition regardless of its texture.

Albitization is an important diagenetic process that is limited to the deeper intervals of the Paleogene section. Incipient albitization occurs in the sandstones overlying the laumontite zone, but complete albitization is present only in those sandstones that are highly zeolitized. This relationship

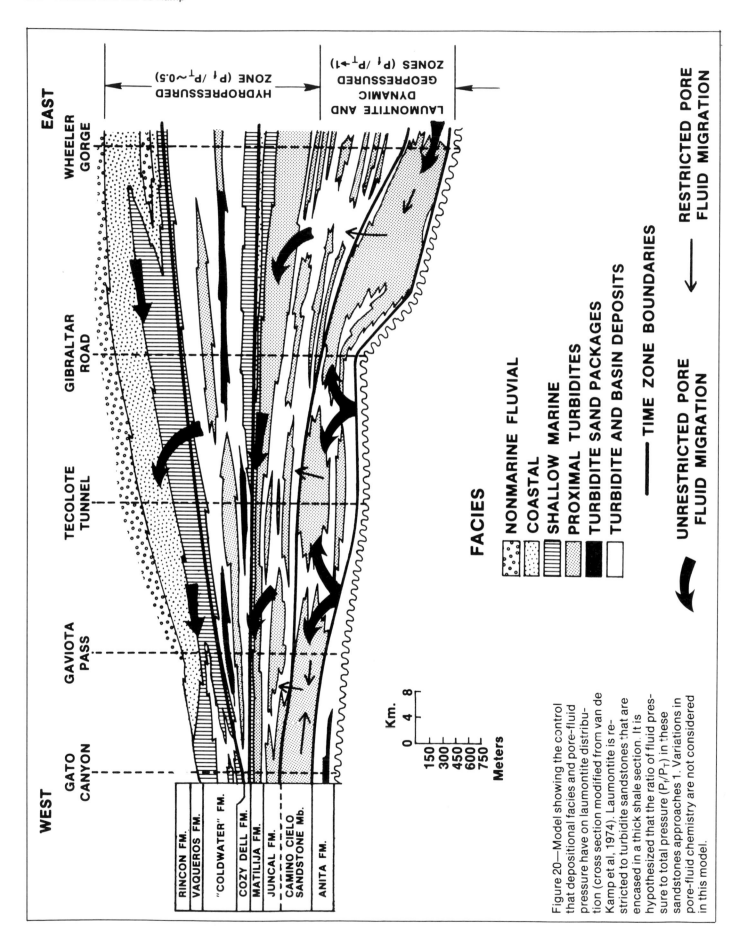

Figure 20—Model showing the control that depositional facies and pore-fluid pressure have on laumontite distribution (cross section modified from van de Kamp et al, 1974). Laumontite is restricted to turbidite sandstones that are encased in a thick shale section. It is hypothesized that the ratio of fluid pressure to total pressure (P_f/P_T) in these sandstones approaches 1. Variations in pore-fluid chemistry are not considered in this model.

along with the chemically complementary nature of the reactions suggests that albitization and zeolitization are coeval.

Level of organic metamorphism, as expressed by vitrinite reflectance, can be used to correlate mineral diagenesis with thermal history of the basin. The top of the albitized zone occurs at 0.9% R_0 (165°C) and 0.75% R_0 (150°C), respectively, in the eastern and central parts of the basin, and 0.5% R_0 (110°C) in the western end. The first appearance of laumontite exhibits a similar pattern occurring in the eastern and central parts of the basin at of 1.0% R_0 (173°C) and 1.3% R_0 (190°C), respectively, while first occurring at 0.5% R_0 (110°C) in the western end of the basin. The top of the laumontite zone occurs at greater burial depths and paleotemperatures in the eastern portion of the basin, and it is, therefore, concluded that temperature and burial depth alone do not play a critical role in the stratigraphic position of its first occurrence.

Laumontite distribution appears to be controlled by pore-fluid chemistry and post-compaction permeability variations, which are responsible for creating differences in fluid pressure between petrologically similar sandstones. "Dynamic" overpressuring may have occurred in the turbidite sandstones whereby pore waters enriched in ions from shale dewatering were continuously supplied for albitization and zeolitization. Overlying sandstones under hydrostatic pressure were not zeolitized and show indications of only incipient albitization.

Compaction, formation of laumontite concurrent with albitization, and precipitation of late-stage calcite are the three principal processes adversely affecting reservoir quality of the sandstones. In light of the relatively new concept of secondary porosity, the extensive calcite cement in many outcrop samples does not necessarily imply that subsurface equivalents will have equally poor reservoir quality. Although recent work suggests that laumontite may break down in the diagenetic environment to yield secondary porosity, this reaction is not considered significant in the laumontite-cemented sandstones of the Santa Ynez Mountains.

ACKNOWLEDGMENTS

We would like to thank W. R. Dickinson for numerous discussions on the regional geology and sandstone petrology of the Santa Ynez Mountains, and for providing continued encouragement during the course of the study. N. H. Bostick gave freely of his time, ideas, and laboratory for performing the vitrinite reflectance analyses. Petrographic thin sections were expertly prepared by R. V. Laniz. X-ray diffraction analyses were performed by W. I. Fugate and B. Storjohann, and E. V. Eslinger provided considerable assistance in the interpretation of XRD patterns of clay minerals. R. A. White provided invaluable technical assistance in the preparation and examination of samples with the scanning electron microscope. All figures were drafted by H. Stephens and the manuscript was typed by J. S. Kline. R. M. Slatt, R. G. Loucks, J. R. Boles, E. V. Eslinger, and M. H. Link reviewed the manuscript and offered many useful suggestions.

The initial part of this work was done by van de Kamp while employed by Shell Development Company. Shell's release of the results of that work for publication is gratefully acknowledged. Portions of this work were supported by the Earth Sciences section of the National Science Foundation with NSF Grant EAR 76-22636. Additional funding was supplied through a Grant-in-Aid of Research from Sigma Xi, the Scientific Research Society of North America. The Geology Department of Stanford University and Cities Service Oil & Gas Corp. also provided financial support for this project.

SELECTED REFERENCES

Albee, A. L., and E.-A. Zen, 1969, Chemical potentials of carbon dioxide and water in the zeolite facies: Ocherki Fizicha-Khimicha Petrologii, v. 1, p. 249–260.

Bailey, T. L., 1952, Geology of southwestern Santa Barbara County, California (rev.): Bulletin of the American Association of Petroleum Geologists, v. 36, p. 176–178.

Boles, J. R., 1982, Active albitization of plagioclase, Gulf Coast Tertiary: American Journal of Science, v. 282, p. 165–180.

Boles, J. R., and D. S. Coombs, 1977, Zeolite facies alteration of sandstones in the Southland syncline, New Zealand: American Journal of Science, v. 277, p. 982–1012.

Boles, J. R., and S. G. Franks, 1979, Clay diagenesis in Wilcox sandstones of southwest Texas: implications of smectite diagenesis on sandstone cementation: Journal of Sedimentary Petrology, v. 49, p. 55–70.

Bostick, N. H., 1979, Microscopic measurement of the level of catagenesis of solid organic matter in sedimentary rocks to aid exploration for petroleum and to determine former burial temperatures—a review, in P. A. Scholle and P. R. Schluger, eds., Aspects of diagenesis: Society of Economic Paleontologists and Mineralogists Special Publication 26, p. 17–43.

Bostick, N. H., and B. Alpern, 1977, Principles of sampling, preparation and constituent selection for microphotometry in measurement of maturation of sedimentary organic matter: Journal of Microscopy, v. 109, p. 41–47.

Bostick, N. H., S. M. Cashman, T. H. McCulloh, and C. T. Waddell, 1978, Gradients of vitrinite reflectance and present temperature in the Los Angeles and Ventura basins, California, in D. F. Oltz, ed., Symposium in geochemistry: low temperature metamorphism of kerogen and clay minerals: Pacific Section Society of Economic Paleontologists and Mineralogists, p. 65–96.

Brown, C. E., and T. P. Thayer, 1963, Low-grade mineral facies in Upper Triassic and Lower Jurassic rocks of the Aldrich Mountains, Oregon: Journal of Sedimentary Petrology, v. 33, p. 411–425.

Castaño, J. R., and D. M. Sparks, 1974, Interpretation of vitrinite reflectance measurements in sedimentary rocks and determination of burial history using vitrinite reflectance and authigenic minerals, in R. R. Dutcher, P. A. Hacquebard, J. M. Schopf, and J. A. Simon, eds., Carbonaceous materials as indicators of metamorphism: Geological Society of America Special Paper 153, p. 31–51.

Coombs, D. S., 1952, Cell size, optical properties and chemical composition of laumontite and leonhardite: American Mineralogist, v. 37, p. 812–830.

———, 1954, The nature and alteration of some Triassic sediments from Southland, New Zealand: Royal Society of New Zealand Transactions, v. 82, p. 65–109.

———, 1961, Some recent work on the lower grades of metamorphism: Australian Journal of Science, v. 24, p. 203–215.

———, 1971, Present status of the zeolite facies: Advances in Chemistry Series 101, American Chemical Society, p. 317–327.

Curran, J. F., K. B. Hall, and R. F. Herron, 1971, Geology, oil fields, and future petroleum potential of Santa Barbara, Channel area, California, in I. H. Cram, ed., Future petroleum provinces of the United States—their geology and potential: American Association of Petroleum Geologists Memoir 15, p. 192–211.

Deer, W. A., R. A. Howie, and J. Zussman, 1962, Rock-forming minerals, v. 1-5: New York, Wiley and Longman, 1788 p.

Dibblee, T. W., 1950, Geology of southwestern Santa Barbara County, California: California Department of Natural Resources, Division of Mines and Geology Bulletin 150, 95 p.

————, 1966, Geology of the central Santa Ynez Mountains, Santa Barbara County, California: California Department of Natural Resources, Division of Mines and Geology Bulletin 196, 99 p.

Dickinson, W. R., 1962, Petrology and diagenesis of Jurassic andesitic strata in central Oregon: American Journal of Science, v. 260, p. 481-500.

————, 1969a, Geologic problems in the mountains between Ventura and Cuyama, in J. C. Taylor and W. R. Dickinson, eds., Upper Sespe Creek field trip guidebook: Pacific Section, Society of Economic Paleontologists and Mineralogists, p. 1-23.

————, 1969b, Miocene stratigraphic sequence on Upper Sespe Creek and Pine Mountain, in J. C. Taylor and W. R. Dickinson, eds., Upper Sespe Creek field trip guidebook: Pacific Section, Society of Economic Paleontologists and Mineralogists, p. 49-55.

Dickinson, W. R., and D. R. Lowe, 1966, Stratigraphic relations of phosphate- and gypsum-bearing Upper Miocene strata, Upper Sespe Creek, Ventura County, California: Bulletin of the American Association of Petroleum Geologists, v. 50, p. 2464-2470.

Dott, R. H., 1964, Wacke, graywacke and matrix—what approach to immature sandstone classification?: Journal of Sedimentary Petrology, v. 34, p. 625-632.

Dow, W. G., 1977, Kerogen studies and geological interpretations: Journal of Geochemical Exploration, v. 7, p. 79-99.

Edwards, L. N., 1972, Notes on the Vaqueros and Rincon Formations, west-central Santa Ynez Mountains, in D. W. Weaver, ed., Central Santa Ynez Mountains, Santa Barbara County, California guidebook: Pacific Section, Society of Economic Paleontologists and Mineralogists, p. 46-54.

Evans, Jr., H. T., D. E. Appleman, and D. S. Handwerker, 1963, The least squares refinement of crystal unit cells with powder diffraction data by an automatic computer indexing method (abs.): American Crystallographic Association Annual Meeting, Cambridge, MA, Program and Abstracts, no. E-10, p. 42-43.

Folk, R. L., 1974, Petrology of sedimentary rocks: Austin, Hemphill Publishing Co., 182 p.

Frost, B., R. C. Surdam, and L. J. Crossey, 1982, Secondary porosity in laumontite-bearing sandstones (abs.): Bulletin of the American Association of Petroleum Geologists, v. 66, p. 569-570.

Galloway, W. E., 1974, Deposition and diagenetic alteration of sandstones in northeast Pacific arc-related basins: implications for graywacke genesis: Geological Society of America Bulletin, v. 85, p. 379-390.

————, 1979, Diagenetic controls of reservoir quality in arc-derived sandstones: Implications for petroleum exploration, in P. A. Scholle and P. R. Schluger, eds., Aspects of diagenesis: Society of Economic Paleontologists and Mineralogists Special Publication 26, p. 17-43.

Ghent, E. D., 1979, Problems in zeolite facies geothermometry, geobarometry and fluid compositions, in P. A. Scholle and P. R. Schluger, eds., Aspects of diagenesis: Society of Economic Paleontologists and Mineralogists Special Publication 26, p. 81-87.

Gibbs, R. J., 1968, Clay mineral mounting techniques for X-ray diffraction analysis: a discussion: Journal of Sedimentary Petrology, v. 38, p. 242-244.

Givens, C. R., 1974, Eocene molluscan biostratigraphy of the Pine Mountain area, Ventura County, California: University of California Publications in Geological Sciences, v. 109, 107 p.

Glover, J. E., 1963, Studies in the diagenesis of some western Australian sedimentary rocks: Journal of the Royal Society of Western Australia, v. 46, p. 33-56.

Hawkins, P. J., 1978, Relationship between diagenesis, porosity reduction, and oil emplacement in late Carboniferous sandstone reservoirs, Bothamsall oil field, E. Midlands: Journal of the Geological Society of London, v. 135, p. 7-24.

Hay, R. L., 1966, Zeolites and zeolitic reactions in sedimentary rocks: Geological Society of America Special Paper 85, 130 p.

Heald, M. T., 1950, Authigenesis in West Virginia sandstones: Journal of Geology, v. 58, p. 624-633.

Heald, M. T., and G. F. Baker, 1977, Diagenesis of the Mt. Simon and Rose Run sandstones in western West Virginia and southern Ohio: Journal of Sedimentary Petrology, v. 47, p. 66-77.

Heald, M. T., and R. E. Larese, 1973, The significance of the solution of feldspar in porosity development: Journal of Sedimentary Petrology, v. 43, p. 458-460.

Helmold, K. P., 1979, Diagenesis of Tertiary arkoses, Santa Ynez Mountains, Santa Barbara and Ventura Counties, California (abs): Bulletin of the American Association of Petroleum Geologists, v. 63, p. 465.

————, 1980, Diagenesis of Tertiary arkoses, Santa Ynez Mountains, California: Unpublished Ph.D. dissertation, Stanford University, 225 p.

Helmold, K. P., and E. H. Snider, 1981, Geothermal gradients, organic metamorphism, and mineral diagenesis in Tertiary rocks, Santa Ynez Mountains, California (abs.): Bulletin of the American Association of Petroleum Geologists, v. 65, p. 937.

Hoffman, J., 1976, Regional metamorphism and K-Ar dating of clay minerals in Cretaceous sediments of the disturbed belt of Montana: Unpublished Ph.D. dissertation, Case Western Reserve University, 266 p.

Hood, A., and J. R. Castaño, 1974, Organic metamorphism: its relationship to petroleum generation and application to studies of authigenic minerals: United Nations Economic Commission for Asia and the Far East, Committee for Coordination of Joint Prospects, Mineral Resources Asian Offshore Areas, Technical Bulletin, v. 8, p. 85-118.

Hower, J., 1981, X-ray diffraction identification of mixed-layer clay minerals, in F. J. Longstaffe, ed., Clays and the resource geologist: Mineralogical Association of Canada, Short Course Handbook, v. 7, p. 39-59.

Hower, J., E. V. Eslinger, M. E. Hower, and E. A. Perry, 1976, Mechanism of burial metamorphism of argillaceous sediment, 1, mineralogical and chemical evidence: Geological Society of America Bulletin, v. 87, p. 725-737.

Iijima, A., and M. Utada, 1971, Present-day zeolite diagenesis of the Neogene geosynclinal deposits in the Niigata Oil Field, Japan: Advances in Chemistry Series 101, American Chemical Society, p. 342-349.

Ingle, Jr., J. C., 1980, Cenozoic paleobathymetry and depositional history of selected sequences within the southern California continental borderland: Cushman Foundation for Foraminiferal Research Special Publication 19, p. 163-195.

Jackson, M. L., 1969, Soil chemical analysis—advanced course: Published by the author, Department of Soil Science, Madison, University of Wisconsin, 895 p.

Kastner, M., 1971, Authigenic feldspars in carbonate rocks: American Mineralogist, v. 56, p. 1403-1442.

Kastner, M., and R. Siever, 1979, Low temperature feldspars in sedimentary rocks: American Journal of Science, v. 279, p. 435-479.

Kerr, P. F., and H. G. Schenck, 1928, Significance of the Matilija overturn: Geological Society of America Bulletin, v. 39, p. 1087-1102.

Land, L. S., and K. L. Milliken, 1981, Feldspar diagenesis in the Frio Formation, Brazoria County, Texas Gulf Coast: Geology, v. 9, p. 314-318.

Laniz, R. V., R. E. Stevens, and M. B.

Norman, 1964, Staining of plagioclase feldspar and other minerals: United States Geological Survey Professional Paper 501-B, p. B152–B153.

Lapham, D. M., 1963, Leonhardite and laumontite in diabase from Dillsburg, Pennsylvania: American Mineralogist, v. 48, p. 683–689.

Link, M. H., 1975, Matilija sandstone: a transition from deep-water turbidite to shallow-marine deposition in the Eocene of California: Journal of Sedimentary Petrology, v. 45, p. 63–78.

Link, M. H., and J. E. Welton, 1982, Sedimentology and reservoir potential of Matilija Sandstone: an Eocene sand-rich deep-sea and shallow-marine complex, California: Bulletin of the American Association of Petroleum Geologists, v. 66, p. 1514–1534.

Liou, J. G., C. Y. Lan, J. Suppe, and W. G. Ernst, 1977, The east Taiwan ophiolite: its occurrence, petrology, metamorphism and tectonic setting: Mining Research and Service Organization Report 2, Industrial Technology Research Institute, Taipei, Taiwan, 195 p.

Lowe, D. R., 1969, Santa Margarita Formation (Upper Miocene), Upper Sespe Creek area, Ventura County, California, in J. C. Taylor and W. R. Dickinson, eds., Upper Sespe Creek Fieldtrip Guidebook: Pacific Section, Society of Economic Paleontologists and Mineralogists, p. 56–62.

Madsen, B. M., and K. J. Murata, 1970, Occurrence of laumontite in Tertiary sandstones of the central Coast Ranges, California: United States Geological Survey Professional Paper 700-D, p. D188–D195.

Maxwell, J. C., 1964, Influence of depth, temperature, and geologic age on porosity of quartzose sandstone: Bulletin of the American Association of Petroleum Geologists, v. 48, p. 697–709.

McCulloh, T. H., 1981, Mid-Tertiary laumontite isograd offset 37 km by left-lateral strike-slip on Santa Ynez Fault, California (abs.): Bulletin of the American Association of Petroleum Geologists, v. 65, p. 956.

McCulloh, T. H., S. M. Cashman, and R. J. Stewart, 1978, Diagenetic baselines for interpretive reconstructions of maximum burial depths and paleotemperatures in clastic sedimentary rocks, in D. F. Oltz, ed., Symposium in geochemistry: low temperature metamorphism of kerogen and clay minerals: Pacific Section, Society of Economic Paleontologists and Mineralogists, p. 18–46.

McCulloh, T. H., V. A. Frizzell, Jr., R. J. Stewart, and I. Barnes, 1981, Precipitation of laumontite with quartz, thenardite, and gypsum at Sespe Hot Springs, western Transverse Ranges, California:

Clays and Clay Minerals, v. 29, p. 353–364.

Merino, E., 1975, Diagenesis in Tertiary sandstones from Kettleman North Dome, California, I, Diagenetic mineralogy: Journal of Sedimentary Petrology, v. 45, p. 320–336.

Nagle, H. E., and E. S. Parker, 1971, Future oil and gas potential of onshore Ventura basin, California, in I. H. Cram, ed., Future petroleum provinces of the United States—their geology and potential: American Association of Petroleum Geologists Memoir 15, p. 254–297.

Nilsen, T. H., and S. H. Clarke, Jr., 1975, Sedimentation and tectonics in the early Tertiary continental borderland of central California: United States Geological Survey Professional Paper 925, 64 p.

O'Brien, J. M., 1973, Narizian–Rufugian (Eocene–Oligocene) sedimentation, western Santa Ynez Mountains, Santa Barbara County, California: Unpublished Ph.D. dissertation, Santa Barbara, University of California, 304 p.

Page, B. M., J. G. Marks, and G. W. Walker, 1951, Stratigraphy and structure of mountains north-east of Santa Barbara, California: Bulletin of the American Association of Petroleum Geologists, v. 35, p. 1727–1780.

Perry, E., and J. Hower, 1970, Burial diagenesis in Gulf Coast pelitic sediments: Clays and Clay Minerals, v. 18, p. 165–177.

Pettijohn, F. J., P. E. Potter, and R. Siever, 1972, Sand and sandstone: New York, Springer-Verlag, 618 p.

Pittman, E. D., 1972, Diagenesis of quartz in sandstones as revealed by scanning electron microscopy: Journal of Sedimentary Petrology, v. 42, p. 507–519.

Redwine, L. E., chairman, 1952, Cenozoic correlation section paralleling north and south margins western Ventura basin from Point Conception to Ventura and Channel Islands, California: American Association of Petroleum Geologists, Pacific Section, Geological Names and Correlation Committee.

Reynolds, Jr., R. C., 1980, Interstratified clay minerals, in G. W. Brindley and G. Brown, eds., Crystal structures of clay minerals and their X-ray identification: Mineralogical Society of London, p. 249–303.

Schmidt, V., and D. A. McDonald, 1979, The role of secondary porosity in the course of sandstone diagenesis, in P. A. Scholle and P. R. Schluger, eds., Aspects of diagenesis: Society of Economic Paleontologists and Mineralogists Special Publication 26, p. 175–207.

Siever, R., 1959, Petrology and geochemistry of silica cementation in some Pennsylvanian sandstones, in H. A. Ireland, ed., Silica in sediments: Society of Eco-

nomic Paleontologists and Mineralogists Special Publication 7, p. 55–79.

Stanley, K. O., and L. V. Benson, 1979, Early diagenesis of high plains Tertiary vitric and arkosic sandstones, Wyoming and Nebraska, in P. A. Scholle and P. R. Schluger, eds., Aspects of diagenesis: Society of Economic Paleontologists and Mineralogists Special Publication 26, p. 401–423.

Stauffer, P. H., 1965, Sedimentation of lower Tertiary marine deposits, Santa Ynez Mountains, California: Unpublished Ph.D. dissertation, Stanford University, 213 p.

————, 1967, Sedimentologic evidence on Eocene correlations, Santa Ynez Mountains, California: Bulletin of the American Association of Petroleum Geologists, v. 51, p. 607–611.

Surdam, R. C., S. Boese, and L. J. Crossey, 1982, Role of organic and inorganic reactions in development of secondary porosity in sandstones (abs): Bulletin of the American Association of Petroleum Geologists, v. 66, p. 635.

Thompson, A. B., 1970, Laumontite equilibria and the zeolite facies: American Journal of Science, v. 269, p. 267–275.

Tompkins, R. E., 1981, Scanning electron microscopy of a regular chlorite/smectite (corrensite) from a hydrocarbon reservoir sandstone: Clays and Clay Minerals, v. 29, p. 233–235.

Towe, K. M., 1962, Diagenesis of clay minerals as a possible source of silica cement in sedimentary rocks: Journal of Sedimentary Petrology, v. 32, p. 26–28.

van de Kamp, P. C., and J. D. Harper, 1969, The Cretaceous and lower Tertiary of the Wheeler Springs-Ojai area, Ventura basin, California, in J. C. Taylor and W. R. Dickinson, eds., Upper Sespe Creek fieldtrip guidebook: Pacific Section, Society of Economic Paleontologists and Mineralogists, p. 24–26.

van de Kamp, P. C., J. D. Harper, J. J. Conniff, and D. A. Morris, 1974, Facies relations in the Eocene–Oligocene in the Santa Ynez Mountains, California: Journal of the Geological Society of London, v. 130, p. 545–565.

van de Kamp, P. C., B. E. Leake, and A. Senior, 1976, The petrography and geochemistry of some Californian arkoses with application to identifying gneisses of metasedimentary origin: Journal of Geology, v. 84, p. 195–212.

Waldschmidt, W. A., 1941, Cementing materials in sandstones and their probable influence on migration and accumulation of oil and gas: Bulletin of the American Association of Petroleum Geologists, v. 25, p. 1839–1879.

Walker, T. R., 1976, Diagenetic origin of continental red beds, in H. Falke, ed., The continental Permian in west, central, and south Europe: Proceedings of the NATO Advanced Study Institute, Mainz, Germany; Dordrecht, Holland, and Boston, D. Reidel Publishing Co., p. 240–282.

Walker, T. R., B. Waugh, and A. J. Crone, 1978, Diagenesis in first-cycle desert alluvium of Cenozoic age, southwestern United States and northwestern Mexico: Geological Society of America Bulletin, v. 89, p. 19–32.

Waugh, B., 1970, Formation of quartz overgrowths in the Pemrith sandstone (lower Permian) of northwest England as revealed by scanning electron microscopy: Sedimentology, v. 14, p. 309–320.

Wilson, M. D., and E. D. Pittman, 1977, Authigenic clays in sandstones: recognition and influence on reservoir properties and paleoenvironmental analysis: Journal of Sedimentary Petrology, v. 47, p. 3–31.

Yoder, H. S., 1955, Role of water in metamorphism, *in* A. Poldervaart, ed., Crust of the earth: Geological Society of America Special Paper 62, p. 505–524.

Zen, E.-A., 1961, The zeolite facies: an interpretation: American Journal of Science, v. 259, p. 401–409.

Formation of Secondary Porosity: How Important Is It?

Knut Bjørlykke
University of Bergen
Bergen, Norway

INTRODUCTION

Evidence of secondary porosity has been reported by many authors, including McBride (1977), Parker (1974), Loucks et al (1977, 1980), Lindquist (1977), Stanton (1977), and Schmidt and McDonald (1979a, 1979b). The common occurrence of corroded and partly dissolved grains and/or cement suggests that post-depositional leaching is an important process that may change the reservoir characteristics of many sandstones. However, we still do not know how important diagenetic leaching processes are in terms of creating new pore space. This paper discusses different processes that can cause mineral dissolution and attempts to quantify the amounts of secondary porosity that can be expected to form under different conditions.

Pore waters in sedimentary basins have two principally different origins: (1) meteoric water flowing into the basin, driven by the head of an elevated ground-water table, and (2) pore water buried with sediments (connate water) and forced upwards by the compaction of the sediments. The flow of compactional water is limited by the amount of water contained in the underlying sedimentary sequence in the basin. The total average upwards flux is on the order of 10^4–10^5 cu cm/sq cm of pore water for a bed of sandstone from the time of deposition to burial to about 3 km (Fig. 1). This calculation is based on an average porosity depth curve and changes in this curve make little difference to the order of magnitude of this pore-water flux. There is, however, no specific limit to the magnitude of meteoric pore-water flux through sediments, although time and flow rates put some constraints on what may seem

ABSTRACT. Secondary porosity in sandstones is created by subsurface dissolution of grains or cement by pore water that is undersaturated with respect to one or more of the major mineral phases. Such undersaturated pore water may be derived from: (1) meteoric water driven by a hydrostatic head; (2) compactional pore water containing CO_2 released from maturing kerogen; (3) clay minerals reactions including the transformation of kaolinite and smectite to illite; and (4) reactions between clay minerals and carbonate releasing CO_2. Calculations of the CO_2 generated from different types of kerogen suggest that few basins will generate enough CO_2 to produce large-scale leaching in thicker sandstones. Dissolution of minerals and removal of aluminum and silica in solution requires that very large volumes of pore water flow through the sandstone. Because leaching often enlarges primary pore space, it is very difficult to estimate the percentage of the pore space that is secondary. Leaching and formation of secondary pore space may also be accompanied by reprecipitation of other minerals so that the net gain in porosity is less than the observed secondary pore space.

possible (Blatt, 1979).

Secondary porosity by dissolution requires a through-flow of pore water that is undersaturated with respect to one or more of the major mineral phases present. The most obvious source of undersaturated pore fluid is meteoric water which, at least initially, is undersaturated with respect to carbonate and feldspar. In sandstone, the most common form of diagenetic secondary porosity is dissolution of carbonate cement and/or grains and dissolution of feldspar (Schmidt and McDonald, 1979a, 1979b). What we observe as secondary porosity may, however, not represent net increase in the porosity, but may, to a large extent, represent a redistribution of the primary porosity. Dissolution of early argonite or high magnesium calcite may result in precipitation of almost

equal volumes of calcite cement. Dissolution of feldspar may lead to the precipitation of kaolinite and quartz:

$$2KAlSi_3O_8 + 2H_2CO_3 + 9H_2O \rightarrow Al_2Si_2O_5(OH)_4 + 4H_4SiO_4 + 2HCO_3^- + 2K^+.$$

If only potassium is removed in solution and the pore water is already saturated with respect to quartz and kaolinite, the volume of the reaction product producing new cement will exceed the amount of secondary porosity gained as a result of dissolution. One volume of feldspar will result in approximately 60% kaolinite and 40% quartz. Similar reactions apply to sodium feldspars.

The total meteoric pore-water flux through sandstones depends on the hydraulic head, sandstone permeability, the degree of connections between

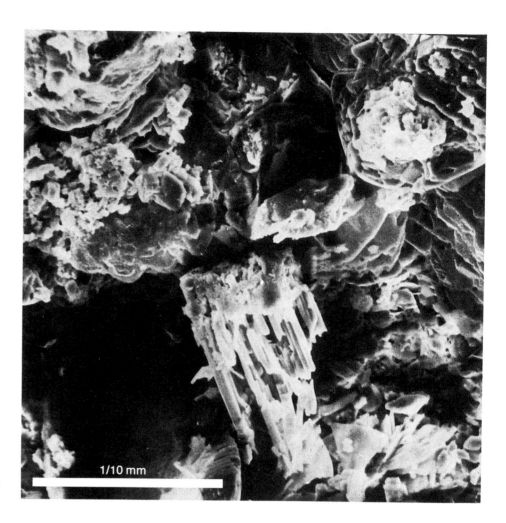

Figure 3—Leached feldspar with subsequent precipitation of authigenic feldspar. From the Etive Formation, Brent Group, Statfjord field.

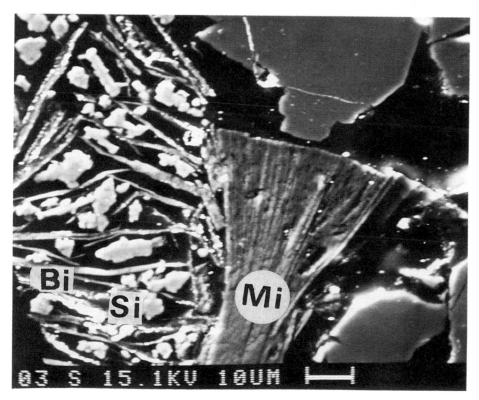

Figure 4—Electron back-scatter picture showing expression of mica (**Mi**) as a result of alteration to kaolinite. Wedged between sheets of biotite are authigenic siderite crystals. Rannock Formation, Brent Group, Statfjord field.

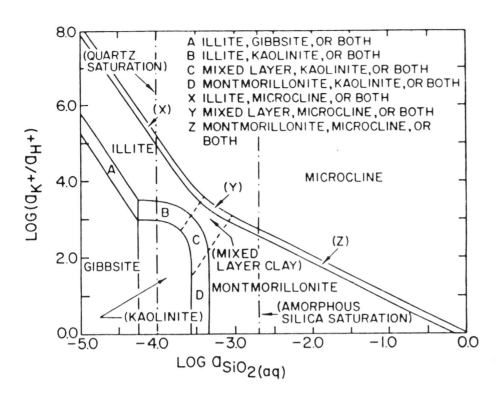

Figure 5—Stability diagram for minerals in aqueous solution at 25°C. From Aagaard (1979). Used with permission of the author.

before reaching the sandstones. It is difficult to see how leaching of carbonate and feldspar by CO_2 released from kerogen can be a general mechanism responsible for the formation of secondary porosity in the wide range of basins. In the Jurassic Brent sandstones of the North Sea, leaching of feldspar and carbonate has previously been interpreted as being the result of late-stage leaching by CO_2 from maturing kerogen (Schmidt and McDonald, 1979a, 1979b). Later studies have, however, produced evidence that the leaching of carbonate and feldspar may have been early and that kaolinite formation seems to predate or at least to be contemporaneous with the main phase of quartz overgrowth (Bjørlykke et al, 1979). Late-stage burial is dominated by feldspar overgrowth and later diagenetic Fe carbonates (Bjørlykke et al, 1979). The Brent sandstones contain layers of unaltered Jurassic volcanic glass (Malm et al, 1979) which suggests that there was little pervasive leaching. Also, the source rocks in the Viking Graben are relatively sapropelic (Bjorøy, personal communication) and will, therefore, release comparatively small amounts of CO_2.

In the Gulf Coast, the evidence for the formation of secondary porosity has been well documented (Loucks et

al, 1977). The sequence of diagenetic events proposed by these authors, however, creates theoretical problems. In their "zone of reservoir development" at 2500–4000 m, they find that porosity is created by leaching of feldspar and volcanic rock fragments and that kaolinite and Fe carbonates are precipitated in the deep subsurface (below 4000 m). This sequence seems to assume that feldspar is totally dissolved and removed without precipitating reaction products, such as kaolinite or quartz. This means that aluminum and silica from the feldspar must be removed in solution and that the kaolinite formed at greater depth has another source of aluminum and silica. Given the low solubilities of quartz and kaolinite, it is very difficult to see how the ions from dissolved feldspar can be removed in solution by the amounts of available water. Less than 100 ppm SiO_2, and even smaller amounts of Al_2O_3, are found in solution in natural pore water below 150°C (White et al, 1963). In explaining cementation of sandstones by pore-water flow from compacting shales, we are faced with the same problem of having insufficient volumes of pore water. The formation of kaolinite from feldspar requires a continuous removal of the released potassium or sodium so that the pore-

water composition remains in the stability field of kaolinite.

Dissolution of plagioclases often results in the precipitation of albite (albitization: Boles, 1982), which releases excess Ca^{+2} and Al^{+2} in the pore water. Alternatively, if aluminum is conserved, silica and sodium may be added from the pore water. At quartz saturation, kaolinite is stable at rather high Na^+/H^+ ratios before the sodium concentration needed to form albite is reached (Kastner and Siever, 1979). The Na^+/K^+ ratio in natural pore water ranges from 50 to 80 (Merino, 1975). Leaching of potassium feldspar and precipitation of kaolinite will increase the K^+ concentration in the pore water so that its composition is no longer in the kaolinite field (Fig. 5) unless potassium is effectively removed (that is, by meteoric water) or there is an effective sink for potassium. Because plagioclase feldspar is more soluble than potash feldspar, it is more likely to be leached during burial below the zone of meteoric water. This selective dissolution results in precipitation of kaolinite and increased porosity. The resulting aluminum and silica ions must be removed in solution. The solubility of quartz at 100°C is only 68 ppm and silica concentrations in natural oil field brines are rarely more than 100 ppm (Merino,

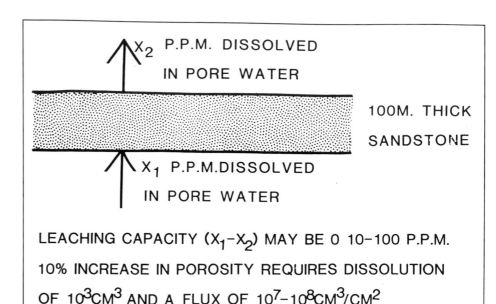

Figure 6—Estimated pore-water flux required to form 10% secondary porosity in a 100 m thick sandstone by leaching. For each 1 cu cm of horizontal cross section there is 10^4 cu cm of sandstone. In the case of quartz where the solubility is 10–100 ppm, 10^4–10^5 cu cm of pore water undersaturated with respect to quartz is required to dissolve 1 cu cm. A total flux of 10^7–10^8 is, therefore, required to produce an increase in porosity of 10%. In the case of carbonate cement with solubility 100–1000 ppm, the requirement is correspondingly lower (10^6–10^7 cu cm/sq cm).

1975; Galloway et al, 1982). Each volume of pore water cannot, therefore, be expected to remove more than 10–100 ppm in solution if it is at all undersaturated with respect to quartz. The solubility of aluminum is even lower than that of silica at pH 5–7 (Garrels and Christ, 1965) and at least 10^4–10^5 volumes of pore water are required to form one volume of pore space by leaching and removal of the reaction products. It has been shown experimentally that Al solubilities can be increased very significantly by organic complexing (Surdam, 1982). So far, however, naturally occurring pore water with very high aluminum concentration has not been reported. Even in very acid ground water (pH 4) influenced by acid rain in Scandinavia, the aluminum concentration is less than 5 mg/liter (Eriksson, 1981; Hultberg and Johansson, 1981). At the moment, therefore, we are unable to account for the observed secondary porosity in the Gulf Coast sandstones in terms of the volumes of water required (Milliken et al, 1981). A 10% increase in porosity in a 100 m thick sandstone requires a pore-water flux at 10^7–10^8 cu cm/sq cm (Fig. 6), or a column of 10–100 km of water, which is very difficult to obtain in a sedimentary basin. Here we are faced with the same problem as in the case of explaining quartz cementation by water from compacting shales, namely that the pore-water flux

requirements are very large (Bjørlykke, 1979).

In the case of carbonate, this figure may be reduced by one order of magnitude (10^6–10^7 cu cm/sq cm). At 3-km depth, the total pore-water flux attributable to compaction in a sedimentary basin is approximately 5.10^4 cu cm/sq cm (Fig. 7). For simplicity, these calculations assume an average vertical pore-water flow. If large volumes of basinal shale are drained through a smaller cross section of sandstones, considerably larger pore-water fluxes and leaching may occur so that the net gain in porosity in the sandstone is smaller than the volume dissolved. It is often difficult to quantify the volume of minerals dissolved, and it may be difficult to estimate the original distribution of leached cement. It is, in most cases, unlikely that a sandstone would have lost most of its primary porosity prior to leaching. In their diagenesis, Schmidt and McDonald (1979) have assumed that nearly all primary porosity is lost by compaction and cementation and that secondary porosity is formed later by leaching. It can be argued, however, that the maximum leaching by pore water will occur in sandstones having the highest primary porosity. In well-cemented tight sandstones, leaching is expected to be controlled by fractures.

Convection currents are driven by the temperature gradient (2–4°C/100

m) in sedimentary basins (Cassan et al, 1981; Wood and Hewett, 1982). Calculations by Wood and Hewett (1982) suggest that pore-flow velocities up to about 1 m/yr occur in porous sandstones. However, convection currents are not likely to flow across low-permeability barriers like thin shales or carbonate-cemented beds. During intervals when pressure gradients are higher than hydrostatic, the pore-water flow must be unidirectional since convection currents cannot flow back against the pressure gradient. Convection currents are, therefore, probably only important in rather permeable sandstones and possibly on sets of intersecting fractures. Many types of sandstone facies, particularly turbidites and fluvial sandstones, have thin interlayered shale beds, which will tend to limit the scale of convection currents.

A model using convection currents as a mechanism for transfer of solids in solution in the subsurface is attractive because it allows pore water to be recycled, thereby overcoming the main problem we are facing in understanding large-scale dissolution of minerals as implied by formation of net secondary porosity. As pointed out by Cassan et al (1981), however, the convection model predicts that carbonate will dissolve near the top of convection cells because its solubility is inversely related to temperature and that silicate minerals will dissolve in the lower part of the

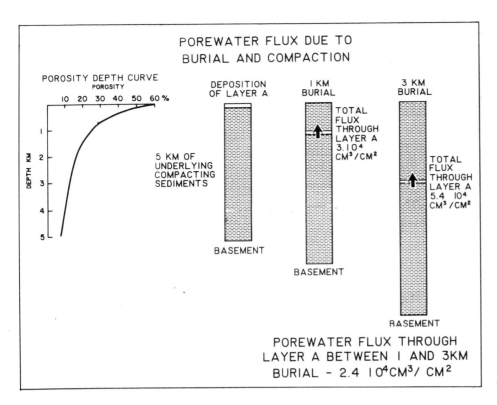

Figure 7—Calculated upwards pore-water flux attributable to burial and compaction.

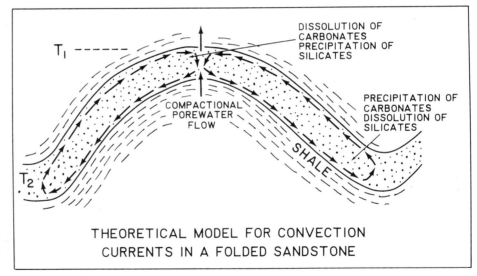

Figure 8—Model for precipitation and leaching, assuming convection occurs in a folded sandstone.

cell. In an anticline structure, we would expect secondary porosity as a result of leaching of carbonate, while in the syncline we would expect leaching of silicates (Fig. 8).

ALTERNATIVE REACTIONS CAUSING DISSOLUTION OF FELDSPAR AND CARBONATE IN THE SUBSURFACE

Since the release of CO_2 from kerogen may be insufficient to explain large-scale leaching of carbonate and feldspar in the subsurface, we should look for alternative mechanisms to explain such leaching processes (Table 1).

As the temperature increases with burial depth, clay minerals that formed during weathering and early diagenesis become unstable. Smectite becomes unstable at 60–100°C and may form illite, and also chlorite and quartz (Hower et al, 1976). As pointed out by Hower et al (1976), there are few other sources of K^+ than potash feldspar in the fine fraction of the shales. Authigenic clay minerals in the sandstones will also be subjected to the same instability, and kaolinite becomes increasingly unstable between 120–150°C (Hower et al, 1976; Hoffman and Hower, 1979). Kaolinite may dehydrate to pyrophyllite $(Al_2Si_2O_5(OH)_2$, but only above 300°C (Velde, 1977). If potassium is available, it will transform to illite at lower temperatures (130–150°C). The following set of reac-

TABLE 1

Sand volume:	1.15×10^{13} cu m
Mudstone volume:	4.2×10^{13} cu m
Thermally mature sandstone:	3.7×10^{13} cu m
Total organic carbon:	0.3% (humic kerogen)
Total kerogen:	1.1×10^{11} cu m $= 2.210^{14}$ kg
Total gas (methane):	2.5×10^{11} cu m
Sand porosity (18%):	2.1×10^{12} cu m

Approach 1

Assume that the CO_2 generated is 50% of the methane volume: 1.25×10^{11} cu m CO_2 gas. This equals 5.4 $\times 10^{12}$ mole (Hunt, 1979). If all the CO_2 reacts with carbonate, 5.4×10^{14} g (1.9×10^{14} cu cm) pore space or less than 0.01% of the sand porosity (2.1×10^{12} cu m).

Approach 2

Assume a CO_2 yield of 50 liters CO_2/kg (50 cu cm/g) of gasprone kerogen (humic kerogen) during heating to 150°C (Tissot and Welte, 1978), 2.210^{14} kg kerogen will produce 1.1×10^{16} liters CO_2 or 4.5×10^{14} moles liters. This amount of CO_2 may dissolve 4.8×10^{16} g $CaCO_3$ or 1.8×10^{10} cu m. This is less than 1% of the observed porosity (2.1×10^{12} cu m).

During maturation, the shales may dewater by 5–10% and expel 1.8–3.7 10^{12} cu m H_2O. This volume of water is sufficient to dissolve the total amount of CO_2 produced from kerogen (2.6×10^3 cu m CO_2) as each cubic meter of water will dissolve 30 cu m of CO_2 at 300 atm 100°C (Hunt, 1979).

Table 1—Calculated CO_2 production and leaching capacity of source rocks associated with the Frio Formation, compared with the porosity in the sandstones. The calculations are based on data from Galloway et al, 1982.

tions indicated by Hower et al (1976) may then occur:

$$3Al_2Si_2O_5(OH)_4 + 2K^+ \quad (1)$$
Kaolinite
$$= 2KAl_3Si_3O_{10}(OH)_2 + 2H^+ + H_2O \ .$$
Illite

This reaction releases protons, lowers the pH, and feldspar may dissolve to provide the necessary potassium.

$$3KAlSi_3O_8 + 2H^+ \longrightarrow \quad (2)$$
K-feldspar
$$KAl_3Si_3O_{10}(OH)_2 + 6SiO_2 + 3H_2O \ .$$
Illite

Combining equations 1 and 2, we obtain:

$$3Al_2Si_2O_5(OH)_4 + 3KAlSi_3O_8$$
Kaolinite   Feldspar

$$= 3KAl_3Si_3O_{10} + 6SiO_2 + 3H_2O$$
Illite

In this calculation we have, for simplicity, used the stoichiometric formula of moscovite for illite (Bjørlykke, 1980). A similar equation can be presented in the case of sodium feldspar. Since illite has a somewhat lower potassium content, the amount of feldspar dissolution relative to kaolinite content may be lower than indicated above. In sandstones, diagenetic kaolinite formed by early diagenetic reaction between meteoric water and feldspar may thus be responsible for further leaching of feldspar and formation of illite. This reaction may explain the leaching of feldspar in cases where burial depth is sufficient for kaolinite to become unstable.

Other possible mechanisms for leaching the subsurface include release of CO_2 from carbonate rocks in areas of high geothermal gradients by metamorphism of limestones at 500–600°C. As pointed out by Hutcheon et al (1980), CO_2 may be released at a much lower temperature by reactions

between carbonate and clay minerals such as: kaolinite + 5 dolomite + quartz + $H_2O \longrightarrow$ Mg-chlorite + 5 calcite + $5CO_2$
15 dolomite + 2 muscovite + 3 quartz + $11H_2O \longrightarrow 3$Mg-chlorite + $2K^+ + 20H^-$ + $15CO_2$. These reactions may be important at temperatures between 180–225°C and may be responsible for leaching at greater burial depth.

CONCLUSIONS

The literature on sandstone diagenesis now contains many good descriptions of secondary porosity documented by evidence of leaching of cements or grains. However, our understanding of the processes that may produce large-scale secondary porosity and are, therefore, of great importance to the reservoir properties of sandstone, is incomplete. This paper discusses the requirements for the formation of secondary porosity. The main conclusions are:

(1) Secondary porosity may form during early diagenesis by meteoric

water flushing, particularly in deltaic and fluvial sandstones.

(2) Late diagenetic secondary porosity is not important in all types of reservoir sandstone. Some of the examples of secondary porosity claimed to be formed by leaching CO_2 from maturing kerogen can be interpreted to be formed by meteoric-water leaching.

(3) The amount of CO_2 produced by maturing kerogen depends on the composition and quantity of kerogen and the shale composition.

(4) In many basins, the CO_2 released by kerogen may be insufficient to explain the observed leaching of carbonate and feldspars, and we should look into other mechanisms involving clay mineral reactions that may account for leaching during deeper burial (2–4 km).

(5) It is very difficult to quantify the volumes of secondary porosity created by leaching. Since some of the material dissolved may be reprecipitated elsewhere in the sandstone, the observed leached porosity cannot always be taken as net gains in the porosity of the sandstone.

(6) Convection currents provide an attractive model for supplying the necessary volume of water for transport of solids in the subsurface. It remains to be seen, however, whether such currents are of great significance for long-distance mass transfer.

(7) Because leaching requires an initial permeability in the sandstone, sandstones with higher primary porosity have the highest potential for leaching. Porosity gained by leaching may, therefore, often represent a redistribution and possibly an enhancement of the primary porosity.

(8) Our knowledge about the formation of secondary porosity is still very limited. Until we have a satisfactory model for the formation of secondary porosity, this concept cannot be used efficiently in exploration and production geology.

ACKNOWLEDGMENTS

Financial support from NTNF Research Council and Statoil is gratefully acknowledged. The author is indebted to Dr. W. E. Galloway, Bureau of Economic Geology, Texas, for discussion and for providing data from the Frio Formation. Drs. J. Boles and D. Pierce have kindly read and suggested improvements on the manuscript.

SELECTED REFERENCES

Aagaard, P., 1979, Thermodynamic and kinetic analyses of silicate hydrolyses of low temperatures and pressures: Ph.D. dissertation, Berkeley, University of California, 126 p.

Aagaard, P., and H. C. Helgeson, 1982, Thermodynamic and kinetic constraints on reaction rates among minerals and aqueous solutions, I, Theoretical considerations: American Journal of Science, v. 282, p. 237–285.

Bjørlykke, K., 1979, Cementation of sandstones: Journal of Sedimentary Petrology, v. 49, p. 1358–1359.

————, 1980, Clastic diagenesis and basin evolution: Revista del Instituto de Investigaciones Geologas, Diputacon Provincial, Universitad de Barcelona, v. 34, p. 21–44.

Bjørlykke, K., O. Malm, and A. Elverhøi, 1979, Diagenesis in the Mesozoic sandstones from Spitsbergen and the North Sea: Geologisches Rundschau, v. 68, p. 1151–1171.

Blanche, J. B., and J. H. M. Whitaker, 1978, Diagenesis of part of the Brent Sand Formation (Middle Jurassic) of the northern North Sea: Quarterly Journal of the Geological Society of London, v. 135, p. 73–82.

Blatt, H., 1979, Diagenetic processes in sandstones: Society of Economic Paleontologists and Mineralogists Special Publication 26, p. 141–157.

Boles, J. R., 1982, Active albitization of plagioclase, Gulf Coast Tertiary: American Journal of Science, v. 282, p. 165–180.

Cassan, J. P., M. C. Garcia Palacios, B. Fritz, and Y. Tardy, 1981, Diagenesis of sandstone reservoirs as shown by petrographical and geochemical analyses of oil bearing formations in Paton basin: Bulletin des Centres de Recherches Exploration-Production Elf-Aquitaine, v. 5, p. 113–135.

Chauvin, A. L., and L. Z. Valachi, 1980, Sedimentology of the Brent and Statfjord Formation of the Statfjord field: Norwegian Petroleum Society Publication, Geilo Meeting, 17 p.

Davies, S. N., and R. J. M. DeWiest, 1965, Hydro geology: New York, John Wiley and Sons, 463 p.

Englund, J. O., and K. F. Meyer, 1980, Groundwater discharge through springs with well-defined outlets: a case study from Jeløya-Moss, S. Norway: Nordic Hydrology, v. 11, p. 145–158.

Englund, J. O., and J. A. Myhrstad, 1980, Groundwater chemistry of some selected areas in southeastern Norway: Nordic Hydrology, v. 11, p. 33–54.

Eriksson, E., 1981, Aluminum in ground-water—possible solution equilibria: Nordic Hydrology, v. 12, p. 43–50.

Fung, P. C., and G. G. Sanipelli, 1980, Repeated leaching of microcline at 25°C, in A. Campbell, ed., Water–rock interaction, 3rd International Symposium on Water–rock Interaction: International Association of Geochemistry and Cosmochemistry, Edmonton, Alberta Research Council, p. 67–71.

Galloway, W. E., D. V. Hobday, and K. Magara, 1982, Frio Formation of the Texas Gulf Coastal Basin—depositional system, structural framework and hydrocarbon origin, migration, distribution and exploration potential: Bureau of Economic Geology, University of Texas (in press).

Garrels, R. M., and C. L. Christ, 1965, Solutions, minerals and equilibria: New York, Harper and Row, 450 p.

Garrels, R. M., and P. Howard, 1957, Reaction of feldspar and mica with water at low temperature and pressure: Clay and Clay Minerals, v. 6, p. 68–88.

Gieskes, J. M., 1981, Deep-sea drilling interstitial water studies: implications for chemical alteration of the oceanic crust, Layer 1 and 4: Society of Economic Paleontologists and Mineralogists Special Publication 32, p. 149–167.

Hancock, J. N., 1978, Diagenetic modelling in the middle Jurassic of the Brent Sand of the northern North Sea: European Offshore Petroleum Conference and Exhibition, EUR 92, p. 275–279.

Hoffman, J., and J. Hower, 1979, Clay mineral assembly as low grade metamorphic Grotlermometers: application to the thrust faulted disturbed belt of Montana, USA: Society of Economic Paleontologists and Mineralogists Special Publication 26, p. 35–79.

Hower, J., E. V. Eslinger, M. E. Hower, and C. A. Perry, 1976, Mechanism of burial-metamorphism of argillaceous sediments, 1, Mineralogical and chemical evidence: Geological Society of America Bulletin, v. 87, p. 725–737.

Hultberg, H., and S. Johansson, 1981, Acid groundwater: Nordic Hydrology, v. 12, p. 51–64.

Hunt, J. M., 1979, Petroleum geochemistry and geology: San Francisco, W. H. Freeman & Co., 716 p.

Hutcheon, I., A. Oldershaw, and E. D. Ghent, 1980, Diagenesis of Cretaceous sandstones of the Kootenay Formation at Elk Valley (Southeastern British Columbia): Geochimica et Cosmochimica Acta, v. 44, p. 1425–1436.

Kastner, M., and R. Siever, 1979, Low temperature feldspars in sedimentary rocks: American Journal of Science, v. 279, p. 435–479.

Land, D. R., and D. Prezbindowski, 1981,

The origin and evolution of saline formation waters: Lower Cretaceous carbonates, south central Texas, USA: Journal of Hydrology, v. 54, p. 51–74.

Lindquist, S. J., 1977, Secondary porosity development and subsequent reduction over pressured Frio Formation Sandstone (Oligocene), south Texas. Gulf Coast Association of Geological Societies Transactions, v. 27, p. 99–107.

Loucks, R. G., D. G. Bebout, and W. E. Galloway, 1977, Relationship of porosity formation and preservation to sandstone consolidation history—Gulf Coast Tertiary, Frio Formation: Bureau of Economic Geology, Austin, University of Texas, Geological Circular 77, p. 109–120.

Loucks, R. G., M. M. Douge, and W. E. Galloway, 1980, Importance of secondary leached porosity in Lower Tertiary sandstone reservoirs along the Gulf Coast: Bureau of Economic Geology, Austin, University of Texas, Geological Circular 80-2.

Malm, O. A., H. Furnes, and K. Bjørlykke, 1979, Volcanoclastics of Middle Jurassic age in the Statfjord Oil Field in the North Sea: N.J. Geol. Paleont. H., v. 10, p. 607–618.

Manheim, F. T., 1967, Evidence for submarine discharge of water on the Atlantic continental slope of southern United States, and suggestions for further research: Transactions of the New York Academy of Sciences, v. 11, no. 29, p. 839–853.

Manheim, F. T., and C. K. Paull, 1981, Patterns of groundwater salinity changes in a deep continental oceanic transect off the southeastern Atlantic Coast of the U.S.A.: Journal of Hydrology, v. 54, p. 95–105.

McBride, E. F., 1977, Secondary porosity—importance in sandstone reservoirs in Texas: Gulf Coast Association of Geological Societies Transactions, v. 27, p. 121–122.

Merino, E., 1975, Diagenesis in Tertiary sandstones from Kettleman North Dome, California, II, Interstitial solutions: distribution of the diagenetic mineralogy: Geochimica et Cosmochimica Acta, v. 39, p. 1629–1645.

Milliken, K. L., L. S. Land, and R. G. Loucks, 1981, History of burial diagenesis determined from isotope geochemistry, Frio Formation, Brazoria County, Texas: Bulletin of the American Association of Petroleum Geologists, v. 65, p. 1397–1413.

Parker, C. A., 1974, Geopressures and secondary porosity in the deep Jurassic of Mississippi: Gulf Coast Association of Geological Societies Transactions, v. 29, p. 69–80.

Schmidt, V., and D. A. McDonald, 1979a, Texture and recognition of secondary porosity in sandstones: Society of Economic Paleontologists and Mineralogists Special Publication 26, p. 175–207.

————, 1979b, The role of secondary porosity in the course of sandstone diagenesis: Society of Economic Paleontologists and Mineralogists Special Publication 26, p. 175–207.

Skarpnes, O., E. D. Ormåsen, K. H. Jacobsen, and G. P. Hamar, 1980, Tectonic development of the northern North Sea, north of the central highs: Norwegian Petroleum Society Publication, Geilo Meeting, 11 p.

Stanton, G. C., 1977, Secondary porosity in sandstones of the lower Wilcox (Eocene) Karnes County, Texas: Gulf Coast Association of Geological Societies Transactions, v. 27, p. 197–207.

Surdam, R. C., 1982, Role of organic and inorganic reaction in development of secondary porosity: American Association of Petroleum Geologists Abstract, v. 66, p. 635.

Tissot, B. P., and D. H. Welte, 1978, Petroleum formation and occurrence: Berlin, Springer-Verlag, 538 p.

Velde, B., 1977, Clays and clay minerals in natural and synthetic systems, in Developments in sedimentology, 2: New York, Elsevier, 218 p.

White, D. E., J. D. Main, and G. A. Waring, 1963, Chemical composition of subsurface waters, in Data of geochemistry: United States Geological Survey Professional Paper 44-F, 67 p.

Wood, J. R., and T. A. Hewett, 1982, Fluid convection and mass transfer in porous sandstones—a theoretical model: Geochimica et Cosmochimica Acta, v. 46, p. 1707–1713.

3

Applications of Clastic Diagenesis in Exploration and Production

Diagenesis of the Frontier Formation, Moxa Arch: A Function of Sandstone Geometry, Texture and Composition, and Fluid Flux

Sharon A. Stonecipher
Robert D. Winn, Jr.
Michele G. Bishop
Marathon Oil Company
Littleton, Colorado

INTRODUCTION

Many of the factors that control diagenesis in siliciclastic rocks (Fig. 1) appear to be related directly or indirectly to processes operating in the original depositional environment. Each depositional environment produces a lithofacies with a limited range of petrophysical characteristics such as grain size and sorting (Visher, 1969; Reed et al, 1975; Anderson et al, 1982), sedimentary textures and structures (Friedman, 1961; Moiola and Weiser, 1968), and size and shape of the sand body (Peterson and Osmond, 1961). These characteristics act as important controls on fluid flux through a sandstone and thereby also indirectly control pore-fluid composition (Potter, 1967; Pettijohn et al, 1972). Detrital mineralogy has also been shown to be process dependent because each detrital component has a characteristic size distribution (Blatt et al, 1972; Odom, 1975; Doe et al, 1976). As a result, hydraulic sorting will generally produce different detrital assemblages in sediment deposited in environments of different energies (see Davies and Ethridge, 1975). Detrital composition also affects pore-fluid compositions. It is likely, therefore, that because of these differences in physical and compositional characteristics, diagenesis should proceed along different paths in rocks of different facies (Taylor, 1978; Blanche and Whitaker, 1978; Stonecipher, 1982). This is illustrated by the

ABSTRACT. Many of the major factors that control diagenesis, such as detrital composition, fluid composition, and fluid flux, can be related directly or indirectly to physical and biological processes operating at the time of deposition. Each depositional environment produces a lithofacies with a specific limited range of physical and compositional characteristics that affect diagenesis. The concept of diagenesis as a function of facies is well illustrated by the second Frontier sandstone of the Moxa Arch, Wyoming.

The lower Frontier Formation on the Moxa Arch comprises sandstones and mudstones deposited in a delta/strand plain system on the western edge of the interior Cretaceous seaway. Depositional environments represented by the rocks include: marine shelf with sand ridges, marine shoreline, fluvial channels, and fluvial flood plain.

Marine sandstones are significantly more quartzose than fluvial units because of sorting within the delta and wave abrasion on beaches. Substantial input of silica-rich water expelled from the underlying Aspen Shale caused nearly complete cementation of the cleaner beach and backshore sandstones by quartz overgrowths. Fluvial sandstones contain less quartz and more chert grains and rock fragments than the marine sandstones, and as a result, were less affected by quartz cementation. In addition, temporary filling of pores by calcite prevented further irreversibly destructive diagenesis. As a result, fluvial sandstones are better reservoirs, even though they are compositionally less mature. Clay-rich sandstones of the lower shoreface, lower sections of the sand ridges, and muddy fine-grained fluvial sandstones have poor present-day porosity and permeability primarily because of compaction.

Fluid flux also appears to have played an important role in determining present-day porosity and permeability profiles. Because of a very low sandstone/shale ratio, fluid channeling in fluvial sandstones on the southern end of the Moxa Arch seems to have caused extreme leaching. The sandstone section here, although it is very thin, is very permeable. To the north, a higher sandstone/shale ratio appears to have permitted a lower fluid throughput per unit volume of sand. As a result, fluid channeling was not as severe, detrital and authigenic clays are more common, and the sandstone section is more homogeneous and of lower overall permeability.

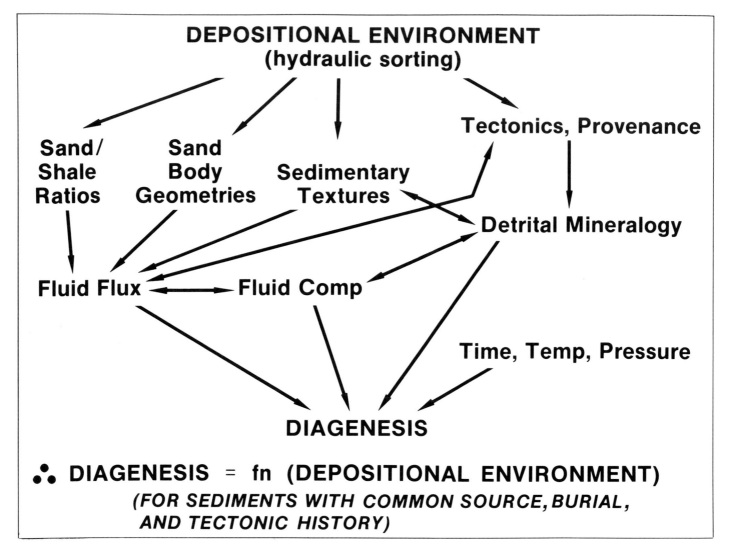

DEPOSITIONAL ENVIRONMENT
(hydraulic sorting)

Sand/ Shale Ratios

Sand Body Geometries

Sedimentary Textures

Tectonics, Provenance

Detrital Mineralogy

Fluid Flux ⟷ **Fluid Comp**

Time, Temp, Pressure

DIAGENESIS

∴ **DIAGENESIS = fn (DEPOSITIONAL ENVIRONMENT)**
(FOR SEDIMENTS WITH COMMON SOURCE, BURIAL, AND TECTONIC HISTORY)

Figure 1—Flow chart of the interrelationship of many of the factors that are primary controls on diagenesis. Most of the factors can be related back to the original depositional environment.

second Frontier sandstones of the Moxa Arch, Wyoming. There, sediment from the same source was sorted into several different facies that have markedly different petrological characteristics and, as a result, different diagenetic histories.

The second Frontier Formation consists of sandstones and mudstones of Late Cretaceous (Cenomanian) age deposited in a wave-dominated delta/ strand plain system that prograded eastwards into the western margin of the interior Cretaceous seaway (Fig. 2);

headwaters of the multiriver system were in the Sevier thrust belt to the west. Facies represented in these rocks include: offshore marine muds and interbedded minor sands, marine sand ridges, marine shoreline sequences, distributary channel point bars and related avulsion bodies, and flood-plain muds (for a more detailed discussion of stratigraphy see Winn and Smithwick, 1980; Winn et al, 1984). The series of cross sections constructed from cores and electric logs in Figure 4 shows the variation in depositional facies from south to north along the Moxa Arch (location of cross sections is shown in Fig. 3). Not all facies are present in all wells. Greater erosion to the south by channel incision and migration has resulted in complete removal of the

shoreline sequence; there fluvial sandstones rest directly on shoreface deposits (cross section A-A'). Further to the north, channels did not incise as deeply and there fluvial sandstones overlie a full shoreline sequence (cross section C-C').

Lower Frontier sandstones consist of subequal amounts of quartz and rock fragments and minor amounts of feldspar and mica (Fig. 5). The rock fragments are almost entirely cherts. Many of the chert grains contain sponge spicules and foraminifera molds typical of the Phosphoria Formation and other Paleozoic units, which are their presumed source. Other rock fragments include polycrystalline quartz grains (thought to be derived from the same source) as well as a few metamorphic

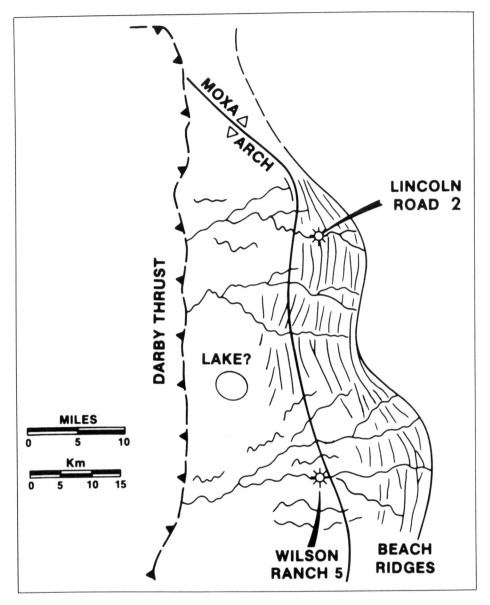

Figure 2 — Schematic reconstruction of depositional environments of the lower Frontier Formation along the Moxa Arch. The two wells shown for reference are located in the cross sections of Fig. 4 and on the map in Fig. 3. The reconstruction has been located with respect to the Darby thrust and the Moxa Arch axis even though these are both later features. Adapted from Winn and Smithwick, 1980.

and volcanic rock fragments (Goodell, 1962). Feldspars are dominantly plagioclase.

Although the same detrital components are present in all of the sandstones of the second Frontier, the relative abundances of components in fluvial and in marine sandstones are very different (Figs. 5, 6). Both shore-line and sand ridge deposits are more quartz-rich and have fewer rock fragments than fluvial units. Marine sandstones have 15–20% less rock fragments than fluvial sandstones because of hydraulic sorting (by grain size and, therefore, composition) within the delta and wave abrasion of mechanically unstable rock fragments along the shoreline (Winn et al, 1984).

Thus, Frontier sandstones show not only the changes in sedimentary textures and structures from facies to facies but also differences in composition. In addition, lateral and vertical variations in porosity and permeability related to petrological characteristics of facies strongly controlled fluid-flow

Figure 3 — Location map of the units on the Moxa Arch and of wells referred to in this paper. Cross sections A-A′, B-B′, and C-C′ are shown in Fig. 4. Inset shows location of the study area in southwestern Wyoming. Adapted from Winn and Smithwick, 1980.

Figure 4 — (a) Cross section A-A′ of the lower Frontier Formation on the southern end of the Moxa Arch. Core depths have been adjusted to log depths. Data are the approximate bases of the marine sand ridges. Cross sections are located in Fig. 2. (b) Cross section B-B′ through the middle of the Arch. Datum is the top of the foreshore and the tops of the fluvial sandstones that cut into the shoreline sequence. (c) Cross section C-C′ through the northern end of the Arch. Datum is the top of the foreshore (beach) zone. After Winn and Smithwick, 1980.

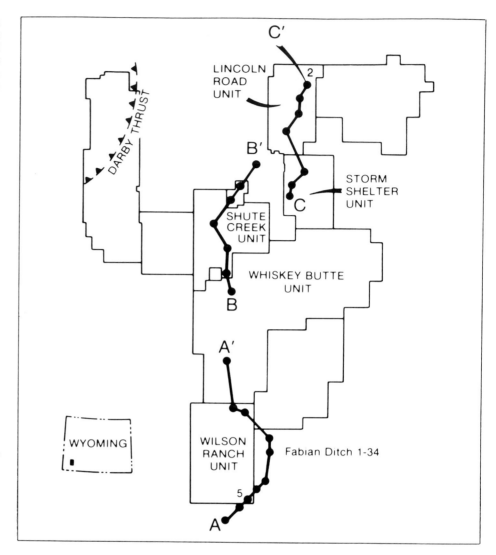

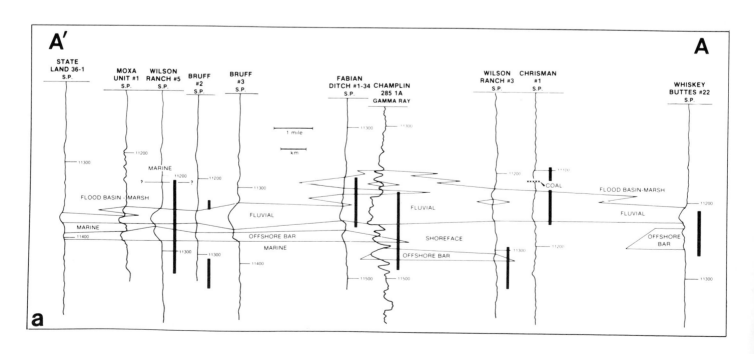

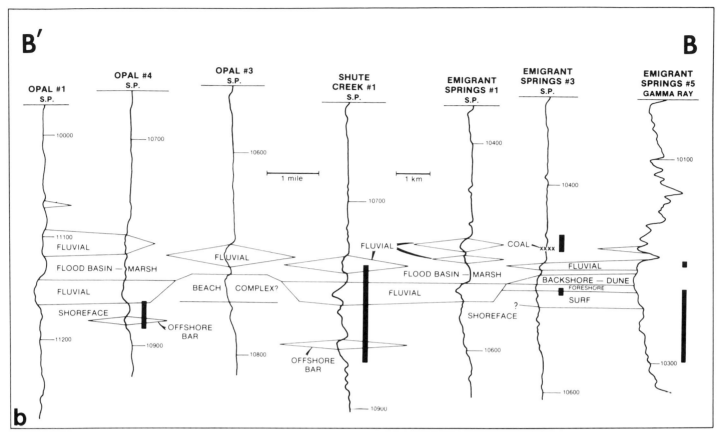

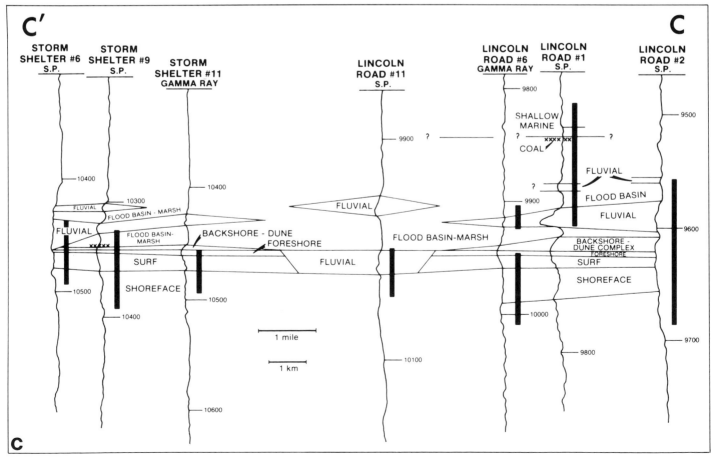

Figure 8—Scanning electron micrograph of clays typical of the lower marine sand ridge sandstones. The flaky material is randomly interstratified illite/smectite mixed-layer clay (I/S MLC). Note how the edges of the flakes exhibit a whiskery illite morphology. It appears that neomorphism (overgrowth and partial transformation) has taken place, and that the overgrowths have a more illitic composition than the original detrital I/S MLC core. Lath-shaped crystals are detrital illite. 2000×; bar is 10 μm.

(Thomas, 1978). Neomorphism (overgrowth and/or recrystallization) of clays (Fig. 8) was the dominant reaction, and the resulting growth of clays in pores and pore throats caused a further reduction in permeability.

Where the crest of the sand ridge was in the shallowest water (for example, Wilson Ranch 5, Figs. 3, 7), the sandstone is extremely clean, well sorted, and quartzose, and most closely resembles reworked beach sands. It is not clear whether this compositional maturity was the result of reworking offshore on the sand bar, or whether it was due to reworking on a nearby beach with subsequent transport to the sand ridges. Whatever the cause, these sandstones have a very high quartz content and contain very little clay. They are also largely cemented by quartz, as a result of several factors. First, the underlying Apsen shale is very siliceous and contains numerous bentonites and porcellanites (Meyers, 1977). As a result, water expressed from this shale during compaction would be expected to be rich in silica, thereby promoting silica cementation in the sandstones of the overlying Frontier Formation. Secondly, numerous authors have shown an inverse relationship between the amount of clay in a sandstone and the extent of cementation (Siever, 1959;

Pettijohn et al, 1972). This is partly due to the decrease in permeability caused by the clay matrix, which limits both fluid flux and fluid access to grain surfaces. In addition, clays have been shown to compete for available dissolved silica with silica precipitates such as quartz overgrowths (Mackenzie et al, 1967; Kastner et al, 1977). Where clay is absent, on the other hand, none of these factors act to inhibit quartz precipitation. In places, quartz overgrowths were inhibited by calcite cementation early in the burial history. Subsequent dissolution of the calcite in these sandstones results in zones of good porosity and permeability, but the zones appear to be patchy and not well connected. In places where calcite precipitation did not occur, nearly all porosity in the clean sand ridge sandstones is occluded by early diagenetic quartz (Fig. 9).

Marine Shoreline Sands

The marine shoreline sequence is interpreted to consist of four facies: lower shoreface, surf, beach, and backshore. The sandstones are largely fine to very fine grained, but they can be as coarse as medium grained. The sequences generally coarsen and become better sorted stratigraphically upwards, although coarse-grained material may

also be found in the surf facies.

Lower shoreface sandstones are largely indistinguishable from lower sections of offshore sandbars. The sandstones are generally poorly sorted and contain abundant detrital I/S MLC because of their deposition under relatively low-energy conditions. Homogenization by burrowing produced textures such as sand grains "floating" in a matrix of I/S MLC (Fig. 10); small discontinuous shale laminae are also common. As with the marine sand ridges, porosity and permeability in these lower shoreface sandstones were originally poor because of the abundance of detrital intergranular clay and were further reduced at an early stage by compaction. Because of the very low permeability, the sandstones acted as closed systems to migrating pore waters and the predominant diagenetic process was neomorphism of clays. Present-day porosity and permeability in these sandstones is, as expected, very poor.

Sandstones deposited in surf, beach, and backshore zones are all very similar. As a result of relatively high energy at the site of deposition (or in the case of the backshore, owing to the source of sand), these sandstones are better sorted, much more quartzose, and much cleaner than lower shoreface sandstones. Transition from the under-

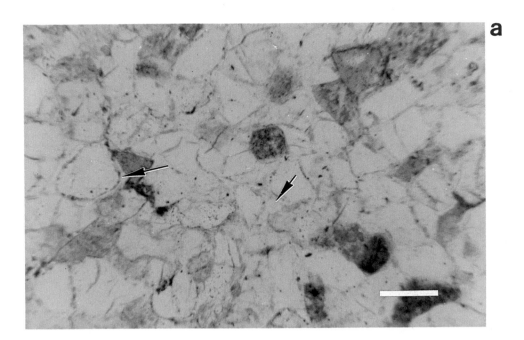

Figure 9—Photomicrograph of sandstone from top of a sand ridge showing nearly complete occlusion of porosity by quartz overgrowths. Outlines of original detrital quartz grains are indicated by thin clay rims (arrows). The unfilled pores (purple patches) are secondary and the result of dissolution of feldspars or calcite (stained pink) replacements of feldspars. (a) plane polarized light; (b) crossed-nicols. Wilson Ranch 5, 11,273 ft; bar equals 0.2 mm.

lying lower shoreface sandstones is gradational and marked by a general decrease upwards in the amount of biogenically introduced clay.

Porosity in all three facies appears to be largely a function of the abundance of sand-to-pebble sized mudclasts and of interstitial clay content. Some of the interstitial clay may be formed from the alteration of detrital biotite, but most of it appears to have been deposited as fecal pellets or small rip-up clasts, both of which were squeezed around sand grains by compaction (Fig. 11). For the reasons discussed previously, the abundance of interstitial clay controls the extent of quartz overgrowths; that is, where clay content is very low, silica

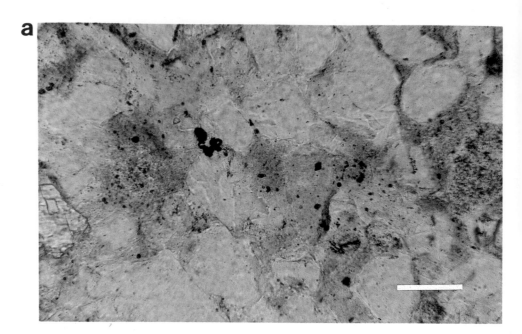

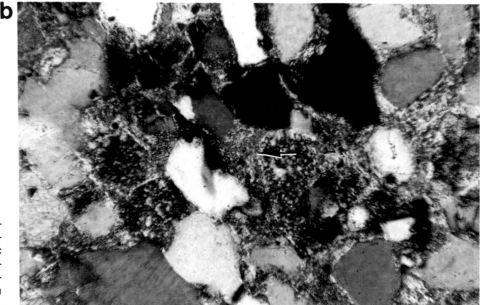

Figure 10—Photomicrograph of abundant I/S MLC filling pore in lower shoreface sandstone. Note characteristic speckled yellowish-brown birefringence of the I/S MLC. (a) plane polarized light; (b) crossed-nicols. Lincoln Road 3, 9578.5 ft; bar equals 0.1 mm.

overgrowths have occluded nearly all porosity. Thus, although in general the cleanest sand makes the best reservoir, in this case it did not. Because of the inferred large input of silica in pore waters, the fact that these shoreline sandstones were originally clean, quartzose, and porous and permeable resulted in their present poor reservoir quality.

Most of the present porosity in shoreline sandstones is secondary in origin, formed primarily by the dissolution of clay clasts (Fig. 12). Where clay clasts were originally abundant, porosity can be moderately high, but permeability is still relatively low because this grain moldic porosity has been iso-

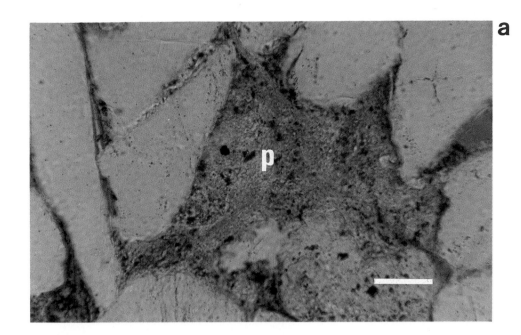

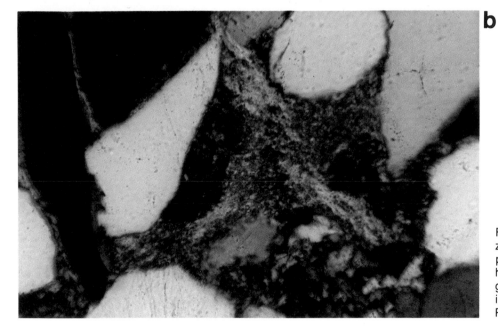

Figure 11—Photomicrograph of surf-zone sandstone showing probable fecal pellet (**P**) composed of I/S MLC which has been squeezed around the adjacent grains by compaction. (a) plane polarized light; (b) crossed-nicols. Lincoln Road 2, 9627 ft; bar equals 0.1 mm.

lated by quartz overgrowths and suture boundaries between the surrounding grains (Fig. 13).

In certain parts of the shoreline section the I/S MLC matrix is of slightly different character. Petrographically the clay exhibits higher birefringence and relief than elsewhere (compare Fig. 14 to Fig. 10), and it occurs as thin well-defined wisps between grains. X-ray diffraction patterns of oriented slides of these samples show the development of a "superlattice" peak ranging from 28–31 Å (Fig. 15), which indicates maximum IM ordering of the mixed-layer clay (Reynolds and Hower, 1970). The development of this peak ranges from a poorly defined shoulder to a sharp,

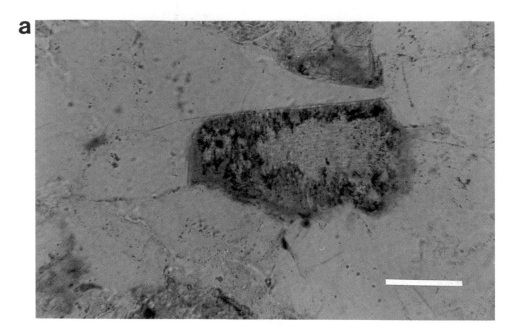

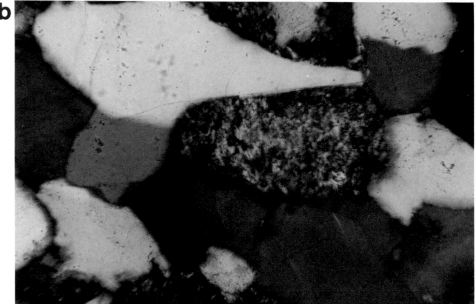

Figure 12—Photomicrograph of partially dissolved mud clast in a backshore sandstone. Note development of porosity (blue) along ragged edges and within etched, honeycombed grain. (a) plane polarized light; (b) crossed-nicols. Lincoln Road 6, 9953 ft; bar equals 0.1 mm.

well-formed peak. Controls on the formation of the ordered I/S MLC itself are not obvious. It appears to be related to a balance between the amount of detrital clay present and the inferred amount of fluid throughput in a given sandstone. If clay content were very high and therefore fluid throughput were low, the sandstone would have acted essentially as a closed system. This case is exemplified by the lower sand ridges and lower shoreface where normal burial diagenesis seems to have taken place and where randomly interstratified, low birefringence I/S MLC is common. If the clay content was low, fluid throughput was probably high, and excessive dissolution of clays seems

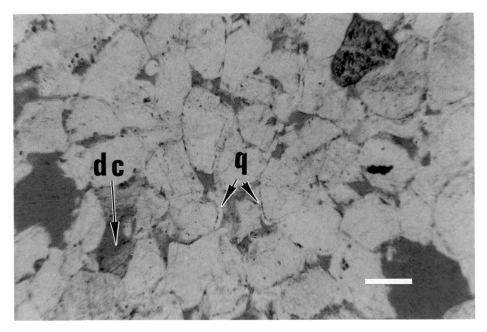

Figure 13—Photomicrograph of backshore sandstone showing secondary pores (large purple areas) formed primarily by the dissolution of clay clasts (**dc**). Quartz overgrowths (**q**) have filled in much of the primary porosity and limit transmissivity of even the large secondary pores; plane polarized light. Lincoln Road 3, 9530 ft; bar equals 0.2 mm.

to have dominated (as occurred in the beach and backshore facies). Ordered I/S MLC formation seems to require the moderate amounts of clay and moderate throughput typical of sandstones in the surf zone facies. Excess sil-

ica in the pore waters, as evidenced by common quartz overgrowths, may also have been a factor in causing a slow transformation of expansible to nonexpansible layers, thereby permitting ordering.

Fluvial Point Bars and Avulsion Bodies

Porosity and permeability profiles through point bars clearly show the effects of hydraulic sorting on grain sizes, detrital compositions, and sedimentary textures and thus on diagene-

Figure 14—Photomicrograph of surf-zone sandstone in which brown patches of IM-ordered illite/smectite mixed layer clay (I/S MLC) fill nearly all the interstitial pore space. Note how the birefringence colors of the well-ordered I/S MLC shown in the crossed-nicols view are orangish-red to yellow. Compare this with the pale yellow to yellowish brown colors typical of the randomly interstratified I/S MLC (Fig. 10). Lincoln Road 2, 9627 ft; bar equals 0.2 mm.

sis. In well-developed point bars, the sand is coarsest at the base and gradually fines upwards. The basal point bar sands are generally medium to coarse grained, although a few are pebble conglomerates. They contain very abundant chert grains and lesser amounts of other lithic fragments and mudclasts, as well as moderately rare feldspars and biotites. The tops of these point bars are fine to very fine grained and more quartzose than the lower part of the point bar. Detrital clay content is variable, although the finer grained sandstones tend to contain more clay. Thicknesses of well-developed point bars are 8–12 m.

Differences in textures and detrital

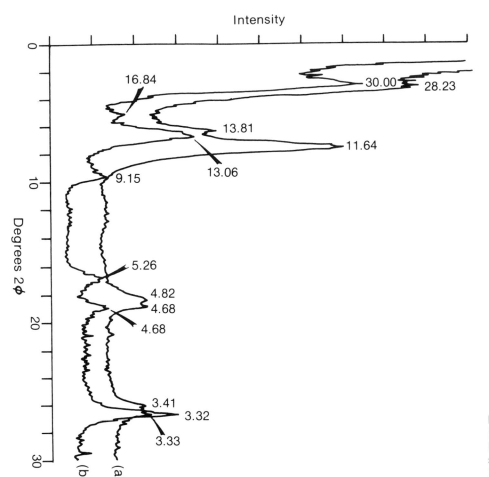

Figure 15—X-ray diffraction patterns of less than 2-μm fraction IM-ordered I/S MLC. (a) Mg-saturated sample; (b) sample saturated with ethylene glycol. Peak labels are d-spacings (Å).

compositions have resulted in different diagenetic histories for different parts of point bars. Because the upper parts of these point bar sequences were finer grained and generally contained more clay, porosity and permeability were originally poor following deposition, and were further reduced at an early stage of burial in clay-rich sections by compaction. Because the point-bar tops were also generally more quartzose, porosity was diminished even in clay-free sections by extensive quartz overgrowth development. Thus, porosity and permeability in upper point bars is relatively poor (Fig. 7).

In basal point bar sandstones, on the other hand, porosity and permeability were good following deposition. These sandstones are coarse grained and contain abundant chert grains and lithic fragments. Because quartz tends to precipitate on monocrystalline quartz grains in preference to polycrystalline or microcrystalline forms (Jonas and McBride, 1977), the presence of fewer monocrystalline quartz grains in fluvial sandstones meant fewer pores were filled by early quartz overgrowths. In addition, lack of abundant, dispersed detrital clay limited the effects of compaction to minor permeability reduction caused by squeezing of rare mud-clasts and by crushing of altered micas. The maintenance of good porosity and permeability through early diagenetic stages probably resulted in these sands being the preferred conduits for fluid flow, which had a profound effect on later diagenesis.

Flushing of large volumes of water through these sandstones is suggested by many lines of evidence. Chemically unstable materials such as clay clasts (Fig. 16), altered micas, and feldspars (Fig. 17) show signs of partial to com-plete dissolution. Poorly to well-developed microstylolitic pressure solution boundaries are common between chert grains and, in some cases, the chert grains are tripolitic. Removal of the seemingly large amounts of material in solution without extensive repre-cipitation in other forms requires maintenance of low solute concentrations. This is most easily accomplished by flushing with large volumes of fluid. In addition, authigenic chlorite is present in the form of patchy rims and partial pore fillings (Fig. 18), which also suggests good fluid mobility for reasons discussed in Winn et al (1983). The combination of high fluid flux with the abundance of chemically unstable materials resulted in extensive development of secondary porosity through clast dissolution. In addition, the input of large quantities of fluid would also provide a renewable source of CO_2

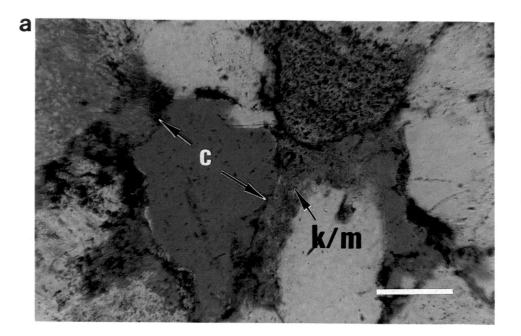

Figure 18—Photomicrograph of nearly complete chlorite rim (**c**). Rim marks outline of probable feldspar grain that was dissolved to form secondary pore. Adjacent space probably represents primary porosity and is now filled with a mixture of kaolinite and I/S MLC (**k/m**). (a) plane polarized light; (b) crossed-nicols. Lincoln Road 2, 9600 ft; bar equals 0.1 mm.

bars, one from the Lincoln Road 2 well (LR 2) and one from the Wilson Ranch 5 well (WR 5) (Fig. 7). The maximum grain size in WR 5 is slightly larger than that in LR 2 (2–2.5 mm for the largest clasts in WR 5 as opposed to 1.5–2 mm in LR 2), but on the average the two sands have roughly the same grain-size distributions. Sandstones fine upward from a pebble conglomerate and pebbly sandstone (WR 5) or from a pebbly sandstone and very coarse-grained sandstone (LR 2). Both basal sandstones also appear to have had the same composition originally. Both were chert-rich with lesser amounts of quartz and moderately common mudclasts, feldspars, and biotites. Diagenesis has

Figure 19—Photomicrograph of basal point-bar sandstone showing feldspar grain (**f**) that has been almost completely replaced by calcite (stained pink by alizarin red). Subsequent dissolution has created small secondary pores in the calcite, which are now filled with kaolinite (**k**). (a) plane polarized light; (b) crossed-nicols. Lincoln Road 6, 9920.2 ft; bar equals 0.1 mm.

caused significant differences in the present composition of these sandstones, however, primarily in the amount and type of clay present. The fluvial sandstone in LR 2 contains significantly more clay, both detrital/ neomorphosed and authigenic, than does the fluvial sandstone in WR 5. Partially dissolved clay clasts abound in LR 2, and pores are commonly filled with a mixture of neomorphosed I/S MLC and authigenic chlorite (Fig. 21). Many pores are also filled with late stage authigenic kaolinite (Fig. 22). In WR 5, on the other hand, interstitial clay and micas are less common. Thin patchy rims of chlorite cover some of the chert grains (Fig. 23) and a few

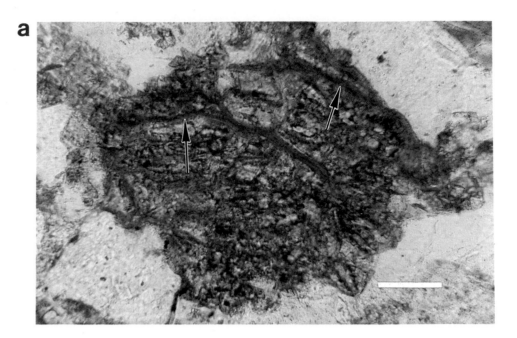

Figure 20—Photomicrograph of sideritized biotite in fine-grained sandstone near top of a point-bar top sequence. Siderite precipitated along and between cleavage planes of the biotite and gradually replaced all but a few remnant shreds of the mica (arrows). (a) plane polarized light; (b) crossed-nicols. Lincoln Road 6, 9920.2 ft; bar equals 0.1 mm.

small pores are filled with a mixture of I/S MLC and chlorite, but these are rare. Kaolinite in any form is totally absent.

The cause of these differences in diagenesis and current petrophysical prop-erties appears not to be related to the original petrological characteristics of these fluvial sands. One possible expla-nation for the observed trends would be that the underlying Mowry shale (the more regionally accepted name for the Aspen) is not chemically or mineralogi-cally homogeneous over the study area. If this were true, compositions of expressed water also could be expected to vary from point to point. Very few data are available on the Mowry shale,

Figure 21—SEM photo of pore in Lincoln Road 2 fluvial sandstone that is filled with a mixture of detrital/neomorphosed I/S MLC (flaky material) and authigenic chlorite (small interpenetrating blades). 800×; bar equals 10 μm.

but what data there are suggest that this explanation is probably not a reasonable one. Schrayer and Zarella (1963, 1966), Davis (1970), and Nixon (1973) discuss regional variations in the organic content and mineralogy of the Mowry shale. These authors indicate that depositional strike for the Mowry is north-northeast to northeast, and that compositional variations along strike are minimal. Because the strike of the Moxa Arch is approximately north-south, one would also expect very small variations in composition of the Mowry along the arch, especially because the study section is so short (~25 mi, Fig. 2).

Another possible, and to us more reasonable, explanation might be that variations in diagenetic history between the fluvial sandstones in the two wells are related to variations in fluid flux. LR 2 is located at the northern end of the Moxa Arch where the fluvial sandstones overlie a thick sequence of shoreline sandstones. On the southern end of the arch, however, where WR 5 is located, erosion has removed nearly all of the shoreline sequence and there

the fluvial sandstones rest directly on muddy lower shoreface deposits and sandstones representing offshore sand ridges (Fig. 4). If we assume: (1) that most of the water that passed through the second Frontier sands, especially early in their history, was the result of compaction and sediment dewatering (see Hayes, 1979), and (2) that because the underlying Aspen shale is of approximately the same thickness over the area of interest (a maximum of 10% variation in thickness; Nixon, 1973; Meyers, 1977), sandstones received the discharge from equal volumes of shale, then the thinner fluvial sandstones to the south should have experienced a far greater throughput or discharge (Q) per unit volume of sand than did the combined fluvial/shoreline sequence to the north. Moreover, because the velocity of flow (V) is a function of the cross-sectional area of flow (V = Q/A, where A is approximated here by the thickness of the sand section), the flow rate or velocity of flow was presumably greater in WR 5 where A is small than it was in LR 2 where A is large (Fig. 24). Thus, not only would the fluvial

sands in WR 5 have seen a much larger total volume of fluid throughput, but also the rate at which fluid passed through the sand would have been much higher. As a result, in WR 5, the amount of time that a given fluid parcel would have to react with the sand would be limited and the buildup of reaction-limiting ions, such as aluminum, in pore waters would be reduced, thus increasing the probability of severe leaching. That leaching did occur in WR 5 is evident from the abundance of honeycombed, partially dissolved, tripolitic chert grains (Fig. 25), the well-developed microstylolitic boundaries between chert grains (Fig. 26), and the presence of clay clasts and altered micas only as rare solution remnants.

In LR 2, on the other hand, the smaller total volume of fluid throughput would reduce the amount of leaching, and the slower flow rate would permit longer contact of pore fluids with the rock, increasing the likelihood of in situ reactions. Thus, in LR 2, rather than being flushed out of the system, ions (especially Al^{+3}) released by the partial dissolution of clay clasts,

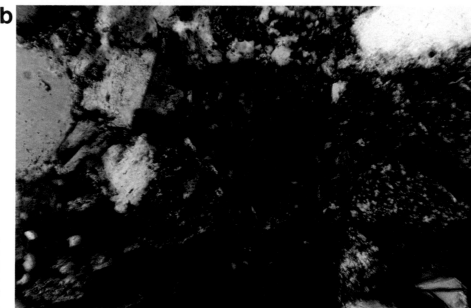

Figure 22—Photomicrograph of secondary pore filled with clean, well-crystallized, authigenic kaolinite (**k**). (a) plane polarized light; (b) crossed-nicols. Lincoln Road 2, 9600 ft; bar equals 0.1 mm.

micas, and feldspars were retained within the system and were available for interaction with reactive minerals. As a result, the products of these interactions, neomorphosed and authigenic clays, are far more abundant in LR 2.

The net effect is, thus, one of rearrangement within the sandstone system rather than dissolution and removal.

Clay mineral diagenesis in LR 2 versus WR 5 may also have differed because of other fluid-flow variables,

such as fluid source and path length. Previously it was noted that, whereas kaolinite is abundant in LR 2, it is totally absent from WR 5. Although the conditions required for the formation of kaolinite are not well known, it

Figure 23—SEM photo of badly corroded remnant of interstitial I/S MLC with small chlorite blades growing on top of it. Note lack of thin wispy edges to the mixed-layer clay (compare this with Fig. 8). Wilson Ranch 5, 750×; bar equals 10 μm.

is generally thought to form in waters that have both a low pH and a low K^+/H^+ ratio (Garrels and Christ, 1965; Pettijohn et al, 1972). Bucke and Mankin (1971) suggested that kaolinite could precipitate from saline formation waters if: (1) degraded detrital illite present in the sandstones were to act as a K^+ acceptor to selectively remove potassium from solution; and (2) organic acids were released in suitable amounts during the maturation of organic matter to maintain the low pH. Whereas detrital illite is fairly common in some facies of the lower Frontier (Winn et al, 1983), Schrayer and Zarella (1963, 1966) and Nixon (1973) indicate that not only is the organic content of the underlying Mowry shale probably very low in the Moxa Arch area, but also the main area of organic deposition was in north-central and northeastern Wyoming, a considerable distance from the arch. Thus, it would be difficult to call upon the latter mechanism to explain the observed kaolinite distribution.

Another possible way to satisfy both requirements for kaolinite formation would be to flush the sandstone system with low-salinity, low-pH fresh water. Influx of fresh water is usually associated with meteoric recharge in areas that have been uplifted and exposed at the surface. Thomaidis (1973) states that during the main period of folding of the Moxa Arch in the Late Cretaceous, basal Mesaverde rocks and upper Hilliard shale were eroded off the crest of the arch. In addition, Stearns et al (1975) indicate from seismic evidence that southward thickening of the uppermost Cretaceous section suggests the present-day southward plunge of the Moxa Arch began at that time with uplift of the northern end of the arch. The Hilliard directly overlies the Frontier, but whether freshwater input along the unconformity in the Hilliard could have reached Frontier sandstones is, at this time, unknown. There is, however, some suggestion that meteoric water may have been responsible for kaolinite development. Kaolinite is most common in LR 2, which is located on the north end of the southward plunging arch; presumably Frontier sandstones at the north end of the arch would have been closer to the source of the fresh water than Frontier sandstones in WR

5. Thus, if the fresh water reached the Frontier in the LR 2 area it may still have been dilute and acidic enough to permit kaolinite formation. By the time the fresh water had traveled the longer distance to the WR 5 area, however, continued reactions with soluble minerals in the rocks may have increased the pH and produced solute levels too high for kaolinite to precipitate. It is likely that further work on determining fluid-flow paths in the Frontier, together with information from oxygen-isotope analyses, will help resolve this question.

The differences in diagenetic history described above are clearly reflected in present-day permeability profiles (Fig. 21). Porosity in the two sands is nearly the same, averaging about 13% in WR 5 and a slightly higher 15% in LR 2. Permeability, however, is quite different in the two wells. The most permeable section in WR 5 (average = 10 md) is an order of magnitude more permeable than the most permeable section in LR 2 (average = 1 md). This difference in permeability may in part be related to slight variations in grain-size profiles for the two wells (Table 1). Both point

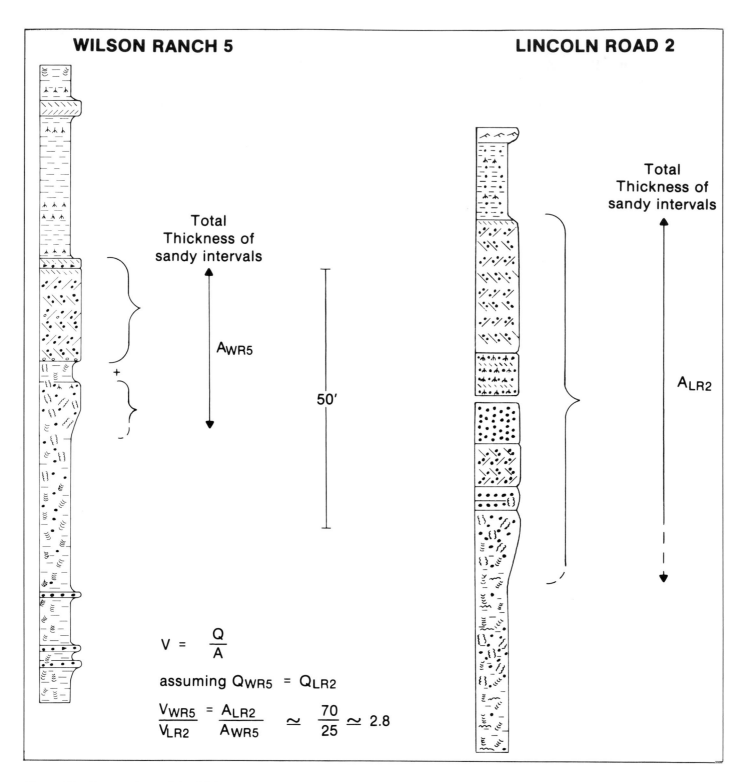

Figure 24—Comparison of the thickness of the permeable sandy intervals in Wilson Ranch 5 and Lincoln Road 2. Assuming that the same volume of water passed through each of these sandy intervals ($Q_{WR5} = Q_{LR2}$), then the rate at which water moved through the sand section in Wilson Ranch 5 was nearly 3 times the rate at which it moved through the sand section in Lincoln Road 2.

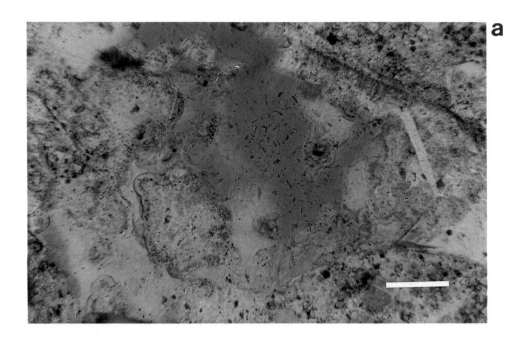

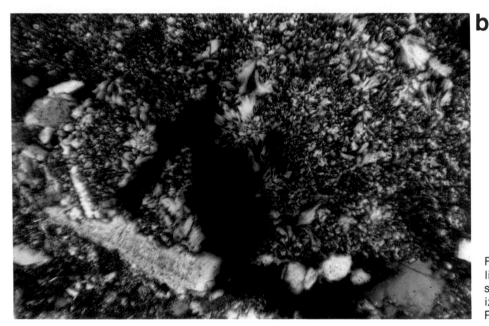

Figure 25—Photomicrograph of tripo-litic chert grain from a basal point-bar sand in Wilson Ranch 5. (a) plane polar-ized light; (b) crossed-nicols. Wilson Ranch 5, 11,267.5 ft; bar equals 0.1 mm.

bars show the expected fining-upward trend, but the coarsest grain size in the LR 2 point bar is one ϕ unit finer than the coarsest grain size in WR 5. In addition, the percent decrease in grain size from point bar base to point bar top is slightly less in LR 2 (60% for LR 2 versus 70% for WR 5). But, it seems unreasonable that these slight differences in grain size could completely account for the marked difference in the permeability profiles in the two wells. Although permeability in the LR

2 fluvial section varies little downhole (about one order of magnitude), the basal point bar in WR 5 is markedly more permeable than the rest of the fluvial section in that well (more than two orders of magnitude). These observations appear to be related to the clay content. As we have shown, extreme fluid channeling and extensive leaching in WR 5 appears to have caused a thin, clean, very high-permeability sand section, whereas the lack of channeling and the abundance of detrital and

authigenic clays in LR 2 resulted in a more homogeneous, low-permeability sand interval.

SUMMARY

Sandstone diagenesis consists of a set of competitive processes or reactions, some of which tend to preserve or enhance porosity and some of which tend to destroy it. The end result of diagenesis in terms of reservoir quality is dependent not only on the detrital

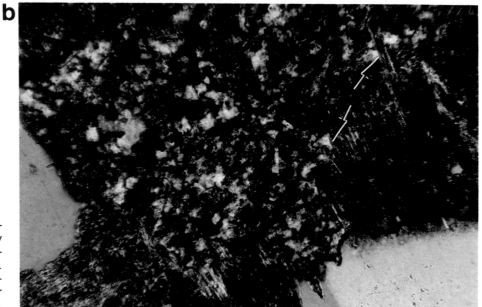

Figure 26—Photomicrograph of well-developed microstylolitic boundary between chert grains. Note concentration of clay along boundary (arrows). (a) plane polarized light; (b) crossed-nicols. Wilson Ranch 5, 11,265.5 ft; bar equals 0.1 mm.

composition of the sandstones, but also on the amount and composition of material input or removed from the system (that is, fluid flux and fluid composition). Each of these factors can be shown to be related in part to the depositional facies. For example, hydraulic sorting will generally result in different detrital assemblages in environments of different energies. Likewise sedimentary textures, sand-body geometries, and sand/shale ratios are important controls on fluid flow through a system. Because of these environmentally controlled differences in original physical and compositional characteristics, diagenesis tends to proceed along different paths in different facies.

ACKNOWLEDGMENTS

The authors would like to thank Marathon Oil Company for permission to publish this work. We would also like to thank E. Pfister, who peformed many of the grain-size analyses, the drafting department at the Denver Research Center, and C. Pedde, who patiently typed the many drafts.

TABLE 1

	Mean[1] (φ)	Standard Deviation	Skewness	Kurtosis
Wilson Ranch 5				
Point bar				
Top (11,251.8 ft)	2.898	0.322	−0.010	0.923
Upper middle (11,256.8 ft)	2.410	0.402	0.122	1.082
Middle (11,259.5 ft)	2.104	0.462	0.188	1.183
Base (11,264.2 ft)	0.803	0.820	−0.167	1.347
Lincoln Road 2				
Point bar				
Top (9582.4 ft)	3.173	0.264	−0.050	0.958
Middle (9600 ft)	2.319	0.438	0.140	1.032
Near base (9603.9 ft)	1.804	0.603	0.112	1.067

[1]All values calculated by method of Folk and Ward (1957).

Table 1—Grain-size profiles of point bars in the WR 5 and LR 2 wells.

SELECTED REFERENCES

Anderson, J. B., C. Wolfteich, R. Wright, and M. L. Cole, 1982, Determination of depositional environments of sand bodies using vertical grain-size progressions: Gulf Coast Association of Geological Societies Transactions, v. 32, p. 565–577.

Blanche, J. B., and J. H. M. Whitaker, 1978, Diagenesis of part of the Brent Sand Formation (Middle Jurassic) of the northern North Sea Basin: Quarterly Journal of the Geological Society of London, v. 135, p. 73–82.

Blatt, H., G. Middleton, and R. Murray, 1972, Origin of sedimentary rocks: Englewood Cliffs, NJ, Prentice-Hall, Inc., 634 p.

Brenner, R. L., 1980, Construction of process-response models for ancient epicontinental seaway depositional systems using partial analogs: Bulletin of the American Association of Petroleum Geologists, v. 64, p. 1223–1244.

Bucke, D. P., and C. J. Mankin, 1971, Clay mineral diagenesis within interlaminated shales and sandstones: Journal of Sedimentary Petrology, v. 41, p. 971–981.

Davies, D. K., and F. G. Ethridge, 1975, Sandstone composition and depositional environment: Bulletin of the American Association of Petroleum Geologists, v. 59, p. 239–264.

Davis, J. C., 1970, Petrology of Cretaceous Mowry Shale of Wyoming: Bulletin of the American Association of Petroleum Geologists, v. 54, p. 487–502.

Doe, T. W., R. H. Dott, Jr., and I. E. Odom, 1976, Nature of feldspar grain-size relations in some quartz-rich sandstones: Journal of Sedimentary Petrology, v. 46, p. 862–870.

Folk, R. L., and W. C. Ward, 1957, Brazos River bar: a study in the significance of grain size parameters: Journal of Sedimentary Petrology, v. 27, p. 3–26.

Friedman, G. M., 1961, Distinction between dune, beach and river sands from their textural characteristics: Journal of Sedimentary Petrology, v. 31, p. 514–529.

Garrels, R. M., and C. L. Christ, 1965, Solutions, minerals, and equilibria: New York, Harper and Row, 450 p.

Goodell, H. G., 1962, The stratigraphy and petrology of the Frontier Formation of Wyoming: Wyoming Geological Association Guidebook, 17th Annual Field Conference, p. 137–145.

Hayes, J. B., 1979, Sandstone diagenesis—the hole truth, in P. A. Scholle and P. R. Schluger, eds., Aspects of diagenesis: Society of Economic Paleontologists and Mineralogists Special Publication 26, p. 127–139.

Irwin, H., and C. Curtis, 1977, Isotopic evidence for source of diagenetic carbonates formed during burial of organic-rich sediments: Nature, v. 269, p. 209–213.

Jonas, E. C., and E. R. McBride, 1977, Diagenesis of sandstone and shale: application to exploration for hydrocarbons: Austin, University of Texas, Department of Geological Sciences Continuing Education Program Publication 1, 165 p.

Kastner, M., J. B. Keene, and J. M. Gieskes, 1977, Diagenesis of siliceous oozes, I, Chemical control on the rate of opal-A to opal-CT transformation—an experimental study: Geochimica et Cosmochimica Acta, v. 41, p. 1041–1059.

Mackenzie, F. T., R. M. Garrels, O. P. Bricker, and F. Bickley, 1967, Silica in seawater: control by silica minerals: Science, v. 155, p. 1404–1405.

Meyers, R. C., 1977, Stratigraphy of the Frontier Formation (Upper Cretaceous), Kemmerer area, Lincoln County, Wyoming: Wyoming Geological Association Guidebook, 29th Annual Field Conference, p. 271–311.

Moiola, R. J., and D. Weiser, 1968, Textural parameters: an evaluation: Journal of Sedimentary Petrology, v. 38, p. 45–53.

Nixon, R. P., 1973, Oil source beds in Cretaceous Mowry Shales of northwestern interior United States: Bulletin of the American Association of Petroleum Geologists, v. 57, p. 136–161.

Odom, I. E., 1975, Feldspar grain-size relations in Cambrian arenites, Upper Mississippi Valley: Journal of Sedimentary Petrology, v. 45, p. 636–650.

Peterson, J. A., and J. C. Osmond, eds., 1961, Geometry of sandstone bodies: Tulsa, American Association of Petroleum Geologists, 240 p.

Pettijohn, F. J., P. E. Potter, and R. Siever, 1972, Sand and sandstone: New York, Springer-Verlag, 618 p.

Potter, P. E., 1967, Sand bodies and sedimentary environments: a review: Bulletin of the American Association of Petroleum Geologists, v. 51, p. 337–365.

Reed, W. E., R. Lefever, and G. J. Moir, 1975, Depositional environment interpretation from settling velocity (psi) distributions: Geological Society of America Bulletin, v. 86, p. 1321–1328.

Reynolds, R. C., and J. Hower, 1970, The nature of interlayering in mixed-layer illite-montmorillonites: Clays and Clay Minerals, v. 18, p. 25–36.

Schrayer, G. J., and W. M. Zarella, 1963, Organic geochemistry of shales, I, Distribution of organic matter in the siliceous Mowry Shale of Wyoming: Geochimica et Cosmochimica Acta, v. 27, p. 1033–1046.

————, 1966, Organic geochemistry of shales, II, Distribution of extractable organic matter in the siliceous Mowry Shale of Wyoming: Geochimica et Cosmochimica Acta, v. 30, p. 415–434.

Siever, R., 1959, Petrology and geochemistry of silica cementation in some Pennsylvanian sandstones, *in* H. A. Ireland, ed., Silica in sediments: Society of Economic Paleontologists and Mineralogists Special Publication 7, p. 55–79.

Stearns, D. W., W. R. Sacrison, and R. C. Hanson, 1975, Structural history of southwestern Wyoming as evidenced from outcrop and seismic, *in* Rocky Mountain Association of Geologists: Symposium on Deep Drilling Frontiers in the Central Rocky Mountains, p. 9–20.

Stonecipher, S. A., 1982, Diagenetic/stratigraphic modeling: a tool for understanding porosity/permeability profiles: Abstracts of the 11th International Association of Sedimentologists Congress, Hamilton, ON, p. 120.

Swift, D. J. P., 1976, Continental shelf sedimentation, *in* D. J. Stanley and D. J. P. Swift, eds., Marine sediment transport and environmental management: New York, John Wiley and Sons, p. 311–350.

Taylor, J. C. M., 1978, Control of diagenesis by depositional environment within a fluvial sandstone sequence in the northern North Sea Basin: Quarterly Journal of the Geological Society of London, v. 135, p. 83–92.

Thomaidis, N. D., 1973, Church Buttes arch, Wyoming and Utah: Wyoming Geological Association Guidebook, 25th Annual Field Conference, p. 35–39.

Thomas, J. B., 1978, Diagenetic sequences in low-permeability argillaceous reservoirs: Quarterly Journal of the Geological Society of London, v. 135, p. 93–100.

Visher, G. S., 1969, Grain-size distributions and depositional processes: Journal of Sedimentary Petrology, v. 39, p. 1074–1106.

Winn, Jr., R. D., and M. E. Smithwick, 1980, Lower Frontier Formation, southwest Wyoming: depositional controls on sandstone compositions and on diagenesis: Wyoming Geological Association Guidebook, 31st Annual Field Conference, p. 137–153.

Winn, Jr., R. D., S. A. Stonecipher, and M. G. Bishop, 1983, Depositional controls on diagenesis in offshore sand ridges, Frontier Formation, Spearhead Ranch field, Wyoming: Mountain Geologist, v. 20, p. 41–58.

————, 1984, Sorting and wave abrasion: controls on composition and diagenesis in the Lower Frontier sandstones, southwestern Wyoming: Bulletin of the American Association of Petroleum Geologists, v. 68, p. 268–284.

Diagenetic History of the Phosphoria, Tensleep and Madison Formations, Tip Top Field, Wyoming

Janell D. Edman
Ronald C. Surdam
University of Wyoming
Laramie, Wyoming

INTRODUCTION

Purpose of Investigation

The relationship of diagenesis to the preservation and development of porosity in sedimentary rocks has received increasing attention during the last five years. In particular, the timing of diagenetic events relative to the maturation and migration of hydrocarbons in an area could critically control the availability of porous reservoir rocks into which the hydrocarbons could migrate. In addition, there appears to be evidence that the maturation reactions of the organic material may affect the inorganic diagenetic reactions and the consequent generation of porosity (Surdam, et al, 1982; Moore and Druckman, 1981). This investigation was designed with these considerations in mind. Specific objectives in this study included:

(1) Relate the timing of porosity destruction and generation in the reservoir rocks to the timing of maturation and migration of hydrocarbons in the source rock.

(2) Identify and delineate the major factors controlling the inorganic diagenesis to facilitate a greater understanding of the diagenetic processes as well as aid in prediction of porosity trends.

(3) Integrate the organic reactions with the inorganic diagenesis to determine the role of the organic system in the generation or destruction of porosity.

The Paleozoic section of the Tip Top

ABSTRACT. Petrographic and geochemical data from cores in the Wyoming thrust belt are used to relate maturation and migration of Phosphoria Formation organic material to the evolution of porosity in the Tensleep and Madison Formations. Observed paragenetic sequences for each formation indicate that all three formations exhibit two distinct phases of diagenesis. These two phases are delineated chronologically by the simple designations of "early" and "late." The period of early diagenesis was responsible for the pore network present in the Tensleep and Madison during migration of Phosphoria-sourced hydrocarbons. This early diagenetic phase exhibits a strong dependence on depositional environment, and is distinct for each formation. In contrast to this early diagenesis, the late diagenesis is consistent and similar for all three formations. This is interpreted to be the result of a relatively uniform sequence of fluids migrating through a pervasive Sevier fracture system that penetrated the entire 2000-ft thick stratigraphic interval of interest. The late diagenesis determined the present porosity configuration in these formations.

The sequence of organic reactions related to the maturation and migration of Phosphoria organic material also influenced late diagenesis in all three formations. Generation of hydrocarbons was the first event in the late diagenetic period of each formation. Abundant organic material in the Phosphoria Formation entered the liquid window just prior to the emplacement of the Darby thrust during the Paleocene. Fracturing associated with thrusting during the Sevier orogeny provided conduits for the migration of Phosphoria hydrocarbons. Subsequent deep burial of the source and reservoir rocks beneath the Darby plate resulted in thermal degradation of the hydrocarbons, and liquid hydrocarbons were altered to solidified bitumen, methane, CO_2, and H_2S. Both the solidified bitumen as well as late dolomite cement severely damaged the early pore systems in the Tensleep and Madison. However, additional by-products of the thermal degradation process (CO_2 and H_2S) dissolved remnant fossil fragments in the dolomite facies of the Madison to produce late, moldic porosity. The end result of this complex sequence of events is that at present the Phosphoria and Tensleep are tight and extremely well indurated, while the Madison possesses late moldic porosity from which to produce the methane reserves in the Tip Top field.

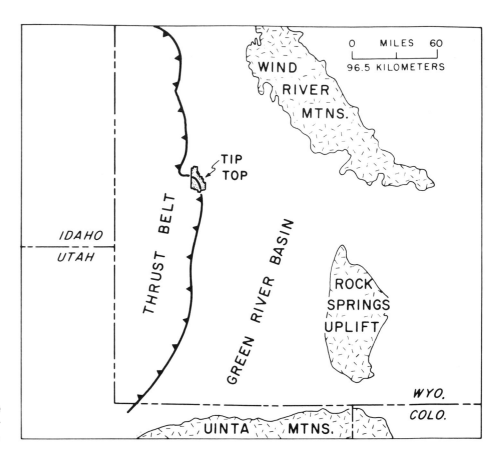

Figure 1—Regional map showing the location of the Tip Top field on the eastern edge of the thrust belt.

field (Figs. 1, 2) in western Wyoming provides an excellent opportunity to implement the program of diagenetic investigation outlined above. In particular, three formations (the Phosphoria, Tensleep, and Madison) provide a unique opportunity to study and integrate the organic and inorganic diagenesis in both varied lithologic and geochemical systems. First, the Tip Top field is an area of active hydrocarbon production where there exists the opportunity to relate the organic and inorganic geochemical systems. Second, diagenetic studies of these three formations, which are located in close stratigraphic proximity to each other (Fig. 2), allow one to compare and contrast the chemical reactions and controlling factors within three different sedimentary/geochemical systems; the organic-rich Phosphoria, the silici-clastic Tensleep, and the carbonate Madison. Thus, this particular stratigraphic interval provides an almost ideal situation for investigating both the control and timing of the organic and inorganic diagenetic systems.

Location of the Study Area and Sample Selection

The Tip Top field (Fig. 1) is located on the La Barge platform, which is a large, positive structural feature situated on the western margin of the Green River Basin at the juncture of the Moxa Arch and the Overthrust Belt. Four wells (Fig. 3) that cored the Paleozoic rocks in this area provided geologic control for this investigation. Most of these cores were taken from the footwall of the Darby thrust as shown in Figure 4. Rocks equivalent to those cored in the Tip Top field do not crop out on the surface, and consequently this study was based entirely on subsurface data. The study was limited to those intervals that were cored.

General Geology

The three formations investigated in this study were all deposited in an extensive miogeosyncline, which dominated deposition along the western margin of North America from the Late Precambrian until the Late Triassic. From the Upper Precambrian

through the Lower Triassic, the Tip Top area was either in a shelf edge or near shelf edge environment. Sedimentation in this environment was predominantly characterized by deposition of marine clastics and carbonates. This vast depositional system was interrupted and eventually destroyed by the onset of uplift and thrusting in the Late Triassic (Mansfield, 1927; Armstrong and Oriel, 1965) and Jurassic (Sippel and Schmitt, 1982).

Thin-skinned thrusting during the Sevier orogeny had a profound impact on the Tip Top region. Emplacement of the Darby thrust during the Paleocene was the major tectonic event influencing the study area. This fault thrust Paleozoic rocks over younger rocks along a bedding plane thrust in the Mesaverde (Fig. 4). Vertical separation on this thrust is approximately 20,000 ft and horizontal movement along the fault has probably been about 10 mi (Wallem, 1981). Both the increased overburden and abundant fracturing associated with emplacement of the Darby thrust had significant effects on

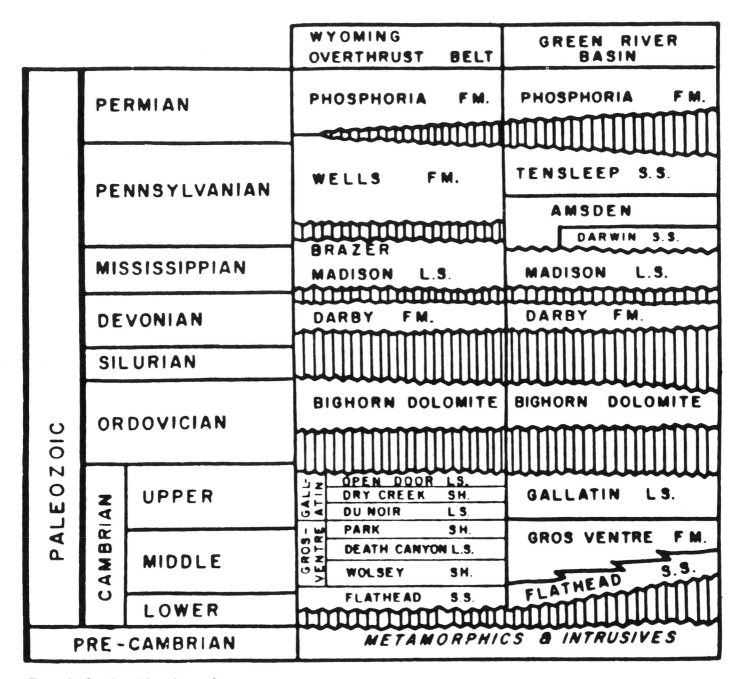

	WYOMING OVERTHRUST BELT	GREEN RIVER BASIN
PERMIAN	PHOSPHORIA FM.	PHOSPHORIA FM.
PENNSYLVANIAN	WELLS FM.	TENSLEEP S.S.
		AMSDEN
		DARWIN S.S.
MISSISSIPPIAN	BRAZER MADISON L.S.	MADISON L.S.
DEVONIAN	DARBY FM.	DARBY FM.
SILURIAN		
ORDOVICIAN	BIGHORN DOLOMITE	BIGHORN DOLOMITE

(PALEOZOIC — CAMBRIAN)

		Wyoming Overthrust Belt	Green River Basin
UPPER	GALLATIN	OPEN DOOR LS. / DRY CREEK SH. / DU NOIR LS.	GALLATIN L.S.
MIDDLE	GROS VENTRE	PARK SH. / DEATH CANYON L.S. / WOLSEY SH.	GROS VENTRE FM. / FLATHEAD S.S.
LOWER		FLATHEAD S.S.	

PRE-CAMBRIAN | METAMORPHICS & INTRUSIVES

Figure 2—Stratigraphic column of Paleozoic rocks in western Wyoming. The formations being investigated in this study are the Mississippian Madison Limestone, the Pennsylvanian Tensleep Sandstone, and the Permian Phosphoria. Modified after WGA stratigraphic column.

the diagenesis of the Phosphoria, Tensleep, and Madison at Tip Top.

Methods and Techniques

Several types of organic and inorganic analyses were performed on rocks from these three Paleozoic formations to facilitate interpretation of the diagenetic history. Organic analyses performed were total organic carbon, pyrolysis, vitrinite reflectance, and visual kerogen analysis. The inorganic studies utilized core analysis, field work, petrography, cathodoluminescence, X-ray diffraction, fine fraction analysis, scanning electron microscope observations, X-ray fluorescence examination, and isotopic analyses.

In addition to these analyses, several other techniques that should have provided more quantitative data to incorporate into the final interpretation were also attempted. These techniques included electron microprobe analyses of the major mineralogic phases in each formation, formation water analysis, observation of hydrocarbon fluid inclusions, and utilization of strontium isotopes as described by Burke et al (1982) to distinguish between marine and

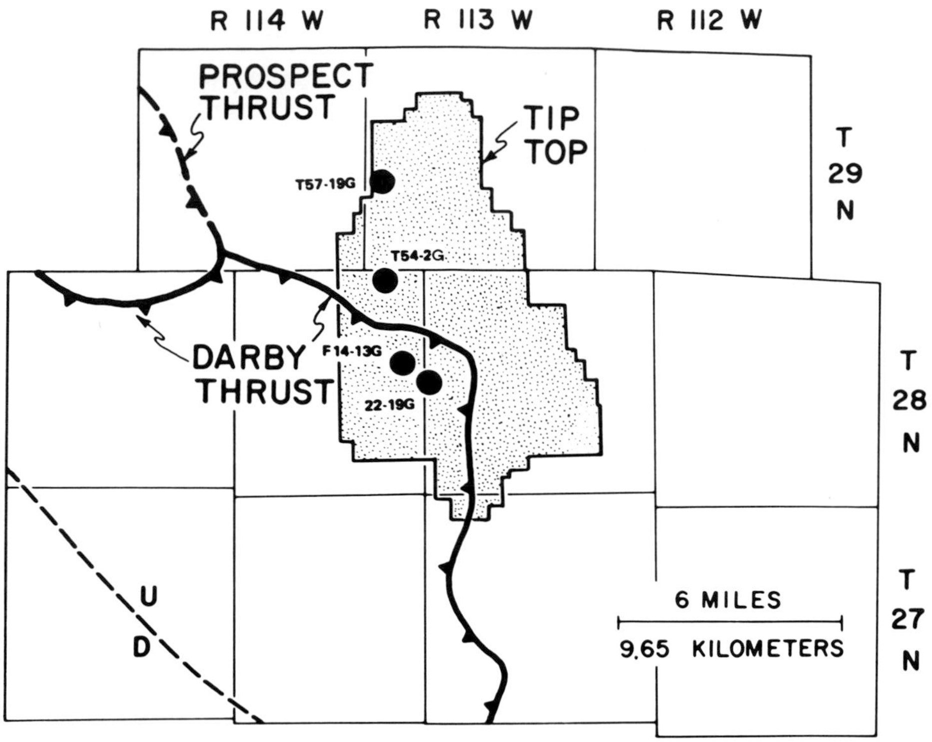

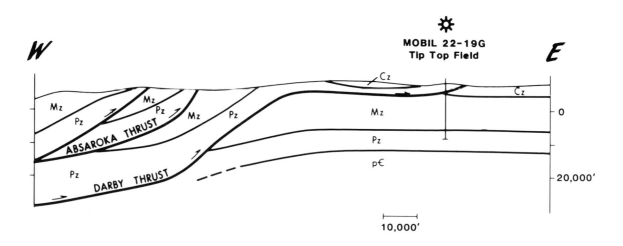

Figure 3—Location of well control within the Tip Top area used in this study. Dashed fault in the southwest corner is in the subsurface.

Figure 4—Cores used in this investigation are located structurally in the footwalls of major thrusts. This schematic cross section through the 22-19G well shows the structural position of the cored Paleozoic intervals in the footwall of the Darby thrust. The cross section is grossly simplified, and some faults are not shown.

fresh-water carbonates. These analyses were unsuccessful. Detailed description of the application and use of all these techniques is given in Edman (1982).

PHOSPHORIA

Stratigraphy and Depositional Environment

The Permian strata in the study area constitute a complex sequence of intricately interbedded lithologies (Sheldon, 1963; Peterson, 1980). These Permian units are underlain by the Tensleep Sandstone of Pennsylvanian age and are overlain by the Triassic Dinwoody Formation. Although the term Phosphoria Formation is often applied to the entire Permian interval, the sequence at this locality is actually comprised of three distinct Permian formations. These are the Phosphoria Formation, the Shedhorn Sandstone, and the Park City Formation. However, because the dominant Phosphoria Formation has frequently been used by the oil industry to describe the entire Permian interval at the Tip Top field, for the purposes of simplicity this same nomenclature will be adopted for the remainder of this report.

The primary constituents of the Phosphoria Formation at Tip Top are dark shale, phosphorite, and chert, which were deposited in the latest Paleozoic expression of the Cordilleran miogeosyncline. These organic-rich sediments (Table 1) accumulated along a continental shelf, which had considerable bathymetric relief (Peterson, 1980). A widespread oxygen-minimum zone coincided either continually or frequently with the sea floor as deposition progressed, and upwelling was both strong and persistent. Both cores and thin sections from the Phosphoria (Figs. 5, 6) display the imprint of this depositional environment. Thin laminae and abundant organic material in Phosphoria samples indicate anoxic conditions in which few bottom-feeding organisms could survive. Soft sediment deformation is also commonly observed, and may be attributed to gravity slumping on an unstable bottom.

Diagenesis

The paragenetic sequence given below comprises a composite sequence as derived from thin section samples

representative of the various Phosphoria lithologies present in the three wells that cored the Phosphoria. It is important to emphasize that the following sequence is a composite sequence determined from all the Phosphoria thin sections, and that no one thin section will contain all the various phases. Also, minor phases present in only a few isolated thin sections have been eliminated in order to simplify and generalize the overall sequence. The composite paragenetic sequence for the Phosphoria Formation is given schematically in Figure 7. The sequence is described below in detail and illustrated in Figures 8, 9, and 10.

Apatite

Chemically, the phosphate minerals in the Phosphoria Formation are fluorapatites with the basic formula $Ca_5(PO_4)_3F$. These minerals occur in a wide variety of forms: oolites, debris comprised of the phosphatic hard parts of various organisms, and authigenic precipitates. The authigenic precipitates include films of apatite coating other minerals, pore-filling material in brachiopod punctae (Figs. 8, 9), and euhedral crystals. In this diagenetic study, only the authigenic phosphate minerals are considered in the paragenetic sequence. The other phosphatic materials are detrital.

The controversy regarding the origin of phosphate minerals—whether they are precipitated directly from sea water or from interstitial waters in the unconsolidated sediment—has only recently been resolved. There is general agreement that most phosphate minerals are formed close to the sediment/water interface by precipitation from interstitial waters (Burnett and Oas, 1979; Bentor, 1980). Many of the chemical reactions that occur during precipitation of calcium phosphates from interstitial waters have yet to be precisely documented. However, several processes related to the precipitation of phosphate material are known. Nissenbaum et al (1972), Sholkovitz (1973), and Bentor (1980) have all observed that under anoxic conditions interstitial waters become greatly enriched in phosphate compared with the concentration of phosphate in bulk sea water. Decomposition of organic phosphates causes the dissolved phosphate concen-

tration in the interstitial water to rise (Berner, 1974). Furthermore, unconsolidated sediments containing both calcareous fossils and abundant carbonate generated by biological processes are probably already saturated with respect to calcite, and as the solubility of one or more of the apatites is exceeded, calcium phosphate precipitates.

Another critical factor that may actually determine whether or not calcium phosphate minerals precipitate is the concentration of Mg^{+2} ion in the interstitial water. Magnesium ions interfere with the precipitation of apatite and the preliminary removal of Mg^{+2} from the pore waters by dolomitization or the uptake of Mg^{+2} during the formation of clay minerals may be crucial to the subsequent precipitation of apatite (Burnett, 1974, 1977; Bentor, 1980). Dolomitization is accelerated by sulfate reduction (Baker and Kastner, 1980; Baker et al, 1981). Thus the association of authigenic phosphate minerals with organic-rich sediments may be a consequence of the chemical reactions that are prerequisites for the precipitation of calcium phosphates. Although the early dolomites or clay minerals that may have preceded apatite precipitation were not observed in any of the Phosphoria thin sections, they could easily have been removed by subsequent diagenetic reactions, leaving the present record incomplete as to the actual sequence of events.

Fossil Dissolution

With the exception of sponge spicules and brachiopod fragments, the fossil fragments in the Phosphoria are so broken and altered that they cannot be identified. The dissolution of fossil fragments immediately following the precipitation of calcium phosphates was probably related to removal of calcium from the pore waters by the calcium phosphate minerals. As precipitation of apatite began, the concentration of calcium ion decreased, which resulted in the dissolution of the calcareous fossils.

Calcite

Most of the calcite occurring at this stage in the paragenetic sequence is a fracture-filling material (Fig. 10). Some calcite is also present as a cement or as an authigenic replacement of pre-

TABLE 1

Sample Description	R₀ Mean	TAI	TOC	Pyrolysis	Thermal Maturity
(1) Twin Creek F14-13G, 8918 ft	1.59	—	0.08	—	Overmature
(2) Dinwoody T54-2G, 13,424 ft	2.42	—	0.10	—	Overmature
(3) Phosphoria F14-13G, 12,776 ft	2.90	—	0.35	—	Overmature
(4) Phosphoria F14-13G, 12,785 ft	2.91	4 to 4+	0.11	—	Overmature
(5) Phosphoria T54-2G, 13,459 ft	2.53	—	0.19	Overmature	Overmature
(6) Phosphoria T54-2G, 13,465 ft	—	—	0.14	—	Overmature
(7) Phosphoria 22-19G, 12,610–12 ft	2.59	—	0.63	Overmature	Overmature
(8) Phosphoria 22-19G, 12,653 ft	2.56	—	0.57	—	Overmature
(9) Phosphoria 22-19G, 12,700–05 ft	2.51	—	4.49	Overmature	Overmature
(10) Phosphoria 22-19G, 12,785–88 ft	2.51	—	0.30	Overmature	Overmature
(11) Tensleep F14-13G, 13,081 ft	—	—	0.08	—	Overmature
(12) Tensleep F14-13G, 13,111 ft	—	—	0.33	—	Overmature
(13) Tensleep T54-2G, 13,772 ft	2.62	—	1.05	Overmature	Overmature
(14) Tensleep T54-2G, 13,803 ft	2.60	—	1.34	Overmature	Overmature
(15) Tensleep 22-19G, 12,263–67 ft	2.43	—	0.58	—	Overmature
(16) Tensleep 22-19G, 13,318–21 ft	2.76	—	0.28	—	Overmature
(17) Madison F14-13G, 13,907 ft	2.59	—	0.07	—	Overmature
(18) Madison F14-13G, 13,941 ft	2.06	—	0.51	—	Overmature
(19) Madison T54-2G, 14,636 ft	2.60	—	0.69	—	Overmature
(20) Madison T54-2G, 14,670–71 ft	—	—	0.11	—	Overmature
(21) Madison 22-19G 13,766–69 ft	—	—	0.08	—	Overmature
(22) Madison 22-19G, 14,010–20 ft	2.67	—	0.43	—	Overmature
(23) Bighorn F14-13G, 15,154 ft	—	—	0.14	—	Overmature
(24) Bighorn 22-19G, 15,060–63 ft	—	—	0.10	—	Overmature

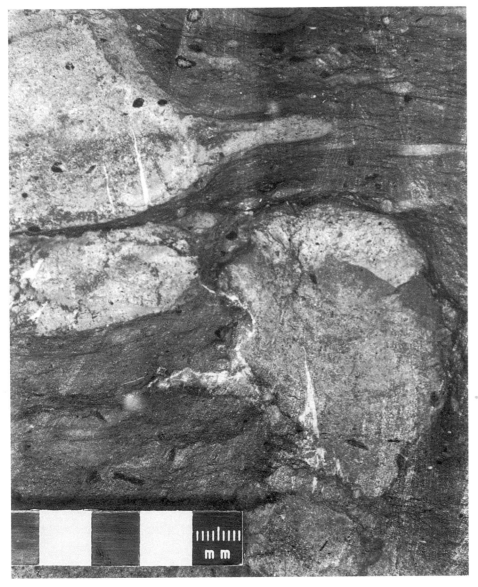

Figure 5—F14-13G, 12,772 ft, photograph of slabbed Phosphoria core illustrating soft sediment deformation. Deformed pods of predominantly light grey spicule material are surrounded by a micritic matrix. In this piece of core, micrite, spicules, shaly material, organic material, apatite, and fossil fragments (mostly bryozoans) are swirled together and deformed.

sive patches of chert when viewing the thin sections through uncrossed-nicols.

In addition to the in situ alteration of sponge spicules to chert, fossils and fossil fragments in the Phosphoria have undergone silicification (Figs. 8, 9). Detailed analysis of the silicification of Permian pelecypods and brachiopods from Wyoming is described in Schmitt and Boyd (1981). It is this silicification of fossil material as well as the introduction of authigenic silica as a cement that comprise the fourth step in the Phosphoria paragenetic sequence.

Although it is not shown in the paragenetic sequence, chalcedony is present as both a fracture- and pore-filling material in the Phosphoria. The abundant sponge spicules in the Phosphoria provided a ubiquitous source of silica, which was frequently mobilized along several different fracture systems. Chalcedony fracture-filling material first formed following the dissolution of fossil fragments, and subsequent episodes of chalcedony fracture filling continued to occur until the final diagenetic event, the precipitation of authigenic pyrite.

Hydrocarbon Migration

A detailed account of the maturation and migration of Phosphoria-sourced hydrocarbons will be given in the final portion of the Phosphoria section. Briefly, the fracturing accompanying the Sevier orogeny provided conduits for the migration of hydrocarbons generated by the progressive maturation of the abundant organic matter in the Phosphoria.

Dolomite

Dolomite exhibits two distinct morphologies in thin section. First, it displays a blocky, massive morphology in large patches associated with fractures. Second, dolomite is also present as rel-

viously existing matrix, but the calcite observed in these two modes was probably also introduced along fractures. The origin of this material is uncertain.

Silicification/Authigenic Silica

Silica occurs within the Phosphoria Formation in a variety of forms. Detrital quartz clasts and siliceous biogenic debris (primarily sponge spicules) were among the depositional components of the Phosphoria. Authigenic silica is

also present, and occurs as chert (microcrystalline quartz), chalcedony, and megaquartz. Authigenic silica is observed as a fracture-filling, authigenic replacement, and as an alteration product. Although much of this silica originated as siliceous biogenic debris, the general mineralogic progression—opal-A to opal-CT to quartz—has gone to completion in the Phosphoria as documented by examination and analysis of the X-ray diffraction patterns. Consequently, many of the original as well as intermediate textures and mineralogies of the silica minerals have been obliterated by diagenesis. It is still possible, however, to identify outlines of the original sponge spicules in mas-

Table 1—Organic maturation data for the Tip Top field.

Figure 6—F14-13G, 12,776 ft, Phosphoria photomicrograph. Swirling mixture of organic matter (dark), shale (white to light gray clasts), calcite (red), and dolomite (yellow-brown). Such assemblages are typical throughout much of the Phosphoria in the study area and are probably caused by both early soft sediment deformation and later diagenesis.

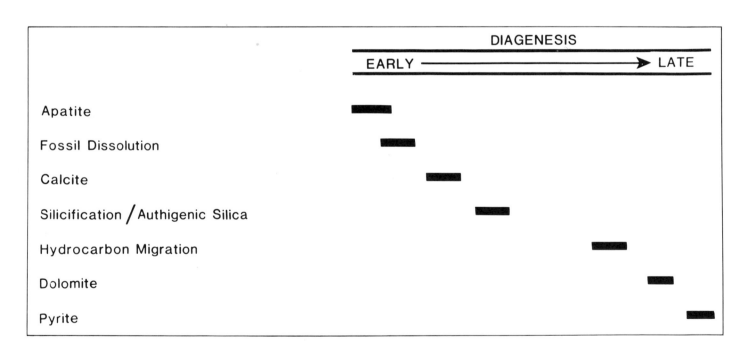

Figure 7—Paragenetic sequence of diagenetic events in the Phosphoria Formation.

atively euhedral rhombs that have replaced chert, silt grains, chalcedony, and apatite. Dolomitization of calcite also occurred during this stage.

Pyrite

The precipitation of euhedral pyrite crystals is the last major step in the paragenetic sequence. Although pyrite is more abundant in the organic-rich facies, it also replaced quartz, dolomite, calcite, and apatite (Figs. 8, 9, and 10). Formation of pyrite probably occurred as the result of a reaction between iron and thermally generated H_2S.

Hydrocarbon Maturation and Migration

Derivation of the maturation and migration history of Phosphoria-sourced hydrocarbons is based on several lines of evidence. This evidence includes geochemical analyses, maturation modeling, and textual/petrographic data. These data were interpreted in conjunction with the tectonic history of the Tip Top field to produce an integrated and consistent record of hydrocarbon migration in the area.

The percentage of total organic carbon (TOC) contained in Phosphoria samples from the Tip Top field is given

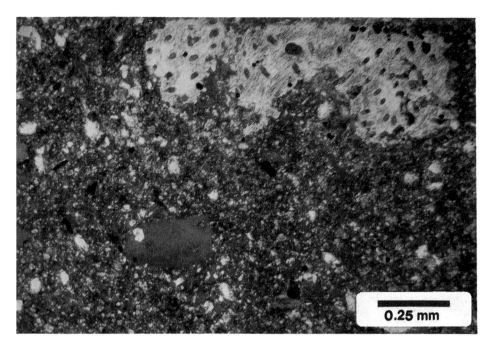

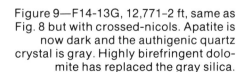

Calc

Pres

Hydr

Dolo

Pyrit

Figure
diagene
Format

Diagen

The
Tenslee
phoria
sequen
from n
basic T
simpler
and we
diagene
parage
Forma
Figure

*Very Ec
Cemen*

Thes
mutual
observe
thin sec
the earl
primar
frequer
these co
grains :
and the
of clast
In plac
been ce
drite o
well-so
cement
tively e
clastic
early c
itation
Pressu
not no

Figure 8—F14-13G, 12,771–2 ft, Phosphoria photomicrograph illustrating part of the Phosphoria paragenetic sequence. The brachiopod fragment in the upper right has been silicified. Prior to silicification, the brachiopod punctures were filled with authigenic apatite. In the lower left center of the photograph, a calcite fossil fragment (stained with alizarin red) is partially replaced by a euhedral quartz crystal. The quartz is then partially replaced by dolomite. Finally, pyrite has replaced the silicified brachiopod fragment along the top right edge of the fragment.

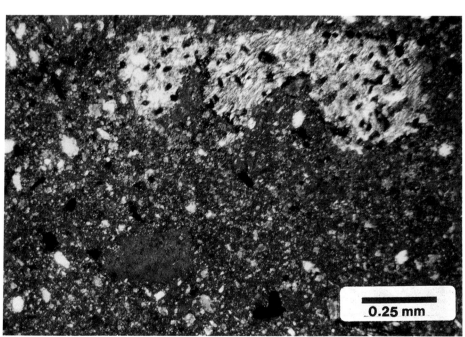

Figure 9—F14-13G, 12,771–2 ft, same as Fig. 8 but with crossed-nicols. Apatite is now dark and the authigenic quartz crystal is gray. Highly birefringent dolomite has replaced the gray silica.

in Table 1. The TOC measured in these samples is only that carbon that still remains after hydrocarbon generation. These thermally mature samples have already released some of their organic carbon as hydrocarbons, and consequently the TOC values reported in Table 1 are minimum values. These TOC data show that some of the Phosphoria facies contained sufficient organic carbon to serve as source beds.

A maturation model for the organic material in the Phosphoria Formation was derived by using the method described by Furlong and Edman (in press). This method incorporates the thermal effects of thrust faulting (Brewer, 1981) with techniques developed by Lopatin (1971). Vitrinite reflectance values (R_0) of 1.59 for the Jurassic Twin Creek Formation at a depth of 8918 ft in the F14-13G well and 2.59 for the Phosphoria at a depth of 12,610–12 feet in the 22-19G well were used to de-

termine the thickness of the overriding sediments in the Darby thrust sheet. Using these R_0 values and a series of thermal models for both values, it appears that 7300 ft is a reasonable thickness for the Darby plate in the Tip Top area.

A modified Lopatin diagram for the study area is shown in Figure 11. It shows that generation of liquid hydrocarbons began slightly before the onset of thrusting and continued for a brief period of time after the emplacement of

Figur
ria §
m
fra
solv
cipit
solu
mol
solv
then
clast
ter of
solut
an

Figure 16—F14-13G, 13,107 ft, Tensleep photomicrograph. Clastic grains in a matrix of highly birefringent early anhydrite cement. Note how well rounded the clastic grains remain. Apparently the anhydrite cement prevented interpenetration of the grains as well as the introduction of additional pore fluids into this section of the rock.

stitial
tiona
majo
perio
of thi
Phos
cons
and p

Th
had c
diage
orgai
enter
the ei
Fract
prob;
tion (
dolo
bene:
thern
hydr
the o
meth
the H
tem l
pyrit
phor
sequ

any significant burial and to have preceded both the silica cement and migration of hydrocarbons. The mutual exclusion of these two early cements is probably related to differences in their original environments of deposition. Calcite precipitation may have been controlled by the water table, whereas anhydrite was probably precipitated as gypsum from brines in interdunal ponds. Besides influencing early porosity distribution in the Tensleep, the anhydrite cement may have also influenced later diagenesis. It is possible that the sulfur in the anhydrite may have been a source for sulfur in the H_2S gas existing throughout the Paleozoic reservoirs. However, the trace amount of anhydrite observed in the Paleozoic cores may not have been volumetrically sufficient to have generated all of the H_2S. Thermal degradation of the liquid hydrocarbons probably contributed sulfur as well.

Pressure Solution/Silica Cementation

Silica cement and intergranular pressure solution of the detrital quartz grains cause the Tensleep Sandstone in the Tip Top field to be extremely well indurated. Commonly the silica overgrowths and interpenetration of grains are so abundant that the outlines of the original grains cannot be seen (Fig. 18). The well-rounded, well-sorted nature of the original clastic grains is completely

obscured by sutured contacts and overgrowths. Even cathodoluminescence did not reveal the original grain outlines (Fig. 19). Silica cementation and intergranular pressure solution took place as the Tensleep underwent continued burial beneath an ever-increasing sedimentary overburden. The silica cement was probably sourced internally within the Tensleep as a result of the pressure solution process. However, there was sufficient primary porosity remaining, even at burial depths greater than 12,000 ft, when hydrocarbons generated in the Phosphoria migrated into the Tensleep (Fig. 20).

Hydrocarbon Migration

The dominant characteristic of much of the Tensleep Sandstone cored in the Tip Top field is the dark gray color (Fig. 17). This is in contrast to the surface outcrops of the Tensleep near Snyder Basin and Middle Piney Creek, which are white to tan in color. Examination of Tensleep thin sections from Tip Top shows the dark gray color is the result of abundant solidified bitumen present throughout the rock (Fig. 20). Much of this solidified bitumen appears to have formed from what was originally a liquid phase rather than from a slurry of liquid and solid. However, as in the Phosphoria, brecciation, injection of breccia slurries, and hydro-

plastic flow associated with hydrocarbon emplacement are observed in both Tensleep cores (Fig. 21) and thin sections.

These observations are interpreted to indicate that hydrocarbons migrated along a pervasive fracture system into the Tensleep. Size limitations imposed by the diameter of the fracture and pore network resulted in fractionation of the denser, primary slurry material as coarser constituents were progressively left behind. The remnant liquid hydrocarbon phase migrated into what was left of the primary porosity in the Tensleep Sandstone: there is no evidence of any dissolution or secondary porosity existing at the time of migration (Fig. 20).

Dolomite

Dolomite replaced both silica and calcite cement. Dolomite rhombs also appear to have grown in pores that were partially filled with oil at one time (Fig. 33) and precipitated in the few isolated open pores that were left after hydrocarbon migration. This late stage dolomite and the solidified bitumen have almost completely destroyed any remnant primary pore space present prior to migration of the hydrocarbons (Figs. 18, 19).

Pyrite

Authigenic, euhedral pyrite crystals replaced calcite, silica, solidified bitu-

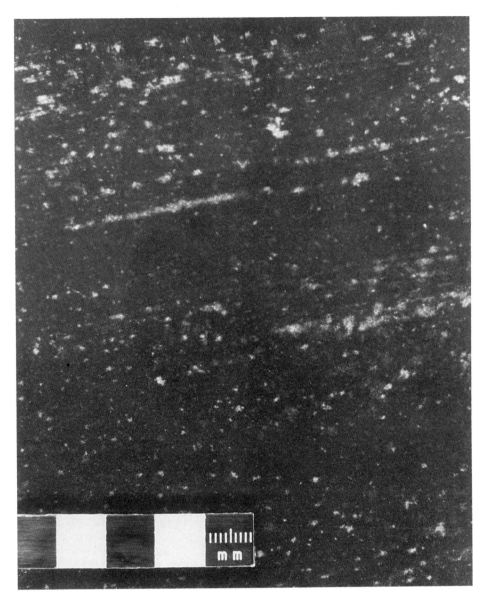

Figure 17—F14-13G, 13,096 ft. Typical Tensleep core observed in the study area. It is a very well indurated, silica-cemented rock that has been stained dark gray by solidified bitumen. The white blebs and laminae are portions of the rock that are cemented by calcite. This material remained light because the hydrocarbons could not penetrate the preexisting calcite cement.

paragenetic sequence (hydrocarbon migration, dolomite, and pyrite) are identical to the three phases of late Phosphoria diagenesis. Late diagenesis dominated development of the pore system that currently exists in the Tensleep. Abundant solidified bitumen and dolomite cement emplaced during late diagenesis have destroyed almost all porosity in the Tensleep, leaving a generally tight sandstone with only very localized traces of porosity.

MADISON

Stratigraphy and Depositional Environment

The Madison Limestone of Mississippian age rests disconformably on the Devonian Darby Formation (Benson, 1966) and is disconformably overlain by the Pennsylvanian Amsden Formation (Sando et al, 1975). Most of the Madison Limestone in the study area was deposited on a carbonate platform located on the western edge of a broad craton. During the interval from the Late Devonian to the Early Mississippian, Antler orogenic activity to the west affected sea level (Gutschick et al, 1980), and the carbonate shelf experienced episodic transgressive and regressive events.

These changes in sea level may have had a significant influence on carbonate sedimentation on the platform. The major portion of the cored interval is an alternating sequence of tight limestone and porous dolomite with the individual limestone and dolomite intervals being approximately 10 to 15 ft thick. The limestones are massive mudstones, wackestones, and packstones that presently have less than 2% porosity. Common allochems in these rocks are ooids, peloids, and fossil fragments (Figs. 22, 23). Most of the identifiable fossil fragments are echinoderms, pelecypods, or solitary rugose corals. The limestones containing fossil

men and dolomite. This relatively minor phase (less than 1% of most of the samples) is the last diagenetic event in this sequence. As in the Phosphoria, pyrite was probably precipitated by the reaction of H_2S with iron.

Tensleep Summary

The Tensleep Sandstone was deposited in the final stages of a regression that dominated sedimentation in the Cordilleran miogeosyncline throughout the Pennsylvanian. This quartzarenite is apparently the product of deposition in a coastal environment—both eolian dune and shallow marine sediments may be present. Like the Phosphoria, it seems to have undergone two distinct episodes of diagenesis. The early

paragenetic sequence consists of (1) precipitation of calcite or anhydrite cement, and (2) silica cementation and pressure solution. The initial phase of diagenesis was influenced by the original detrital particles and their interaction with fluids encountered both near the earth's surface and during burial. Early diagenetic processes controlled development of the pore system present at the time of migration of hydrocarbons from the Phosphoria during the Sevier orogeny. The early pore system provided a reservoir for the Phosphoria-sourced hydrocarbons. It is the remnants of this migration event that have stained the Tensleep dark grey.

The three stages in the late Tensleep

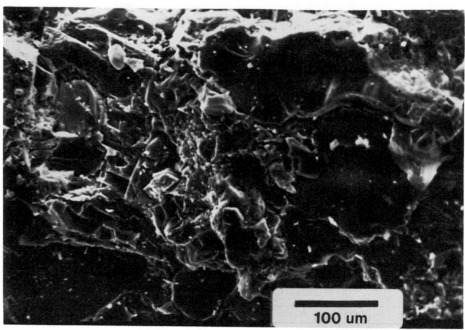

Figure 18—22-19G, 13,784 ft, Tensleep SEM photomicrograph of a sandstone cemented with silica and dolomite. It is difficult to identify individual clastic grains because of the pervasive silica cementation and pressure solution. The grains appear to be "welded" together. A patch of dolomite cement can be identified in the upper left corner. Note the distinctive dolomite cleavage. There is little, if any, porosity remaining in this sample.

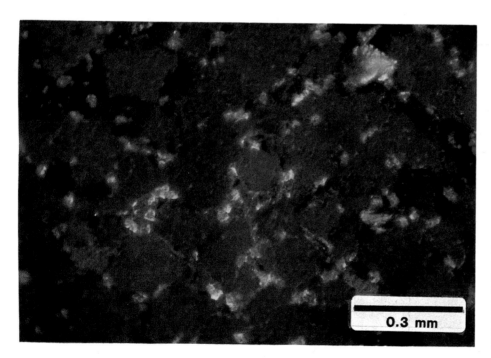

Figure 19—22-19G, 13,142–46 ft, Tensleep cathodoluminescence photomicrograph. Quartz grains luminesce a dull blue-gray while the feldspars luminesce light blue. Late dolomite cement luminesces bright red and has partially replaced some of the quartz grains. The black material is solidified bitumen.

fragments are generally greater than 50% lime micrite. The limestones containing ooids and peloids are usually well sorted and have little carbonate mud matrix. Sparry calcite is the dominant cement in the oolitic and peloidal facies.

The dolomites are very homogeneous and uniform. They are finely crystalline and commonly have 12% or greater porosity according to core analysis. The dolomites contain only minor fossil debris and are massive to vaguely laminated or mottled. Cryptalgal structures, rip-up clasts, and burrows indicative of intertidal deposition are present in the dolomite portions of the cores. Both the dolomites and the limestones are predominantly medium gray.

The alternating sequence of limestones and dolomites can be interpreted to be the result of a series of shallowing upward carbonate intervals (Edman, 1982). Utilizing the shallowing upward model of James (1980), the oolitic facies would represent high-energy, surf-zone deposition and the fossil fragment wackestones would be subtidal open marine or lagoonal facies. In this same model the dolomites repre-

Figure 20—T54-2G, 13,803 ft, Tensleep photomicrograph. This sandstone thin section has been impregnated with blue epoxy to show porosity. The distribution of the black bitumen approximates what the pore configuration must have been in the Tensleep at the time of hydrocarbon migration. This solidified bitumen is responsible for the dark color of the Tensleep Sandstone at the Tip Top field.

sent intertidal deposition. The alternation of deeper water subtidal limestones with intertidal dolomites could be related either to the previously mentioned tectonically induced sea level changes or to progradation of the carbonate sediments themselves causing a series of shallowing upward sequences (Edman, 1982).

Diagenesis

Composite paragenetic sequences for the Madison Limestone were determined primarily through the use of thin sections stained with alizarin red and potassium ferricyanide, and sample examination with the scanning electron microscope. The paragenetic sequence for the limestones is distinct from the sequence for the dolomites, and so the two sequences will be described separately. The limestones sequence, which will be considered first, is illustrated schematically in Figure 24.

Allochem and Lime Mud Deposition

The original depositional material of the Madison limestones appears to have been allochems (primarily ooids, peloids, and fossil fragments) and varying amounts of lime mud (Figs. 22, 23). Most of the recognizable fossil fragments are either pelecypod, echinoderm, or coral debris. The oolitic facies have the least amount of original mud

matrix, while the fossil fragment facies are frequently a 50–50 mixture of fragments and mud. The variation in the amount of lime mud is probably related to the original depositional environment.

Sparry Calcite Cement and Conversion of Fossil Fragments to Spar

Sparry calcite is the dominant cement in the well-sorted oolitic and peloidal lime grainstones. It is not observed in the lime wackestones and mudstones. The sparry calcite completely fills the pore spaces between particles and is not present in thin sections as a meniscus cement. Only slight deformation and compaction of the particles are observed to have occurred prior to precipitation of sparry calcite. These textural relationships are interpreted to indicate emplacement of sparry calcite cement in the phreatic zone.

Sparry calcite that has replaced the original fossil fragments is characterized by uniform extinction in contrast to the more undulatory type of extinction present in the original fossil material (Halley, 1982, personal communication). It is difficult to determine from the available textural evidence whether replacement of the fossil fragments took place before or after precipitation of sparry calcite cement. All that can be

stated is that both happened after deposition but before calcite fracture filling.

Calcite Fracture Filling

Fractures filled with calcite cut across both cement and sparry calcite fossil fragments.

Hydrocarbon Migration

The Madison limestones had little porosity at the time of hydrocarbon migration. The mudstones and wackestones had almost no primary porosity, and sparry calcite cement eliminated most of the primary porosity in the grainstones. As a result, hydrocarbons exist only in a few hairline fractures in the limestone portion of the Madison (Fig. 32). As in the Phosphoria and Tensleep Formations, these hydrocarbons have undergone thermal degradation so that the remaining material is solidified bitumen.

Silicification

This phase consists primarily of chert (microcrystalline quartz) replacing fossil fragments. In a few thin sections, patches of calcite spar and ooids are also partially replaced by chert. Unlike the euhedral authigenic silica precipitated in the next step, the replacement chert has an irregular morphology lacking any crystalline structure.

Figure 21—T54-2G, 13,797 ft, Tensleep core photograph illustrating fracturing and brecciation caused by the migration of hydrocarbons through the Tensleep. The migrating hydrocarbon material has ripped off fragments of the country rock and incorporated them into the migrating slurry.

Authigenic Silica

Small, euhedral authigenic quartz crystals replaced calcite spar and micrite (Fig. 23). Silica for both this phase and the silicification phase is interpreted to have been introduced along fractures.

Dolomite

Dolomite replaced both calcite and authigenic silica (Fig. 23). As with the silica phases, the dolomite may have been introduced along fractures.

Pyrite

Precipitation of small, euhedral pyrite crystals seems to have been the last major diagenetic phase.

The paragenetic sequence for the Madison dolomites is more complex than that for the limestones. However, much of the dolomite paragenetic sequence can be explained by two phenomena. First, there is the progression of the carbonate minerals toward more stable mineralogical phases. According to Bathurst (1971), carbonate minerals progress from high-Mg calcite or aragonite to calcite and finally to dolomite in order to achieve the most stable configuration. Thus, high-Mg calcite and aragonite will be most vulnerable to early alteration. Second, dolomitization is a highly fabric-selective process, and fine-grained material is preferentially dolomitized before grains and fragments composed of relatively pure calcite crystals (Davies, 1979). With these considerations in mind, the Madison dolomite paragenetic sequence is shown schematically in Figure 25.

Deposition of Lime Mud and Fossil Fragments

It is difficult to reconstruct accurately the original material that has undergone dolomitization. Estimates can be made, however, by examining thin sections in which the dolomitization process has not gone to completion, through depositional environment analysis, and by considering the stratigraphic position of the dolomitized sequence. Most of the dolomitized material is composed of a finely crystalline subhedral dolomite matrix with a few fossil fragments scattered throughout (Fig. 26). Almost all of the recognizable fossil fragments are crinoid plates. No remnant ooids or peloids are seen in the dolomites. There is both moldic (Fig. 27) and intercrystalline porosity. Some of the molds have a sufficiently characteristic shape to make identification possible. Many of these identifiable molds seem to have origi-

nated by the dissolution of pelecypod fragments (Fig. 28). The present dolomites seem to indicate an original material of lime mud and fossil fragments. It is difficult to estimate the original proportions of these constituents because it is impossible to determine the number of highly unstable fossil fragments that may have dissolved before dolomitization began. It seems likely though, because crinoid plates are commonly preserved even through late burial, and that these fragments are much more abundant in the limestones than in the dolomites, that there were probably fewer fossil fragments in the original dolomitized lime mud than in the sections that were

Figure 22—T57-19G, 15,252 ft, Madison limestone core photograph. This limestone is a massive, coarsely crystalline wackestone. Most of the white patches and blebs in this photograph are fossil fragments.

cal data. Textural evidence shows that dolomitization exhibits a strong fabric selectivity whereby finer-grained material is preferentially dolomitized first. This is observed in thin sections where the dolomitization process has not gone to completion. In such thin sections, dolomitization began after the sparry calcite cementation stage mentioned in the limestone paragenetic sequence. Dolomitization occurred first in the micritic material and extended into the sparry calcite only after dolomitization of the mud was near completion. Even at this stage, however, dolomitization of sparry calcite was not complete and a few calcite fossil fragments remained in the dolomitized matrix.

Examination of the entire Madison stratigraphic interval provides additional information regarding the dolomitization process. The boundary between the dolomitized lime mud and the remnant original limestone is sharp and abrupt. This is observed in cores and in thin sections. In one thin section, from 13,942 ft in the 22-19G well, the transition from limestone to dolomite occurs in an interval approximately one inch wide. This is interpreted to indicate that the zone containing the dolomitizing fluids must also have had very abrupt and distinct boundaries. According to P. K. Scott (1981, personal communication), the study area contains 12 such dolomitized intervals, which can be correlated across the entire Tip Top field. This can be used to infer that, although the boundaries of the zones undergoing dolomitization were very sharp, the zones of dolomitization were themselves fairly regional and areally extensive.

Carbon-isotope data are also helpful in interpreting the zones of dolomitization. The data included in Table 2 demonstrate that the $\delta^{13}C$ values for limestone and dolomite pairs are very similar. This is related to the vast reservoir of carbon in the original carbonate rocks compared with the minimal carbon present in the formation fluids. It

not dolomitized. It appears the primary depositional material was predominantly lime mud with minor fossil fragments. Such material may have been deposited in an intertidal lagoon or mud flat where occasional storms carried stenohaline marine fossil fragments into the lagoon or mud flat.

Dissolution of Least Stable Fossil Fragments

This step is somewhat hypothetical and, although no evidence of this event remains in the rock record, its occurrence is reasonable based upon what is known regarding the stabilization of carbonate minerals. It is possible that

fossil fragments comprised primarily of high-Mg calcite and aragonite may have dissolved prior to dolomitization, leaving no record of their original presence.

Dolomitization of Lime Mud

Dolomitization of primary lime micrite and fossil fragments was responsible for the generation of the porous zones that existed within the Madison during hydrocarbon migration. Several lines of evidence can be utilized to document and model the dolomitization process. These include textural, stratigraphic, and geochemi-

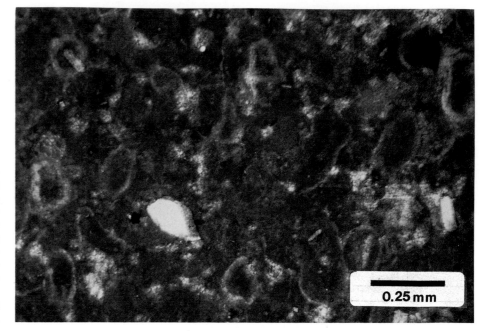

Figure 23—22-19G, 13,940 ft, photomicrograph of the Madison limestone facies. Fossil fragments are common within the micritic matrix of this limestone, which has been stained with alizarin red. Euhedral silicate crystals (white) have replaced both fossil fragments and the micritic matrix. Gray-green dolomite then replaced the authigenic silicates. Later, pyrite (dark) apparently replaced both the silicates and the dolomite.

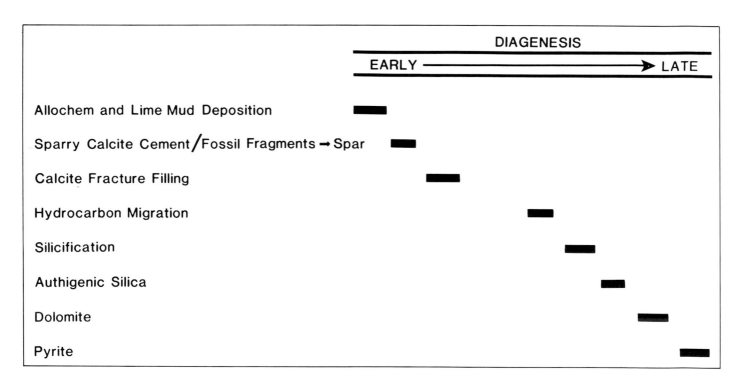

Figure 24—Paragenetic sequence of diagenetic events in the Madison limestone facies.

would have taken considerable pore volumes of formation waters to alter the $\delta^{13}C$ values fixed at the time of formation of these carbonates. Consequently, the $\delta^{13}C$ values for both the limestone and the dolomite are similar and represent artifacts of the original shallow marine deposition.

Oxygen-isotope values for several limestone/dolomite pairs within the Madison are also given in Table 2. The $\delta^{18}O$ values included here are within the range of values given by Land (1980) for ancient dolomites. However, considering the lack of precise geochemical models for explaining oxygen-isotope data in ancient dolomites (Land, 1980), it is not possible to use these data to constrain either the temperature of formation or fluid chemistry in the dolomitization model.

The data listed above can be added to information mentioned previously in the Madison section to delineate con-

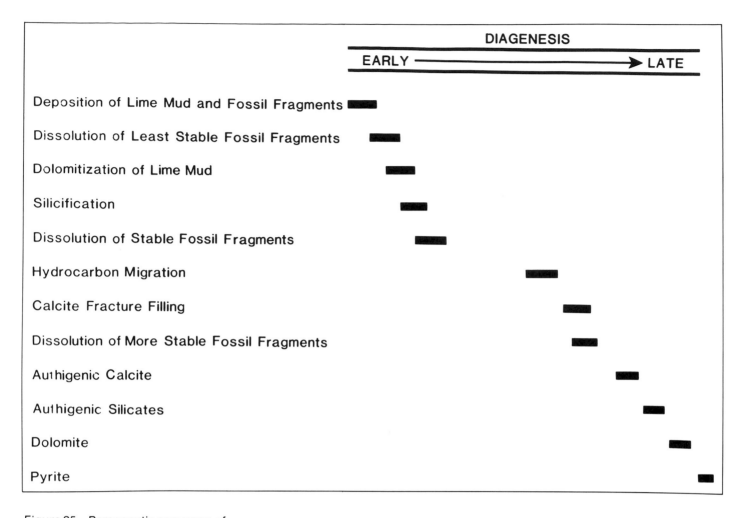

Figure 25—Paragenetic sequence of diagenetic events in the Madison dolomite facies.

straints on a dolomitization model. Any model for dolomitization must:

(1) Provide a mechanism that can generate the alternating dolomite/limestone intervals;

(2) Explain the sharp contacts between successive alternating layers of limestone and dolomite;

(3) Be a regional phenomenon because the porous zones can be correlated across the Tip Top field;

(4) Be consistent with the narrow range of $\delta^{13}C$ values, which indicate a marine origin for the primary carbonate material;

(5) Allow some burial of the lime ooids and peloids to occur prior to dolomitization;

(6) Explain why subtidal carbonates were not dolomitized while intertidal carbonates were.

(7) Be consistent with the lack of features characteristic of evaporites or a sabkha environment in the dolomitized interval.

Considering all of these constraints, a mixed-water dolomitization model most adequately fits the data. For this model to work, the intertidal carbonates must have been subaerially exposed during regional regression and become sites for freshwater recharge. The zone of freshwater recharge provides an area where fresh water interacts with connate sea water to form replacement dolomite (Folk and Land, 1975). The deeper water subtidal carbonates were never subaerially exposed and were not subjected to mixed-water dolomitization. The regressive events responsible for subaerial exposure could have been controlled by tectonic movement related to the Antler orogeny or by progradation of the carbonate sediments themselves. A regressive

phase would produce an interval consisting of dolomitized intertidal deposits, while a transgressive phase would result in deposition of subtidal limestones. Alternation of sea level would result in the observed alternation of lithologies. The mixed-water model satisfies most of the enumerated constraints for the dolomitization process.

Silicification

This process consists of replacement of the dolomitized lime mud by chert (microcrystalline quartz). It appears to be a localized phenomenon and is observed only in a few thin sections from the T57-19G well. The chert seems to act selectively to replace the dolomitized mud but not the fossil fragments. These fossil fragments are removed by a subsequent diagenetic process, leaving moldic porosity in the chert matrix. The source of the silica is uncertain.

TABLE 2

Sample Description	Carbonate Content μmoles/g of rock	$\delta^{13}C$ PDB in o/oo	$\delta^{18}O$ PDB in o/oo
(1) Madison T57-19G, 15,064 ft dolomite	10.49	+2.42	−1.99
(2) Madison T57-19G, 15,068 ft calcite	9.83	+1.51	−8.40
(3) Madison T57-19G, 15,214 ft dolomite	9.96	+2.42	−3.51
(4) Madison T57-19G, 15,217 ft calcite	10.07	+2.25	−6.33
(5) Madison 22-19G, 13,939 ft dolomite	10.08	+3.37	+2.33
(6) Madison 22-19G, 13,940 ft calcite	9.55	+1.19	−8.67
(7) Madison 22-19G, 13,953 ft calcite	9.36	+1.70	−12.11
(8) Madison 22-19G, 13,957 ft dolomite	10.54	+1.99	−2.90

Table 2—Carbon and oxygen isotope data.

step (Figs. 29, 30). Either the fluid chemistry changed from that existing during dissolution of the fossil fragments or this calcite—like the calcite of the remnant crinoid plates—is more resistant to dissolution by CO_2 and H_2S.

Authigenic Silicates

Both quartz and feldspar precipitated in the late molds (Figs. 28–30) and replaced the previously described authigenic calcite. Like the authigenic calcite, the crystals of these authigenic silicates are euhedral. In particular, the quartz crystals are doubly terminated.

Dolomite

Late dolomite apparently precipitated in pores that were previously partially filled with oil, in moldic porosity, and also replaced both authigenic calcite and authigenic silicates (Figs. 29–33). This late dolomite phase generally appears to be more euhedral and has slightly larger crystals than does the matrix dolomite. All of the post-hydrocarbon migration mineralogic phases mentioned above (calcite, silicates, and dolomite) are interpreted to have been introduced as fluids moving along fractures into the stratigraphic interval of interest.

Pyrite

As in the Phosphoria, Tensleep, and Madison limestone, the last main diagenetic event is the precipitation of

small, euhedral pyrite crystals from the reaction of H_2S with iron.

Madison Summary

The Mississippian Madison carbonate sequence was deposited on a shallow platform located on the western edge of a broad craton. Diagenesis in the Madison carbonates is generally more complex than diagenesis observed in the Phosphoria and Tensleep Formations. The paragenetic sequence for the limestones differs from that in the dolomites, and both sequences exhibit numerous phases. Despite the apparent complexity, diagenesis for both facies can be broadly divided into two periods —an early phase prior to migration and a later stage including migration and extending through the precipitation of pyrite.

Figure 29—T57-19G, 15,247 ft, Madison photomicrograph illustrating part of the paragenetic sequence in the dolomite facies. In this thin section, late moldic porosity (pale blue) has formed in a matrix of dolomicrite. Authigenic calcite (stained pink by alizarin red) has precipitated into the mold. Later, the regular lath-shaped silicate crystal in the center of the photomicrograph replaced this calcite. Late-stage dolomite replaced both the calcite and the silicate. The large black irregular pyrite crystal is the last major diagenetic phase present in this slide.

Figure 30—Same as Fig. 29 but with crossed-nicols. The authigenic silicate shows up clearly as a white lath in the center of the photomicrograph. Late dolomite that replaced the earlier authigenic calcite is the lighter, more birefringent material in the almost extinct calcite crystal in the lower right.

Diagenesis prior to hydrocarbon migration can be related to the initial detrital components and paleogeography of deposition. Subtidal limestones containing abundant allochems underwent gradual burial and compaction and were never in contact with dolomitizing fluids. Instead, the primary porosity present in the lime grainstones was destroyed relatively early in the burial history by precipitation of sparry calcite, and subsequently the limestones were tight and impermeable at the onset of migration. In contrast, the dolomite facies were predominantly lime mud with relatively fewer allochems. These intertidal lime muds underwent mixed-water dolomitization, which was responsible for generating the porosity and permeability present in the dolomites during hydrocarbon migration.

Later diagenesis in the Madison, beginning with hydrocarbon migration and extending through the precipitation of pyrite, was apparently influenced by Sevier fracturing. Most of the

Figure 31—22-19G, 13,784 ft, SEM photomicrograph illustrating moldic and intercrystalline porosity in the Madison dolomite. Note how some of the molds have late authigenic dolomite precipitated in them. This authigenic dolomite probably changes the shapes of the molds so that in the cores they appear as vugs rather than distinct, identifiable molds.

constituents observed in this phase of diagenesis (hydrocarbons, calcite, silicates, dolomite, and pyrite) were probably introduced into the Madison carbonates as liquids migrating along a pervasive fracture system. Many of these late phases are present in both the limestone and dolomite facies.

Dolomitization appears to have been the critical step in the diagenesis of the Madison carbonates. It seems to have controlled development of both early and, indirectly, late porosity. Subtidal limestones that were buried beyond the zone of mixed-water dolomitization were tight during migration and hydrocarbons were present in this facies only along small, hairline fractures. However, the dolomitized muds provided excellent reservoirs for the Phosphoria-sourced hydrocarbons and abundant oil filled the intercrystalline and early moldic porosity. After the emplacement of the liquid hydrocarbons in the dolomite, deep burial of these reservoir rocks resulted in the thermal degradation of the hydrocarbons. By-products of these degradation reactions were solidified bitumen, methane, CO_2, and H_2S. The CO_2 and H_2S dissolved remnant calcite fossil fragments in the dolomite, creating late, deep secondary porosity. Because the impermeable limestones had never contained abundant liquid hydrocarbons, they were

not exposed to the degradation products, CO_2 and H_2S. Thus the opportunity for similar development of late secondary porosity in the limestones never existed.

DISCUSSION

The Phosphoria, Tensleep, and Madison Formations were deposited in an extensive miogeosyncline that dominated deposition along the western margin of North America from the Late Precambrian until the Late Triassic. These formations, even though they are located in close stratigraphic proximity, were deposited in diverse environments. The Phosphoria represents deposition in an organic-rich marine system influenced by coastal upwelling. The Tensleep contains siliciclastic rocks that were deposited in an eolian dune/shallow marine system. The Madison carbonates were deposited on a broad shallow shelf. Despite these obvious differences, some generalizations can be made regarding the diagenesis of these three formations.

First, in each formation the diagenesis can be divided into two broad periods. These periods are an early phase prior to hydrocarbon migration and a late phase beginning with hydrocarbon migration and ending with the precipitation of pyrite. Early diagenesis con-

trolled the porosity present at the time hydrocarbons were migrating, late diagenesis strongly influenced the porosity configuration observed in the rocks today. Examination of the paragenetic sequences for each formation shows that the early period of diagenesis was different for each formation, while the late phase exhibits similarities in late diagenesis between the three formations.

During early diagenesis many of the observed processes involved interactions between initial depositional components and interstitial fluids as the sediments underwent lithification and burial. These early processes include such reactions as precipitation of apatite in the Phosphoria, pressure solution and silica cementation in the Tensleep, conversion of fossil fragments to spar in the Madison limestones, and mixed-water dolomitization of lime mud in the Madison dolomites. The diversity of early diagenetic reactions in the three formations is directly related to differences in primary depositional constituents as well as variations in early burial history, which were functions of paleogeography. In the most general sense, this period of early diagenesis appears to have been controlled by factors dependent upon depositional environment.

By comparison, the late diagenetic

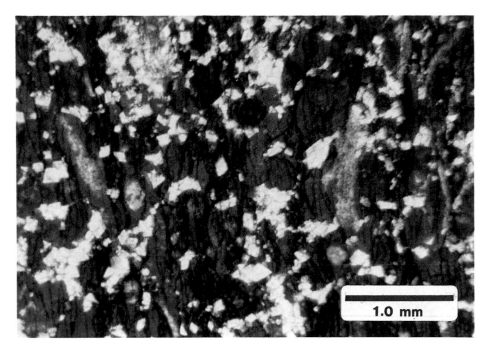

Figure 32—22-19G, 13,968 ft, Madison limestone photomicrograph where the hydrocarbon-filled fractures cut across a matrix of lime micrite and fossil fragments. After emplacement of the hydrocarbons, dolomite replaced both the lime matrix and the fossil fragments. Growth of these dolomite crystals has clearly taken place after hydrocarbon migration, as is documented by numerous euhedral dolomite rhombs extending across the hydrocarbon-filled fractures.

processes in these formations seem to have been very similar. The late paragenetic sequences for the Phosphoria and Tensleep are the same: migration of hydrocarbons, dolomite, and pyrite. These same three phases are also present in both the Madison limestones and dolomites although the paragenetic sequences in the carbonates are complicated by intervening calcite and silicate phases. The correspondence of late diagenetic processes across 2000 vertical stratigraphic feet of diverse lithologies that display distinctly different early paragenetic sequences is too consistent to be coincidental. The simplest explanation would be for some pervasive, widespread mechanism to have influenced and controlled late diagenesis in all three formations.

The initiation of late diagenesis in each formation began with the migration of hydrocarbons. This migration event was associated with fracturing and brecciation related to emplacement of the Darby thrust. Temporally, the beginning of late diagenesis is very close to the onset of thrusting in the study area. The coincidence of thrusting with the initiation of late diagenesis provides the most reasonable model for explaining the similarity in late diagenesis among the three formations.

Observations document the migration of hydrocarbons along a pervasive fracture system into the stratigraphic interval of interest. Correlation between late diagenetic events in the three formations is interpreted to indicate the continued presence of a widespread fracture system throughout much of the late diagenetic history of the area. This fracture system would provide a mechanism for introducing a relatively consistent and similar sequence of fluids into the Phosphoria, Tensleep, and Madison. The first fluid pulse was predominantly hydrocarbons followed by dolomitic liquids, and then fluids containing iron, which subsequently reacted with H_2S to precipitate trace amounts of pyrite. This model of tectonic control over late diagenesis is consistent with textural and petrographic observations as well as the regional structural evolution of the area.

Finally, in addition to generating the fracture system discussed above, tectonics appear to have influenced late diagenesis in yet another manner. Burial of the stratigraphic sequence beneath the Darby thrust had a significant impact on the hydrocarbon material that had migrated into the Tensleep and Madison reservoirs. Increased temperatures and pressures resulting from an additional 7300 ft of overburden caused thermal degradation of the once liquid hydrocarbons. The products of these degradation reactions were solidified bitumen, methane, CO_2, and H_2S. The solidified bitumen greatly reduced porosity and permeability in both the Tensleep and Madison. Furthermore, solidified bitumen in conjunction with late dolomite eliminated most porosity in the Tensleep and destroyed its reservoir capacity. The Madison would have undergone a similar post-migration history had it not been for the interaction of H_2S and CO_2 with remnant fossil fragments in the Madison dolomite. These gases dissolved fossil fragments in the dolomite to produce late moldic porosity in the dolomite facies. Although generation of moldic porosity did not enhance permeability in the dolomites, it did increase porosity and in places core measurements indicate up to 20% porosity. The low permeability, however, is not detrimental to gas production, and the Madison dolomites are currently the best reservoir for producing the methane present at Tip Top.

CONCLUSIONS

The three formations that have been examined for this investigation exhibit two distinct phases of diagenesis. These are given the chronological designations "early" and "late." The early phase of diagenesis was primarily con-

Figure 33—22-19G, 13,927–32 ft, Madison dolomite photomicrograph showing a euhedral dolomite crystal in a pore partially filled with solidified bitumen. This dolomite is believed to be the same post-hydrocarbon migration phase of dolomite as illustrated in Fig. 32. Absence of both embayments and other irregularities filled with oil in the surface of the crystal, as well as the apparent lack of oil staining of the dolomite are used to infer precipitation of dolomite after hydrocarbon emplacement. Precipitation of the dolomite crystal may have been initiated in an open space generated by shrinkage of the liquid hydrocarbon material during the thermal degradation process. Continued crystal growth produced an unstained dolomite rhomb with clean, sharp edges. Similar textural relationships are observed in the Phosphoria and Tensleep Formations.

0.25 mm

trolled by the depositional environment in which the sediments accumulated. The term depositional environment is defined here in the broadest sense and includes detrital lithology, paleogeography, depositional setting, and the chemical environment. This early diagenetic phase is different for each of the formations examined—the variation in the early diagenesis being a direct reflection of the diversity in depositional environments that exists among the three formations. The period of early diagenesis was responsible for the pore network present in the rocks at the time of hydrocarbon migration.

The organic material in the Phosphoria entered the liquid window just prior to the emplacment of the Darby thrust. Tectonic activity associated with the onset of the Sevier orogeny in the study area had a dramatic effect on both the migration of hydrocarbons and the late episode of diagenesis. Fractures generated during the Sevier orogeny provide conduits for the migrating hydrocarbons. These hydrocarbons filled the porous Tensleep Sandstone and Madison dolomite. Subsequent tectonic movements provided additional conduits for formation waters, which controlled the late diagenetic phases. The late phase is relatively uni-

form throughout the entire stratigraphic interval investigated. The uniformity of the late diagenesis is believed to result from the pervasive and widespread migration of a relatively constant sequence of fluids along fractures that penetrated all three formations. The late phase of diagenesis determined the porosity configuration that currently exists in these rocks.

Finally, although it is more difficult to document, this study indicates that the organic diagenetic reactions did influence the inorganic system. Of greatest importance in this study is the impact the organic reactions had on the pore systems in the reservoir rocks. Thermal degradation of the liquid hydrocarbons, which occurred as the result of deep burial, produced solidified bitumen, methane, CO_2, and H_2S as by-products of the degradation process. The solidified bitumen, in conjunction with late dolomite cement, almost completely destroyed the porosity in the Tensleep Sandstone and eliminated much of the intercrystalline and early moldic porosity in the Madison. However, the CO_2 and H_2S generated in thermal degradation reactions caused dissolution of calcite fossil fragments in the Madison dolomites creating late, deep, secondary moldic porosity. As a consequence of this sequence of events,

the Madison is currently the best reservoir rock for producing the methane present in the Tip Top field today.

ACKNOWLEDGMENTS

We would like to thank Mobil for providing the major support for this investigation. Numerous individuals within Mobil also generously donated their time, information, and support for this project. In particular, we would like to thank Phyllis Scott, Dr. Eve Sprunt, Maury Mendenhall, Selena Dixon, Aus Melker, and Dr. Dave Eby. Chevron U.S.A. aided in the vitrinite reflectance analyses. Financial aid to J. D. Edman during the course of this study was provided by the Department of Geology and Geophysics at the University of Wyoming. Drs. Donald L. Blackstone, James I. Drever, Kevin P. Furlong, and Duncan Harris reviewed a preliminary copy of this manuscript.

SELECTED REFERENCES

Armstrong, F. C., and S. S. Oriel, 1965, Tectonic development of Idaho–Wyoming Thrust Belt: Bulletin of the American Association of Petroleum Geologists, v. 49, no. 11, p. 1847–1866.

Baker, P. A., and M. Kastner, 1980, The origin of dolomite in marine sediments: Geological Society of America Abstracts, 93rd Annual Meeting, Atlanta, p. 381–382.

Baker, P. A., M. Kastner, and G. E. Anderson, 1981, Mechanism and kinetics of sulfate inhibition on dolomitization of calcium carbonate: Bulletin of the American Association of Petroleum Geologists, v. 65, p. 893–894.

Bathurst, R. G. C., 1971, Carbonate sediments and their diagenesis, in Developments in sedimentology 12: New York, Elsevier Publishing Co., 658 p.

Benson, A. L., 1966, Devonian stratigraphy of western Wyoming and adjacent areas: Bulletin of the American Association of Petroleum Geologists, v. 50, no. 12, p. 2566–2603.

Bentor, Y. K., 1980, Unsolved problems, in Y. K. Bentor, ed., Marine phosphorites: Society of Economic Paleontologists and Mineralogists Special Publication 27, p. 3–18.

Berner, R. A., 1974, Kinetic models for the early diagenesis of nitrogen, sulfur, phosphorus, and silicon in anoxic marine sediments, in E. D. Goldberg, ed., The sea, v. 5: New York, John Wiley and Sons, p. 427–450.

Brewer, J., 1981, Thermal effects of thrust faulting: Earth and Planetary Science Letters, v. 56, p. 233–244.

Burke, W. H., R. E. Denison, E. A. Hetherington, R. B. Koepnick, H. F. Nelson, and J. B. Oho, 1982, Variation of seawater 87Sr/86Sr throughout Phanerozoic time: Geology, v. 10, no. 10, p. 516–519.

Burnett, W. C., 1974, Phosphorite deposits from the sea floor off Peru and Chile: radiochemical and geochemical investigations concerning their origin: Ph.D. dissertation, University of Hawaii, 196 p.

———, 1977, Geochemistry and origin of phosphorite deposits from off Peru and Chile: Geological Society of America Bulletin, v. 88, p. 813–823.

Burnett, W. C., and T. G. Oas, 1979, Environment of deposition of marine phosphate deposits off Peru and Chile (abs.), in P. J. Cook, and J. H. Shergold, eds., Proterozoic and Cambrian phosphorites: Report of the 1st International IGCP Project 156, Canberra, ANU Press, p. 54–56.

Davies, G. R., 1979, Dolomite reservoir rocks: processes, controls, porosity development: American Association of Petroleum Geologists Short Course on Carbonate Porosity.

Edman, J. D., 1982, Diagenetic history of the Phosphoria, Tensleep and Madison Formations, Tip Top Field, Wyoming: Unpublished Ph.D. dissertation, University of Wyoming, 229 p.

Folk, R. L., and L. S. Land, 1975, Mg/Ca ratio and salinity: two controls over crystallization of dolomite: Bulletin of the American Association of Petroleum Geologists, v. 59, no. 1, p. 60–68.

Fryberger, S. G., 1979, Eolian–fluviatile (continental) origin of ancient stratigraphic trap for petroleum in Weber Sandstone, Rangely oil field, Colorado: Mountain Geologist, v. 16, no. 1, p. 1–36.

Furlong, K. P., and J. D. Edman, in press, Geographical approach to the determination of hydrocarbon maturation in overthrust terrains: Bulletin of the American Association of Petroleum Geologists.

Gutschick, R. C., C. A. Sandberg, and W. J. Sando, 1980, Mississippian shelf margin and carbonate platform from Montana to Nevada, in T. D. Fouch and E. R. Magathan, eds., Paleozoic paleogeography of west-central United States: West-Central United States Paleogeography Symposium, Denver.

James, N. P., 1980, Facies models 10: shallowing upward sequences in carbonates, in R. G. Walker, ed., Facies models: Geoscience Canada, Reprint Series 1, p. 109–119.

Land, L. S., 1980, The isotopic and trace element geochemistry of dolomite: the state of the art, in D. H. Zenger, J. B. Dunham, and R. L. Ethington, Society of Economic Paleontologists and Mineralogists Special Publication 28, p. 87–110.

Lopatin, N. V., 1971, Temperature and geologic time as factors in coalification (in Russian): Akademiya Nauk SSR Izvestiya, Seriya Geologicheskaya, n. 3, p. 95–106.

Mankiewicz, D., and Steidtmann, J. R., 1979, Depositional environments and diagenesis of the Tensleep Sandstone, eastern Big Horn basin, Wyoming, in P. A. Scholle and P. R. Schluger, eds. Aspects of diagenesis: Society of Economic Paleontologists and Mineralogists Special Publication 26, p. 319–336.

Mansfield, G. R., 1927, Geography, geology and mineral resources of part of southeastern Idaho: United States Geological Survey Professional Paper 152, 453 p.

Moore, C. H., and Y. Druckman, 1981, Burial diagenesis and porosity evolution, Upper Jurassic Smackover, Arkansas and Louisiana: Bulletin of the American Association of Petroleum Geologists, v. 65, p. 597–628.

Nissenbaum, A., B. J. Presley, and I. R. Kaplan, 1972, Early diagenesis in a reducing fjord, Saanich Inlet, British Columbia, I, Chemical and isotopic changes in major components of interstitial water: Geochemica et Cosmochimica Acta, v. 36, p. 1007–1027.

Orr, W. L., 1977, Geologic and geochemical controls on the distribution of hydrogen sulfide in natural gas, in R. Campos and J. Goni, eds., Proceedings of the 7th International Meeting on Organic Geochemistry: Revista Espanola de Micropaleontologia, p. 571–597.

Peterson, J. A., 1980, Depositional history

and petroleum geology of the Permian Phosphoria, Park City and Shedhorn Formations, Wyoming and southeastern Idaho: United States Geological Survey, Open File Report 80-667, 42 p.

Sando, W. J., M. Gordon, Jr., and J. T. Dutro, Jr., 1975, Stratigraphy and geologic history of the Amsden Formation (Mississippian and Pennsylvanian) of Wyoming: United States Geological Survey Professional Paper 848-A, 78 p.

Schmitt, J. G., and D. W. Boyd, 1981, Patterns of silicification in Permian pelecypods and brachiopods from Wyoming: Journal of Sedimentary Petrology, v. 51, p. 1297–1308.

Sheldon, R. P., 1963, Physical stratigraphy and mineral resources of Permian rocks in western Wyoming: United States Geological Survey Professional Paper 313-B, 273 p.

Sholkovitz, E. R., 1973, Interstitial water chemistry of the Santa Barbara Basin sediments: Geochimica et Cosmochimica Acta, v. 37, p. 2043–2073.

Sippel, K. N., and J. G. Schmitt, 1982, Early Cretaceous depositional and structural development of Wyoming–Idaho–Utah foreland basin (abs.): Bulletin of the American Association of Petroleum Geologists, v. 66, p. 631.

Surdam, R. C., S. Boese, and L. J. Crossey, 1982, Role of organic and inorganic reactions in development of secondary porosity in sandstones (abs.): Bulletin of the American Association of Petroleum Geologists, v. 66, p. 635.

Wallem, D. B., 1981, Environmental, diagenetic and source rock analysis of the Bear River Formation, Western Wyoming: unpublished M.S. thesis, University of Wyoming, 101 p.

Lithofacies, Diagenesis and Porosity of the Ivishak Formation, Prudhoe Bay Area, Alaska

John Melvin
Angela S. Knight
Sohio Petroleum Company
San Francisco, California

INTRODUCTION

The Prudhoe Bay field is the largest oil field in North America, with recoverable reserves of 9.6 billion barrels. It is located in the northeastern part of Alaska's North Slope, in close proximity to Prudhoe Bay (Fig. 1). The field was discovered in 1968 with the drilling of the ARCO-Humble Prudhoe Bay State 1 well. Since that time a large number of wells have been drilled in the field area, particularly following commencement of production in 1977. In addition, the number of exploration wells in the onshore and offshore environs of Prudhoe Bay field continues to grow.

The geological setting of Alaska's North Slope in general, and of the Prudhoe Bay field reservoirs in particular, has been discussed by a number of authors (Rickwood, 1970; Detterman, 1970; Morgridge and Smith, 1972; Jones and Speers, 1976; Jamison et al, 1980). The regional geological setting of the field will be described only briefly herein. The purpose of this paper is to discuss the depositional and diagenetic histories of the main reservoir at Prudhoe Bay, namely, the Ivishak Formation. These aspects will be demonstrated as having a controlling effect on reservoir parameters.

The number of wells in the Prudhoe Bay area provides a large data base for study of the reservoir. The present study is the result of examination of a number of selected wells that had full reservoir penetration by core. The cores were analyzed with particular regard to lithofacies distribution, and the petrography of rock material representative of those lithofacies.

ABSTRACT. The Permo-Triassic Ivishak Formation is the main reservoir interval of the Prudhoe Bay field, North Slope, Alaska. Studies of cores from the field area reveal that porosity development within the Ivishak Formation has a complex relationship dependent on both depositional (lithofacies) and post-depositional (diagenetic) history.

Four dominant lithofacies are identified: (1) interbedded very fine sandstones and mudstones; (2) parallel laminated carbonaceous fine sandstones; (3) multistory upward-fining medium sandstones; and (4) conglomerates. These lithofacies occur everywhere as upward-coarsening to conglomerate sequences. In the main field area the coarsening sequence is overlain by a gross upward-fining sequence of gravelly to medium-grained multistory sandstones, which thins dramatically to the north. Consideration of lithofacies and thickness variation leads to an interpretive model concerning evolution of the basin with respect to tectonics and sedimentation. Thus initial progradation of an active alluvial fan–delta system from the northeast was replaced by progressive transgression from the south of more distal upon proximal facies.

Petrographic characteristics of the rocks reveal that porosity development is related to the diagenetic history of each lithofacies. Porosity within the medium-grained sandstones is predominantly secondary because of dissolution of grains and early grain-replacement calcite. Porosity within the conglomeratic intervals appears to be much more of a primary (textural) origin.

GEOLOGICAL SETTING

The main structural features of the North Slope are the Barrow Arch, which parallels the Arctic coastline and plunges eastward from Point Barrow, and the Colville Trough to the south. The latter is a strongly asymmetric syncline with its axis close to the front of the central and western Brooks Range (Morgridge and Smith, 1972) (Fig. 2). The Prudhoe Bay field is located on the easterly extension of the Barrow Arch.

The stratigraphy of the North Slope has been discussed by a number of authors. Pre-Mississippian rocks are considered to be economic basement (Jamison et al, 1980). Beds ranging in age from Mississippian to earliest Cretaceous are assigned to the Ellesmerian Sequence of Lerand (1973). The Brookian Sequence (Lerand, 1973) comprises the Late Cretaceous and Tertiary succession. The Ellesmerian Sequence is in most places overlain unconformably by the Brookian sequence. Jones and

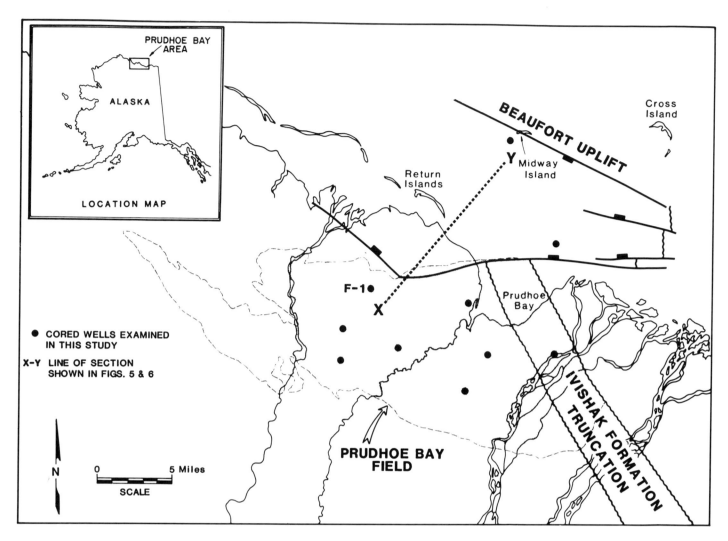

Figure 1—Location map of Prudhoe Bay
field, Alaska.

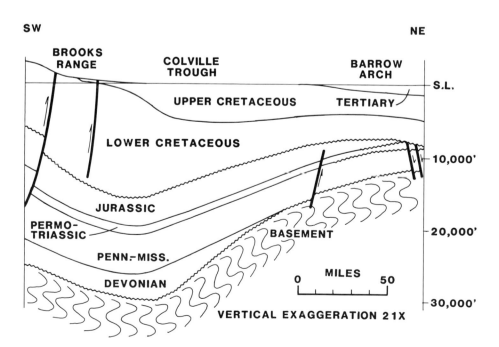

Figure 2—Southwest-to-northeast
generalized cross section from Brooks
Range to Prudhoe Bay.

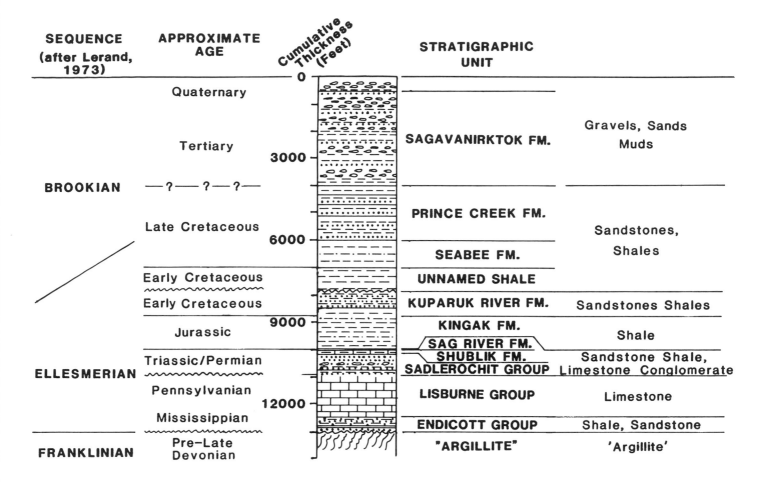

SEQUENCE (after Lerand, 1973)	APPROXIMATE AGE	Cumulative Thickness (Feet)		STRATIGRAPHIC UNIT	
BROOKIAN	Quaternary	0			
	Tertiary	3000		SAGAVANIRKTOK FM.	Gravels, Sands Muds
	—?—?—?—				
	Late Cretaceous	6000		PRINCE CREEK FM.	Sandstones, Shales
				SEABEE FM.	
	Early Cretaceous			UNNAMED SHALE	
	Early Cretaceous			KUPARUK RIVER FM.	Sandstones Shales
	Jurassic	9000		KINGAK FM.	Shale
				SAG RIVER FM.	
ELLESMERIAN	Triassic/Permian			SHUBLIK FM. SADLEROCHIT GROUP	Sandstone Shale, Limestone Conglomerate
	Pennsylvanian	12000		LISBURNE GROUP	Limestone
	Mississippian			ENDICOTT GROUP	Shale, Sandstone
FRANKLINIAN	Pre–Late Devonian			"ARGILLITE"	'Argillite'

Figure 3—Generalized stratigraphic column at Prudhoe Bay field.

Speers (1976) named this important break the "Lower Cretaceous Unconformity." Hydrocarbons within the Ellesmerian Sequence are trapped in Laramide structures as well as stratigraphically beneath the Lower Cretaceous unconformity (Jones and Speers, 1976).

The main reservoir within the Ellesmerian Sequence at Prudhoe Bay is the Ivishak Formation. This formation occurs within the Sadlerochit Group, which on palynological evidence is considered to be Middle to Late Permian and/or Early Triassic in age (Jones and Speers, 1976). The Ivishak Formation was defined by Jones and Speers (1976) from a type section in British Petro-

leum 19-10-15 well (Sec. 19, T10N, R15E). It is gradationally underlain by the Kavik Shale, and its upper contact is commonly associated with a thin, phosphatic radioactive conglomerate overlain by mudstones of the Shublik Formation. The formation thickness ranges from approximately 650 ft in the south to 300 ft in the northeast. Jones and Speers (1976) informally subdivided the Ivishak reservoir into five members, in ascending order, A, B, C, D, and E. In the far western part of the field, the Shublik–Sadlerochit contact is less distinct. An interval of calcareous sandstones and shales occurs between "typical" Shublik and Ivishak lithologies in this area (Jones and Speers, 1976). That sequence is absent over much of the main body of the field and will not be included in the following discussion. A generalized stratigraphic section of the Prudhoe Bay area is presented in Figure 3.

SEDIMENTOLOGY

The lithic content and depositional environment of the Ivishak Formation have been described in general terms and with little significant disagreement by most previous authors (Rickwood, 1970; Morgridge and Smith, 1972; Jones and Speers, 1976; Eckelmann et al, 1975; Jamison et al, 1980). A sequence comprising mudstones, sandstones, and conglomerates was deposited within a fluvio-deltaic environment, having been derived from uplands to the north of the present-day coastline (the Beaufort Uplift: Morgridge and Smith, 1972; see Fig. 1). The sedimentological interpretations presented here are consistent with this previous work and provide refinements based on detailed lithofacies and vertical sequence analysis of cores from a number of wells within and to the north of Prudhoe Bay field (Fig. 1).

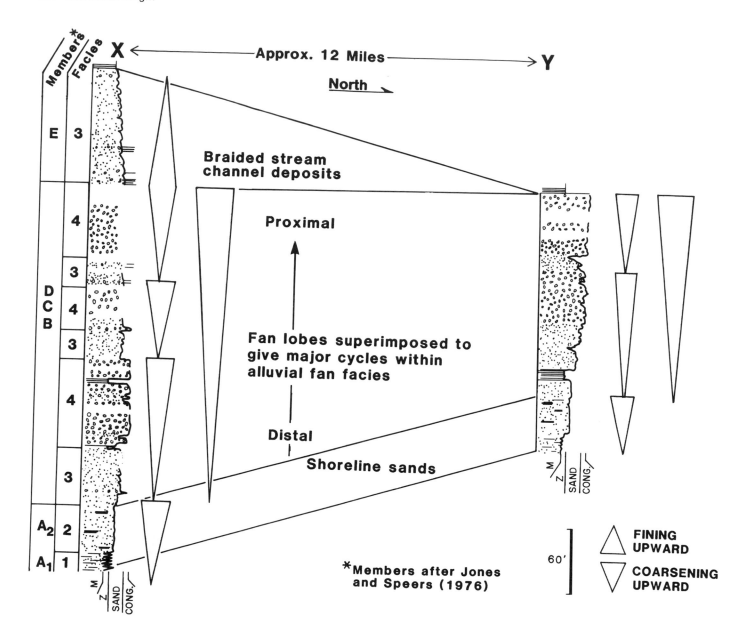

Figure 5—Schematic correlation of lithofacies within Ivishak Formation from south to north. Note: (1) persistence of basal "shoreline sands" comprising Lithofacies 1, 2; (2) overall thinning of upper part of sequence to north; (3) loss of uppermost Lithofacies 3 sandstones to the north.

Lithofacies Associations

The entire Ivishak sequence can be described as a single succession of specific lithofacies. However, real differences in the proportion of those lithofacies relative to each other are observed laterally, as noted by Jones and Speers (1976). Those variations in the distribution of lithofacies are discussed below. Figure 5 illustrates generally the vertical lithofacies associations and their lateral variation in the Ivishak Formation.

Lithofacies 1 and 2 only occur in the lowermost part of the Ivishak, and Lithofacies 1 in all places coarsens upwards into Lithofacies 2 (Fig. 5). This basal upward-coarsening sequence is broadly equivalent to Member A (Submembers A1 and A2) of Jones and Speers (1976). It is present in all the wells examined, and its thickness (50 to 60 ft) varies little across the entire study area.

Overlying this basal association of Lithofacies 1 and 2, the sequence continues to coarsen upward through medium- to coarse-grained sandstones of Lithofacies 3 into an interval dominated by conglomerates of Lithofacies 4. In detail, this coarse-grained section comprises a number of units, each of which coarsens upwards from Lithofacies 3 sandstones into Lithofacies 4 conglomerate (Fig. 5). However, the uppermost conglomerate of the highest

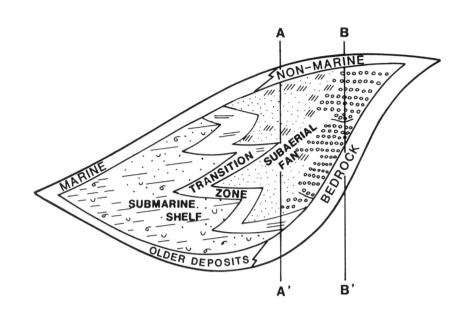

coarsening unit in any well generally contains the coarsest detritus, and this approaches cobble size in the north-northeast. Jones and Speers (1976) believed that to be the direction from which the sediment was derived. Members B, C, and D (Jones and Speers, 1976) are together more or less equivalent to this interval of interstratified Lithofacies 3 sandstones and Lithofacies 4 conglomerates. The thickness of this interval is about 350 to 400 ft in the main body of Prudhoe Bay field, and thins to 200 ft towards the north-northeast (Fig. 5).

In Prudhoe Bay field, the overall upward-coarsening from Lithofacies 1 through Lithofacies 4 is ultimately overlain by a sequence dominated by Lithofacies 3 sandstones and mudstones. The presence of finer-grained detritus upward results in a distinct upper sand unit in the main field area, which itself displays a gross upward-fining trend (Fig. 5). This uppermost sandstone sequence, dominated by Lithofacies 3, is roughly equivalent to Member E of the Ivishak Formation as defined by Jones and Speers (1976). As stated by them (Jones and Speers, 1976), Member E thins in a northeasterly direction from more than 200 ft thick in the southwest part of the field, to nil in the extreme northeast. This is represented diagrammatically in Figure 5.

Depositional History

The Ivishak Formation comprises a lower sequence of basal shoreline deposits overlying deeper water sediments. These basin margin sediments are overlain everywhere by a sequence of braided alluvial stream channel deposits. These, in turn, exhibit features characteristic of varying degrees of proximity to source. This stratigraphic sequence is interpreted to be the product of fan–delta sedimentation.

Wescott and Ethridge (1980) described stratigraphic models for fan-delta sedimentation building onto either (1) island or continental slopes or (2) a continental or island shelf. In the latter model lateral facies variations are characterized by gravelly proximal, braided stream deposits that grade seaward into sandy, distal braided stream deposits, well-laminated sands of the beach and near-shore zone, and finally into burrowed shoreface muds. This model is represented in Figure 6a. The vertical section A-A' in Figure 6a

Figure 7—Schematic diagram illustrating depositional development of Ivishak Formation through time: (**a**) shoreline deposits develop at edge of low-relief, stable hinterland; (**b**) initial uplift in source area produces deposition by sandy braided streams; (**c**) maximum uplift produces gravelly detritus closest to source in alluvial fans, with continued distal deposition by sandy braided streams; (**d**) uplift ceases, and denudation of source area results in sandy braided stream deposition over the alluvial fan gravels.

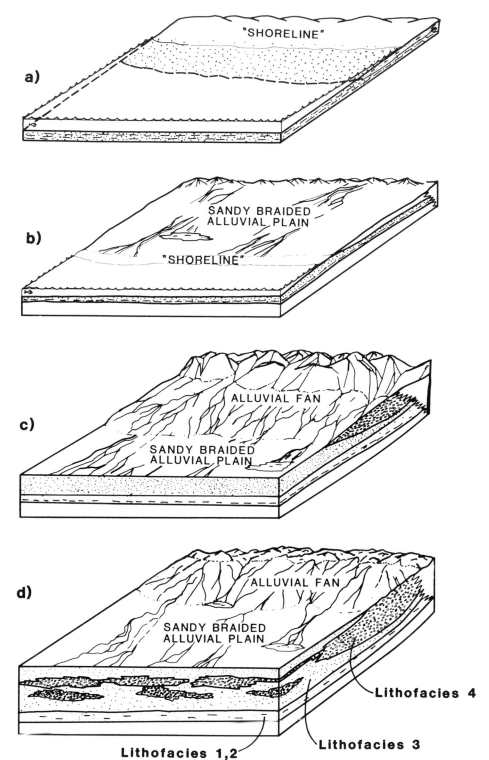

has been superimposed on the Wescott and Ethridge (1980) model and is clearly similar to the sedimentary succession in the Ivishak Formation.

The geometric configuration of lithofacies within the Ivishak Formation is shown in Figure 6b, and it is clear that important differences exist between these rocks and the model illustrated in Figure 6a. Figure 6a indicates that although normal fan–delta progradation would in most places produce a classic upward-coarsening sequence, in areas closest to source the vertical sequence would lack any basal facies representative of initially more distal environments. This is represented by the vertical section B-B′ in Figure 6a. Figure 6b, however, shows that the

basal fine-grained deposits of Lithofacies 1 and 2 in the Ivishak Formation maintain their aggregate thickness more or less across the entire area, even into very proximal parts of the basin.

This suggests changing tectonic influences on the basin history during Ivishak times.

The depositional history of the Ivishak Formation can be discussed in

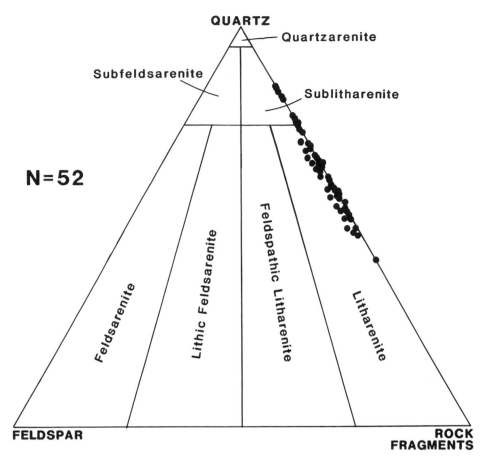

Figure 8—Ternary plot illustrating detrital grain composition of sandstones from Ivishak Formation. Point-count data (250 points per sample) from selected number of representative samples; classification scheme after Folk et al (1970). Reproduced from New Zealand Journal of Geology and Geophysics.

terms of a number of stages of sedimentation, each of which reflects the tectonic conditions prevailing at the time (Fig. 7). Figure 7a shows deposition across the entire area of the lowest upward-coarsening sequence (Lithofacies 1 to 2). This deposit is attributed to progradation into a standing body of water of shoreline deposits. The widespread occurrence of this sequence and its limited variation in thickness, in association with the textural maturity of the shoreline sands themselves, implies deposition over a wide area, from a provenance of subdued relief in an environment of relative tectonic stability.

A major change occurred with the introduction of coarser (medium-grained) braided stream channel sandstones over the shoreline deposits (Fig. 7b). These represent the beginning of

alluvial fan deposition in response to uplift of the source area.

With increased tectonic activity in the source area, higher topographic relief produced coarse (gravelly) detritus and progradation of alluvial fans into the basin (Fig. 7c). Fan, or fan-lobe switching produced different episodes of lobe development as represented by the superimposed upward-coarsening sequences of Lithofacies 3 into Lithofacies 4 described previously (Fig. 5). This phase of coarsest-grained alluvial fan-lobe development represents the phase of maximum input of sediment into the basin during a time of most active tectonism.

Subsequent denudation of the source area with minimal tectonic rejuvenation is reflected by the superposition of braided stream sands over the relatively

proximal alluvial fan surface conglomeratic deposits (Fig. 7d).

PETROGRAPHY

Over 500 thin sections of sandstones and conglomerates from the Ivishak Formation were examined petrographically as part of this study. More than 50 of those sections were point counted for detrital grain components, intergranular material, and porosity. All samples were impregnated with a blue-dyed epoxy resin to facilitate identification of porosity and the slides were stained with Alizarin Red-S and potassium ferricyanide solutions to aid in identification of carbonate cements.

Generally, the rocks are moderately sorted to moderately well-sorted sandstones and conglomerates: chert litharenites and chert lithrudites (Fig. 8)

Figure 9—Photomicrograph illustrating general features of sandstone (Lithofacies 3) in Ivishak Formation. Note: chert grains (light brown), quartz with well-developed overgrowth (arrowed), and patchy poikilotopic siderite (dark brown). The siderite exhibits abundant evidence of dissolution. Scale bar = 900 μ. Plane-polarized light, porosity in blue.

Figure 10—Photomicrograph of fine-grained sandstone from Lithofacies 2. The rock contains very little porosity owing to quartz cementation and compactional deformation of matrix and labile grains (brown material). Scale bar = 500 μ. Plane-polarized light.

according to the classification scheme of Folk et al (1970). A typical example of Ivishak Formation sandstone is shown in Figure 9. The predominant detrital grain type is monocrystalline quartz (27 to 63%; avg. 47%). Chert grains are abundant, however, and two major types are observed. One of these is a dense chert (6 to 44%; avg. 19%) comprising tightly interlocking crystals of microcrystalline quartz and containing various amounts of argillaceous material. The other chert type is microporous and occupies 1 to 12% (avg. 5%) of the bulk rock. It is identified by a greenish-blue hue observed in plane polarized light. This is due to impregnation by the blue-dyed resin of a micropore system that pervades the chert. Accessory detrital grains (avg. 9%) include polycrystalline quartz (that is, metaquartzite and vein quartz), sedimentary rock fragments (mudstone, siltstone, sandstone), metasedimentary rock fragments and extremely rare feldspar (both potassium feldspar and plagioclase). A fine-grained clay matrix occurs in various amounts. In general, however, the highest amounts occur in the very fine-grained sandstones of Lithofacies 1.

Original depositional textures within the sandstones vary considerably in grain size and sorting. Inhomogeneous packing density within the sandstones is interpreted to be a result of post-

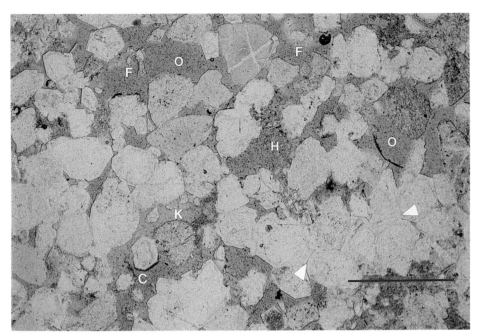

Figure 11a—Photomicrograph illustrating porosity development in Lithofacies 3 sandstones: well-developed secondary porosity (blue) as revealed by occurrence of oversized pores (**O**), floating grains (**F**), channelized pores (**C**), honeycombed grains (**H**). Note local occurrence of kaolinite (**K**) and patchy quartz overgrowth cement (arrowed), creating impression of inhomogeneous packing. Scale bar = 625 μ. Plane-polarized light.

Figure 11b—Different sample from Lithofacies 3 showing abundance of features similar to Fig. 11a. Note abundance of well-developed quartz overgrowth. Symbols same as Fig. 11a. Scale bar = 600 μ. Plane-polarized light.

depositional dissolution of the rock and is discussed more fully below. Within the conglomerates, variation in sorting in the inter-pebble sandy matrix is pronounced and proved to be significant in the appraisal of the Ivishak Formation as a reservoir rock. This is also discussed more fully below.

The authigenic mineralogy of the rocks includes siderite, pyrite, ferroan carbonate, kaolinite, and quartz. Of these, siderite, pyrite, and quartz are the dominant cements. Siderite forms poikilotopic rhombs (Fig. 9) or clusters of rhombs up to a few millimeters in diameter. These commonly impart a "spotted" appearance to the rock in hand specimen. Pyrite occurs as dispersed cubes that replace matrix and organic debris, or as nodules enclosing and replacing detrital grains. Such pyrite nodules are commonly observed in the core in excess of 1-inch diameter. Other carbonate cements include ferroan dolomite and rare ferroan calcite. Kaolinite forms a patchy pore-filling cement and may represent neomorphic replacement of preexisting detrital lithic grains. It is not a quantitatively significant pore-occluding cement (avg. 2–3%). Quartz forms as syntaxial overgrowth on detrital quartz. It is very common (Fig. 9) but rarely forms a significant pore-occluding cement.

The primary objective of this petrographic study was to describe and

Figure 12a—Photomicrograph illustrating dissolution features in Lithofacies 3 sandstones. Patchy poikilotopic siderite with abundant leached porosity (arrowed). Note also honeycombed rock fragments in upper left (**H**). Scale bar = 600 μ. Plane-polarized light, porosity in blue.

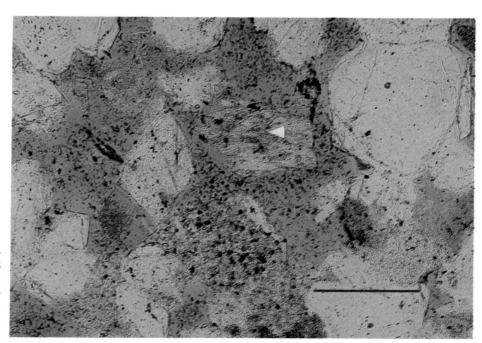

Figure 12b—Photomicrograph illustrating dissolution features in Lithofacies 3 sandstones. Detrital siliceous rock fragment "floating" in pore and displaying intragranular porosity (arrowed). Scale bar = 375 μ. Plane-polarized light, porosity in blue.

understand the nature of the pore system within this important reservoir rock. In particular, it was considered important to elucidate the relationship, if any, which the present-day pore system may have to lithofacies distribution and diagenesis. It was observed that each lithofacies can be characterized by a specific set of petrographic relationships. A discussion of those relationships follows.

Lithofacies 1 and 2

The fine- to very fine-grained sandstones that occur interbedded with mudstones in Lithofacies 1 are petrographically similar to the laminated fine-grained "shoreline" sandstones of Lithofacies 2. Thus, the rocks are very fine- to fine-grained, well-sorted chert litharenites that contain a higher proportion of heavy minerals (zircon, hornblende, tourmaline, and rutile) and

carbonaceous debris than the other lithofacies.

The Lithofacies 2 sandstones that occur at the top of the basal upward-coarsening shoreline deposits have undergone significant cementation by quartz (Fig. 10). This quartz cement also occurs in Lithofacies 1 sandstones, but there it is accompanied by an increase in intergranular clay matrix, which increases in abundance strati-

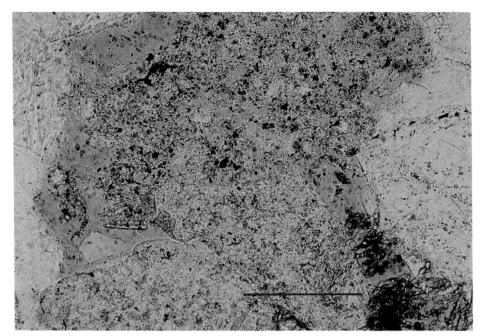

Figure 12c—Photomicrograph illustrating dissolution features in Lithofacies 3 sandstones. Detrital argillaceous chert grain with extremely skeletal appearance as a result of pervasive development of intragranular porosity. Scale bar = 125 μ. Plane-polarized light, porosity in blue.

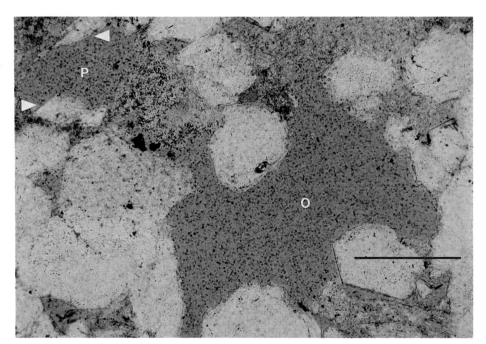

Figure 12d—Photomicrograph illustrating dissolution features in Lithofacies 3 sandstones. Quartz fragments (arrowed), which are in optical continuity, indicating they were once part of a single grain, now represented by porosity (**P**). The oversized pore (**O**) may have once contained a number of such grains. Scale bar = 250 μ. Plane-polarized light, porosity in blue.

graphically downward as the Ivishak Formation grades into the underlying Kavik Shale. This clay is commonly associated with very fine-grained organic detritus and forms an intergranular pore fill that dominates these lithofacies. Clay probably inhibited the introduction of other pore-filling cements and consequently resulted in a higher degree of compaction.

Lithofacies 3

The braided stream sandstones of Lithofacies 3, which occur directly below and above the conglomeratic interval of the Prudhoe Bay reservoir (see Fig. 5), have petrographic characteristics that prove most important in understanding the development of reservoir quality.

They are medium- to coarse-grained,

moderately sorted to moderately well-sorted chert litharenites with only small amounts of primary intergranular clay matrix. Kaolinite is the most abundant authigenic clay (3%) and is characterized by its vermicular habit. It occurs either as a dispersed partial pore filling or as replacement of preexisting detrital grains. It does not form a significant pore-occluding cement within this

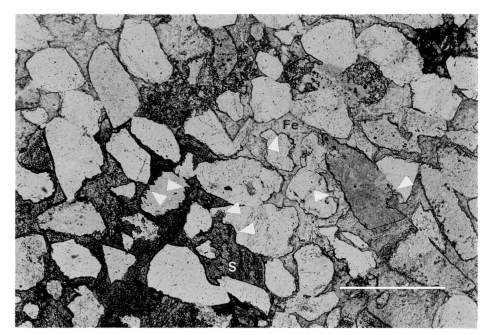

Figure 13a—Photomicrograph illustrating relationship of carbonate cements to detrital grains and porosity development in Lithofacies 3 sandstones. Poikilotopic development of both siderite (**S**) and blue-stained ferroan calcite (**Fe**), both of which are aggressively corroding and replacing the detrital grains (arrowed). Scale bar = 500 μ. Plane-polarized light.

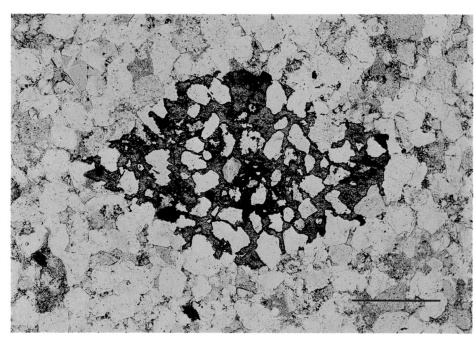

Figure 13b—Photomicrograph illustrating relationship of carbonate cements to detrital grains and porosity development in Lithofacies 3 sandstones. Poikilotopic development of rhombic siderite showing the extent to which detrital grain replacement affects grain shape and apparent sorting, and creates illusion of open packing. Scale bar = 400 μ. Plane-polarized light.

lithofacies, and, where present, contains substantial microporosity. This is identified in thin section by the greenish-blue hue imparted to the kaolinite by the blue-dyed resin impregnating the pore system (Fig. 11a).

The most significant petrographic feature of Lithofacies 3 is that a large proportion of the porosity is secondary. Many of the criteria described by Schmidt et al (1977) and Schmidt and McDonald (1979a) as being diagnostic of dissolution of preexisting grains and

cements are recognized within Lithofacies 3 of the Ivishak Formation in Prudhoe Bay field. Figures 11a and 11b illustrate inhomogeneous packing, oversized pores, floating grains, channelized pores, and honeycombed grains as they occur in this lithofacies, and all of which reflect an aggregate pore system of excellent quality.

It has been argued that a pore-system characterized by such features is produced by the mesogenetic leaching of the carbonate minerals calcite, dolo-

mite, and siderite (Schmidt and McDonald, 1979b). Figure 12a shows that dissolution of poikilotopic siderite has occurred within the Ivishak. However, there is abundant petrographic evidence showing that dissolution features occur not only in carbonate but also in rock fragments (Fig. 12b), chert (Fig. 12c), and indeed quartz (Fig. 12d). In attempting to understand the diagenetic processes that produced the pore system observed within the Ivishak Formation, we must consider the ques-

Figure 13c—Photomicrograph illustrating relationship of carbonate cements to detrital grains and porosity development in Lithofacies 3 sandstones. Poikilotopic siderite on left side similar to Fig. 13b. Note, in particular, the rhombic reentrants on the detrital grain boundaries (arrowed). Right side shows grains within a porous area (blue) exhibiting similar textural relationships to grains enclosed by carbonate. Scale bar = 600 μ. Plane-polarized light.

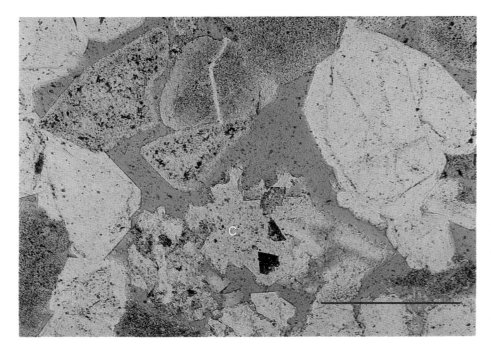

Figure 13d—Photomicrograph illustrating relationship of carbonate cements to detrital grains and porosity development in Lithofacies 3 sandstones. Detrital chert grain (**C**) in porous area, exhibiting well-developed intragranular rhombic porosity, and abundant rhombic reentrants around the grain boundary. The brown rhombic areas within the grain are hydrocarbon-filled pores. Scale bar = 250 μ. Plane-polarized light.

tion: Did the porosity result from direct dissolution of siliceous detrital grains, as well as carbonate cements, or were those detrital grains first replaced by cements prior to dissolution? The petrographic evidence in Lithofacies 3 argues strongly that a great number of the grains were first replaced by carbonate prior to dissolution.

Figure 13a shows that at least two phases of carbonate cement occur at the present time within Lithofacies 3 of the Ivishak Formation in the Prudhoe

Bay area. One of these is siderite, the other is a ferroan calcite. Both of these phases are clearly highly replacive of the detrital grains (Fig. 13a). The grain replacement produces textural relationships that imply open packing and poor sorting of the detrital grains within the carbonate-cemented area (Fig. 13b). Figure 13c shows poikilotopic siderite strongly replacing detrital grains, producing corroded grain margins with rhombic reentrants, and an apparent poor sorting of the grains similar to

Figure 13b. However, in the same illustration (Fig. 13c), there occurs adjacent to the poikilotopic siderite an area of porosity within which the detrital grains clearly exhibit evidence of dissolution. Furthermore, the textural relationships of those grains suggest that they are quite poorly sorted, at least in that localized area. It can be inferred that the area presently occupied by porosity in Figure 13c was at one time the site of a grain-replacive cement. Figure 13d shows a detrital chert grain

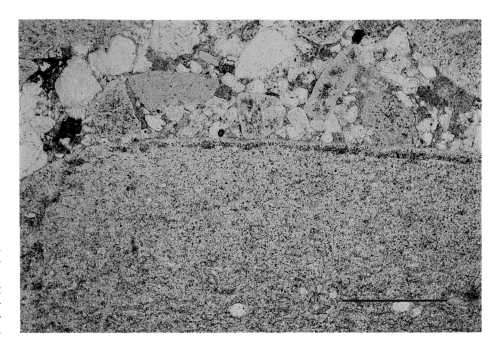

Figure 14a—Photomicrograph illustrating textural variation within sandy matrix of Lithofacies 4 conglomerates. Poorly sorted matrix, and concomitant severe reduction of primary intergranular porosity. Scale bar = 700 μ. Plane-polarized light.

that unequivocally exhibits intragranular rhombic porosity, and a grain margin that is severely indented with rhombic reentrants, similar to those described on the grains poikilotopically enclosed by siderite in Figure 13c. The shape of this grain is too fragile to have been transported in that form as a sedimentary particle. Its present form results from an episode of severe grain replacement by an unknown cementing mineral, which has subsequently been removed by dissolution. The rhombic aspect of the pores and reentrants suggests that the mineral was a carbonate. At least two phases of grain-replacive carbonate cement (siderite and ferroan calcite) have been described from this lithofacies in the Ivishak (Fig. 13a). Those minerals occur within the rock, with only minor evidence of having been dissolved (Figs. 12a, 9). The grain-replacive carbonate phase, which has been removed to produce much of the enhanced porosity within Lithofacies 3 of the Ivishak, is inferred to have been relatively soluble to the dissolving fluids permeating the rock than the siderite or ferroan calcite. The solubility of carbonate minerals relative to each other cannot be defined in a simple sense, since it will be dependent in a complex way on the physical chemistry of the formation fluids, including variables such as temperature, pressure, pH, Eh, cation concentrations, etc.

Thus the carbonate that has been dissolved out of these sandstones cannot be unequivocally identified. However, circumstantial considerations outlined below suggest that it may have been calcite.

The sandstones of Lithofacies 3 have been interpreted above as the deposits of a distal alluvial fan/braided alluvial plain. Furthermore, the conglomeratic deposits of those inferred alluvial fans have been interpreted as comprising, in part, debris flow deposits. The occurrence of such deposits suggests abrupt, short-lived depositional events, such as might occur in flash floods after heavy storms. The preservation of such deposits results from a lack of reworking: the alluvial fan sedimentation was thus most likely ephemeral, and occurred in a semi-arid climate, in which prolonged surface run-off was not common. Such an environment leads to the development of nodular and layered calcite deposits (caliche) in many areas (Reeves, 1976).

It is thus possible that early calcite cementation in the form of caliche occurred within the braided stream sandstones (Lithofacies 3) of the Ivishak Formation. Certainly, the petrographic aspects of the areas presently cemented by siderite and/or ferroan calcite bear a close resemblance to caliche deposits (both modern and ancient) described by Nagtegaal (1969)

and Steel (1974). Nagtegaal (1969) noted that more than 50% (volume) of the rock was occupied by caliche in both recent and ancient examples, following grain replacement in each case. It is not easy to prove unequivocally that the Ivishak sandstones supported widespread caliche development. Nagtegaal (1969) indicated the dual problems associated with identifying caliche profiles in ancient rocks. Those problems are (1) a high depositional rate that prevents the development of complete and thick profiles and (2) the occurrence of much later replacement and cementing carbonate. Steel (1974) has noted that (caliche) carbonate may be partly or completely replaced by dolomite or silica. Possibly the siderite and/or ferroan calcite observed in Lithofacies 3 sandstones at the present time represent replacement of an earlier calcite caliche deposit.

The development of secondary porosity within the braided stream sandstones (Lithofacies 3) of the Ivishak Formation is widespread over Prudhoe Bay field. If such porosity does indeed result from dissolution of an early calcite cement (caliche), then that cement was also widespread. Steel (1974) has observed that areas containing relatively thin deposits of alluvium (and which are thus inferred to have subsided relatively slowly) contain a much higher percentage of caliche than areas

Figure 14b—Photomicrograph illustrating textural variation within sandy matrix of Lithofacies 4 conglomerates. Coarser, and better sorted matrix, resulting in much improved primary intergranular porosity. Scale bar = 600 μ. Plane-polarized light.

that subsided rapidly, accumulating thick sedimentary successions. The Ivishak Formation is rarely thicker than about 650 ft, which is relatively thin as an alluvial succession compared with some. For example, the Devonian alluvium of the Hornelen Basin in Norway exceeds 25,000 m in thickness (Steel et al, 1977). Thus it should be expected that caliche development in such a thin depositional unit as the Ivishak would be widespread.

Lithofacies 4

The conglomerates of Lithofacies 4 are generally pebble-sized, bimodally sorted chert lithrudites. The degree of sorting within the interpebble sandy matrix varies considerably.

Carbonate cement occurs in these rocks but it does not appear to be, or to have been, abundant. The reservoir characteristics of Lithofacies 4 appear to be controlled, not by secondary diagenetic phenomena but by primary textural variation in the interpebble sandy matrix. Figure 14 illustrates this variation. In some cases, the interpebble matrix is poorly sorted (Fig. 14a) and the accommodation of the varying sizes of detrital grains reduces severely the primary intergranular porosity. In other examples (Fig. 14b), the intergranular sandy matrix is coarser (generally concomitant with a grain-

size increase in the gravel-size mode), and in addition exhibits significantly improved sorting. This leads to an improvement in the overall primary intergranular porosity of these rocks. In places the porosity within this coarser and better-sorted interpebble matrix may be slightly enhanced by secondary dissolution processes. This is nowhere as marked as the secondary porosity in Lithofacies 3, and does not alter the argument concerning textural porosity control within the conglomerates. Poorly sorted interpebble matrix would thus have had low permeability to any cement-bearing fluids and subsequent solvents; the coarser, better-sorted sediments would have had a higher permeability to such fluids. Any secondary porosity enhancement in these sediments would nonetheless be a reflection of the better initial, texturally related primary porosity. The porosity of the conglomerates is thus only moderate at best, and inherently primary intergranular in nature.

CONCLUSIONS

The Permo-Triassic Ivishak Formation in the Prudhoe Bay oil field comprises a sequence of sandstones and conglomerates with mudstones that was deposited in an alluvial fan-delta depositional environment.

Porosity diagenesis in the Ivishak Formation is clearly lithofacies dependent. Specifically, dissolution of an early carbonate cement (inferred to be caliche) which formed within the sandy braided stream deposits of the distal alluvial fan apron led to the occurrence of well-developed secondary porosity, which was thus generally limited to that lithofacies only. Porosity distribution in the Ivishak Formation can thus be discussed with respect to the distribution of lithofacies occurring within it (Fig. 15).

The basal upward-coarsening sequence of the Ivishak fan-delta exhibits the poorest reservoir quality. Porosity characteristics vary from poor to moderately good, worsening downward into the interbedded sediments of Lithofacies 1 (Fig. 15). This reflects primary depositional characteristics of the rocks, and in particular the intergranular clay content related to finer depositional grain size.

The best reservoir quality is found within the sandstones of Lithofacies 3. These occur most commonly directly above and below the interval dominated by Lithofacies 4 conglomerates (Fig. 15). Local occurrences of Lithofacies 3 sandstones within the conglomerates are reflected as high-porosity streaks.

The occurrence of conglomerate

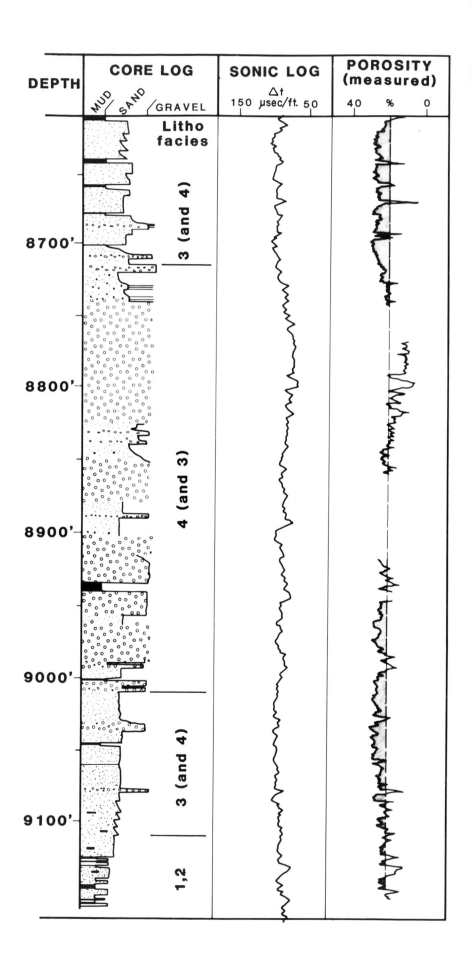

Figure 15—Vertical logs of lithofacies, sonic transit time and measured porosity from Sohio Drill Pad F-1 well (see Fig. 1), Prudhoe Bay field.

shows marked correlation with reduced porosity (Fig. 15). This is attributed to the lack of significant secondary porosity enhancement within this lithofacies. Variation within the conglomerate porosity profile reflects variable sorting in the interpebble matrix of the conglomerates.

Porosity development in this important reservoir thus results from an interplay of variables that depend on both the primary depositional and secondary diagenetic history of the rocks. The fullest understanding of reservoir quality and its distribution in any reservoir can only be achieved by the integration of studies concerning not only the burial history of the rocks but also the variations that occur in textural characteristics related to primary depositional lithofacies.

ACKNOWLEDGMENTS

The writers thank the management of Sohio Petroleum Company for allowing them to publish this paper. The subject matter was originally presented at the AAPG convention in Calgary, 1982. Aspects of this work have benefited from discussion and comment from Drs. D. L. Bremner, M. J. Mayall, and R. J. Steel. In particular, we appreciate the constructive criticism of our colleagues at Sohio Petroleum, namely Drs. D. A. Barnes and H. G. Bassett; the opinions and interpretations expressed are nonetheless those of the writers alone. Particular thanks go to C. D. Noyes for his invaluable assistance with our petrographic studies, Leta Blome for her expert drafting of the figures, Mike Joyce for his help in preparing the photographs, and Dotti Campbell and Theresa Villezar for their patience with us while typing the manuscript.

SELECTED REFERENCES

Bluck, B. J., 1967, Deposition of some Upper Old Red Sandstone conglomerates in the Clyde area: a study in the significance of bedding: Scottish Journal of Geology, v. 3, p. 139–167.

Detterman, R. L., 1970, Sedimentary history of the Sadlerochit and Shublik Formations in northeastern Alaska, in Proceedings of the Geological Seminar on the North Slope of Alaska: Pacific Section, American Association of Petroleum Geologists, p. O1–O13.

Eckelmann, W. R., R. J. DeWitt, and W. L. Fisher, 1975, Prediction of fluvial-deltaic reservoir geometry, Prudhoe Bay Field, Alaska: 9th World Petroleum Congress Proceedings, v. 2, p. 223–227.

Folk, R. L., P. B. Andrews, and D. W. Lewis, 1970, Detrital sedimentary rock classification and nomenclature for use in New Zealand: New Zealand Journal of Geology and Geophysics, v. 13, p. 937–968.

Goldring, R., and P. Bridges, 1973, Sublittoral sheet sandstones: Journal of Sedimentary Petrology, v. 43, p. 736–747.

Hooke, R. L., 1967, Processes on arid-region alluvial fans: Journal of Geology, v. 75, p. 438–460.

Jamison, H. C., L. D. Brockett, and R. A. McIntosh, 1980, Prudhoe Bay—a 10-year perspective, in M. T. Halbouty, ed., Giant oil and gas fields of the decade: 1968-1978: American Association of Petroleum Geologists Memoir 30, p. 289–310.

Jones, H. P., and R. G. Speers, 1976, Permo–Triassic reservoirs of Prudhoe Bay field, North Slope, Alaska, in J. Braunstein, ed., North American oil and gas fields: American Association of Petroleum Geologists Memoir 24, p. 23–50.

Larsen, V., and R. J. Steel, 1978, The sedimentary history of a debris-flow dominated, Devonian alluvial fan—a study of textural inversion: Sedimentology, v. 25, p. 37–59.

Lerand, M., 1973, Beaufort Sea, in R. G. McGrossan, ed., The future petroleum provinces of Canada—their geology and potential: Canadian Society of Petroleum Geologists Memoir 1, p. 315–386.

Morgridge, D. L., and W. B. Smith, Jr., 1972, Geology and discovery of Prudhoe Bay field, Eastern Arctic Slope, Alaska, in R. E. King, ed., Stratigraphic oil and gas fields—classification, exploration methods, and case histories: American Association of Petroleum Geologists Memoir 16, p. 489–501.

Nagtegaal, P. J. C., 1969, Microtextures in recent and fossil caliche: Leidse Geologische Mededelingen, v. 42, p. 131–142.

Pettijohn, F. J., 1957, Sedimentary rocks, New York, Harper and Row, 718 p.

Reeves, C. C., 1976, Caliche—origin, classification, morphology and uses: Lubbock, TX, Estacado Books.

Rickwood, F. K., 1970, The Prudhoe Bay field, in Proceedings of the Geological Seminar on the North Slope of Alaska: Pacific Section, American Association of Petroleum Geologists, p. L-1–L-11.

Schmidt, V., and D. A. McDonald, 1979a, Texture and recognition of secondary porosity in sandstones, in P. A. Scholle and P. R. Schluger, eds., Aspects of diagenesis: Society of Economic Paleontologists and Mineralogists Special Publication 26, p. 209–225.

———, 1979b, The role of secondary porosity in the course of sandstone diagenesis, in P. A. Scholle and P. R. Schluger, eds., Aspects of diagenesis: Society of Economic Paleontologists and Mineralogists Special Publication 26, p. 175–207.

Schmidt, V., D. A. McDonald, and R. L. Platt, 1977, Pore geometry and reservoir aspects of secondary porosity in sandstones: Bulletin of Canadian Petroleum Geology, v. 25, p. 271–290.

Steel, R. J., 1974, Cornstone (fossil caliche)—its origin, stratigraphic, and sedimentological importance in the New Red Sandstone, Western Scotland: Journal of Geology, v. 82, p. 351–369.

Steel, R. J., R. Nicholson, and L. Kalander, 1975, Triassic sedimentation and paleogeography in Central Skye: Scottish Journal of Geology, v. 11, p. 1–13.

Steel, R. J., S. Maehle, H. Nilsen, S. L. Roe, and A. Spinnangr, 1977, Coarsening-upward cycles in the alluvium of Hornelen Basin (Devonian) Norway: sedimentary response to tectonic events: Geological Society of America Bulletin, v. 88, p. 1124–1134.

Wescott, W. A., and F. G. Ethridge, 1980, Fan-delta sedimentology and tectonic setting—Yallahs Fan Delta, southeast Jamaica: Bulletin of the American Association of Petroleum Geologists, v. 64, p. 374–399.

Diagenesis and Evolution of Secondary Porosity in Upper Minnelusa Sandstones, Powder River Basin, Wyoming

John C. Markert
Cities Service Oil and Gas Corp.
Oklahoma City, Oklahoma

Zuhair Al-Shaieb
Oklahoma State University
Stillwater, Oklahoma

INTRODUCTION

The Raven Creek field was discovered in March 1960 by the Kewanee Oil Company No. 1 Norman (Sec. 14, T48N, R69W, Campbell County, Wyoming). The field trends northwest and appears to be genetically related to the adjacent Reel field (Fig. 1). Production is from a sandstone of the Permian Upper Minnelusa. The trap is formed where the productive sandstone is truncated updip against the Opeche Shale at the post-Minnelusa unconformity. The structure, stratigraphy, and trapping mechanism at the Raven Creek field are described in detail by Tranter (1963).

The purpose of this study is to investigate the diagenetic history and the evolution of secondary porosity of the Upper Minnelusa sandstones at Raven Creek and Reel fields. In so doing, we attempt to correlate major tectonic events in the Powder River basin with the changes in the geochemistry of formation water and the sequence of diagenetic events observed. Utilization of the WATEQF computer program (Plummer et al, 1976) proved to be helpful in evaluating chemical analyses of formation waters.

ABSTRACT. Lower Permian sandstones of the Minnelusa Formation have produced a significant number of hydrocarbon reservoirs in the Powder River basin. At the Raven Creek and Reel fields the Upper Minnelusa consists of interbedded sandstones, dolomites, and anhydrites. These sediments represent shoaling-upward cycles consisting of three facies: subtidal, intertidal, and supratidal. Complete cycles are interrupted by well-sorted, fine- to coarse-grained sandstones.

Quartzarenites, the dominant sandstone type, have framework constituents of quartz, feldspars, micas, and heavy minerals. Diagenetic minerals include anhydrite, dolomite, mixed-layer illite–smectite, kaolinite, quartz, chert, and pyrite. A progressive sequence of diagenetic events from oldest to youngest evident in Upper Minnelusa sandstones is: (1) precipitation of poikilotopic anhydrite and quartz overgrowths, (2) dissolution of anhydrite cement resulting in the formation of secondary porosity, (3) precipitation of dolomite rhombs, mixed-layer illite–smectite, kaolinite, and chert in both primary and secondary pore space, and (4) precipitation of pyrite and the accumulation of hydrocarbons in the reservoirs, halting further chemical diagenesis. Dissolution of anhydrite is related to tectonism during Jurassic or Early Cretaceous times. Hydrocarbons migrated into the reservoirs in response to the Laramide orogeny.

Subsurface formation waters from the Upper Minnelusa were analyzed using the WATEQF computer program. Results indicate that the waters are undersaturated to slightly supersaturated with respect to anhydrite. There is a positive correlation between the zones of low anhydrite saturation and the zones of high porosity. This suggests that the development of secondary porosity is attributable to the dissolution of anhydrite. The WATEQF program may be used in delineating secondary porosity fairways in similar geologic settings.

METHODOLOGY

Twenty-eight cored sections of the Upper Minnelusa from wells in, or near, the trend of the Raven Creek and Reel fields were selected for this study. Whole cores were described in detail with emphasis placed on lithology, texture, sedimentary structures, and diagenetic mineralization.

Fifty-two representative samples of sandstone were selected for petrographic analysis. The thin sections were impregnated with blue epoxy to show porosity and provide a basis for inferring permeability, and were stained with potassium ferrocyanide to distinguish iron-bearing dolomite. Modal analyses were based on counting 400 points on each slide.

One-hundred-thirty-eight bulk samples representative of sandstones from

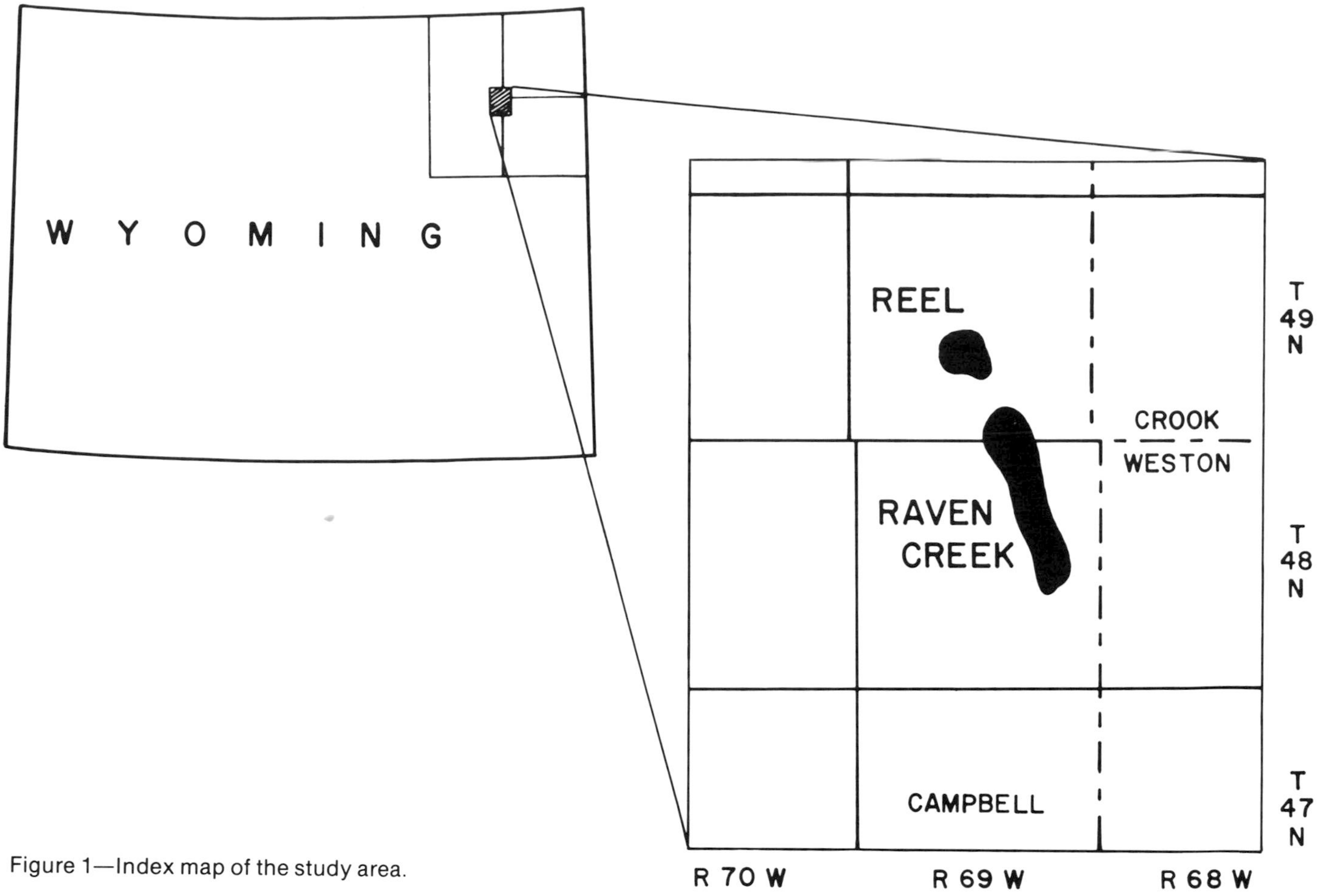

Figure 1—Index map of the study area.

27 cored sections of the Upper Minnelusa were collected according to a stratified random sampling plan and analyzed by semi-quantitative X-ray diffraction techniques. Clay minerals from two cores were separated and identified using X-ray diffraction techniques.

Several samples were examined with a scanning electron microscope coupled with energy dispersive X-ray analysis (SEM and EDAX) to identify diagenetic minerals and determine their textural relationships and composition. Samples of primary and secondary minerals were selected for isotopic analysis. These samples provided three $\delta^{13}C$ and $\delta^{18}O$ values for primary dolomite and seven $\delta^{13}C$ and $\delta^{18}O$ values for dolomite cement.

TECTONIC SETTING

The Raven Creek and Reel fields are located on the eastern flank of the Powder River basin (Fig. 2). The basin trends northwesterly across northeastern Wyoming into southeastern Montana. The basin is bounded on the north by the Miles City Arch and on the east by the broad arch of the Black Hills uplift. The southern part of the basin is bounded by the Hartville uplift to the southeast and the Laramie Range to the southwest. The Casper Arch and Big Horn Mountains form the western boundary.

The Powder River basin is markedly asymmetric with the synclinal axis adjacent to the Casper Arch and Big Horn Mountains. The eastern and northern flanks of the basin are devoid of strong structural features. On the northeastern flank, the beds are relatively undisturbed and dip westward off the Black Hills uplift, at approximately one degree. The basin is bounded by zones of stronger deformation along the margins of the Laramie, Hartville, and Black Hills uplifts on the south,

and the Casper Arch and Big Horn Mountains on the west.

The present configuration of the Powder River basin is the result of Late Cretaceous and Early Tertiary deformation. The first major indication of Laramide orogeny is suggested by relatively strong subsidence of the basin during latest Cretaceous times (Glaze and Keller, 1965). During the Paleocene, the Big Horn Mountains, Black Hills, and Laramie Range began to rise, delineating the major basin features (Love et al, 1963). Most of the present structures in the basin were formed by strong folding and faulting during the Early Eocene. Westward tilting resulted from renewed uplift in the Black Hills in Late Eocene times (Glaze and Keller, 1965). Uplift, downwarping, and regional tilting extended into the Miocene (Curtis et al 1958).

Although the present form of the Powder River basin is principally the result of Laramide deformation, the general configuration of the basin developed as a result of pre-Late Cretaceous structural movements occurring episodically in the form of localized deformation accompanying broad regional downwarps (Curtis et al, 1958). Isopach maps of Mississippian, Pennsylvanian and Lower Permian, Upper Jurassic and Upper Cretaceous rocks reflect gentle warpings with northwest trends, which were recurrent with marked persistence through geologic time (Curtis et al, 1958; Strickland, 1958). Blackstone (1963) suggested that Paleozoic and Upper Cretaceous movements were minor, but that a major arching occurred during the Jurassic. The Raven Creek and Reel fields are located on the northeast-trending Belle Fourche Arch. Uplift of the arch was initiated during the deposition of Lower Cretaceous sediments (Slack, 1981).

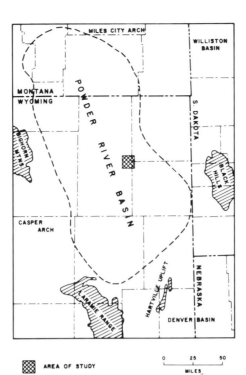

AREA OF STUDY

0 25 50
MILES

Figure 2—Powder River basin and surrounding structural features.

STRATIGRAPHIC SETTING

The major uplifts flanking the Powder River basin have different nomenclatures applied to the outcropping Pennsylvanian and Permian strata. These include the Minnelusa Formation in the Black Hills, the Hartville Formation in the Hartville uplift, the Casper Formation in the Laramie Range, and the Amsden and Tensleep Formations in the Big Horn Mountains. These strata were correlated and described in detail by Mallory (1967) and Maughan (1978) and are summarized in Figure 3. Because of their lithologic and sequential similarities, the Casper, Hartville, and Minnelusa Formations should be identified as a single formation (Maughan, 1978). The name, "Minnelusa," has gained wide acceptance by the petroleum industry for Pennsylvanian and Lower Permian strata in most of the Powder River basin, with the exception of the west flank and the Casper Arch.

Foster (1958) divided the Minnelusa into three members separated by regional unconformities. The lower member overlies the Mississippian Pahasapa Limestone and consists of red mudstone, interstratified with fine-grained sandstone and thin beds of limestone (Maughan, 1978) of predominantly Atokan age (Foster, 1958). The middle member is Desmoinesian, Missourian, and Virgilian in age (Foster, 1958), and is an interbedded sequence of dolomite, argillaceous dolomite, quartzose sandstone, and thin but persistent, radioactive black shales (Foster, 1958; Maughan, 1978).

A red arenaceous to argillaceous mudstone termed the "red shale marker" is at the base of the Upper Minnelusa in most of eastern Wyoming (Foster, 1958). The marker overlies an erosional contact between Pennsylvanian and Permian rocks. The top of the upper member is marked by an unconformity that separates the Minnelusa from the overlying Opeche Shale Member of the Permian and Triassic Goose Egg Formation. The Upper Minnelusa is composed of sandstone, dolomite, and anhydrite. Sandstone predominates on the periphery of the basin (Tenney, 1966). The percentage of carbonate increases basinward (Foster, 1958) and grades laterally toward the south through penesaline sediments into halite deposits (Tenney, 1966). Paleontological evidence derived from

the Hartville uplift (Love et al, 1963), Laramie Range (Agatston, 1954), and Big Horn Mountains (Verville, 1957) indicates that the Upper Minnelusa is Wolfcampian in age.

The productive sandstones in the Raven Creek and Reel fields are in the upper member of the Minnelusa Formation. In this area, Wolfcampian age strata are composed of sandstone beds interbedded with microcrystalline dolomite, anhydrite, and thin beds of red and black shale (Fig. 4). The sandstones, which vary in thickness, are white to light gray, fine- to very fine-grained, rounded to subrounded, and are cemented by anhydrite and dolomite. Dolomite beds are gray to pink, stylolitized, fractured, and contain chert and anhydrite inclusions. Anhydrite beds are white to red, and textures range from microcrystalline to sucrosic.

The Minnelusa is overlain by the red, anhydritic, and locally arenaceous shales of the Opeche Member of the Goose Egg Formation. Locally, a rubble zone composed of chert, dolomite, sandstone, and anhydrite pebbles in a red shale matrix is present at the base of the Opeche (Hudson, 1963).

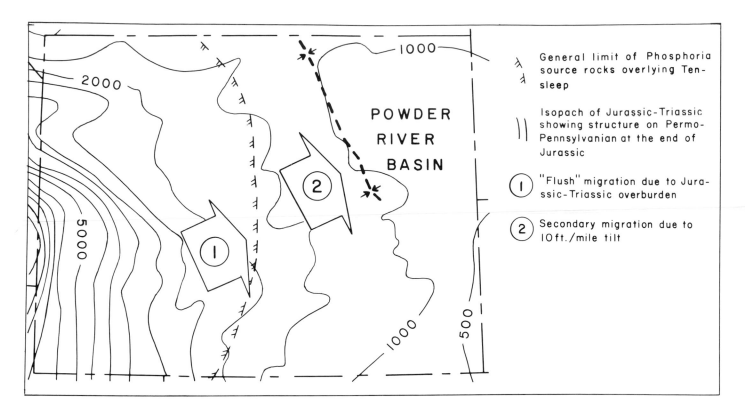

Figure 5—Regional geologic conditions that permitted migration of Phosphoria-sourced oils toward eastern Wyoming. Adapted from Barbat (1967).

Phosphoria Formation. Cheney and Sheldon (1959) speculated that the phosphatic shale beds of the Phosphoria were sources of petroleum found in the Park City Formation in southwestern Wyoming and northern Utah. Stone (1967) extended this concept to include most of the oil in Paleozoic reservoirs in the Big Horn basin. A similar source for Paleozoic accumulations in the Wind River basin has been suggested by Keefer (1969). Barbat (1967), Sheldon (1967), and Momper and Williams (1979) concluded that the oil in Upper Paleozoic rocks in the Powder River basin migrated eastward from the Phosphoria Formation. The Phosphoria and its partial stratigraphic equivalents of marine origin cover much of Montana, Wyoming, Colorado, Idaho, Utah, and Nevada (McKelvey et al, 1959). Maughan (1975) suggested that the principal area of oil generation in the Phosphoria would have been eastern Idaho, adjacent central-western Wyoming and southwestern Montana.

Hydrocarbons were expelled, possibly in response to overburden pressures, from Phosphoria source rocks into adjacent porous and permeable Permian and Pennsylvanian conduit beds, with the vertical and lateral limits of migration controlled by impermeable sealing beds (Sheldon, 1967). Gussow (1954) suggested that approximately 2000 ft of overburden are required before hydrocarbons are expelled from source rocks and primary migration can occur. Much of the Phosphoria source area was being subjected, or had already been subjected, to a critical load of 2000 to 3000 ft of burial by the end of the Jurassic (Barbat, 1967). Sheldon (1967), Stone (1967), and Keefer (1969) concluded that the geologic conditions in the Phosphoria source area were such that primary migration was initiated during Early Jurassic time (Fig. 5). Primary migration probably was complete by Early Cretaceous time (Barbat, 1967).

Differences in overburden pressure, as well as the regional westward dip of strata across Wyoming, would have induced an eastward migration of hydrocarbons toward central Wyoming (Sheldon, 1967). Barbat (1967) suggested that secondary migration of oil

from the Tensleep to the laterally contiguous Minnelusa may have occurred in response to the development during Late Jurassic time of regional southwestward tilting across the broad shelf of central Wyoming (Fig. 5). Momper and Williams (1979) suggested that oil derived from the Phosphoria entered northeastern Wyoming across the Casper Arch before the Powder River basin was formed. Initially, the oil migrated through Pennsylvanian Tensleep sandstones. Then, where Lower Permian Tensleep sandstones were preserved, oil migrated through conduit beds of that age. The emergence of the Ancestral Wind River Range in Maastrichtian time probably halted the uninterrupted eastward secondary migration of Phosphoria-derived oil (Sheldon, 1967).

Two types of oil are present in Upper Minnelusa reservoirs at the Raven Creek and Reel fields. This suggests that two cycles of migration are responsible for accumulation. The sandstones contain streaks and patches of thick, heavy black oil, while the sand grains are evenly saturated with a producible brown oil. The trap at Raven Creek is formed by the truncation of the reservoir rock against the Opeche Shale at the post-Minnelusa unconformity (Tranter, 1963). Tranter and Kerns (1972) suggested that oil could have been trapped in a pre-Opeche structure and seeped out when the sandstone was exposed by erosion. In this case, the black oil is the residue of an inspissated pre-Opeche accumulation, and the brown oil migrated into the trap after deposition of the Opeche. Primary migration of oil from the Phosphoria probably did not begin until the Jurassic. Therefore, if oil was in place in the reservoir before post-Minnelusa erosion, it was probably derived from a source other than the Phosphoria.

An alternative to this hypothesis is that the two types of oil represent two stages of migration of Phosphoria-sourced oil into the reservoir. The first cycle of oil may have migrated into the Raven Creek and Reel fields as early as the Late Jurassic. Considering the history of tectonic instability in the area, it is not improbable that an influx of meteoric water interfered with the migration pathways and altered the composition of the oil by nonthermal alteration processes, such as biodegradation and water washing. Basinward tilting of strata as early as latest Cretaceous time may have resulted in the remigration of Phosphoria-sourced oil from other parts of the basin (Momper and Williams, 1979) into the Raven Creek and Reel fields.

COMPOSITION

Minnelusa cores from the Raven Creek and Reel fields consist of sandstone, shale, dolomite, and anhydrite. Primary attention in this study was directed to analysis of sandstones in the Upper Minnelusa, which total approximately 65% of the rock types within the cores studied. Shales are the least abundant rock type, constituting less than 5% of the Upper Minnelusa. Dolomite and anhydrite make up the remainder of the Upper Minnelusa section. Opeche shale constitutes the remainder of the rock types in the cores.

Detrital Constituents

Although some variation exists in the relative mineralogic proportions (Table 1), the detrital mineralogy of Upper Minnelusa sandstones is essentially the same in all cores examined. The sandstones are predominantly quartzarenites (classification of Folk, 1974). A few of the samples are subarkoses. The majority of the framework grains are quartz, with the remainder mostly alkali feldspar. Rounded chert grains are present but volumetrically negligible. Heavy and accessory minerals are generally present but rarely exceed 1% of the total composition. Zircon, the most abundant heavy mineral, occurs as well-rounded grains. Other minor accessory minerals include biotite and rounded grains of tourmaline. Thin illitic clay seams occur in many of the sandstones.

Detrital quartz grains range from 44 to 94% and average 77% of the total mineralogy of the sandstones. The types of quartz observed are unstrained monocrystalline grains, strained monocrystalline grains, and composite or polycrystalline grains. Strained single grains make up approximately two-thirds of the total quartz. Polycrystalline quartz is the least abundant variety. There is generally a bimodal distribution of grain sizes. The finer mode of moderately to well-sorted, subangular to subrounded, fine- to very fine-grained sand is commonly mixed, and occasionally interstratified with a well-sorted, subrounded to rounded, medium- to coarse-grained sand fraction. The grains are commonly frosted.

Feldspar constitutes, on the average, 2% of the total composition of the sandstones. Untwinned orthoclase is the most common variety, followed in abundance by rare grains of twinned microcline. No plagioclase was observed. Feldspar grains are subrounded, and are invariably larger than the mean grain size of samples in which they occur. Two distinct populations of feldspar are recognized. Orthoclase grains show signs of dissolution or chemical alteration. Microcline grains are commonly unaltered.

Diagenetic Constituents

The reservoir quality of Upper Minnelusa sandstones is closely related to diagenesis, which in turn is dependent upon composition. The dominant diagenetic minerals in the sandstones are anhydrite and dolomite. Less common authigenic constituents include

mixed-layer illite–smectite, kaolinite, quartz, chert, and pyrite.

Anhydrite

Anhydrite is the most conspicuous diagenetic mineral in Upper Minnelusa sandstones (Fig. 6), ranging from 1 to 54% and averaging 13%. The mode of occurrence of anhydrite ranges from small scattered or layered nodules to pervasively cemented beds. In thin section it occurs as poikilotopic masses enclosing and replacing detrital (Fig. 7) or authigenic quartz. A stairstep boundary was formed where anhydrite replaced quartz. Later dissolution of anhydrite left discontinuous quartz boundaries.

The precipitation of the calcium sulfate cement may have been contemporaneous with the deposition of the sand in a coastal sabkha environment. Achauer (1982) noted that nodular anhydrite cements in Upper Minnelusa sandstones are similar to nodular anhydrite cements in the eolian facies of Persian Gulf coastal sabkha depositional cycles. The early formation of calcium sulfate cement is evidenced by low compaction in sandstones completely cemented by it. In such samples the detrital grains are "floating" in an anhydrite matrix. The calcium sulfate cement may have been precipitated as anhydrite or as gypsum. Later burial below 2000 to 3000 ft would have dehydrated gypsum to anhydrite (Pettijohn et al, 1973), and resulted in a cement volume loss and greater compaction (Mou and Brenner, 1982).

Much of the anhydrite cement in Upper Minnelusa sandstones may have formed after deposition of the sands. The overlying Goose Egg and Spearfish Formations were deposited under evaporitic conditions. Supersaturated calcium sulfate brines originating in these formations may have percolated down into the Upper Minnelusa sands and cemented them. Dissolution of the supratidal facies in the Upper Minnelusa and the secondary precipitation of calcium sulfate from laterally migrating solutions that were supersaturated with respect to calcium sulfate may be another mechanism by which this cement was formed (Stone, 1969).

TABLE 1

Well Number	Depth (ft)	Detrital Constituents[1]				Diagenetic Constituents[1]			
		Quartz	Feldspar	Mica	Misc.	Anhydrite	Dolomite	Clays	Pyrite
Government F12-6-G	8058	76	tr	—	tr-Z	2	22	tr	—
Krause	8352	78	2	tr	—	12	8	tr	—
F12-2-P	8340	79	1	tr	—	3	17	tr	tr
Krause	8349	75	1	tr	—	19	5	tr	tr
F21-2-P	8530	69	1	tr	tr-Z	13	17	tr	—
	8595	71	3	tr	tr-Z	2	24	—	—
Krause	8348	83	1	tr	—	3	13	—	—
F23-2-P	8370	82	1	tr	—	5	12	—	tr
Krause	8373	80	1	tr	tr-C	7	12	tr	—
F41-3-P	8390	75	2	tr	tr-C	8	15	tr	—
Krause	8385	71	1	tr	—	5	23	—	—
F32-3-P	8394	80	1	—	tr-Z	1	18	—	—
	8415	83	3	tr	—	4	10	—	—
Talley 1	8655	80	2	tr	—	1	17	—	tr
	8670	85	1	tr	tr-Z	3	11	—	—
Krause	8425	67	4	tr	—	17	12	—	—
C-1	8471	85	1	—	—	2	12	—	tr
Krause	8370	76	5	—	tr-Z	5	14	tr	—
F12-11-P	8388	69	2	—	tr-Z	25	4	tr	—
Krause F34-11-G	8340	76	4	—	—	15	5	—	—
Clark	8491	85	1	tr	—	2	12	—	tr
F-48-69-13-C4	8520	80	3	tr	—	3	14	tr	—
Norman 1	8350	68	3	—	—	23	6	tr	—
	8390	77	4	tr	tr-Z	17	2	tr	—
	8430	84	2	tr	tr-Z	13	1	tr	—
Government	9380	76	2	tr	tr-Z	21	1	—	—
Neuen. 1	9420	80	2	—	—	5	13	—	—
State F34-16-S	8505	76	1	—	—	14	9	—	—
Reel	8530	75	1	tr	—	15	9	—	—
F14-21-P	8550	79	1	tr	—	8	11	tr	—
Reel 1	8467	79	1	tr	tr-Z, T	16	3	—	1
	8484	79	4	tr	—	2	15	—	—
Wolfe 1	8345	77	4	tr	tr-Z	5	14	tr	—
	8385	75	8	—	—	5	12	tr	—
Bryant P-49-69-27-A4	8459	79	2	tr	tr-C	3	16	tr	—
Reel	8555	77	1	—	—	8	14	—	—
Government 23-28A	8575	77	6	tr	—	10	7	tr	—
Reel	8530	76	1	tr	—	19	4	—	—
Government 21-28A	8550	75	3	tr	—	5	17	tr	—
Krause F14-34-P	8480	77	1	tr	tr-Z	7	15	tr	—
Krause	8405	79	2	tr	tr-Z	3	16	tr	tr
F34-34-P	8435	68	tr	tr	—	30	2	—	—
Government 14-1	9055	68	tr	tr	—	30	2	—	—

[1](tr = trace, Z = zircon, T = tourmaline, C = chert).

Table 1—Summation of thin-section data.

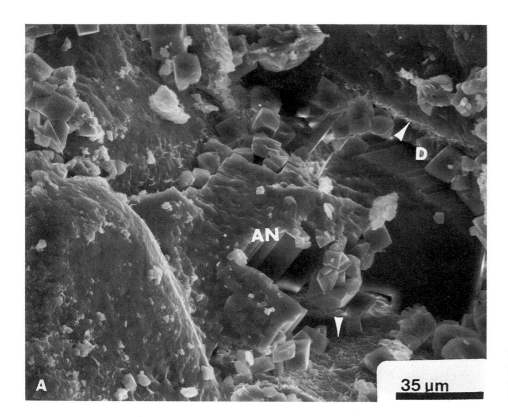

35 µm

Figure 6a—SEM photomicrograph of authigenic minerals. Anhydrite (**AN**), dolomite (**D**), and pore-lining mixed-layer illite–smectite (arrows).

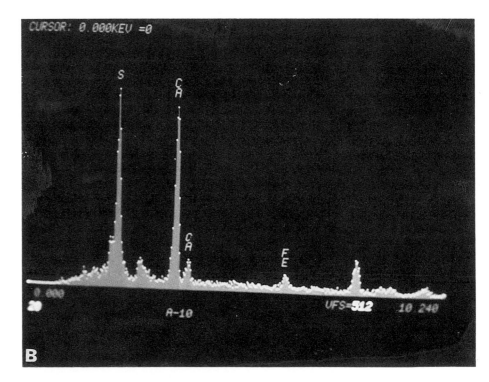

Figure 6b—EDAX of anhydrite cement.

0.1mm

Figure 7—Light photomicrograph of poikilotopic anhydrite (**AN**). Crossed polarizers.

Dolomite

The dolomite content of Upper Minnelusa sandstones ranges from 1 to 35% and averages 8%. It occurs as euhedral rhombs (Fig. 8), suggesting a diagenetic origin. The rhombs are generally much smaller than the quartz grains. They occur as aggregates that occupy significant portions of the pore space (Fig. 9) or as single crystals that line pores (Fig. 10) or replace anhydrite.

The ferroan nature of the dolomite is indicated by the blue color developed when stained with potassium ferrocyanide (Fig. 11). An increase in the Fe^{+2}/Mg^{+2} ratio in solution and a corresponding minimal level of magnesium ion activity (Katz, 1971) are required to form ferroan dolomite. The maintenance of an elevated concentration of Fe^{+2} in solution is dependent on the pH and Eh regime. Recognizing that the Powder River basin has a long history of tectonic instability that probably resulted in a complex history of fluid exchange, the assumption is made that the present pH of the formation water, approximately 6.8 (Wells et al, 1979), is representative of water character during the precipitation of authigenic dolomite. If this supposition is correct, a strongly reducing environment was required to maintain elevated concentrations of Fe^{+2} ions in solution (Hem and Cropper, 1959; Katz, 1971;

Al-Shaieb and Shelton, 1978).

Dolomite beds and diagenetic dolomite from the sandstones were analyzed for carbon- and oxygen-isotope ratios to determine the origin of the carbon and the isotopic nature of the depositional waters. The results are summarized in Figure 12. The mean values of $\delta^{13}C$ and $\delta^{18}O$ for the dolomite beds are -4.2 o/oo (PDB) and $+34.3$ o/oo (SMOW), respectively. The mean values of $\delta^{13}C$ and $\delta^{18}O$ for diagenetic dolomite are $+2.0$ and $+25.5$ o/oo, respectively. The differences in the mean values indicate that the two types of dolomite were formed in waters having markedly different isotopic compositions.

The average $\delta^{13}C$ value of marine carbonates is $+0.6 \pm 2.8$ o/oo (Keith and Weber, 1964). Diagenetic dolomite in Upper Minnelusa sandstones is enriched in ^{13}C and has approximately the same $\delta^{13}C$ value as marine carbonate. Hudson (1975) suggested that late diagenetic carbonate cement forms during the compactional phase of burial diagenesis. This process involves the dissolution of carbonates originating in marine environments and their reprecipitation. The difference in the mean values of $\delta^{13}C$ for the two types of dolomite in the Upper Minnelusa indicates that the dolomite beds were not the source of the ^{13}C found in the

diagenetic dolomite.

The equilibrium value of $\delta^{18}O$ for dolomite forming at 25°C probably ranges from $+31.9$ to $+38.1$ o/oo; however, these values should decrease with increasing temperature (Land, 1980). Choquette (1971) noted the occurrence of ferroan dolomite cement with a $\delta^{18}O$ value of $+24.5$ o/oo. It was suggested that the cement originated late in diagenesis at advanced stages of induration and burial, and in the presence of heated and reducing formation waters. The $\delta^{18}O$ value, $+25.5$ o/oo, of ferroan dolomite in Upper Minnelusa sandstones suggests that the dolomite may have precipitated in a similar environment from water with approximately the same isotopic composition.

Authigenic Clay Minerals

Authigenic clay minerals in representative samples of Upper Minnelusa sandstones were identified as mixed-layer illite–smectite and kaolinite. The delicate projections on the mixed-layer illite–smectite and the platy morphology of the kaolinite indicate their authigenic nature. Mixed-layer illite–smectite is the most abundant authigenic clay mineral. It occurs as pore linings, pore bridgings, and pore fillings (Figs. 13, 14). Kaolinite occurs as booklets filling pores. Both clays were generally observed on top of other

9 µm

Figure 8a—SEM photomicrograph of authigenic minerals and secondary porosity. Dolomite (**D**), mixed-layer illite–smectite (arrow), and secondary porosity.

cements or on exposed detrital grain surfaces, suggesting contemporaneous precipitation.

The precipitation of authigenic clay minerals reflects changes in the chemistry of the formation water. As a result of this new equilibrium regime, detrital illite, either in the detrital clay seams or as transported clay cutans on detrital grains, and detrital feldspar became unstable. This instability may have resulted in the transformation of minor quantities of detrital illite and feldspar to authigenic mixed-layer illite–smectite and kaolinite, respectively.

Silica

Silica is a minor cementing agent in Upper Minnelusa sandstones. Syntaxial quartz overgrowths (Fig. 15) occur sporadically throughout the cored sections. The initial prism faces were lost as overgrowths from adjacent detrital grains merged to form crystallographically irrational compromise boundary surfaces that may be mistaken for pressure solution features (Pittman, 1972). The paucity of sutured grain contacts indicates that much of the silica was

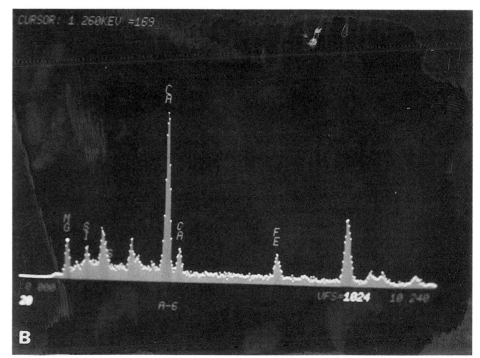

Figure 8b—EDAX of euhedral dolomite.

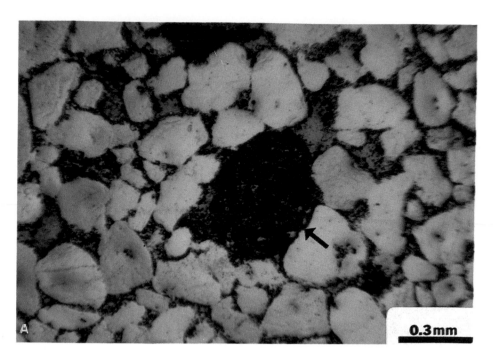

Figure 9a—Light photomicrograph of pore-filling dolomite aggregates (**D**). Plane light.

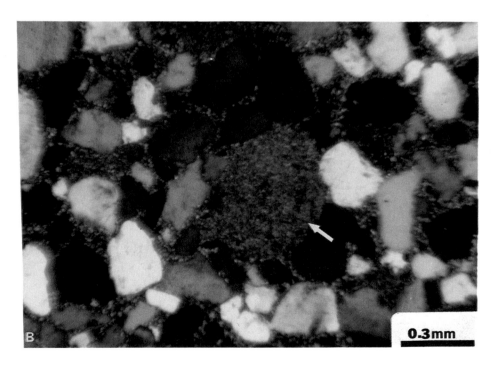

Figure 9b—Crossed polarizers.

precipitated from migrating fluids rather than as a result of local pressure solution. Traces of diagenetic chert were also observed.

Pyrite

Hydrocarbon migration had a limited influence on diagenetic mineralization in Upper Minnelusa sandstones. Alteration of the sandstones by hydrogen sulfide gas associated with petroleum resulted in the precipitation of disseminated euhedral pyrite crystals (Markert, 1982). Pyrite can be formed in red beds by the reduction of iron by hydrogen sulfide gas associated with hydrocarbons (Lilburn, 1979). In this process the iron is reduced from the ferric to ferrous state, with the sulfur in the pyrite provided by the hydrogen sulfide gas. The iron needed for the reaction originated either from the hematite associated with the Opeche shale, or by base exchange from clay minerals (Parker, 1973).

DIAGENETIC HISTORY AND SECONDARY POROSITY

Diagenetic modifications of the original depositional fabric have resulted in the present-day porosity and permeability in Upper Minnelusa sandstones. Although the pore geometry within these sandstones is a combination of

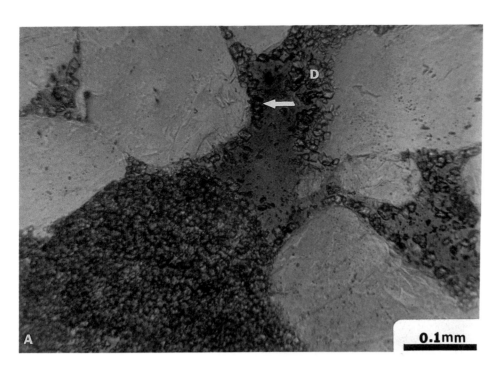

Figure 10a—Light photomicrograph of pore-lining (arrow) dolomite (**D**). Plane light.

both primary and secondary porosity, the volume of secondary porosity exceeds that of primary porosity. Compaction and the precipitation of authigenic anhydrite, quartz, dolomite, clay minerals, and miscellaneous diagenetic products acted to reduce reservoir potential. Most of the secondary porosity was developed by the dissolution of calcium sulfate cement. The following discussion of the nature, degree, and sequence (Fig. 16) of diagenesis of Upper Minnelusa sandstones is based on the textural relationships of detrital and diagenetic minerals. Correlation of these events with the tectonic history of the Powder River basin is based on less direct lines of evidence.

The presence of detrital grains that appear to "float" without any self-supporting framework in an anhydrite matrix suggests that calcium sulfate may have been introduced into the sandstones during deposition in a sabkha environment. Early calcium sulfate cement in Minnelusa sandstones may have absorbed the overburden pressure and prevented mechanical compaction at depths of burial less than 10,000 ft (Moore, 1975).

Quartz overgrowths occur sporadically throughout the core samples. The Upper Minnelusa in the Raven Creek and Reel fields was probably never sub-

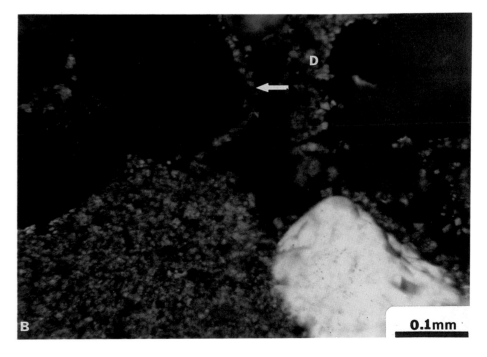

Figure 10b—Crossed polarizers.

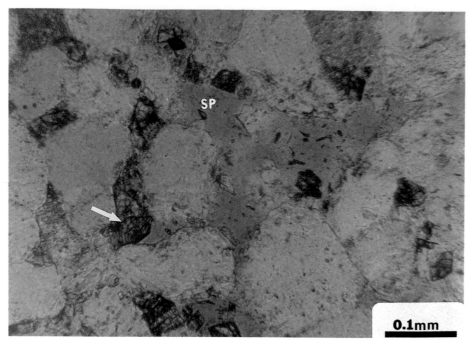

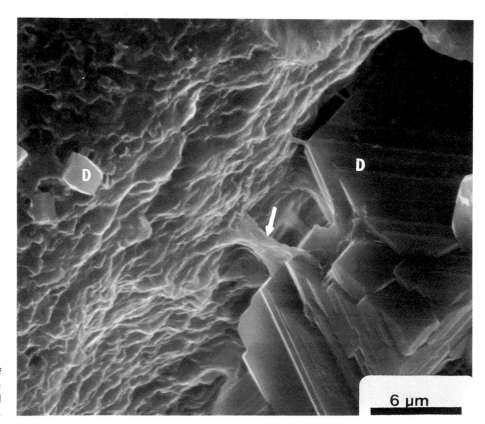

0.1mm

Figure 11—Light photomicrograph of ferroan dolomite (arrow) and secondary porosity (**SP**). Plane light.

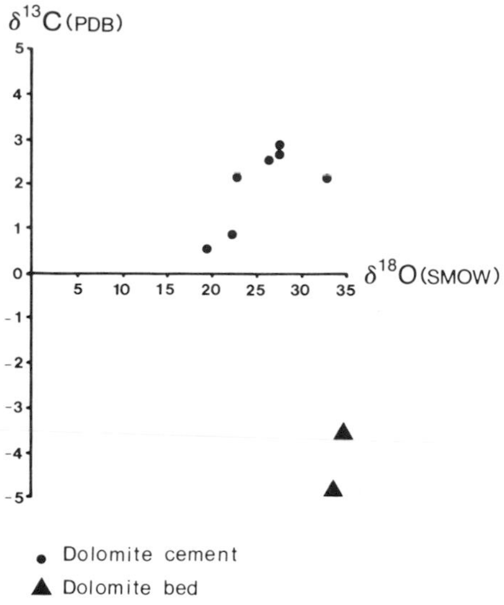

$\delta^{13}C$ (PDB)

$\delta^{18}O$ (SMOW)

• Dolomite cement
▲ Dolomite bed

Figure 12—Relationship between $\delta^{13}C$ and $\delta^{18}O$ values of dolomite in the Upper Minnelusa.

6 µm

Figure 13—SEM photomicrograph of mixed-layer illite–smectite (arrow) bridging between detrital quartz and euhedral dolomite.

Figure 14—SEM photomicrograph of pore-bridging (arrow) and pore-filling mixed-layer illite–smectite.

jected to depths as great as 10,000 ft. Therefore, much of the silica may have been precipitated from migrating fluids, rather than from oversaturated solutions resulting from pressure solution. The presence of poikilotopic masses of anhydrite enclosing and replacing unabraded quartz overgrowths, and the absence of quartz overgrowths in some samples that are well cemented by anhydrite suggests that quartz may have precipitated between two episodes of calcium sulfate precipitation. The anhydrite units in the Upper Minnelusa or the overlying Goose Egg and Spearfish Formations are the more probable sources of the precipitate associated with the later episode of calcium sulfate cementation.

The dissolution of calcium sulfate cement (Fig. 17) was an important diagenetic event in reservoir development of the Upper Minnelusa sandstones. The abundance of authigenic anhydrite in the cored sections and the frequency with which anhydrite dissolution was observed throughout the sandstones suggests that calcium sulfate cementation occluded most of the primary intergranular porosity. With-

Figure 15—Light photomicrograph showing extreme silicification. Note quartz overgrowths and "dust" rims (arrows). Crossed polarizers.

outline of original grain which has grown by secondary overgrowths.

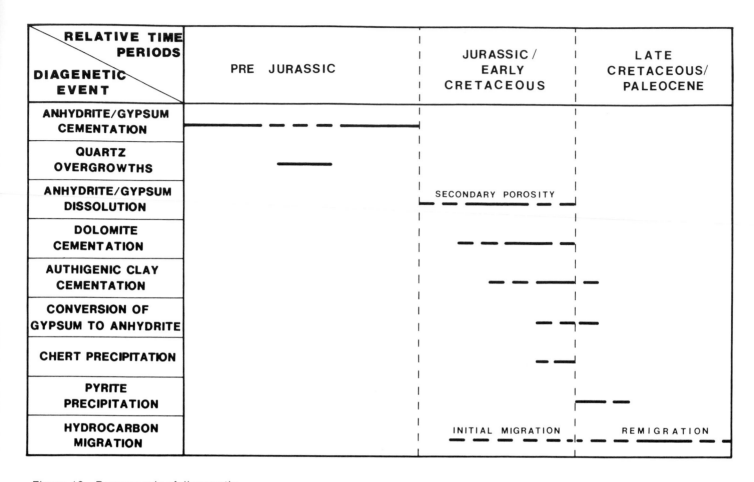

RELATIVE TIME PERIODS / DIAGENETIC EVENT	PRE JURASSIC	JURASSIC / EARLY CRETACEOUS	LATE CRETACEOUS/ PALEOCENE
ANHYDRITE/GYPSUM CEMENTATION	———— – – – –		
QUARTZ OVERGROWTHS	—		
ANHYDRITE/GYPSUM DISSOLUTION		SECONDARY POROSITY – – – –	
DOLOMITE CEMENTATION		– – – —	
AUTHIGENIC CLAY CEMENTATION		– – –	—
CONVERSION OF GYPSUM TO ANHYDRITE		– –	—
CHERT PRECIPITATION		– –	
PYRITE PRECIPITATION			—
HYDROCARBON MIGRATION		INITIAL MIGRATION – – –	REMIGRATION – – – – —

Figure 16—Paragenesis of diagenetic events.

out this dissolution the present-day porosity in the sandstones might have been extremely low. High topographic relief resulting from warping in the Jurassic (Blackstone, 1963) or uplift in the Early Cretaceous (Slack, 1981) may have forced meteoric water into the deeper subsurface under a high hydrostatic head. The encroachment of meteoric water resulted in the dilution of formation fluids and subsequently, the dissolution of calcium sulfate cement.

Authigenic dolomite represents a late-stage diagenetic product. This is evidenced by the presence of dolomite replacing anhydrite, and the occurrence of pore-filling and pore-lining dolomite in Upper Minnelusa sandstones with abundant secondary porosity. Authigenic mixed-layer illite–smectite (Fig. 18) and kaolinite may have formed contemporaneously with diagenetic dolomite. However, the occurrence of

mixed-layer illite–smectite on the surface of dolomite cement suggests that the initiation of dolomite cementation may have preceded the precipitation of authigenic clays. The precipitation of authigenic clay minerals and dolomite reflects changes in the physiochemical environment, possibly as a result of an influx of meteoric waters that were responsible for anhydrite dissolution.

After Early Cretaceous times the Upper Minnelusa would have been buried to depths exceeding 2000 to 3000 ft, and any gypsum cement present in the sandstones would have dehydrated to anhydrite. Moore (1975) estimated that calcium sulfate cement may have constituted an average of 25% or more of the Minnelusa sandstones at the time of deposition. This number represents the difference between the intergranular and effective porosities at zero depth of burial. Recrystallization during dehydration of gypsum to anhydrite results in a 38% volume reduction (Mou and Brenner, 1982). Therefore, if the Minnelusa sandstones contained 25% or

more gypsum cement, dehydration would result in sandstones with approximately 16% anhydrite cement. The loss in volume is accompanied by an increase in compaction.

Early calcium sulfate cementation may have precluded the accumulation of hydrocarbons in Upper Minnelusa sandstones in the Raven Creek and Reel fields before the Jurassic times. Phosphoria-sourced oils possibly migrated into the Upper Minnelusa reservoirs as early as Late Jurassic times. Accumulation of producible hydrocarbons in the reservoirs may have occurred during latest Cretaceous or Early Tertiary times by remigration from other parts of the basin, in response to basinward tilting of Minnelusa strata.

Secondary chert appears to have formed late in the diagenetic history of the Minnelusa. Pyrite was possibly the last authigenic mineral to form in the sandstones. The precipitation of pyrite is related to hydrocarbon migration. Chemical diagenesis was inhibited by

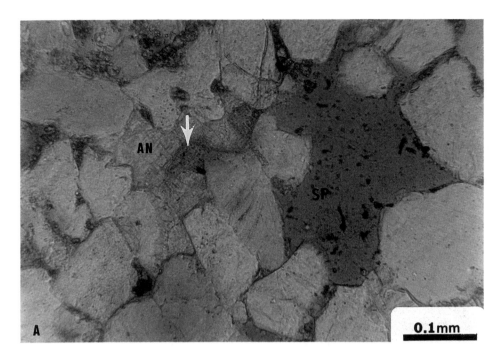

0.1mm

Figure 17a—Light photomicrograph showing the dissolution of anhydrite (**AN**) and the resulting secondary porosity (**SP** and arrow). Plane light.

the accumulation of hydrocarbons in the reservoirs.

WATEQF ANALYSIS OF FORMATION WATERS

Petrographic evidence suggests that secondary porosity is the dominant type of porosity in Upper Minnelusa sandstones in the Raven Creek and Reel fields. Dissolution of anhydrite (Fig. 17) is observed throughout the section. In order to propose a possible mechanism for the dissolution of anhydrite and the generation of secondary porosity, Minnelusa formation waters were analyzed using the WATEQF computer program (Plummer et al, 1976). Chemical analyses of formation waters by Wells et al (1979) provided information on the chemical species, calcium, magnesium, sodium, potassium, iron, bicarbonate, carbonate, sulfate, and chloride, and the pH of the formation waters. The formation temperatures used were obtained from Head et al (1979).

The WATEQF program provides information on the dissociated and associated species, activity coefficients of dissolved species, and the degree of saturation of these species with respect to inorganic solid compounds. Precipitation or dissolution of a specific solid

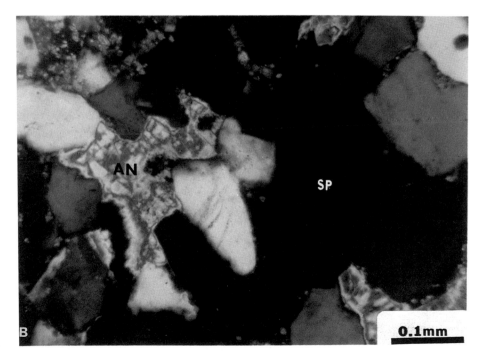

0.1mm

Figure 17b—Crossed polarizers.

Figure 18a—SEM photomicrograph showing textural relationships of detrital and authigenic minerals.

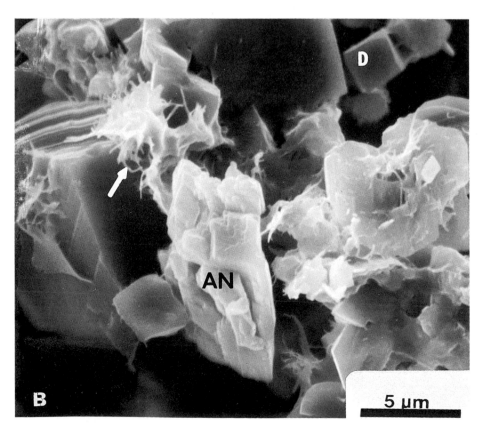

Figure 18b—Enlargement of 18a (arrow) showing textural relationships of authigenic minerals, anhydrite (AN), dolomite (D), and mixed-layer illite–smectite (arrow).

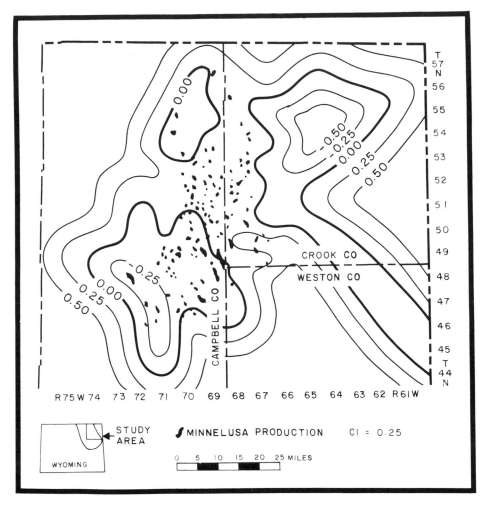

Figure 19—Computed average anhydrite saturation in Upper Minnelusa formation waters.

species (that is, anhydrite) is evaluated by comparing the ion activity product (IAP) with the solubility product (K_T) of this particular species. The term, IAP/K_T, is defined as the saturation index (S). Solutions may be considered to be supersaturated with respect to a specific mineral phase if $S > 1$, undersaturated if $S < 1$, and saturated if $S = 1$.

To determine if a relationship exists between the chemistry of Minnelusa formation waters and the spatial distribution of anhydrite in Upper Minnelusa sandstones, a contour map (Fig. 19) of the anhydrite saturation indices (Table 2) was constructed. This map was compared to published contour maps (Head and Merkel, 1977) of the average porosity distribution (Fig. 20) and the percent volume of sandstone (Fig. 21) in the Upper Minnelusa. There is a distinct correspondence between the zones of high porosity and

high sandstone percentages. There is a positive, but less pronounced, correlation between the zones of low anhydrite saturation and the zones of high porosity and sandstone percentages. The porosity network in the sandstones is a combination of both primary and secondary intergranular porosity, with the second porosity being dominant. Therefore, the latter correlation indicates that the development of secondary porosity in Upper Minnelusa sandstones is attributable to the dissolution of anhydrite, and that the highest secondary porosity development may be encountered in those areas where the formation waters are undersaturated with respect to anhydrite.

The aforementioned observations suggest that the WATEQF computer program may be used as an exploration tool in delineating secondary porosity fairways in geologic settings that are dominated by evaporitic environments.

SUMMARY

The Upper Minnelusa is a multifacies rock system consisting of subtidal and intertidal dolomites, supratidal anhydrite, and sandstones representing deposition in eolian and associated sabkha environments. The present textures and fabric of the sandstones are primarily the product of diagenesis. Although secondary porosity is the dominant type of porosity, a certain amount of primary porosity has been retained.

The dominant modifications of the original depositional fabric are the destruction of primary porosity by the precipitation of authigenic anhydrite, dolomite, clay minerals, chert and pyrite, and porosity enhancement by the dissolution of anhydrite. An understanding of the nature and sequence of diagenesis, when integrated with WATEQF analysis of formation

TABLE 2

Sample Location	Log IAP/K$_T$[1]	Sample Location	Log IAP/K$_T$[1]	Sample Location	Log IAP/K$_T$[1]
Campbell Co.		T51N-R71W	0.11	T51N-R66W	−0.02
T45N-R71W	−0.41	T51N-R72W	−0.05	T51N-R68W	0.15
T46N-R69W	0.12	T52N-R69W	0.07	T52N-R68W	0.21
T46N-R70W	0.22	T52N-R70W	0.16	T52N-R69W	0.20
T47N-R69W	−0.19	T52N-R72W	0.05	T53N-R65W	−0.57
T47N-R70W	0.00	T53N-R69W	0.12	T53N-R67W	−0.12
T47N-R71W	−0.41	T53N-R70W	0.09	T53N-R68W	0.18
T48N-R69W	−0.02	T54N-R70W	−0.02	T54N-R63W	−0.50
T48N-R70W	−0.17	T54N-R71W	−0.08	T54N-R65W	−1.24
T48N-R71W	−0.17	T55N-R69W	−0.04	T54N-R66W	−0.14
T48N-R72W	−1.97	T55N-R71W	0.17	T54N-R67W	0.14
T49N-R69W	0.01	T56N-R69W	0.02	T54N-R68W	0.06
T49N-R70W	−0.14	T57N-R69W	−0.01	T55N-R67W	−0.18
T49N-R71W	−0.07			T55N-R68W	0.06
T49N-R72W	−0.12	Crook Co.		T56N-R61W	0.09
T49N-R74W	−0.014	T49N-R67W	0.08	T57N-R64W	0.12
T50N-R69W	0.07	T49N-R68W	0.18		
T50N-R70W	−0.09	T50N-R65W	−0.07	Weston Co.	
T50N-R71W	0.02	T50N-R66W	0.08	T45N-R61W	−0.05
T51N-R69W	0.02	T50N-R67W	0.09	T47N-R68W	−0.17
T51N-R70W	−0.01	T50N-R68W	0.29	T48N-R68W	0.03

[1]Each value represents the mean saturation index for all samples in the township.

Table 2—Anhydrite saturation indices.

waters, enables the delineation of secondary porosity fairways in Upper Minnelusa sandstones.

ACKNOWLEDGMENTS

We would like to express our appreciation to Shell Development Company, Geophysical Research Department, for supporting this study and giving us the permission to publish our findings. The support and guidance of Alan Rosenthal, Jim Roberts, and Charles Walker were of great help. Special thanks to Gary Stewart for his critical review of this manuscript.

SELECTED REFERENCES

Achauer, C. W., 1982, Sabkha anhydrite: the supratidal facies of cyclic deposition in the Upper Minnelusa Formation (Permian) Rozet fields area, Powder River basin, Wyoming: Society of Economic Paleontologists and Mineralogists Core Workshop 3, p. 193–209.

Agatston, R. W., 1954, Pennsylvanian and Lower Permian of northern and eastern Wyoming: Bulletin of the American Association of Petroleum Geologists, v. 38, p. 508–583.

Al-Shaieb, Z., and J. W. Shelton, 1978, Secondary ferroan dolomite rhombs in oil reservoirs, Chadra sands, Gialo field, Libya: Bulletin of the American Association of Petroleum Geologists, v. 62, p. 463–468.

Barbat, W. N., 1967, Crude-oil correlations and their role in exploration: Bulletin of the American Association of Petroleum Geologists, v. 51, p. 1255–1292.

Berg, R. A., and C. S. Tenney, 1967, Geology of Lower Permian Minnelusa oil fields, Powder River basin, Wyoming: Bulletin of the American Association of Petroleum Geologists, v. 51, p. 705–709.

Blackstone, Jr., D. L., 1963, Development of geologic structure in central Rocky Mountains: American Association of Petroleum Geologists Memoir 2, p. 160–179.

Cheney, T. M., and R. P. Sheldon, 1959, Permian stratigraphy and oil potential, Wyoming and Utah: Intermountain Association of Petroleum Geologists 10th Annual Field Conference, p. 90–100.

Choquette, P. W., 1971, Late ferroan dolomite cement, Mississippian carbonates, Illinois basin, in O. P. Bricker, ed., Carbonate cements: Johns Hopkins University Studies in Geology 19, p. 339–346.

Curtis, B. F., J. W. Strickland, and R. C. Busby, 1958, Patterns of oil occurrence in the Powder River Basin, Wyoming, in Habitat of oil: American Association of Petroleum Geologists, p. 268–292.

Folk, R. L., 1974, Petrology of sedimentary rocks: Austin, Hemphill Publishing Co., 182 p.

Foster, D. I., 1958, Summary of the stratigraphy of the Minnelusa Formation, Powder River basin, Wyoming: Wyo-

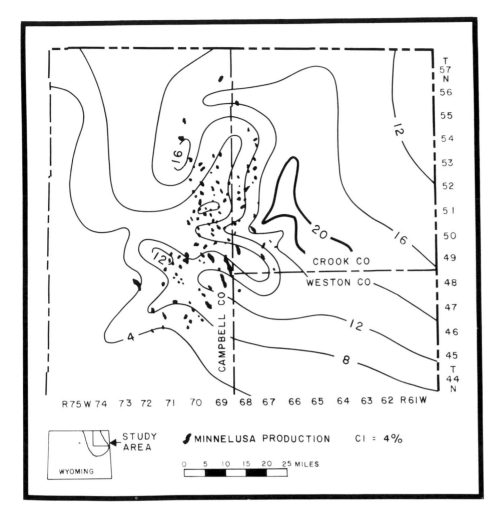

Figure 20—Computed average porosity for the top major sand-carbonate-sand sequence of the upper part of the Minnelusa Formation. Adapted from Head and Merkel (1977).

ming Geological Association Guidebook, 13th Annual Field Conference, p. 39–44.

Glaze, R. E., and E. R. Keller, (co-chairmen Wyoming Geological Association Technical Studies Committee), 1965, Geologic history of the Powder River basin: Bulletin of the American Association of Petroleum Geologists, v. 49, p. 1893–1907.

Gussow, W. C., 1954, Differential entrapment of oil and gas, a fundamental principle: Bulletin of the American Association of Petroleum Geologists, v. 38, p. 816–853.

Head, W. J., and R. H. Merkel, 1977, Hydrologic characteristics of the Madison Limestone, the Minnelusa Formation, and equivalent rocks as determined by well-logging formation evaluation, Wyoming, Montana, South Dakota, and North Dakota: United States Geological Survey Journal of Research, v. 5, p. 473–485.

Head, W. J., K. T. Kilty, and R. K. Knottek, 1979, Maps showing formation temperatures and configurations of the tops of the Minnelusa Formation and the Madison Limestone, Powder River basin, Wyoming, Montana, and adjacent areas: United States Geological Survey Miscellaneous Geologic Investigations Series, Map I-1159.

Hem, J. D., and W. H. Cropper, 1959, Survey of ferrous–ferric chemical equilibria and redox potentials: United States Geological Survey Water-Supply Paper 1459-A, 31 p.

Hudson, J. D., 1975, Carbon isotopes and limestone cement: Geology, v. 3, p. 19–22.

Hudson, R. E., 1963, Halverson Ranch field, Minnelusa production, Campbell County, Wyoming: Billings Geological Society and Wyoming Geological Association Guidebook, 1st Joint Field Conference, p. 123–124.

Katz, A., 1971, Zoned dolomite crystals: Journal of Geology, v. 79, p. 38–51.

Keefer, W. R., 1969, Geology of petroleum in Wind River basin, central Wyoming: Bulletin of the American Association of Petroleum Geologists, v. 53, p. 1839–1865.

Keith, M. L., and J. N. Weber, 1964, Carbon and oxygen isotopic composition of selected limestones and fossils: Geochimica et Cosmochimica Acta, v. 28, p. 1787–1816.

Land, L. S., 1980, The isotopic and trace element geochemistry of dolomite: the state of the art: Society of Economic Paleontologists and Mineralogists Special Publication 28, p. 87–110.

Lilburn, R. A., 1979, Mineralogical, geochemical, and isotopic evidence of diagenetic alteration, attributable to hydrocarbon migration, Cement-Chickasha field, Oklahoma: M.S. thesis, Oklahoma State University, 88 p.

Love, J. D., P. O. McGrew, and H. D. Thomas, 1963, Relationship of latest Cretaceous and Tertiary deposition and deformation to oil and gas in Wyoming:

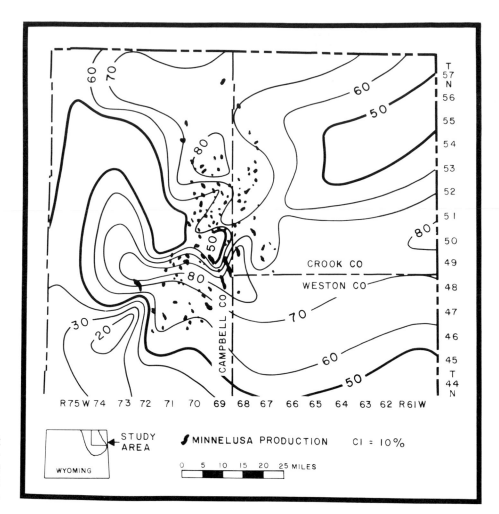

Figure 21—Computed average percent sand in the top major sand–carbonate–sand sequence of the upper part of the Minnelusa Formation. Adapted from Head and Merkel (1977).

American Association of Petroleum Geologists Memoir 3, p. 196–208.

Mallory, W. W., 1967, Pennsylvanian and associated rocks in Wyoming: United States Geological Survey Professional Paper 554-G, 31 p.

Markert, J. C., 1982, Mineralogical, geochemical, and isotopic evidence of diagenetic alteration, attributable to hydrocarbon migration, Raven Creek and Reel fields, Wyoming: M.S. thesis, Oklahoma State University, 126 p.

Maughan, E. K., 1975, Organic carbon in shale beds of the Permian Phosphoria Formation of eastern Idaho and adjacent states—A summary report: Wyoming Geological Association Guidebook, 27th Annual Field Conference, p. 107–115.

————, 1978, Pennsylvanian (Upper Carboniferous) System in Wyoming: United States Geological Survey Open-File Report 78-377, 32 p.

McKelvey, V. E., et al, 1959, The Phosphoria, Park City, and Shedhorn Formations in the western phosphate field: United States Geological Survey Professional Paper 313-A, 47 p.

Momper, J. A., and J. A. Williams, 1979, Geochemical exploration in the Powder River basin: Oil and Gas Journal, December 10, p. 129–134.

Moore, W. R., 1975, Grain packing–porosity relationships of Minnelusa sandstones, Powder River basin, Wyoming: Mountain Geologist, v. 12, p. 45–53.

Mou, D. C., and R. L. Brenner, 1982, Control of reservoir properties of Tensleep Sandstone and diagenetic facies: Lost Soldier field, Wyoming: Journal of Sedimentary Petrology, v. 52, p. 367–381.

Parker, C. A., 1973, Geopressures in the deep Smackover of Mississippi: Journal of Petroleum Technology, v. 25, p. 971–979.

Pettijohn, F. J., P. E. Potter, and R. Siever, 1973, Sand and sandstone: New York, Springer-Verlag, 618 p.

Pittman, E. D., 1972, Diagenesis of quartz in sandstones as revealed by scanning electron microscopy: Journal of Sedimentary Petrology, v. 42, p. 507–519.

Plummer, L. N., B. F. Jones, and A. H. Truesdell, 1976, WATEQF-A-FORTRAN IV version of WATEQ, a computer program for calculating chemical equilibria of natural waters: United States Geological Survey Water Resources Investigations 76-13, 61 p.

Sheldon, R. P., 1967, Long-distance migration of oil in Wyoming: Mountain Geologist, v. 4, p. 53–65.

Slack, P. B., 1981, Paleotectonics and hydrocarbon accumulation, Powder River basin, Wyoming: Bulletin of the American Association of Petroleum Geologists, v. 65, p. 730–743.

Stokes, W. L., 1968, Multiple parallel-truncation bedding planes—A feature of wind deposited sandstone formations: Journal of Sedimentary Petrology, v. 38, p. 510–515.

Stone, D. S., 1967, Theory of Paleozoic oil

and gas accumulation in Big Horn basin, Wyoming: Bulletin of the American Association of Petroleum Geologists, v. 51, p. 2056–2114.

Stone, W. J., 1969, Stratigraphy of the Minnelusa Formation along western and northern flanks of the Black Hills, Wyoming and South Dakota: M.S. thesis, Kent State University, 254 p.

Strickland, J. W., 1958, Habitat of oil in the Powder River basin: Wyoming Geological Association Guidebook, 13th Annual Field Conference p. 132–147.

Tenney, C. S., 1966, Pennsylvanian and Lower Permian deposition in Wyoming and adjacent areas: Bulletin of the American Association of Petroleum Geologists, v. 50, p. 227–250.

Tranter, C., 1963, Raven Creek field, Campbell County, Wyoming: Billings Geological Society and Wyoming Geological Association Guidebook, 1st Joint Field Conference, p. 143–146.

Tranter, C. E., and C. W. Kerns, 1972, Raven Creek field, Campbell County, Wyoming: American Association of Petroleum Geologists Memoir 16, p. 511–519.

Tranter, C. E., and C. K. Petter, 1963, Lower Permian and Pennsylvanian stratigraphy of the northern Rocky Mountains: Billings Geological Society and Wyoming Geological Association Guidebook, 1st Joint Field Conference, p. 45–53.

Verville, G. J., 1957, Wolfcampian fusulinids from the Tensleep Sandstone in the Big Horn Mountains, Wyoming: Journal of Paleontology, v. 31, p. 349–352.

Wells, D. K., J. F. Busby, and K. C. Glover, 1979, Chemical analyses of water from the Minnelusa Formation and equivalents in the Powder River basin and adjacent areas, northeastern Wyoming: Wyoming Water Planning Program Report 18, 27 p.

Reservoir Property Implications of Pore Geometry Modification Accompanying Sand Diagenesis: Anahuac Formation, Louisiana

Michael T. Holland
Terra Tek Research
Salt Lake City, Utah

INTRODUCTION

In the U.S. Gulf Coast, diagenetic processes active during the burial history of Tertiary sandstones are responsible for modifying reservoir properties. Knowledge of when, and to what degree, these diagenetic processes have affected reservoir development can be helpful in predicting the areal development of reservoir quality and inferring potential production characteristics.

Secondary porosity is now recognized to be the predominant and sometimes exclusive form of porosity found in many deep sandstone reservoirs worldwide including the U.S. Gulf Coast region (Schmidt and McDonald, 1980; Lindquist, 1977; Loucks et al, 1977; Milliken et al, 1981; Hayes, 1981). When attempting to produce fluids from reservoirs of high porosity, one must consider the mechanical competence of the grain fabric surrounding this secondary porosity as an important parameter controlling the reservoir's stress sensitivity. As the effective stress in a reservoir increases with production by reducing pore pressure with constant overburden stress, both porosity and permeability decrease. Inelastic rock deformation mechanisms play a significant role in reducing pore space and permeability (Schatz et al, 1982). However, reservoir intervals in quartzose sandstones containing secondary porosity may be more resistive to compaction than those of similar mineralogy containing an equal amount of primary porosity (Schmidt and McDonald, 1980; Hayes, 1981). To determine which portions of

This paper substantially revised and modified from copyright © SPE-AIME Paper #10991 presented at the 57th Annual Fall Technical Conference in New Orleans, Louisiana, Sept. 26–29, 1982.

ABSTRACT. This paper represents a study of reservoir pore modification accompanying diagenetic secondary porosity development within a deep (13,400 ft) overpressured Anahuac Formation sandstone in southern Louisiana. Secondary porosity formed by dissolution of carbonate cement, detrital grains, and other soluble minerals comprises a significant portion of porosity formed in U.S. Gulf Coast Tertiary reservoir sands. The primary pore system within this reservoir is believed to have been significantly enlarged (by up to 32% porosity) by acidic fluids generated during hydrocarbon maturation and dewatering of adjacent shales. Subsurface secondary porosity development within sandstones is significant in influencing the development of potential reservoir porosity after much of the primary porosity has been destroyed by mechanical and chemical compaction. Properties of the reservoir pore system that affect fluid flow and mechanical resistance of the reservoir to compaction accompanying production will also be influenced.

Characteristics of the reservoir pore system were established by study of whole core samples using scanning electron microscopy, petrographic examination, mercury injection, and simulated in-situ reservoir condition core testing. Secondary pore size and distribution was found to be influenced by sandstone mineralogy, grain size, sorting and angularity, the pore matrix content, and by sedimentary structures and resulting textural components that may hinder fluid flow.

Changes in the mechanical resistance to compaction caused by the development of secondary porosity in sandstone reservoirs is important when considering reservoir stress sensitivity. Keystone bridging relationships between grains can be established during the initial phases of compaction so that when leaching of cement and soluble grains occurs, a less soluble quartz grain matrix is left to support porosity development. Special core tests were performed at simulated in-situ reservoir conditions of pressure and temperature to examine porosity and permeability reduction as a function of effective stress generated by pore pressure reduction (simulated fluid production). Observed volumetric strain to uniaxial compaction at reservoir conditions was determined within portions of the sand containing high (25–30%) porosity. Test results exhibited less than 1% reduction in total bulk volumes accompanying a 60% reduction in pore pressure. Permeabilities measured at in-situ conditions were commonly an order of magnitude less than those measured at ambient conditions. However, with increased effective stress applied to the rock fabric, data suggest that permeabilities decrease at a much slower rate, reflecting constriction of pore throats rather than constriction of stress-induced microfactures thought to exist in core samples at ambient conditions.

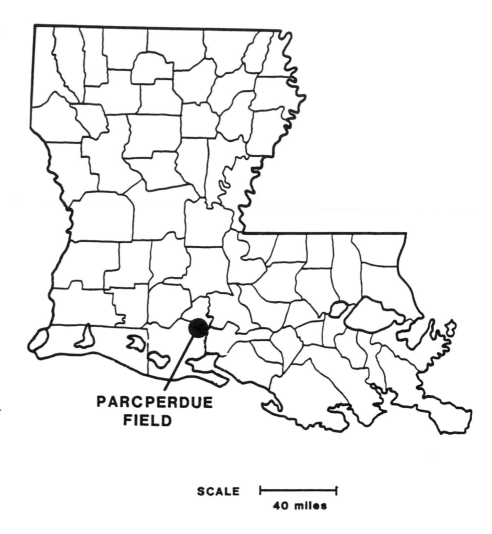

Figure 1—Index map of Louisiana and location of Parcperdue Field.

PARCPERDUE FIELD

SCALE |————————|
40 miles

the reservoir will be most affected by secondary porosity development and their resistance to compaction, one should consider an examination of sandstone mineralogy and understand the factors that control porosity distribution.

The distribution of secondary dissolution porosity within a reservoir will depend upon the movement of acidic and undersaturated pore fluids within the reservoir interval. If pore-fluid movement is minimal, isolated secondary pores may result in low permeability. However, some reservoirs experience removal of soluble grains and cements in such quantities that flow channels are created. If suitable pore fluids are available in sufficient quantity, flow-channel development creates more pore surface area, reducing capillary pressures, thereby enhancing dissolution of soluble minerals. This process may eventually lead to production of good effective porosity within the reservoir.

The case study presented in this paper represents a study of "hybrid" pores whose diagenetic history is applicable to many quartz sandstones (Schmidt and McDonald, 1980). The target sand interval studied by core analyses is called the Cibb Jeff Sand, which occurs in the Upper Oligocene Anahuac Formation in southern Louisiana. Preservation of effective secondary porosity in this predominantly quartz sandstone was dependent upon the ability of the grain fabric to support overburden pressure after cement and grain dissolution. During early diagenesis, mechanically unstable grains were squeezed around more competent grains by compaction. Primary porosity was reduced such that bridging and keystone relations between rigid grains were developed. The amount of compaction was strongly influenced by depositional factors such as grain sorting, size, and angularity.

During the burial history of the Cibb Jeff Sand, the sandstone entered an overpressured regime (pore pressure gradient much greater than 0.465 psi/ft). Presently, overpressure is encountered at approximately 11,000 ft

in this area. It is probable that the sandstone possessed the lowest porosity of its burial history before the advent of these abnormal pore pressures. This assumes that overburden stress was much greater than pore pressure within the reservoir, the difference of which had been increased with continued burial. Generation of hydrocarbons and CO_2 within shales creates abnormal pore pressure (Momper, 1978, 1980; Meisner, 1980), which may result in the formation of fractures, allowing movement of acidic pore waters into the more permeable, normally pressured sandstones in adjacent lithologies. Increased pore pressure in the sandstones reduced effective stress applied to the grain fabric, so that when carbonate cement and soluble grains were dissolved, the grain fabric was partially supported by higher pore pressures. Pressure solution at tangential point contacts also increased the

well-sorted grain fabric strength, thereby supporting porosity development and preventing pore collapse.

STUDY PROCEDURE

One-hundred-and-twenty feet of core (13,340 to 13,460 ft) were recovered from the DOW/DOE L. R. Sweezy No. 1 test well at the Parcperdue Geopressure/Geothermal site located in Vermilion Parish, Louisiana (Fig. 1). The target sand interval represented by this core is called the Cibb Jeff Sand, which occurs in the Upper Oligocene Anahuac Formation. The reservoir area is estimated to be approximately 940 acres with an average total sand thickness of 50 ft giving a reservoir volume of approximately 4.7×10^5 acre-ft (Hamilton and Wilson, 1981).

Geologic analysis of the reservoir interval involved core description, petrographic description, and scanning

Figure 2—High temperature in-situ testing system.

electron microscopy. Grain-size distribution, lithology, and sedimentary structures within the 70-ft total sand interval were determined by a combination of visual examination and low power optical microscopy. Petrographic analysis involved the study of 72 thin sections taken from core in the total sand interval. Sand mineralogy and porosity were determined in 32 sections by 300 point counts. A Cambridge Mark IIA scanning electron microscope equipped with an energy dispersive analyzer was used for detailed microstructural analysis.

All in-situ condition core tests were performed using a system designed and built at Terra Tek, Inc. for high-temperature creep measurements (Fig. 2). The system uses gas-backed, thermally stabilized hydraulic accumulators for applying the axial, confining, and pore pressures. The gas-backed accumulators are designed to maintain constant confining pressure and axial load over long time periods. Figure 3a shows a schematic of the axial load, confining pressure, and pore pressure units. The pore pressure unit is also used for flow (permeability) and ejected fluid volume (pore volume changes) measurements. The axial and transverse strain measurements are made using, respectively, linear variable differential transformers (LVDTs) and

strain-gaged cantilever fixtures. The strain gages have some drift over prolonged measurement durations, especially at elevated temperature. The LVDTs, on the other hand, are extremely stable and have high sensitivity for prolonged measurements on most rocks. Test specimens are heated internally within the pressure cell by convecting hot fluid contained within a ceramic shroud. Temperatures are measured by thermocouples attached to the test specimen at several positions. Figure 3b shows a view of the entire test stack assembly. The system is linked to a PDP-11 computer system that records pressure, temperature, stress and strain transducer outputs in real time.

Pore pressure reduction tests and tests consisting of several hydrostatic, triaxial and uniaxial loading and unloading cycles were designed to provide data for calculating mechanical and transport properties of the core samples. Uniaxial compaction tests only are discussed here. An earlier study by Sinha, et al (1981) describes the triaxial and creep test procedures and results.

Jacketed test specimens are heated and loaded to simulate in-situ reservoir conditions of temperature (107° C), confining pressure (13,000 psi), total overburden stress (13,500 psi), and pore

pressure (12,000 psi). The confining pressure and the pore pressure were increased simultaneously (maintaining a constant differential of approximately 50 psi to prevent jacket failure). After the pore pressure reaches its maximum value, the confining pressure is increased to the reservoir level, followed by deviatoric stress application.

In the compaction tests, the pore pressure is lowered to approximately 60% of its maximum value while the confining pressure and the deviatoric stress values are constantly adjusted to obtain a uniaxial compaction (no lateral strain change) while maintaining a constant total axial stress (representing the overburden). Steady-state permeability to brine of estimated formation salinity (the formation is brine saturated with a trace of oil) was measured after in-situ reservoir conditions were stabilized. A second permeability measurement was made after the one-dimensional compaction phase at maximum drawdown to assess the influence of fluid production on the porosity and permeability of the reservoir rock.

CORE ANALYSIS RESULTS

Macroscopic Core Analysis and Environment of Deposition

A vertical profile for the cored interval is presented in Figure 4. The sand

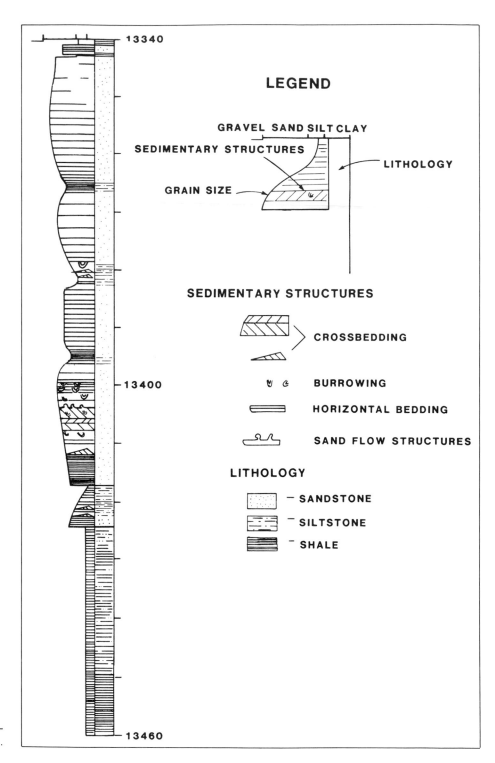

Figure 4—Vertical profile of cored interval for Sweezy No. 1 test well.

tion porosity. Rare patches of carbonate cement and grain surface dissolution features suggest that the upper sand may have originally contained extensive carbonate cement development which wholly or partially replaced feldspar, volcanic lithic fragments, and quartz grains. Carbonate has since been dissolved resulting in partial grain surface solution of detrital quartz and near total dissolution of plagioclase feldspars grains, which appear as honeycombed remnants with intragranular porosity (Fig. 6). Portions of the upper sand also appear to have undergone pressure solution and some welding at quartz grain contacts (Lowry, 1956), with euhedral quartz overgrowths developing in open pores adjacent to pressure solution contacts. Further evidence for pressure solution in this zone is the presence of stress fractures within individual adjacent grains where one grain has penetrated another at tangential point contacts. Figure 7 exhibits shear fractures emanating from grain contacts. These features exemplify how

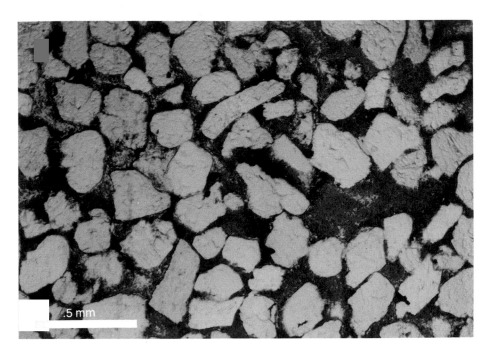

.5 mm

Figure 5—Thin section photomicrograph of the well-sorted upper sand zone rock fabric (beach facies). Porosity (blue) has been enhanced by cement and grain dissolution developing enlarged pores and pore throats (depth 13,346 ft).

overburden stresses may be distributed in the rock framework grains through grain contacts, thereby propping open pores and supporting the rock matrix even when carbonate cement is dissolved. Pores are relatively clean and free of authigenic clays. Pore-throat enlargement improves permeability and has been shown experimentally to improve the recovery efficiency of nonwetting fluid phases (Wardlaw and Cassan, 1979).

Figure 8 represents a chronological sequence of pore-space alteration based upon petrographic observation of samples. Although this diagram is generalized, it portrays the major diagenetic changes responsible for pore-geometry modification in the upper sand zone, culminating in the features observed in actual core samples.

The lower sand interval exhibits fine grain size (average 0.15 mm), fair-to-poor sorting, subangular grains, and porosities of 15–20% (Fig. 9). Mineralogically, detrital grains in this interval are predominately quartz (75%), with feldspar and volcanic lithic fragments (5–10%), and minor mafics (<1%). Matrix components include detrital muscovite and biotite (2%), authigenic kaolinite (5–7%), and mixed-layer smectite/chlorite/illite (1–2%) clays. Quartz overgrowth rims are common (0.003–0.01 mm thick). Glauconite is

.2 mm

commonly dispersed with detrital grains and detrital clays have possibly been dispersed by bioturbation within this interval. Carbonate cement also occurs in scattered patches, occasionally completely filling pores.

Lower sand porosity is unevenly distributed owing to selective secondary porosity development in zones of maximum prefluid flow. Macroporosity

Figure 6—Thin section photomicrograph of honeycombed feldspar grain. Porosity (pink) is both intergranular and intragranular (depth 13,363 ft).

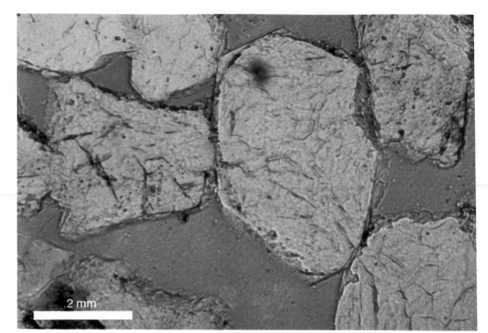

Figure 7—Thin section photomicrograph showing tangential point contacts between grains. Note the tangential shear fractures in the two grains in the lower right corner. Adjacent to the pressure solution contact between these grains, silica overgrowths have developed in open pore space (depth 13,346 ft).

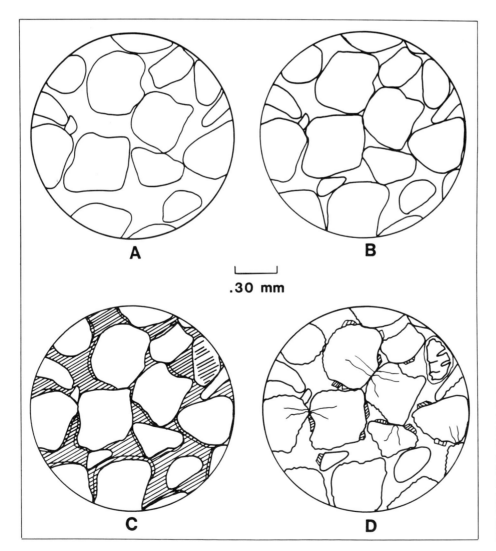

.30 mm

Figure 8—Generalized schematic of diagenetic changes that have occurred in the Cibb Jeff upper sand during burial: (**A**) Depositional texture, unconsolidated sand; (**B**) Initial compaction with shallow burial; (**C**) Carbonate precipitation in primary pores and partial grain replacement; (**D**) Present grain and pore configuration observed after carbonate dissolution, pressure solution at grain contacts, and development of silica overgrowths.

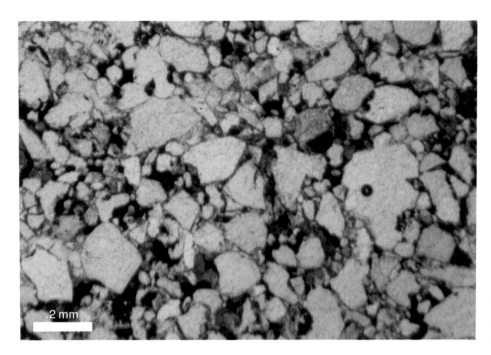

.2 mm

Figure 9—Thin-section photomicrograph from lower sand interval (intertidal to subtidal facies). Note poorer sorting, angular grains, uneven distribution of porosity (dark blue), and increased clay matrix content (depth 13,400 ft).

(pores with diameters $>5~\mu$m) has been increased by the development of oversize pores resulting from detrital grain (plagioclase feldspar and volcanic lithic fragments) and carbonate cement dissolution. However, primary pores and pore throats adjacent to macropores are commonly filled with authigenic clays, creating microporosity (pores with diameters $<5~\mu$m) around individual clay crystals (Fig. 10). Plagioclase feldspars and carbonate cement isolated from large pores by authigenic clays appear unaffected by dissolution. Porosity of this type has been termed "fabric selective" by other workers (Schmidt and McDonald, 1980), describing pores whose shapes generally follow fabric elements of the host rock.

Figure 11 represents a schematic of petrographic features observable in the lower sand zone. These features may be compared with those presented for the upper sand zone in Figure 8.

Scanning Electron Microscope Analysis

Selected samples from the total sand interval were examined using the scanning electron microscope (SEM) to determine pore geometries, type and habit of clay minerals, extent of quartz overgrowths, and dissolution features.

The upper sand exhibits extremely large and clean pores (Fig. 12). Pore shapes are highly variable and display

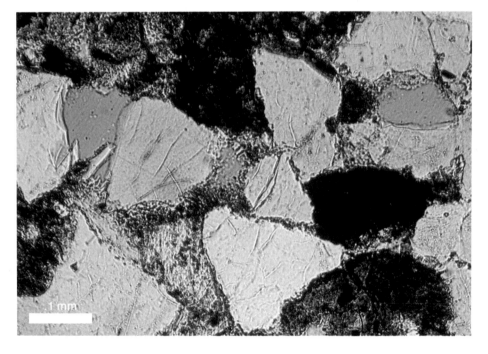

.1 mm

dissolution features in some portions of the sand (that is, fluted groove textures in pore throats [Fig. 13], pitted grains, and rounded quartz overgrowths at pressure solution contacts). Quartz overgrowths are abundant in different stages of development (that is, discrete individual microquartz overgrowths and coalescing overgrowths) (Pittman,

Figure 10—Thin-section photomicrograph from lower sand zone showing how macro-pores are isolated by authigenic clay development (depth 13,400 ft).

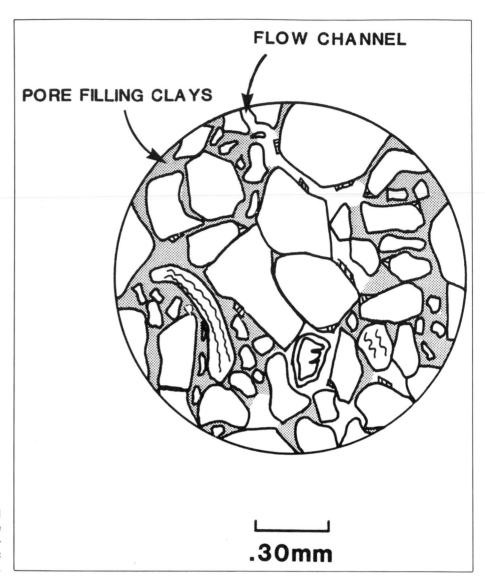

Figure 11—Schematic of lower sand grain fabric that exhibits fabric selective porosity. Shaded area represents porosity filled with detrital and authigenic clays.

1972). Adjacent grains have tangential point contacts (Fig. 14) and some contain sutured contacts (Fig. 15). Grains such as these may appear to "float" in thin section. Feldspars are commonly honeycombed from dissolution (Fig. 16) and partially dissolved volcanic lithic fragments have undergone near-complete alteration to chlorite and mixed-layer illite/smectite (Fig. 17). These grains may also have been partially replaced with carbonate, explaining the large dissolution voids.

The lower sand interval displays a rock fabric much different from that seen in the upper sand (Fig. 18). Authigenic clays are abundant and occur as grain coats, pore linings, and pore fills (Neasham, 1977). Much of the porosity

observed is microporosity (pores less than 5μm in diameter) adjacent to these clay particles. Mixed-layer illite/smectite and chlorite commonly coat grains, while kaolinite exclusively fills pores and constricts pore throats. Kaolinite occurs as discrete crystals often engulfed by quartz overgrowths in pores. Microquartz or zeolites also occur coating these overgrowths (Fig. 19). Quartz overgrowths appear euhedral in shape and do not exhibit the replacement/dissolution textures observed in the upper sand interval, suggesting less dynamic pore-fluid flow (Fig. 20). Dissolved feldspars and lithic fragments appear to have left grain-sized voids infrequently scattered in the rock fabric.

In-Situ Condition Core Testing Results

Uniaxial compaction tests were performed on selected core samples from the cored interval. Uniaxial compaction was induced at in-situ reservoir conditions. Upon establishing reservoir confining pressure (13,000 psi), pore pressure (12,000 psi), overburden pressure (13,500 psi), and temperature (107° C), pore pressure was reduced gradually from 12,000 psi to approximately 7200 psi, keeping lateral strain constant. The deformation behavior of samples may be represented by a mean effective pressure (σ_m) versus volumetric strain (ϵ_v) relationship.

Figure 21 exhibits the stress–strain curve followed in arriving at reservoir conditions in one whole-core sample.

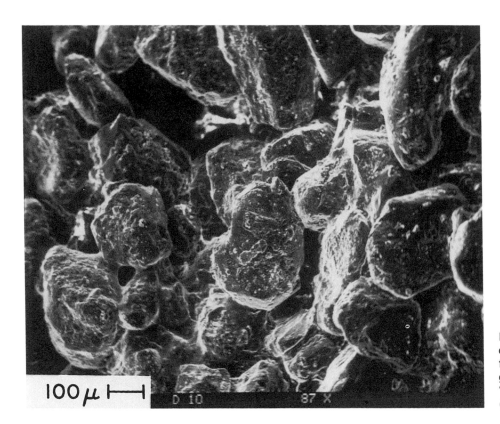

Figure 12—Scanning electron photomicrograph of upper sand zone rock fabric. Grains have predominantly tangential point contacts, which create "floating" appearance in thin section (depth 13,346 ft).

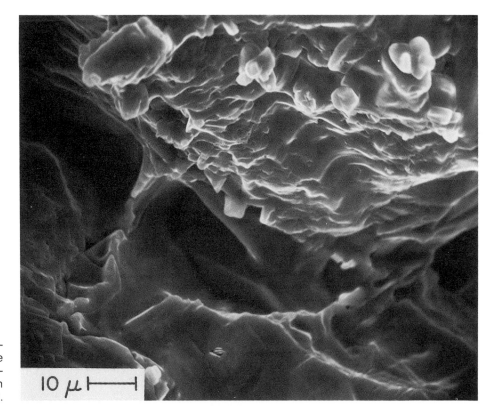

Figure 13—Scanning electron photomicrograph of variable dissolution texture within upper sand facies in close proximity to upper shale boundary (depth 13,343 ft).

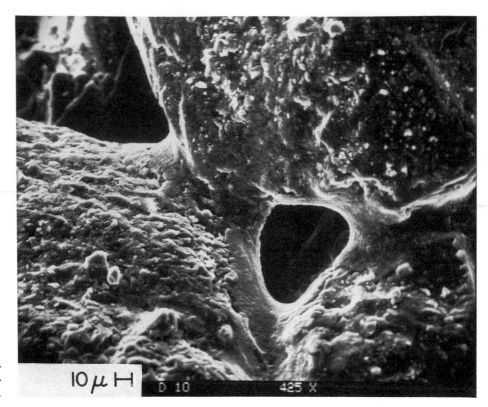

Figure 14—Scanning electron photomicrograph of tangential quartz grain contacts (depth 13,346 ft).

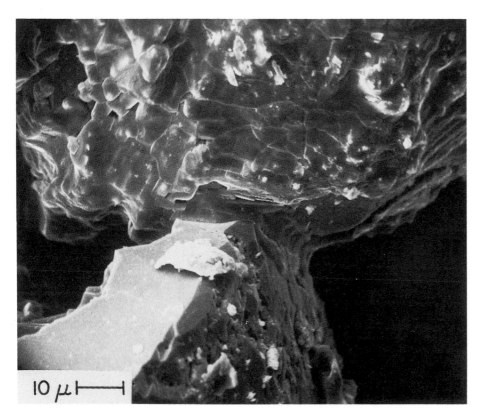

Figure 15—Scanning electron photomicrograph of pressure solution contact between two grains. Note sutured contact between grains and flat, smooth surface where a grain has been plucked.

Figure 16—Scanning electron photomicrograph of partially dissolved feldspar in rock matrix of upper sand facies (depth 13,343 ft).

Figure 17—Scanning electron photomicrograph of partially dissolved volcanic lithic fragment that has undergone alteration to clay minerals (depth 13,343 ft).

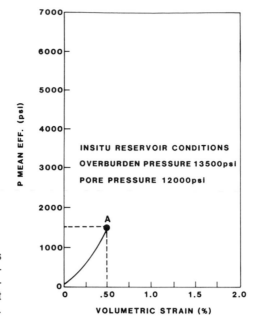

INSITU RESERVOIR CONDITIONS
OVERBURDEN PRESSURE 13500psi
PORE PRESSURE 12000psi

P MEAN EFF. (psi)

VOLUMETRIC STRAIN (%)

Figure 21—Uniaxial compaction stress versus strain curve for obtaining reservoir conditions in one core sample. In-situ permeability was measured at point A.

experienced less exposure to pore-fluid flow and thus less effective porosity development has occurred. These differences have led to variations in reservoir characteristics such as porosity, permeability, and pore-to-throat size ratios.

The porosity, its distribution, and type therefore varies significantly within the cored interval. The best effective porosity occurs within the upper sand zone and is related to its medium grain size, good sorting, the lack of matrix material and authigenic mineral pore fills, and the enlargement of individual pores and pore throats by dissolution of silica, carbonate cement, and unstable detrital grains. Core porosities in this zone determined from bulk and grain density measurements on one-inch core plugs, average 25–35% (Fig. 25). Similar, though slighly lower, values were obtained from thin-section point counts.

Within this upper sand zone, medium grain size coupled with good sorting preserves larger pores and pore throats. The subrounded shape of grains coupled with silica overgrowths and pressure solution at tangential point contacts help prevent grain rotation and reduction in pore size under stress. Pore-to-throat size ratios are small compared with those observed in the lower sand grain fabric. As a result, the recovery efficiency of nonwetting phase fluids should be excellent in this zone.

Porosities in the lower sand average 20–25% (Boyle's Law) and 15–20% when measured in thin section. The difference in these values reflects the microporosity, which is uncounted in thin section. This occurs between authigenic pore-filling clay crystals. Oversized pores exist in pore-fluid migration pathways, these being connected by pore apertures of variable size. When producing a reservoir with this type of pore system, capillary effects in individual pore apertures have been shown experimentally to reduce recovery efficiency (Wardlaw and Cassan, 1979). Nonwetting phase pore fluids are likely to follow the secondary pore pathways with large pore apertures, and bypass many smaller pores that may be isolated as a result of capillary forces in smaller pore-throat diameters. Production-induced compaction of these smaller pore apertures should be much greater than for the upper sand. Poor grain sorting, the presence of labile or deformable grains, and matrix pore fills will also reduce the effective porosity and increase the stress sensitivity.

Ambient and in-situ permeabilities also differ significantly between the upper and lower Cibb Jeff Sand interval. The best permeability occurs in the upper sand zone and can be related to the good grain sorting and roundness, massive bedding, absence of authigenic clays or cements in pore throats, and the enlarged pore apertures.

Figure 26 represents a portion of the Cibb Jeff interval where bench condition, in-situ and post-fluid production permeability measurements were obtained. The upper sand in-situ permeabilities ranged from 43–200 md and the lower sand in-situ permeabilities ranged from 6–26 md. As can be seen, the differences between the bench condition and in-situ values in all samples were commonly one order of magnitude or greater. These differences are believed to be due to the presence of microfractures, or sheetlike pores, induced in core samples during core-retrieval operations. The differences in these values are important when calculating recoverable reserves from the reservoir.

Resistance to collapse of a dissolution pore network under reservoir conditions is of key importance in preserving the low pore-to-throat size ratios and hence maintaining hydrocarbon recovery efficiency. Wardlaw and Cassan (1979) state that hydrocarbon or nonwetting phase recovery efficiency is increased with reduced pore-to-throat size ratio. Results from uniaxial compaction tests in the upper Cibb Jeff Sand evidenced less than 1% volumetric strain corresponding to a 60% reduction in pore pressure at in-situ conditions. Even with overpressured reservoir conditions, this is considered very small for a sandstone of 25–30% porosity.

Overpressuring in the Parcperdue Field area is encountered at approximately 11,000 to 12,000 ft. If it is assumed that compaction of the sandstone proceeded to these depths under a normal hydrostatic formation pressure gradient (0.465 psi/ft), then the formations were subjected to a greater net effective stress during this initial compaction than that seen under overpressured conditions. It is likely that much of the Cibb Jeff sandstone's porosity was filled with carbonate cement at this time, which helped to support the rock fabric at higher stresses. With the advent of overpressuring in adjacent shales, fluids were expelled from these overpressured sediments into the more porous normally pressured sands, thereby increasing their pore pressure, and reducing the effective stress in the rock fabric.

Hydrocarbon maturation also occurred in adjacent shales, producing CO_2, which may dissolve in the pore brines since it is known that interstitial pore waters may become undersaturated with respect to carbonates as the zone becomes overpressured (Schmidt and McDonald, 1980). Assuming flow of pore fluids, these factors support the theory that maximum dissolution of soluble constituents in the sandstone accompanied overpressuring of the reservoir.

The dissolution of supportive carbonate, silica, and unstable grains, would lead to an increase of stress on the quartz grains leading to further compaction and rearrangement of the detrital grain matrix. This would favor pressure solution at tangential point contacts. Grain sorting and pressure solution develop a stronger grain fabric, which has in turn helped to preserve larger dissolution pores for fluid movement within the reservoir interval. In the case of the Cibb Jeff upper sand zone, the quartz grains exhibit tangential point contacts, which distribute stress in the grain fabric. Constriction of pore throat volume between adjacent pores is minimized.

By contrast, the lower sand interval contains variable, yet smaller pores and variable pore-to-throat size ratios, which are affected by pore-filling authigenic clays, grain sorting, and dissolution. The smaller pore apertures, coupled with the fact that there is a greater percentage of labile or deformable framework grains, suggest that if more compaction occurred in this zone, greater sensitivity of pore apertures would be observed. This is supported by in-situ core testing where lower sand zone samples experienced greater permeability reduction during in-situ condition simulation of maximum drawdowns (Fig. 23).

The presence of extensive plugging by authigenic clays within this zone suggests a sealing mechanism whereby

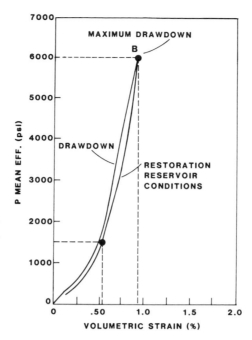

Figure 22—Continuation of Fig. 21 test stress versus strain curve where pore pressure is reduced by 60% (~6000 psi effective stress) at point B. A second permeability measurement was obtained at maximum drawdown. The sample was then reequilibrated to reservoir conditions by increasing pore pressure.

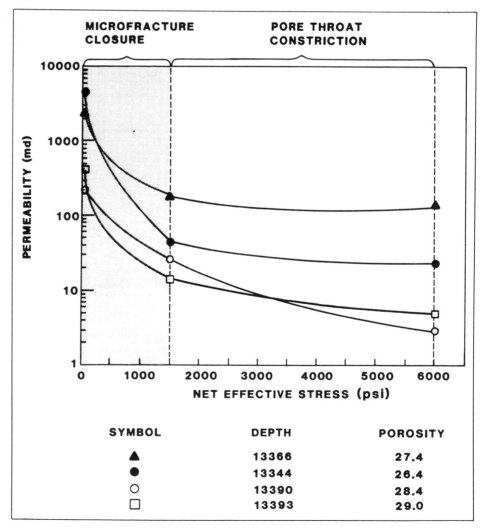

Figure 23—Stress versus permeability plot for permeabilities measured at ambient, in-situ, and maximum effective stress. Curves are drawn as index lines only, however the trend and slope of these curves suggest closure of microfractures from 0–1500 psi, then constriction of pore apertures at greater stresses from 1500–6000 psi net effective stress.

SYMBOL	DEPTH	POROSITY
▲	13366	27.4
●	13344	26.4
○	13390	28.4
□	13393	29.0

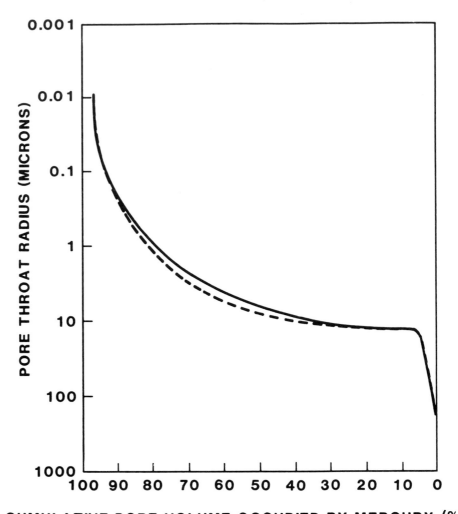

Figure 24—Cumulative pore size distribution curves for a pre- and post-test sandstone sample subjected to uniaxial compaction and simulated production (measured on unstressed sample) (depth 13,366 ft).

hydrocarbons could be trapped. Wilson (1977) observed similar types of reservoirs where hydrocarbon accumulations had prevented diagenetic mineral precipitation in oil-saturated zones. However, below the oil/water contact, diagenetic plugging by precipitation of clay minerals had effectively sealed the reservoirs. The Cibb Jeff Sand represents a similar type of trap but also shows evidence that carbonate cement prevented major authigenic clay accumulations in the upper sand. These types of diagenetic traps may be found in formations thought to be structurally unsuitable for exploration.

Reservoir production can be greatly dependent upon reservoir compressibility. As effective stress increases with production, the stress sensitivity of the reservoir will influence the long-term prediction of reservoir performance. Schatz et al (1982) have shown how inelastic rock deformation mechanisms play a significant role in changing porosity and permeability during production. Inelastic pore collapse can be the dominant mechanism in reducing pore space and permeability even when reservoir conditions are hydrostatic, especially in reservoirs containing a large proportion of matrix-supported

grains. However, for the range of effective stress that would be generated with production in the overpressured Cibb Jeff Sand, core tests suggest that the reservoir compressibility is not as great as could be expected for a sand with 30% porosity.

CONCLUSIONS

I. Study of the Cibb Jeff Sand shows how depositional textures and the timing of diagenetic events are critical factors in determining reservoir properties. The post-depositional movement of acidic

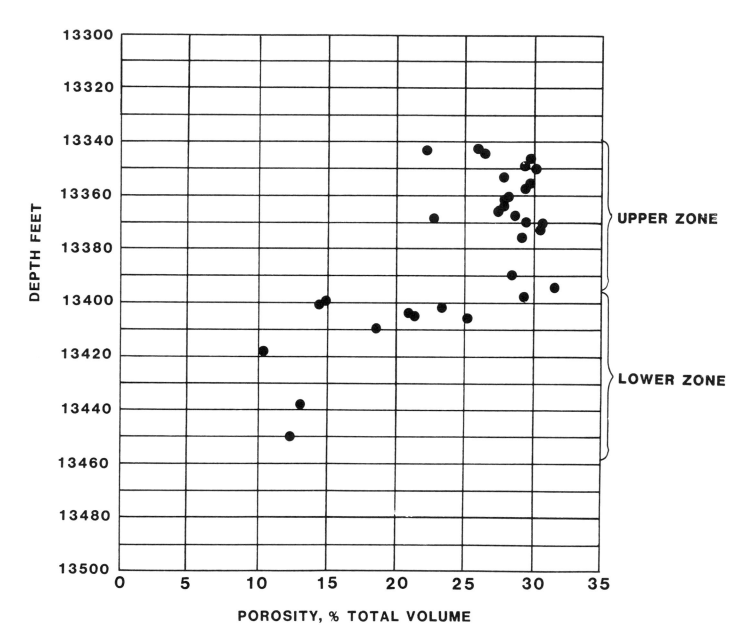

Figure 25—Porosity plot from core plug measurements within the total Cibb Jeff Sand cored interval. The porosity plot is separated into two zones, upper and lower, which reflect facies distribution of porosity and textural features.

and undersaturated fluids within the Cibb Jeff Sand, and the resulting porosity development was dependent upon:

1. the mineralogy of detrital grains;
2. the pore matrix content (carbonate cement and authigenic clays);
3. the sedimentary structures and

resulting textural components that retarded fluid flow;
4. the increase in formation pore pressure as a result of overpressuring, which reduced the effective stress in the reservoir;
5. the development of pressure solution and silica overgrowths in the well-sorted grain matrix, providing a stronger rock

A Review of Artificial Diagenesis During Thermally Enhanced Recovery

Ian Hutcheon
The University of Calgary
Calgary, Alberta

INTRODUCTION

The Cretaceous oil sand deposits of Alberta contain approximately 200×10^9 cu m of heavy crude oil, nearly twice the world's known conventional reserves. Devonian and Mississippian carbonate rocks contain an additional 200×10^9 cu m of heavy crude oil (Outtrim and Evans, 1977). Of this total only 7.5×10^{10} barrels can be recovered by surface mining; the remainder is accessible only by in-situ recovery techniques (Mossop et al, 1979). The addition of heat, usually by steam injection or in-situ combustion, is presently the favored method of enhancing recovery.

Extensive oil sand and heavy oil deposits include Athabasca, Cold Lake, Wabasca, Peace River, and Lloydminster (Fig. 1). Athabasca, the largest deposit, has significant surface exposures suitable for mining but most of the deposit is not accessible by mining. The other deposits range from 75 to 750 m in depth and thermal stimulation is the favored technique for in-situ recovery of these oils, which range from 8–15° API gravity and viscosities of approximately 1×10^3 to 1×10^5 m Pa·sec at 25°C. These high viscosity oils require that heat, solvents, or both be added to reduce the viscosity to levels at which the oil will flow. Two fundamentally different processes are most often used to supply heat to the reservoir: steam injection and in-situ combustion. The temperatures available for the steam process vary, depending on the pressure the reservoir can withstand, and are generally in the 200 to 300°C range. Temperatures during in-situ combustion may reach 700 to 800°C, but the water-to-rock ratio is lower and the time the reservoir is at

ABSTRACT. The tar sand and heavy oil deposits of Alberta and Saskatchewan represent a huge resource, most of which has to be recovered by in-situ methods, rather than surface mining. Oil viscosities are extremely high at normal reservoir temperatures and thermal methods of enhanced recovery, primarily steam injection and in-situ combustion, have been successfully employed on a pilot scale. Lithic sands in the Cold Lake area have been subjected to steam injection and pre- and post-steam cores are available for examination of mineral alteration reactions. A core from quartzose sands in the Lloydminster area, which was cut after in-situ combustion, is also available and the nature of the mineralogical reactions in these compositionally distinct rocks, subjected to a physically very different recovery scheme, can be compared and contrasted with the Cold Lake samples.

Important factors in controlling the extent of mineral alteration include the original composition and mineral distribution of the sands, the temperature and time of exposure to elevated temperature and the water-to-rock ratio. Oil recovery may be affected by mineral reactions if the timing is such that porosity-reducing reactions occur before there has been significant oil displacement. Mineral reactions may also increase porosity and produce CO_2, both of which are potentially beneficial to ultimate oil recovery.

very high temperatures is less than in steam injection.

The extreme temperatures of steam and combustion in-situ recovery processes ensure that chemical reactions will take place betweeen the injected fluids, formation fluids, and the minerals in the rocks. In addition to the obvious effects of increased temperature and invasion by very fresh water (steam has very low dissolved solids content), the composition of the reservoir rocks must play a major role in determining mineral reactions. The period of time during which the rocks are at elevated temperatures is geologi-

cally quite short, probably on the order of years during combustion and steam flooding. One would expect the most important mineral reactions to occur within the pore space, involving authigenic phases and framework grains that suffer dissolution.

The Athabasca deposits are fine sands, with quartz grains comprising 90% (Carrigy, 1973) and only minor amounts of feldspars and clay minerals (Nelson and Glaister, 1978). The Peace River and Cold Lake deposits both contain abundant clay minerals in the less than 44-μm size fraction, in addition to siderite and calcite. In Peace

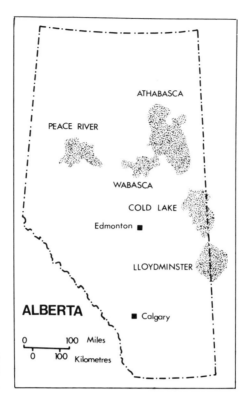

Figure 1—Location map showing the Cold Lake and Lloydminster deposits examined in this study.

River the clay minerals are dominantly kaolinite, chlorite, and illite, while the Cold Lake deposit also contains smectite (Nelson and Glaister, 1978). The framework grains in the Cold Lake deposits are dominated by chert, shale, and volcanic rock fragments and detrital ferromagnesian minerals (Oldershaw et al, 1981). The Lloydminster heavy oil deposits are either dominantly quartzose or lithofeldspathic (Putnam, 1982a).

Of all the deposits, one would expect that the Cold Lake and Peace River sands and the lithofeldspathic portions of the Lloydminster deposits would be the most chemically reactive rocks. This is probably the case, but there are few published papers documenting the alteration of mineralogy during in-situ recovery, so comparisons based on literature data are not possible.

This review paper relies heavily on previously published data on mineral transformations at Cold Lake, published data of other authors and some new data from Lloydminster. Two specific examples, steam flooding in the lithofeldspathic sands at Cold Lake, and in-situ combustion of quartzose Lloydminster deposits are discussed.

The potential positive and negative effects of mineral reactions on the oil recovery also are summarized.

CONDITIONS DURING THERMAL RECOVERY

Steam injection and in-situ combustion are the most commonly employed thermal recovery techniques, although other methods have been proposed. The methods of application of steam or in-situ combustion differ and this greatly affects the temperature of the process, the time the reservoir is at elevated temperature, and the way in which heat is transferred to the reservoir.

Steam Injection
Steam may be applied to a reservoir either in a series of alternating cycles of steam injection and oil production from the same well (referred to as cyclic steam injection, steam stimulation, or "huff and puff") or by injection into one well within a pattern and recovery from surrounding wells (steam flooding or steam drive). In the oil sand deposits in Alberta, steam stimulation is generally used first, to establish communication between wells, before steam flooding begins.

The steam injection process relies on the heat released during condensation of steam to heat the reservoir. The temperature of the steam is a function of the pressure, and thus formation temperatures during recovery depend on the operating reservoir pressure. This implies that there is a minimum depth, above which steam injection cannot generate sufficient heat to significantly reduce oil viscosity. Doscher (1967) reports measured temperatures of 115° C for a pilot site at Cold Lake. During the operation of the Cold Lake pilot, temperatures were maintained at about 200° C for 2 years.

Because steam is much less dense than oil or water, it tends to ride over the oil bank by gravity segregation, forming a chamber the full height of the reservoir at the injection well and becoming progressively narrower away from the injector. This phenomenon is described in texts on reservoir engineering (see Dake, 1978, p. 389, for example). Figure 2 shows that when the steam breaks through to the recovery well, a significant volume of the reservoir has not been swept by steam. As the steam condenses it drains into the oil bank, effecting some recovery as hot water and this is the first recovery fluid to contact the rocks.

In-Situ Combustion
During in-situ combustion, air or oxygen is injected into the formation and the burning oil supplies the heat for the recovery process. Ideally the heat is produced by burning coke, laid down in the formation by heating the heavy oil, and the lighter hydrocarbon compounds are pushed ahead of the flame front. Water may be injected with air to transmit the heat behind the combustion zone to the oil in front of the combustion zone (wet combustion). Temperatures may range from 300 to 700° C (Penberthy and Ramey, 1966) and the combustion front, and its associated heat, moves on the order of 0.7 to 3 m per day (Latil, 1980). Only the zone surrounding the combustion interface is very hot, but a significant portion of the reservoir is at steam temperature. Figure 3 is a generalized diagram showing the various zones present during fire flooding. During in-situ combustion, temperatures are considerably higher than steam injection tempera-

INJECTION   RECOVERY

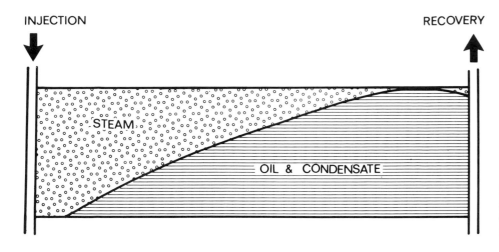

Figure 2—The probable shape of the steam–oil and condensate interface during steam flooding. The interface is not vertical owing to gravity override.

Figure 3—The zones developed during in-situ combustion are: (**1**) injected air and water, steam produced near flame front; (**2**) flame front (300–700° C); (**3**) steam and coke (200° C+); (**4**) gases and light hydrocarbons; (**5**) hot water (20–100° C); and (**6**) oil bank and cold oil in place (reservoir temperature). Adapted from Chierici. Copyright © 1980 Agip Production Development San Donato Milanese.

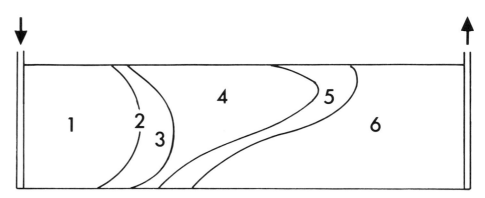

tures but, unlike steam injection, the bulk of the reservoir is not at this very elevated temperature for the entire recovery process. Steam injection has been used more frequently, even though combustion has some obvious advantages.

MINERAL REACTIONS

There are a number of factors, inherent in thermal recovery processes, which would be expected to influence the nature and extent of mineral reactions. These include: (1) temperature, (2) residence time, (3) water-to-rock ratio, and (4) rock and fluid composition.

These parameters will change between and within reservoirs, during the recovery process, and as a function of the way in which the recovery process proceeds. In the following section

two examples, a steam injection pilot in lithofeldspathic rocks (Cold Lake) and an in-situ combustion pilot in quartzose sands (Lloydminster) will be used to show the net effects of some of the factors listed above.

Steam Injection—Cold Lake

The Cold Lake and Lloydminster deposits are both within the Upper Mannville Group, which contains quartzose and lithofeldspathic units (Putnam, 1982a, 1982b). The Cold Lake deposit studied is in the Clearwater Formation; the Lloydminster deposit studied is in the Sparky Formation.

Perry and Gillott (1979, 1982) have examined actual and potential mineral alterations produced in Cold Lake oil sands during wet and dry combustion laboratory experiments. These experiments are used primarily to produce

engineering data on fuel and air requirements and are potentially misleading in two ways. Firstly, the samples are cleaned, disaggregated, packed into the combustion apparatus, re-imbibed with water and hydrocarbons, and ignited. The original texture, mineralogy and structure of the pore system is lost. Secondly, some of the experiments of Perry and Gillott (1982) deal only with the reaction of a single mineral, isolated from the rock and pore fluid environment. This is an unrealistic situation, as mineral reactions usually take place among assemblages of minerals and involve pore fluids. Some of the transformations studied by Perry and Gillott (1982) are not likely to be preserved (for example, the $\alpha-\beta$ quartz transition) and are thus not particularly useful.

More realistically designed experiments have been performed by Boon et al (in press) in static autoclaves and indicate an overall decrease in quartz, kaolinite, and dolomite with a corresponding increase in analcime, chlorite, smectite, and calcite. In addition, flow experiments by these authors show that dispersal of fine-grained material may be an important process during steam injection. Mineral reactions in which dolomite and kaolinite react with the fluid to produce both smectite and chlorite are also documented. Boon et al have chosen to write these reactions using HCO_3^- as one of the products, but it is possible that the solubility of CO_2 could be exceeded during the experiment, or in the field, and that CO_2 would be produced in the vapor phase. In addition, at realistic pressures, the solubility of CO_2 increases with increasing temperature above about 200°C. As these fluids cooled they would evolve CO_2. The experiments of Boone et al very clearly show the type of mineral reactions possible and their potential for restricting porosity and permeability.

A study of the Great Plains pilot project, near Cold Lake (Oldershaw, et al, 1981; Hutcheon et al, 1981) confirms that similar mineral reactions to those studied by Boon et al (in press) do occur and do restrict porosity. Within the Great Plains pilot area at Cold Lake, the framework of the rocks is composed of volcanic fragments, feldspars, chert, shale fragments, and quartz grains. Pore-space minerals include smectite, illite, chlorite, zeolites, and siderite. Kaolinite, which occurs as an alteration product in feldspars and volcanic rock fragments and in shale fragments, is available to the pore fluids through dissolution at the surfaces of grains (Oldershaw et al, 1981).

The Great Plains pilot at Cold Lake was in operation as a steam flood for approximately two years. Before injection started, a core was cut ("pre-steam") in a hole subsequently used as an observation well; 15 m northeast of the proposed injection well, within a five-spot pattern. After two years of steam injection, a second well was cored ("post-steam") 30 m northeast of the injector and in a line with the injec-

tor and the pre-steam well. Details of well locations and injected intervals are given in Oldershaw et al (1981). Based on thermocouple measurements made in the field and interpolated both in time and distance, temperatures in the post-steam well ranged from formation temperature to 250°C during the operation of the pilot. Temperatures in the zone of interest (410–420 m) were about 200°C when the post-steam core was cut (Hutcheon et al, 1981).

The oil sands in Alberta are unconsolidated, with the bitumen and the interlocking nature of the grains generally helping to bind the grains together. After solvent extraction in the laboratory a loose sand is produced. It is possible that early migration of hydrocarbons into the Mannville prevented cementation. Water washing may have degraded the oil, causing it to lose viscosity and be trapped (Deroo et al, 1974a, 1974b). Figure 4 shows one box of core from the pre-steam well and one from the post-steam well, at approximately the same stratigraphic depth interval. It is obvious that the post-steam core has been cemented into larger pieces and is relatively free of hydrocarbons. In some zones, recovery rates ranged up to 60% of the original oil in place, and high recovery rates correlate with zones of high porosity in the pre-steam well (Hutcheon et al, 1981).

Thin-section examination has shown that the framework of the rock remains relatively unaffected, but the pore space mineralogy has been altered. A fine-grained (1–4 μm) early diagenetic assemblage (Fig. 5) of smectite, illite, chlorite, zeolite (clinoptilolite?) and siderite has been replaced by a relatively more coarse-grained (4–20 μm) assemblage of smectite and analcime (Oldershaw et al, 1981). In addition to minor evidence of silica dissolution and transport during steam injection, as postulated by Boon (1977), there is some textural evidence to indicate a reaction relationship between quartz and smectite (Fig. 5g). Textural relationships also show that analcime (sodium zeolite) is the last mineral to form during the pilot operation.

There are good correlations between

decreasing porosity, kaolinite content and increasing smectite content (less than 2 μm size fraction). Absolute decreases in porosity are on the order of 10% and although the actual porosity measurements from well logs may be inaccurate, the difference between measurements should be quite accurate. Without more pre- and post-steam wells it is not possible to determine exactly how material has moved within the reservoir to cause this porosity decrease. If all the change in porosity results from mass transported from the pre-steam well to the post-steam well, a porosity change of 10% would require the solution and redeposition of about 25 g of solid material for every 100 cu cm of the original reservoir. Some of the porosity decrease undoubtedly results from the higher volume occupied by smectite. The dissolution of the precursor kaolinite, in very compact form in framework grains, and deposition of smectite with an open texture and large effective volume in the pore spaces contributes to porosity reduction. Because of the osmotic swelling capability of smectite in fresh water, steam injection may cause even more severe reductions in effective porosity.

Thermochemical Calculations

From the X-ray diffraction data portrayed in Figure 6 and the scanning electron microscopy data summarized in Figure 5 (for more detail see Oldershaw et al, 1981), a mineral reaction between kaolinite and pore fluids to produce smectite may be inferred. The elevated temperatures (250°C) of steam injection indicate that such mineral reactions may approach a metastable equilibrium state. The time scale is on the order of years and Matthews (1980) indicates reaction rates for analcime to albite on the order of two years, at

Figure 4 — Two core boxes from the pre-steam (on the right) and post-steam (on the left) wells. Note the well-consolidated and oil-free nature of the post-steam core. The light band marked **s** is a shale layer. The pre-steam core is blank because it is oil saturated. Scale bar = 10 cm.

Figure 5a–h—Scanning electron micrographs of pre-steam (micrographs a, b, c, and d) and post-steam (micrographs e, f, g, and h) samples from Cold Lake. Scale bar is in micrometers. Note: All micrographs were taken by Alan Oldershaw.

Figure 5a—General view of pre-steam samples shows very limited diagenetic coatings on grains. **Q** is quartz; **Ct** is chert; **Sh** is shale; and **F** is feldspar.

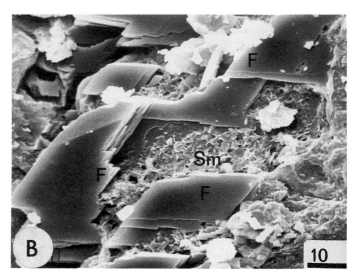

Figure 5b—Euhedral diagenetic feldspar overgrowths (**F**) have fine-grained smectite (**Sm**) mixed with illite on the surface of altered rock fragments. (Pre-steam.)

Figure 5c—Siderite (**Sid**) is rare but usually occurs on chert (**Ct**) grains. (Pre-steam.)

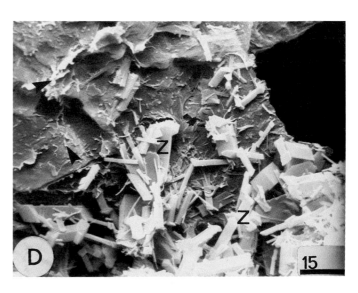

Figure 5d—Zeolites (**Z**), possibly clinoptilolite, are very fine grained in the pre-steam samples.

Figure 5e—Post-steam samples show very extensive grain coatings compared with pre-steam samples (**A**).

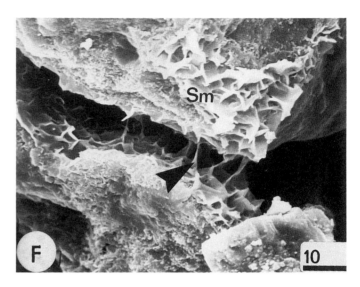

Figure 5f—Post-steam smectites (**Sm**) tend to be more coarse grained (see b) and show pore-bridging structures.

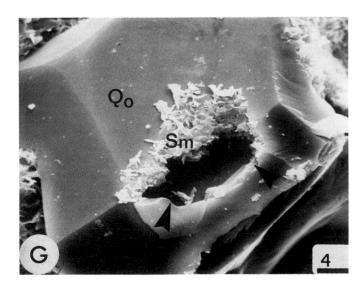

Figure 5g—Quartz overgrowths (**Qo**), present in the pre-steam samples, have reacted with the injected steam to produce smectite (**Sm**). The quartz overgrowths have undergone dissolution (black arrows). (Post-steam.)

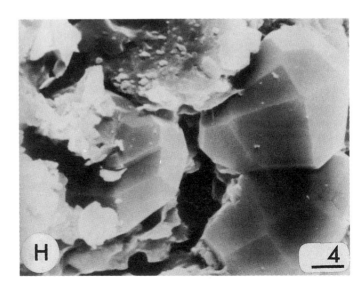

Figure 5h—Coarse-grained (10 μm) analcime crystals are generally the last observed stage of mineral growth in the post-steam core samples.

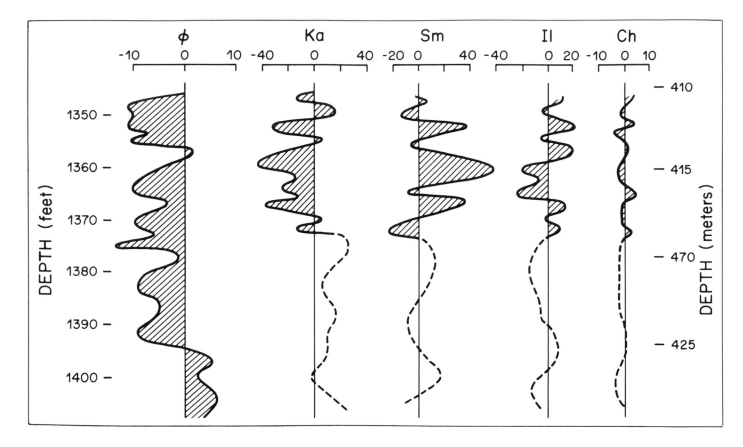

Figure 6—Marked decreases in porosity (ϕ) are associated with decreasing kaolinite (**Ka**) and increasing smectite (**Sm**) contents in the less than 2-μm fraction. Positive deviations indicate an increase from pre-steam to post-steam cores; negative deviations indicate a decrease. The abbreviations **Il** and **Ch** stand for illite and chlorite, respectively. Figure from Hutcheon et al. Copyright © 1981 Canadian Society of Petroleum Geologists Calgary, Alberta. Reprinted with permission.

these temperatures. The clay minerals have much higher surface-area-to-volume ratios and could be expected to react at least as quickly as more equant minerals like analcime. As a first approximation to the mineral reactions concerned, we may employ calculations based on equilibrium thermodynamics. Thermodynamics deals with energy balances (chemical energy in this case), not mass balances, and this allows us to write equilibria that represent real mineral reactions but do not "look like" they involve real minerals. This procedure is theoretically sound. Hutcheon et al (1981) chose the following equilibrium to represent the mineral reaction between kaolinite and sodium smectite:

where the chemical formulae represent thermodynamic components dissolved in, or comprising all of, the mineral phase. There is no thermodynamic requirement that the mineral phase and the thermodynamic component have a one-to-one compositional correspondence, as long as the activity of the thermodynamic component can be estimated. For example, the smectite in the rocks examined clearly contains more than sodium. With as much thermodynamic justification, one could consider the aluminum–silicon framework in smectite as a thermodynamic component, represented by a formula similar to that for pyrophyllite and adjusting the activity of the aluminum–silicon component to account for compositional variations in smectite. This allows us to write the equilibrium:

$$7Al_2Si_2O_5(OH)_4 + 8\ SiO_2 + 2Na^+ \leftrightharpoons \quad (1)$$
$$\text{Kaolinite} \qquad \text{Quartz}$$

$$2NaAl_7Si_{11}O_{30}(OH)_6 + 7H_2O + 2H^+$$
$$\text{Smectite}$$

$$Al_2Si_2O_5(OH)_4 + 2SiO_2 \leftrightharpoons \quad (2)$$
$$\text{Kaolinite} \qquad \text{Quartz}$$

$$Al_2Si_4O_{10}(OH)_2 + H_2O$$
$$\text{Smectite}$$

In (2) the $Al_2Si_4O_{10}(OH)_2$ component is considered to be dissolved in the smectite phase. In both (1) and (2) it is clear that a source of silica is required to form smectite from kaolinite. The scanning electron micrographs indicate that quartz and smectite may not be stable together (Fig. 5f). Rewriting (2) using aqueous silica gives:

$$Al_2Si_2O_5(OH)_4 + 2H_4SiO_4 \leftrightharpoons \quad (3)$$
$$\text{Kaolinite}$$

$$Al_2Si_4O_{10}(OH)_2 + 5H_2O$$
$$\text{Smectite}$$

and considering the dissolution of quartz gives:

$$SiO_2 + 2H_2O \leftrightharpoons H_4SiO_4 \quad (4)$$
$$\text{Quartz}$$

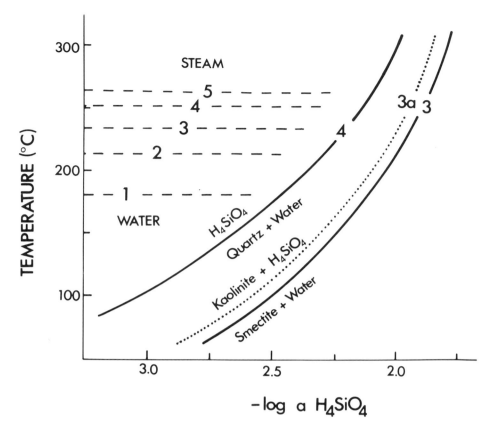

Figure 7—Stabilities of the kaolinite-to-smectite equilibrium using pyrophyllite as the smectite component (**3**), adjusting the activity of pyrophyllite component in smectite to a reasonable value (**3a**), and the dissolution of quartz (**4**) are shown in terms of temperature and activity of H_4SiO_4 at 5 MPa (50 bars). Liquid water–steam temperatures are shown by the horizontal lines at values from 1 to 5 MPa.

Using data from Cobble et al (1982), Helgeson et al (1978), Keenan et al (1969), and Robie et al (1978) the stabilities of (3) and (4) can be calculated as a function of temperature and activity of H_4SiO_4. The calculations are shown at 0.1 MPa (1 atm) but increasing pressure to 4–5 MPa, at which steam injection pilots operate, would have a negligible effect on these equilibria. To make the free energy change for (3) more closely approximate the conditions in the field, an activity term for the activity of $Al_2Si_4O_{10}(OH)_2$ in smectite was calculated using an ideal solution model. Justification for this procedure can be found in Helgeson et al (1978). The calculated activity is 0.71, which changes the stability of (3) to the position shown by (3a) in Figure 7. Both (3) and (3a) are above the quartz solubility and it would require a drastic decrease in the activity of $Al_2Si_4O_{10}(OH)_2$ in smectite to move (3) and (3a) into the H_4SiO_4 stability field. This implies that the textural relationship between smectite and quartz (Fig. 5f) is probably more related to chemical kinetics. Boon et al (in press) note that

in their static autoclave experiments it did not appear that equilibrium had been established, even after 26 days.

Hutcheon et al (1981) present a phase diagram relating the stabilities of smectite, kaolinite, albite, and analcime (sodium zeolite) as a function of temperature. The phase diagram, reproduced here as Figure 8, should be drawn for appropriate activity values of H_4SiO_4, rather than at quartz saturation; however the topology of the diagram does not change significantly. The phase relationships shown in Figure 8 explain the appearance of analcime as the last-formed mineral, as shown by SEM micrographs. When steam injection stops and pressure drops, the formation will be reinvaded by more saline formation water. At constant pH this will increase the sodium-to-hydrogen-ion activity ratio and stabilize analcime at the expense of smectite. Only two months elapsed between the time injection stopped and when the core was cut, implying that analcime nucleation and growth must be very rapid.

There is general agreement between the laboratory experiments by Boon et

al (in press) and the observations in the field (Oldershaw et al, 1981; Hutcheon et al, 1981). One major difference is that the laboratory samples had originally high carbonate mineral contents but the field samples did not, and thus CO_2-producing reactions were unlikely in the field. In addition Boon et al note the growth of feldspar which, in spite of longer reaction times, was not observed in the field.

In-Situ Combustion—Lloydminster

Samples were obtained from a Lloydminster well that had been subjected to in-situ combustion. Temperatures exceeded 700°C. The details of well location and operation are confidential. The stimulated horizon is a quartzose sand, similar to those described by Putnam (1982b). Thin section petrography (Fig. 9) shows that quartz, chert, and feldspars are the major framework grains, with minor coal fragments. Cements and authigenic minerals include siderite, calcite, and kaolinite in samples from unburned portions of this well and other wells that were not subjected to

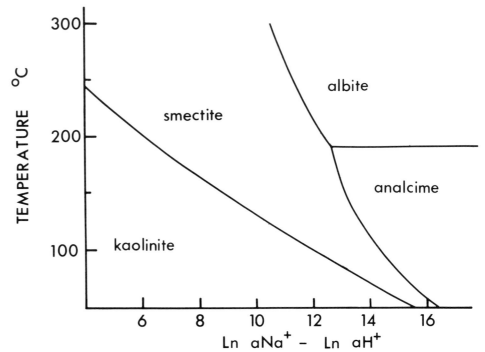

Figure 8—Phase relationships for a Ca/Na = 1.0 smectite at 10 MPa are shown for kaolinite, smectite, albite, and analcime. Note that smectite is the "high-temperature" phase compared with kaolinite and that analcime stability requires increasing the activity of Na at constant pH.

combustion. Minor amounts of authigenic Mg-rich chlorite (by anomalous birefringent colors) were observed in close association with siderite. In spite of these cements, the sediment is unconsolidated. Resistivity and gamma ray logs are shown in Figure 10 for a well cut before combustion commenced, to give some indication of the nature of the reservoir.

Samples recovered from a core cut after the combustion zone had passed the well location are a brick red and show the reaction of siderite and kaolinite to a fine-grained matted coating (Fig. 11) of chlorite. Some intervals were not burned during the combustion process. The color probably results from oxidation of iron minerals, as hematite was identified in trace amounts in X-ray diffractograms of most burned samples. No hematite was observed in virgin samples.

Figure 12 shows weight percent of clay minerals in the sample in the less than 2-μm size fraction, porosity, oil saturation, and weight percent calcite plus siderite in the bulk sample. No core was recovered within the interval from 6 to 12 m* because the unit was

poorly consolidated and apparently not burned. Porosity and oil saturations were determined by a modified Dean Stark analysis while the mineral percentages were determined by X-ray diffraction. The zone marked "interface" on Figure 12, at 15 m, contains a very tight matrix of fine particles and the amount of clay minerals and residual oil saturations (as percent of pore volume) are correspondingly high in this zone, while porosities are low. Near the bottom of the well (22–24 m) the increase in calcite corresponds to a calcite-cemented zone that does not appear to be affected by the combustion process. In contrast to the burned part of the well, siderite and kaolinite are present in, and below, this zone. The tightly clay-cemented zone at 14 m also appears to have affected the combustion process, as a very marked interface can be seen between samples with strong hydrocarbon staining below the interface and clean samples above.

Scanning electron microscopy implies a reaction between siderite and kaolinite to produce iron chlorite.

$$5FeCO_3 + Al_2Si_2O_5(OH)_4 + SiO_2 \quad (5)$$
Siderite   Kaolinite   Quartz

$$+ 2H_2O \leftrightharpoons Fe_5Al_2Si_3O_{10}(OH)_8 + 5CO_2$$
Chlorite

Reliable thermochemical data for iron chlorite are not available and phase relationships for (5) cannot be calculated. Hutcheon et al (1980) show phase relationships for a similar reaction involving dolomite, magnesium chlorite, and calcite, which intersects the miscibility surface between CO_2 and H_2O and is thus capable of producing CO_2 at temperatures at, and above, 180° C (depending on pressure). Equilibrium (5) will probably have a similar topology and could generate CO_2 well below the thermal decomposition temperatures of most carbonate minerals. A cartoon (Fig. 13) of the probable relationship between equilibrium (5) and the CO_2–H_2O miscibility surface shows how CO_2 can be generated by this process.

Pre- and post-burn samples were not as closely located, as in the case of the Cold Lake samples, so direct comparisons could not be made. Thin-section examination indicates that porosity probably increases as a result of the combustion process. This is supported by examination of portions of the interval that were not burned. The effect of mineral reactions on recovery-related parameters is not as great as in the Cold Lake example. Diversion of the combustion front can apparently be

*Note: Depths are given relative to an arbitrarily selected datum.

caused by tight zones on the order of less than 0.1-m thickness. Original geology can, in many cases, play a greater role in influencing recovery than mineral reactions do.

POTENTIAL EFFECTS ON RECOVERY

For the examples considered, the most obvious differences are: (1) different recovery methods were employed; and (2) the Cold Lake sands are lithofeldspathic with substantial clay contents and are chemically very reactive, while the quartzose Lloydminster rocks contain less clay and are chemically less reactive.

Steam injection processes operate at lower temperatures than in-situ combustion but more of the reservoir volume is at elevated temperature for longer times. Because the heat is generated right at the combustion interface, during in-situ combustion, only a small volume of reservoir is heated at any one time. The obvious oil-wetting shown in some of the photomicrographs could be explained by all the pore water being vaporized before all the oil, which was obviously hot enough to have a very low viscosity (Fig. 10), had been displaced, burned, or converted to coke. This implies low water-to-rock ratios during combustion when compared to steam injection.

Effect on Porosity

It is clear from the Cold Lake study that porosity is directly affected by the mineral reactions that produce smectite from kaolinite and quartz. The timing of mineral reactions, relative to oil displacement, is a very important and unknown factor. If mineral growth and porosity occlusion are rapid, the effect on recovery will be negative. During cyclic steam injection the same flow pathways are used for injection and recovery, and continuous porosity occlusion will eventually impair both injectivity and recovery. High recovery rates in the Cold Lake pilot are associated with zones of high initial porosity (Hutcheon et al, 1981); low recovery rates are not associated with high reductions in initial porosity, indicating that in this steam flood pilot porosity-reducing mineral growth postdated most of the recovery.

Figure 2 shows that, as a result of gravity effects, the interface between the steam chamber and the oil bank is not vertical. Once the steam zone establishes communication between the injection and recovery wells, steam will bypass the unrecovered oil in the lower part of the reservoir. Reduction of porosity and permeability in the small zone near the top of the reservoir where steam breakthrough occurs would cause the steam–oil interface to move to a lower level in the reservoir, allowing production of a greater volume of the original oil-in-place (Fig. 14).

Pre-combustion samples directly comparable to post-combustion samples are not available for the Lloydminster area. The reaction between kaolinite and siderite to produce chlorite results in a volume decrease of 57 cu cm (or 21%) for reaction 5 as written. Smectite was not observed in significant quantities in the Lloydminster samples; therefore, porosity decrease by swelling is not an important factor. Kaolinite, a mineral subject to dispersal and pore-throat blocking under high flow rates, is consumed in the mineral reaction. The fine-grained chlorite produced is present as a rimming cement and probably helps to consolidate the rock slightly (perhaps accounting for the good core recovery above the interface at 14 m and the poor recovery below). In this case it appears that the effects of mineral reactions on porosity and reservoir performance are beneficial, if anything. Recovery problems are more obviously related to the pre-combustion state of the rock (the calcite-cemented interval, for example) than mineral reactions.

Production of Carbon Dioxide

Ejiogu et al (1978) noted the production of CO_2 during wet combustion of Cardium Formation sands from the Pembina field in Alberta. They show, using X-ray diffraction data, the presence and decomposition of siderite, calcite, and dolomite. No scanning electron microscopy was done so the mineral reaction is difficult to document. From Tables 4 and 5 in Ejiogu et al, an increase in CO_2 from 21.89 to 25.97 mole % (the remaining gas phase was dominantly nitrogen) was produced by the decomposition of 2.2 wt % of carbonate mineral. This supports the

idea that mineral reactions similar to reaction 5, and documented by Hutcheon et al (1980), are capable of producing significant amounts of CO_2 during thermal recovery.

Solution of CO_2 in crude oil may improve recovery in at least two ways—by reducing viscosity or increasing oil volume. Reducing oil viscosity decreases the viscosity contrast between displaced and displacing fluids, minimizing the tendency for unstable displacement (or "viscous fingering"). Since residual oil saturations are functions of volume, expanding a given mass of oil by CO_2-induced swelling means more oil is recovered. Simon and Graue (1965) show that an increase of 30 mole % CO_2 in a CO_2-crude oil mixture decreases the viscosity of a 13° API oil at 100° C from approximately 60 to 10 centipoises. A temperature increase of 50° C would be required to produce the same viscosity reduction.

CONCLUSION

The elevated temperatures accompanying thermally enhanced in-situ recovery methods cause mineral reactions to occur, which primarily involve pore-space minerals. In a steam flood pilot, these reactions were observed to consolidate the previously unconsolidated rock and occlude porosity, but did not severely limit oil recovery. The sands in the steam flood pilot area were lithoclastic and probably quite chemically reactive. These factors, combined with longer periods at elevated temperature and higher water-to-rock ratios, caused more dramatic mineral reaction effects in the steam flood example, compared with an example of in-situ combustion. Although temperatures were higher during in-situ combustion, the less chemically reactive nature of the quartzose rocks, short periods of time at elevated temperature, and lower water-to-rock ratios all contribute to a minimal effect of mineral reactions on rock petrophysics. While these conclusions are generally reliable, the specific observations cannot be applied to other reservoirs without detailed examination of the mineralogy before and after high-temperature treatment.

The mineral reactions themselves can be studied using equilibrium thermodynamics, but only qualitative results

Figure 9a–d—Thin-section photomicrographs of unburned and burned samples from Lloydminster.

Figure 9a — Cleaned unburned samples contain rosettes or siderite **(s)** and coarse-grained kaolinite **(k)**. These are not visible in the unburned samples, except in areas of high residual oil saturations. Magnification = 63X.

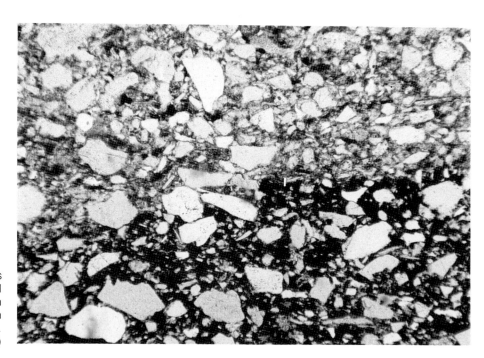

Figure 9b — The fine-grained interval is responsible for some high residual oil saturations. The interface **(l)** between highly oil-saturated and relatively clean rocks can be directly observed. Magnification = 63X. (Burned.)

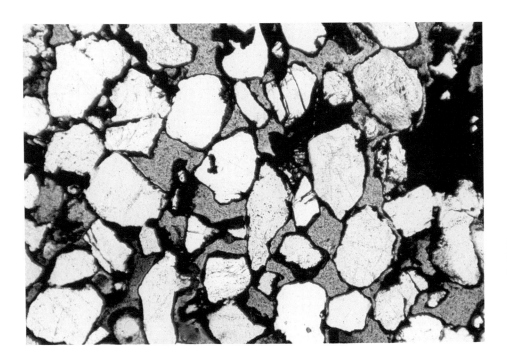

Figure 9c — Below the interface, the burned samples have developed a strongly oil-wet texture, the coke (produced by the heat of the burn) resembling a meniscus cement. The sample could have had all the connate water evaporated by the combustion process. Magnification = 63X.

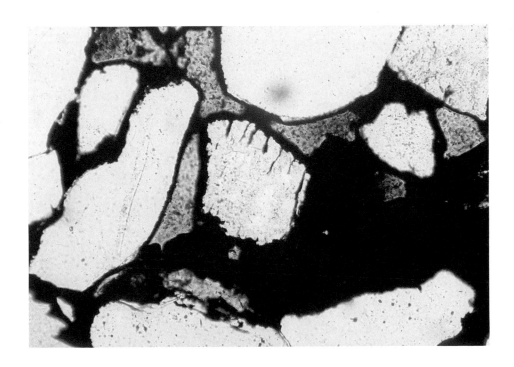

Figure 9d — The oil must have been very hot when the rock became oil-wet as even microfractures in chert grains are oil-wetted. The oil was subsequently coked by the heat of the combustion process. Magnification = 160X.

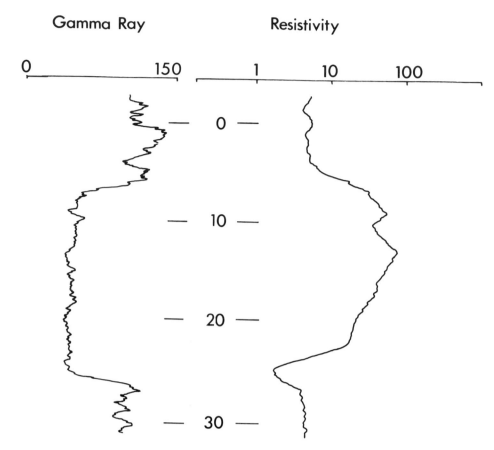

Figure 10—Gamma ray and resistivity logs from a pre-combustion well within the Lloydminster-area pilot. Depths are in meters below an arbitrary datum. Gamma ray log is in API units, resistivity is in ohm sq m/m.

can be obtained. Properly understanding the thermal recovery process and its effect on mineralogy is a complex process, which should include chemical kinetics and mass transfer information.

The mineral reactions observed may occlude or enhance porosity. Depending on the timing of mineral reactions and the nature of the recovery scheme, increases or decreases in porosity may be either beneficial or detrimental to enhanced hydrocarbon recovery. Each specific case must be examined separately. The mineral reactions themselves may provide an in-situ source of CO_2 which may, in itself, improve the recovery process.

ACKNOWLEDGMENTS

I am indebted to Alan Oldershaw for taking, and allowing met to use, the SEM micrographs shown in this paper. I am also grateful to Roger Butler, whose chemical engineering course in thermal recovery helped me to understand some of the intricacies of steam flooding. Much of the work on Cold Lake was supported by Norcen Energy and I am grateful for their permission to publish some of the data. Additional costs were covered by an NSERC grant to the author.

SELECTED REFERENCES

Boon, J. A., 1977, Fluid–rock interactions during steam injection, *in* D. A. Redford and A. G. Winestock, eds., The oil sands of Canada—Venezuela 1977: Toronto, Canadian Institute of Mining and Metallurgy, p. 133–138.

Boon, J. A., T. Hamilton, L. Holloway, and B. Wiwchar, In press, Reaction between rock matrix and injected fluids in Cold Lake oil sands: Journal of Canadian Petroleum Technology.

Carrigy, M. A., 1973, Mesozoic geology of the Fort McMurray area, *in* M. A. Carrigy, ed., Guide to the Athabasca Oil Sands Area: Alberta Research Council Information Series 65, p. 77–101.

Chierici, G. L., 1980, Enhanced oil recovery processes: a state-of-the-art review, 2nd edition: Agip Production Development, San Donato Milanese, 90 p.

Cobble, J. W., R. C. Murray, Jr., P. J. Turner, and K. Chen, 1982, High temperature thermodynamic data for species in aqueous solution: Palo Alto, CA, Electric Power Research Institute Report NP-2400 Project 1167-1.

Dake, L. P., 1978, Fundamentals of reservoir engineering: Amsterdam, Elsevier Publishing Co., 343 p.

Deroo, G., B. Tissot, R. G. McCrossan, and F. Der, 1974a, Geochemistry of the heavy oils of Alberta, *in* L. V. Hills, ed., Oil sands fuel of the future: Canadian Society of Petroleum Geologists, p. 148–167.

————, 1974b, Reply to discussion by D. S. Montgomery, *in* L. V. Hills, ed., Oil sands fuel of the future: Canadian Society of Petroleum Geologists, p. 168–189.

Doscher, T. M., 1967, Technical problems in in-situ methods for recovery of bitumen from tar sands: Panel discussion 13(6), 7th World Petroleum Congress, Mexico City.

Ejiogu, G. C., D. W. Bennion, R. G. Moore, and J. K. Donnely, 1978, Wet combustion—a tertiary recovery process for the Pembina Cardium reservoir: 29th Annual Technological Meeting of the Petroleum Society of the Canadian Institute of Mining and Metallurgy Paper 78-29-32.

Helgeson, H. C., J. M. Delany, H. W. Nesbitt, and D. K. Bird, 1978, Summary and

Figure 11a–d—Scanning electron micrographs of unburned and burned samples from Lloydminster. Scale bars are in micrometers. Note: All micrographs by Alan Oldershaw.

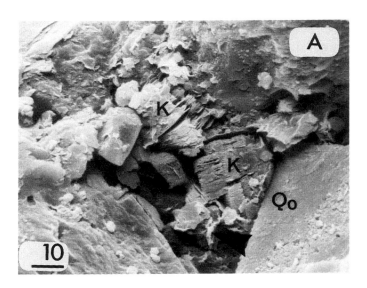

Figure 11a—Kaolinite (**K**) and quartz overgrowths (**Qo**) are common in the unburned samples.

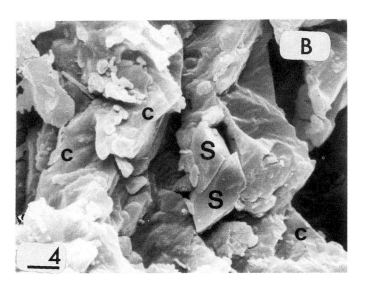

Figure 11b—Calcite (**c**) and siderite (**S**) are found as authigenic phases. (Unburned.)

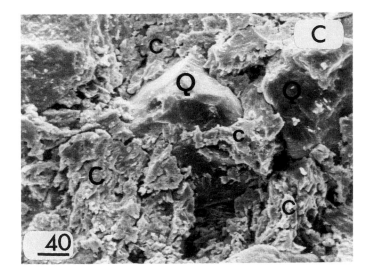

Figure 11c—Chlorite (**c**) forms matted coatings on framework quartz grains (**Q**) in burned samples.

Figure 11d—The reaction of siderite (**s**) and kaolinite to chlorite (**c**) does not consume all the siderite, indicating either that the reaction does not go to completion or kaolinite is consumed before the siderite. Some kaolinite is observed in other burned samples and it seems likely that the reaction does not go to completion. (Burned.)

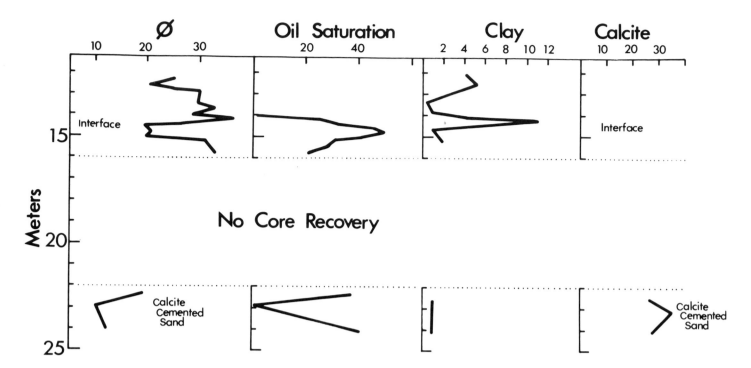

Figure 12—Porosity (ϕ, as vol %), oil saturation as % of pore space, wt % clay in the bulk rock (that is, all material less than 2 μm), and carbonate as wt % (shown as calcite) are plotted as a function of depth in the post-combustion well. Note that oil saturations and clay percentage both increase at the observed oil interface, while the porosity decreases markedly. Depths are given below an arbitrary datum.

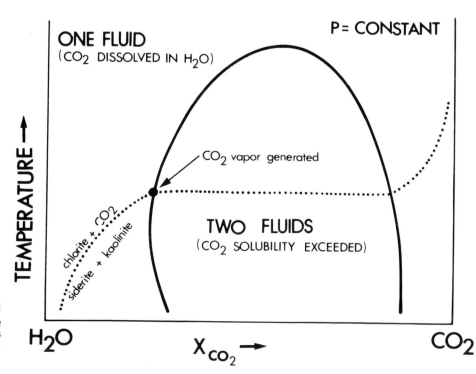

Figure 13—The reaction of siderite and kaolinite to chlorite can generate CO_2 once the solubility of CO_2 in water is exceeded.